Springer Series on
Atoms+Plasmas

3

Guest Editor: J. Peter Toennies

Springer Series on
Atoms+Plasmas

Editors: G. Ecker   P. Lambropoulos   H. Walther

I. Lindgren   J. Morrison

Atomic
Many-Body Theory

Second Edition

With 96 Figures

Springer-Verlag Berlin Heidelberg New York
London Paris Tokyo

Professor Ingvar Lindgren · Dr. John Morrison *
Chalmers University of Technology, Department of Physics
S-41296 Göteborg, Sweden
* Present address: Department of Physics, University of Louisville
Louisville, Kentucky 40292, USA

Guest Editor:
Professor Dr. J. Peter Toennies
Max-Planck-Institut für Strömungsforschung, Böttingerstraße 6−8
D-3400 Göttingen, Fed. Rep. of Germany

Series Editors:
Professor Dr. Günter Ecker
Ruhr-Universität Bochum, Institut für Theoretische Physik, Lehrstuhl I, Universitätsstraße 150
D-4630 Bochum-Querenburg, Fed. Rep. of Germany

Professor Peter Lambropoulos, Ph.D.
University of Crete, P.O. Box 470, Iraklion, Crete, Greece, and
Department of Physics, University of Southern California, University Park
Los Angeles, CA 90089-0484, USA

Professor Dr. Herbert Walther
Sektion Physik der Universität München, Am Coulombwall 1
D-8046 Garching/München, Fed. Rep. of Germany

The first edition was published in 1982 in Springer Series in Chemical Physics, Vol. 13;
Atomic Many-Body Theory

ISBN 3-540-10504-2 Springer-Verlag Berlin Heidelberg New York
ISBN 0-387-10504-2 Springer-Verlag New York Heidelberg Berlin

ISBN 3-540-16649-1 2. Auflage   Springer-Verlag Berlin Heidelberg New York
ISBN 0-387-16649-1 2nd edition   Springer-Verlag New York Berlin Heidelberg

Library of Congress Cataloging-in-Publication Data. Lindgren, Ingvar, 1931 –. Atomic many-body theory. (Springer series in atoms + plasmas ; 3) "The first edition was published in 1982 in Springer series in chemical physics, vol. 13" –– T.p. verso. Bibliography: p. Includes indexes. 1. Atomic theory. 2. Many-body problem. I. Morrison, J. (John), 1940 –. II. Title. III. Series. QD461.L72 1986 539.7 86-10082

Offest printing and bookbinding: Konrad Triltsch, Graphischer Betrieb, Würzburg
2153/3150-543210

Preface to the Second Edition

In the new edition only minor modifications have been made. Some printing errors have been corrected and a few clarifications have been made. In recent years the activity in *relativistic* many-body theory has increased considerably, but this field falls outside the scope of this book. A brief summary of the recent developments, however, has been included in the section on "relativistic effects" in Chap. 14. In addition, only a very limited number of references have been added, without any systematic updating of the material.

Göteborg, December 1985 *I. Lindgren · J. Morrison*

Preface to the First Edition

This book has developed through a series of lectures on atomic theory given these last eight years at Chalmers University of Technology and several other research centers. These courses were intended to make the basic elements of atomic theory available to experimentalists working with the hyperfine structure and the optical properties of atoms and to provide some insight into recent developments in the theory.

The original intention of this book has gradually extended to include a wide range of topics. We have tried to provide a complete description of atomic theory, bridging the gap between introductory books on quantum mechanics — such as the book by Merzbacher, for instance — and present-day research in the field. Our presentation is limited to static atomic properties, such as the effective electron-electron interaction, but the formalism can be extended without major difficulties to include dynamic properties, such as transition probabilities and dynamic polarizabilities.

In order that our presentation of atomic theory be accessible to as wide a group of students as possible we have made a clear decision as to what the reader is and is not expected to know. We assume that the reader has a basic understanding of quantum mechanics, but we do not assume that he is familiar with atomic physics. So the book should be suitable for a first- or second-year graduate student, and it is indeed for such a student that the book is mainly intended. Our assumption as to the reader's background has as a consequence that the level of difficulty varies somewhat throughout the book. For some readers the most difficult chapter may well be the second, which summarizes the angular-momentum and tensor-operator theory we shall use throughout the book. This chapter includes both a summary and an extension of results found in quantum-mechanics books, and — after all — entire books have been written on tensor-operator methods. We have included only those topics which are essential for our purposes, however, and tried to motivate each new idea. Angular-momentum graphs are introduced in the third chapter and developed in a straightforward way in Chap. 4. The fifth and the sixth chapters begin our treatment of atomic physics in a systematic way. Much of the material in these chapters is quite elementary. Afterwords, our treatment is fairly uniform, although, of course, eventually the scope of the material itself becomes quite large.

The first part of the book represents a self-contained presentation of atomic theory at an intermediate level. This part can be of value by itself for the nonspecialist who would like to become familiar with such a widely used method as the Hartree-Fock model.

The general many-body theory presented in the second part of the book is to a large extent independent of the atomic theory presented in the first part and can be applied to other interacting many-fermion systems, such as molecules, nuclei and − to some extent − also to solids. It is only in our applications of the formalism that we have taken up specifically atomic problems.

The presentation in the book is based largely on graphical methods. In the first part angular-momentum graphs are used to represent the coupling of the spin and orbital angular momenta of the electrons. In the second part Feynman-like diagrams − so-called Goldstone diagrams − are used to represent the different terms in the perturbation expansion. These diagrams are evaluated using the angular-momentum graphs developed in the early part of the book. One of the main aims of the book is to make the reader familiar with graphical methods, which are clear and powerful and are now becoming widely used in different branches of physics and chemistry.

A large numer of people have made important contributions to this book at various stages of its development. In particular, we would like to mention our scientific colleagues Barry Adams, Bruce Barrett, Rodney Bartlett, Jacques Bauche, Jíri Čížek, Sten Garpman, Karol Jankowski, Brian Judd, Hugh Kelly, Johannes Lindgren, Stig Lundqvist, Ann-Marie Mårtensson, Joe Paldus, Kathy Rajnak, Arne Rosén, Pat Sandars, Sten Salomonson, Tony Starace, Sune Svanberg, Andy Weiss and Nicholas Winter. In addition, we wish to express our deep appreciation to Marianne Thureson for patiently typing and correcting the many versions of the manuscript and to Bertil Oscarsson for skilfully making the numerous ink drawings, which form an essential part of our presentation.

Göteborg, November 1981   *I. Lindgren · J. Morrison*

Contents

Part II Perturbation Theory and the Treatment of Atomic Many-Body Effects

Part I

Angular-Momentum Theory
and the Independent-Particle Model

1. Introduction

Atoms differ from other microscopic systems by the presence of a well-defined center of force. Because of their attraction to the nucleus, the electrons are confined to a localized region of space and they move in a field which is approximately spherical. This elementary property of the atom explains the importance of angular-momentum theory to atomic physics, and it provides the physical basis of the shell model of the atom. There is, of course, an enormous amount of experimental evidence which supports this approximate description. The regularities of the chemical properties of the elements, for instance, strongly indicate that the properties of an atom are to a large extent determined by the angular-momentum properties of its outer electrons. The first direct evidence of atomic shell structure was given by optical spectroscopy and x-ray absorption data. Nowadays, modern techniques, such as x-ray photoelectron spectroscopy and laser spectroscopy, provide us with a wide variety of accurate information about the structure of the atom.

Because of its obvious relevance, angular-momentum theory was developed very early in a form suitable for atomic problems, particularly by Wigner and Racah. Racah introduced the idea of a spherical tensor, which essentially extends the angular-momentum classification to operators and allows one to treat states and operators on the same footing. The electrostatic and the magnetic interactions among the electrons can be written in terms of tensor operators and their matrix elements readily evaluated in this way. In the second chapter we discuss the elementary properties of angular momentum states and spherical tensors. Graphical methods of representing vector-coupling coefficients and evaluating matrix elements are given in the third and fourth chapters. These early chapters include a number of physical examples, which we shall refer to later. Our main purpose here, though, is to develop the basic techniques of angular momenta and their graphical representation, which we shall use throughout the book.

We begin to treat atomic physics in a systematic way in the fifth chapter. For an atom having more than a few electrons the atomic Hamiltonian is very complex, and it is only possible to find its eigenfunctions within the framework of some approximation scheme. A natural approximation is to assume that each electron moves in an *average* potential due to the nucleus and the other electrons. This assumption leads to the independent-particle model, which essentially reduces the many-electron problem to the problem of solving a number of single-particle equations. For an arbitrary potential, the solutions of the single-electron equations can have quite a complicated form, and they do not serve as a convenient basis set for further calculations. For this reason, one normally assumes also that the potential is spherically symmetric. This is the central-field approximation, which is treated in Chap. 6. The approximate Hamiltonian then describes a number of electrons moving in an average central field. The validity of this approximation depends largely upon the attraction of the electrons to the nucleus and, as we have said, the success of the central-field model explains the importance of angular-momentum theory to atomic physics. It would in a sense be more logical to introduce the central-field approximation before talking at all about angular momentum. Pedagogically, though, we feel that it is better to develop the basic techniques that we shall need first and then to approach the physical problems in earnest. The flow of the book is smoother that way. In Chap. 7 we discuss the restricted Hartree-Fock method, which in some sense can be regarded as the best central-field model. This method is sufficiently accurate to give a qualitative understanding of some of the physical properties of the atom, such as its size and the magnitude of the interactions of the electrons.

The second part of the book deals with the problem of describing the departures from the simple independent-particle model. The methods we shall use for this purpose are based mainly upon perturbation theory, which is described in Chap. 9. In the following three chapters we develop the graphical representation of the perturbation expansion for closed-shell systems, leading to the so-called linked-diagram expansion. Here we follow the basic ideas of Brueckner and Goldstone, although we use a completely time-independent approach based on Rayleigh-Schrödinger perturbation theory. Rules for evaluating Goldstone diagrams are derived by taking advantage of the close analogy between the Goldstone diagrams and the angular-momentum diagrams introduced in the third chapter. Chapters 10–12 thus give a systematic order-by-order description of the departures from the single-particle model for closed-shell systems.

In Chap. 13 the diagrammatic formulation of perturbation theory is extended to open-shell systems, and the electrostatic term structure of the atom is discussed in detail as an application of the formalism. In order to avoid the need of generating a complete set of virtual states, one- and two-particle equations are derived for the particular linear combination of excited states which contribute to the Goldstone diagrams. This enables us to carry out accurate many-body calculations readily without constructing an extensive basis set or generating virtual orbitals.

The open-shell formalism is then applied to the hyperfine interaction in Chap. 14. This provides an example of an "additional" weak perturbation and it illustrates the important concept of an effective operator, which can also be used, for instance, to describe the interaction of the atom with an external field and to include relativistic effects.

Up to this point we have considered mainly second- and third-order contributions to the energy, which can be described by means of first-order one- and two-particle functions. In the final chapter we extend this treatment by considering the iterative solution of the one- and two-particle equations. Solving these equations self-consistently is equivalent to generating large classes of diagrams to all orders of the perturbation theory. This approach is extended further by considering the exp(S) or coupled-cluster procedure, which is now frequently used in quantum chemistry. This formalism, which we present in a form applicable to open-shell systems, includes the important effect of disconnected pair excitations and gives a much more accurate description of such phenomena as molecular dissociation. For atomic and molecular systems, the coupled-cluster formalism leads to significantly improved accuracy and it is generally regarded as the best scheme that is presently available. We also discuss a number of simplified schemes which have been reported in the literature.

2. Angular-Momentum and Spherical Tensor Operators

We begin this book by reviewing the theory of angular-momentum and spherical tensor operators—emphasizing the analogy between them. The material which appears in the present chapter is treated more extensively in several monographs that are devoted entirely to angular-momentum theory [*Rose* 1957; *Edmonds* 1957; *Fano* and *Racah* 1959; *Brink* and *Satchler* 1968] and the reader is referred to these books for further details. We have tried, however, to include all the material we need for the following treatment of atomic systems. In the third chapter graphical methods of representing angular-momentum theory are introduced, and this technique is used in the fourth chapter to develop tensor-operator methods further. In later chapters angular-momentum graphs will be used frequently to evaluate the angular part of interaction matrix elements.

2.1 Elementary Properties of Angular-Momentum and Spherical Tensor Operators

In this section we introduce the commutation relations of the angular-momentum operators and use them to derive the eigenvalues and the matrix elements of these operators. Spherical tensor operators are also introduced. The equations used to define these operators are entirely analogous to the equations satisfied by the angular-momentum states. This close analogy between angular-momentum states and spherical tensor operators, which is a consequence of their basic definitions, will be used frequently in the following.

2.1.1 Angular-Momentum Operators

In classical mechanics, the angular momentum of a particle (m) with respect to a certain reference point (0) is defined as

$$l = r \times p,\tag{2.1}$$

where r is the position vector and $p = m\,v$ is the linear momentum of the particle as indicated in Fig. 2.1. The corresponding quantum-mechanical operator can be obtained by means of the usual replacements

$$r \to r \quad \text{and} \quad p \to -i\boldsymbol{V},$$

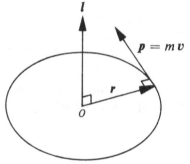

Fig. 2.1. Classical view of the angular momentum

using atomic units (Appendix A). Here, $\boldsymbol{\nabla}$ is the gradient operator. In Cartesian coordinates it is

$$\boldsymbol{\nabla} = \hat{\boldsymbol{x}}\frac{\partial}{\partial x} + \hat{\boldsymbol{y}}\frac{\partial}{\partial y} + \hat{\boldsymbol{z}}\frac{\partial}{\partial z}, \tag{2.2}$$

where $\hat{\boldsymbol{x}}$, $\hat{\boldsymbol{y}}$, $\hat{\boldsymbol{z}}$ are the unit vectors in the directions of the coordinate axes. This gives the quantum-mechanical angular-momentum operator

$$\boldsymbol{l} = -\mathrm{i}\boldsymbol{r} \times \boldsymbol{\nabla} \tag{2.3}$$

with the Cartesian components

$$\begin{cases} l_x = -\mathrm{i}\left(y\dfrac{\partial}{\partial z} - z\dfrac{\partial}{\partial y}\right) \\[2mm] l_y = -\mathrm{i}\left(z\dfrac{\partial}{\partial x} - x\dfrac{\partial}{\partial z}\right) \\[2mm] l_z = -\mathrm{i}\left(x\dfrac{\partial}{\partial y} - y\dfrac{\partial}{\partial x}\right). \end{cases} \tag{2.4}$$

It is easy to show that these components satisfy the commutation relations

$$[l_x, l_y] = \mathrm{i}l_z; \quad [l_y, l_z] = \mathrm{i}l_x; \quad [l_z, l_x] = \mathrm{i}l_y.$$

Here as well as in (2.4), the last two relations can be obtained from the first one by a cyclic permutation of the coordinates, $x \to y \to z \to x$.

It is well known from elementary quantum mechanics that the definition (2.3) is adequate for describing the *orbital* motion of particles, but not, for instance, the *spin* motion of an electron. The commutation relation above, however, is more general than (2.3) and is normally used as the definition. Following that approach here, we define a general angular-momentum operator

$$\boldsymbol{j} = j_x\hat{\boldsymbol{x}} + j_y\hat{\boldsymbol{y}} + j_z\hat{\boldsymbol{z}}$$

by the commutation relation

$$[j_x, j_y] = ij_z \quad \text{(cyclic)},$$ (2.5)

where "cyclic" stands for the above-mentioned cyclic permutation. The angular momentum is a physical observable, and, hence, the operators are assumed to be hermitian.

The square of the operator j is

$$j^2 = j_x^2 + j_y^2 + j_z^2,$$

and it follows from (2.5) that this operator commutes with the components j_x, j_y, and j_z. In particular, since j^2 and j_z are commuting, hermitian operators, they have a complete set of simultaneous eigenstates (see Theorem B.2 in Appendix B). Denoting the eigenvalues by μ and μ_z, respectively, we can write the eigenvalue equations

$$\begin{cases} j^2 \Psi(\gamma, \mu, \mu_z) = \mu \Psi(\gamma, \mu, \mu_z) & \text{(2.6a)} \\ j_z \Psi(\gamma, \mu, \mu_z) = \mu_z \Psi(\gamma, \mu, \mu_z). & \text{(2.6b)} \end{cases}$$

Here, γ represents the additional labels we may need to distinguish between eigenfunctions with the same eigenvalues of j^2 and j_z.

The commutation relations (2.5) are sufficient to determine the spectrum of eigenvalues of j^2 and j_z, and they also serve to determine the matrix elements of these operators. In the interest of completeness, we derive these basic results here—even though they are found in many introductory books on quantum mechanics.

We introduce new operators j_+ and j_-, defined by

$$j_\pm = j_x \pm ij_y.$$ (2.7)

In contrast to j_x and j_y, the operators j_+ and j_- do not correspond to physical observables, and they need not be hermitian. In fact, j_+ and j_- are hermitian adjoints of each other

$$j_+^\dagger = j_-, \quad j_-^\dagger = j_+.$$ (2.8)

From the definition (2.7) and from (2.5) the operators j_+ and j_- can be shown to satisfy the commutation relations

$$\begin{cases} [j_z, j_\pm] = \pm j_\pm & \text{(2.9a)} \\ [j^2, j_\pm] = 0 & \text{(2.9b)} \end{cases}$$

and to have the property

$$j_-j_+ = j^2 - j_z^2 - j_z \tag{2.10a}$$

$$j_+j_- = j^2 - j_z^2 + j_z . \tag{2.10b}$$

Operating on (2.6a) with j_+ from the left and using the fact that this operator commutes with j^2, gives immediately

$$j^2 j_+ \Psi(\mu, \mu_z) = \mu j_+ \Psi(\mu, \mu_z) . \tag{2.11a}$$

(For simplicity, we have left out the additional label γ, which is unaffected by the angular-momentum operators). Similarly, by operating on the left-hand side of (2.6b) with j_+ and using the commutation relation (2.9a), we obtain

$$j_+j_z \Psi(\mu, \mu_z) = (j_z j_+ - [j_z, j_+]) \Psi(\mu, \mu_z) = j_z j_+ \Psi(\mu, \mu_z) - j_+ \Psi(\mu, \mu_z) .$$

Operating on the right-hand side of (2.6b) with j_+, gives

$$\mu_z j_+ \Psi(\mu, \mu_z)$$

and equating with the previous relation we get

$$j_z j_+ \Psi(\mu, \mu_z) = (\mu_z + 1) j_+ \Psi(\mu, \mu_z). \tag{2.11b}$$

So operating on $\Psi(\mu, \mu_z)$ with j_+ gives a new eigenfunction belonging to the same eigenvalue of j^2 but to the eigenvalue $(\mu_z + 1)$ of j_z. By repeatedly operating with j_+ on $\Psi(\mu, \mu_z)$, we can generate a whole set of eigenfunctions belonging to the eigenvalues μ_z, $\mu_z + 1$, $\mu_z + 2$, ... and all belonging to the eigenvalue μ of j^2. For this reason j_+ is called a *step-up* operator.

Equations that are analogous to (2.11a,b) may also be derived in a similar fashion for the operator j_-

$$j^2 j_- \Psi(\mu, \mu_z) = \mu j_- \Psi(\mu, \mu_z) \tag{2.12a}$$

$$j_z j_- \Psi(\mu, \mu_z) = (\mu_z - 1) j_- \Psi(\mu, \mu_z) . \tag{2.12b}$$

It then follows that $j_- \Psi(\mu, \mu_z)$ is an eigenfunction of j^2 corresponding to the eigenvalue μ and an eigenfunction of j_z with the eigenvalue $(\mu_z - 1)$. Thus, j_- serves as a *step-down* operator. j_+ and j_- are also called *ladder operators*.

We shall now determine the possible values of μ and μ_z. For a definite value μ, it is easy to show that μ_z must have an upper and a lower bound. Denoting the maximum value by j, it then follows from (2.11b) that

$$j_+ \Psi(\mu, j) = 0 . \tag{2.13}$$

Otherwise $j_+ \Psi(\mu, j)$ would be an eigenfunction of j_z corresponding to the eigenvalue $j + 1$. Operating on (2.13) from the left with j_-, we obtain

$$j_-j_+ \Psi(\mu, j) = 0 .$$

Using (2.10a) we then get

$$(\mu - j^2 - j) \, \Psi(\mu, j) = 0 \,, \tag{2.14}$$

and since the wave function cannot vanish identically at all points, it follows that

$$\mu = j(j + 1) \,. \tag{2.15}$$

Similarly, let $(j - r)$ be the least eigenvalue of j_z. Then it follows that

$$j_- \Psi(\mu, j - r) = 0$$

and

$$j_+ j_- \Psi(\mu, j - r) = 0.$$

In analogy with (2.14) this leads to

$$\mu - (j - r)^2 + (j - r) = 0 \,.$$

By eliminating μ by means of (2.15), we get the quadratic equation

$$r^2 - r(2j - 1) - 2j = 0 \,,$$

which has only one positive root, $r = 2j$. Thus, the least eigenvalue of j_z is equal to $j - r = -j$. This means that for a particular eigenvalue $\mu = j(j + 1)$ of $\mathbf{j}^2$ there are $2j + 1$ eigenfunctions $\Psi(\mu, m)$ of j_z, corresponding to the eigenvalues

$$m = j, j - 1, \dots, -j + 1, -j.$$

It is also clear from this argument that $2j$ is an integer, which means that *j must be either an integer or a half-integer*. This is a consequence of using the commutation rules (2.5) as the definition of the angular momentum, instead of the explicit expression (2.3), which leads to integral quantum numbers only. The latter expression can, of course, still be used for describing the orbital motion, and we shall refer to (2.3) as the *orbital* angular-momentum operator.

We shall in this book make frequent use of the Dirac notation and write the simultaneous (normalized) eigenstates (2.6) of $\mathbf{j}^2$ and j_z simply as $|\gamma jm\rangle$ (see Appendix B). These states have the property

$$\begin{cases} \mathbf{j}^2|\gamma jm\rangle = j(j + 1)|\gamma jm\rangle & \left(j = 0, \dfrac{1}{2}, 1, \dfrac{3}{2}, \dots \right) & \text{(2.16a)} \\[2mm] j_z|\gamma jm\rangle = m|\gamma jm\rangle & (m = -j, -j + 1, \dots +j), & \text{(2.16b)} \end{cases}$$

and, as we have said, they form a complete (orthonormal) set. Any single-electron state can be expanded in terms of states of this kind.

We would like now to construct the matrices, which correspond to the operators j_+ and j_-. According to (2.11), $j_+|jm\rangle$ is an eigenfunction of j^2 and j_z, corresponding to the eigenvalues $j(j+1)$ and $m+1$, respectively. This state is not necessarily normalized, however. Denoting the normalization factor by α_{jm}, we can write

$$j_+|jm\rangle = \alpha_{jm}|jm+1\rangle. \tag{2.17}$$

The complex conjugate of this equation is

$$(j_+|jm\rangle)^* = \langle jm|j_+^\dagger = \langle jm|j_- = \alpha_{jm}^* \langle j\,m+1|, \tag{2.18}$$

using (2.8). Multiplying (2.17) by (2.18) and integrating, we obtain the *scalar product*

$$\langle jm|j_-j_+|jm\rangle = |\alpha_{jm}|^2 \langle jm+1|jm+1\rangle.$$

With (2.10a) and (2.16) this becomes

$$[j(j+1) - m(m+1)]\langle jm|jm\rangle = |\alpha_{jm}|^2 \langle jm+1|jm+1\rangle.$$

Since the functions $|jm\rangle$ and $|jm+1\rangle$ are assumed normalized, we obtain in this way an expression for α_{jm}

$$|\alpha_{jm}|^2 = j(j+1) - m(m+1).$$

The phase of α_{jm} is not determined by this equation, and it is in fact arbitrary. We shall here follow the phase convention, adopted by *Condon* and *Shortley* [1935], of taking the phase to be $+1$. We then get

$$j_+|jm\rangle = [j(j+1) - m(m+1)]^{1/2}|jm+1\rangle \tag{2.19a}$$

and similarly

$$j_-|jm\rangle = [j(j+1) - m(m-1)]^{1/2}|jm-1\rangle. \tag{2.19b}$$

The nonvanishing matrix elements of j_+, j_-, and j_z are thus given by the equations

$$\langle jm\pm1|j_\pm|jm\rangle = [j(j+1) - m(m\pm1)]^{1/2}$$
$$\langle jm|j_z|jm\rangle = m. \tag{2.20}$$

The corresponding expressions for j_x and j_y are obtained from these equations using the definitions (2.7).

The point we would like to emphasize here is that the relations (2.20) and the spectrum of eigenvalues of j^2 and j_z follow directly from the commutation relations (2.5). They are thus true for any kind of angular momentum. The spin operators s_x, s_y and s_z also satisfy these relations, and the spin part of a single-electron wave function may be written in terms of two spinors, α and β. Each spinor is an eigenfunction of s^2, corresponding to the eigenvalue $(1/2 + 1)/2$, and they have eigenvalues $+1/2$ and $-1/2$ with respect to s_z

$$\begin{cases} s^2\alpha = 3/4\alpha \\ s_z\alpha = +1/2\alpha \\ s^2\beta = 3/4\beta \\ s_z\beta = -1/2\beta . \end{cases} \tag{2.21}$$

Problem 2.1. Show that the matrices for the components of spin angular momentum s are

$$s_x = \frac{1}{2}\begin{pmatrix} 0 & 1 \\ 1 & 0 \end{pmatrix}, \quad s_y = \frac{1}{2}\begin{pmatrix} 0 & -i \\ i & 0 \end{pmatrix}, \quad s_z = \frac{1}{2}\begin{pmatrix} 1 & 0 \\ 0 & -1 \end{pmatrix}, \tag{2.22}$$

where the first row and column correspond to $m_s = 1/2$. These are the so-called *Pauli spin matrices*. Show that they satisfy the commutation rules (2.5).

Problem 2.2 Construct the matrices for the components of an (orbital) angular momentum operator, corresponding to $j = l = 1$.

2.1.2 Spherical Tensor Operators

Thus far, we have discussed the effect of angular-momentum operators acting upon *states*. We would like now to introduce a set of *operators* t_q^k ($q = k, k - 1, \cdots - k$), which satisfy analogous relations when acted upon by the angular-momentum operators. Such operators are called *spherical tensor operators*. In atomic problems they form a convenient basis for operators in the same way as do the angular-momentum eigenstates for states.

We shall define spherical tensor operators of rank k by means of the commutation relations

$$[j_z, t_q^k] = q t_q^k \tag{2.23a}$$

$$[j_\pm, t_q^k] = [k(k + 1) - q(q \pm 1)]^{1/2} t_{q\pm1}^k \tag{2.23b}$$

In order to motivate this definition, we first recall an elementary result from quantum mechanics. The z component of the orbital angular-momentum operator (2.4) is in spherical polar coordinates (see Problem 2.8)

$$l_z = -i\frac{\partial}{\partial\phi}. \tag{2.24}$$

Conventionally, an operator acts on everything that stands to its right. For instance, assuming f and Ψ to be ordinary functions, we have

$$l_z f \Psi = l_z (f \Psi) = -i\left(\frac{\partial}{\partial \phi} f\right)\Psi - i f\left(\frac{\partial}{\partial \phi}\Psi\right),$$

using the ordinary product rule. The last equation can be written

$$l_z f \Psi = \overset{\sqcap}{l_z f}\Psi + f l_z \Psi.$$

Here, the notation $\overset{\sqcap}{l_z f}$ indicates that l_z operates *only* on the function f. Solving for $\overset{\sqcap}{l_z f}\Psi$, we obtain

$$\overset{\sqcap}{l_z f}\Psi = (l_z f - f l_z)\Psi.$$

Since the function Ψ is arbitrary, this leads to the identity

$$\overset{\sqcap}{l_z f} \equiv [l_z, f]. \tag{2.25}$$

Therefore, in order to get the effect of l_z on f alone, we should form the commutator of l_z and f. This argument holds also if f is an operator, and this explains the appearance of the commutators in the defining equations of the spherical tensor operators above. Equations (2.23) are the analogues of the corresponding relations (2.20) for the angular-momentum operators. This implies that the components of a spherical tensor operator transform among themselves as do the corresponding angular-momentum states when operated upon by the angular-momentum operators. This is the underlying reason for the close analogy between spherical tensor operators and angular-momentum eigenstates.

It follows from the definition (2.23) that *the angular-momentum operators themselves are spherical tensor operators of rank one* (see Problem 2.3).

Problem 2.3. Using the commutation relations (2.5), show that the operators

$$j_{+1} = -\frac{1}{\sqrt{2}} j_+ = -\frac{1}{\sqrt{2}}(j_x + i j_y)$$

$$j_0 = j_z$$

$$j_{-1} = \frac{1}{\sqrt{2}} j_- = \frac{1}{\sqrt{2}}(j_x - i j_y)$$

satisfy the relations (2.23) and hence are the components of a spherical tensor operator of rank 1.

2.2 Rotations in Space

In the previous section we defined angular-momentum and spherical tensor operators by means of commutation relations. This definition of spherical tensor operators agrees with that employed by *Racah* [1942] in introducing these operators. In the present section we consider the relation between angular-momentum operators and rotations in space, and we shall see that this leads to an alternative definition of angular-momentum and spherical tensor operators.

2.2.1 Relation Between Angular-Momentum Operators and Infinitesimal Rotations in Space

We study first the effect of rotations upon functions, and for simplicity we restrict the introductory discussion to two dimensions.

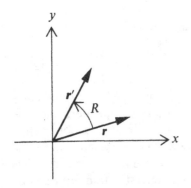

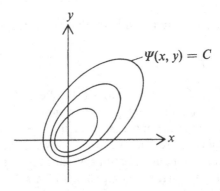

Fig. 2.2 The rotation of a vector in two dimensions

Fig. 2.3. Representation of a function by means of contour curves on which the function has a constant value

Let r and r' be two vectors represented by the column vectors

$$r = \begin{pmatrix} x \\ y \end{pmatrix}; \quad r' = \begin{pmatrix} x' \\ y' \end{pmatrix}.$$

Assuming that they have the same length, these vectors are related by means of a rotation R, as shown in Fig. 2.2,

$$r' = Rr .\tag{2.26}$$

The rotation operator R can be represented by a matrix, which transforms the column vector r into r'.

We would like now to introduce rotation operators which act upon functions. A function $\Psi(x,y)$ can be represented graphically by means of a family of contour curves, along which the function has a constant value,

$$\Psi(x, y) = C,$$

as illustrated in Fig. 2.3.

We consider a single contour curve and imagine that there is a vector attached to each point of the curve. We can then rotate the contour in the same way as the vectors above (Fig. 2.4). In a fixed coordinate system the new contour will be represented by another function, which we denote by

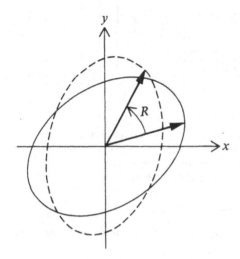

Fig. 2.4. The rotation of a function can be illustrated by rotating its contour curves

$$\Psi'(x, y) = P(R)\Psi(x,y). \tag{2.27}$$

The operator $P(R)$ transforms the original function into the new one.

As a simple illustration we consider the effect of a rotation by 90° in a positive (counter-clockwise) direction upon the function

$$\Psi(x, y) \equiv x.$$

The contours of this function are straight lines, perpendicular to the x axis (see Fig. 2.5a). A positive rotation by 90° turns these lines into lines perpendicular to the y axis, in such a way that positive values of x correspond to positive values of y (Fig. 2.5b). These lines represent the function

$$\Psi'(x, y) \equiv y\,.$$

In analogy with (2.27), we can write this as

$P(90°)x = y$.

Another rotation by 90° in the same direction leads to lines again perpendicular to the x axis, but now so that *positive* values of y correspond to *negative* values of x (Fig. 2.5c). This implies that

$P(90°)\, y = -x$.

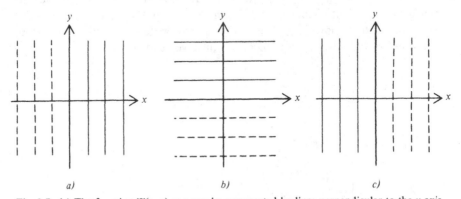

a)   b)   c)

Fig. 2.5. (a) The function $\Psi(x, y) \equiv x$ can be represented by lines perpendicular to the x axis. The solid lines represent positive and the dotted lines negative values of the function. (b) After a rotation by 90° in a positive direction, this function is transformed into $\Psi'(x, y) \equiv y$. (c) A second rotation by 90° in the same direction yields $\Psi''(x, y) \equiv -x$

Problem 2.4 Find how the functions x and xy transform under a rotation by 45° in a positive direction about the z axis.

Using polar coordinates we can express the effect of a general two-dimensional rotation by an angle α as

$\Psi'(r, \phi) = P(\alpha)\Psi(r, \phi)$.

The operator $P(\alpha)$ rotates the contours of the function $\Psi(r, \phi)$ by an angle α in a positive direction. Then the new function has the same value at the angle ϕ as the original function has at the angle $(\phi - \alpha)$,

$P(\alpha)\Psi(r, \phi) = \Psi(r, \phi - \alpha)$.

An analogous result holds in three dimensions for a rotation about the polar axis (z axis)

$$P_z(\alpha)\, \Psi(r, \theta, \phi) = \Psi(r, \theta, \phi - \alpha), \qquad (2.28)$$

when spherical polar coordinates are used (see Fig. 2.6).

Polar axis

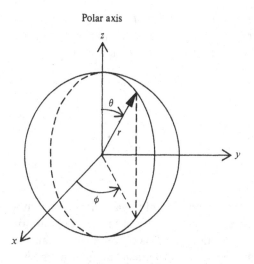

Fig. 2.6. Definition of spherical polar coordinates

$$\begin{cases} x = r \sin \theta \cos \phi \\ y = r \sin \theta \sin \phi \\ z = r \cos \theta \end{cases} \quad \begin{cases} r = (x^2 + y^2 + z^2)^{1/2} \\ \theta = \arccos (z/r) \\ \phi = \arctan (y/x) \end{cases}$$

We are now in a position to give a definition of the angular-momentum operators in terms of rotation operators. In order to do this, we first cast the orbital angular-momentum operator (2.24)

$$l_z = -\,\mathrm{i}\frac{\partial}{\partial\phi} \tag{2.29}$$

into a more general form by expressing the derivative $\partial/\partial\phi$ as a limit of a finite difference ratio

$$\frac{\partial}{\partial\phi}\,\Psi(r, \theta, \phi) = \lim_{\delta\phi\to 0}\frac{\Psi(r, \theta, \phi) - \Psi(r, \theta, \phi - \delta\phi)}{\delta\phi} .$$

Using the definition (2.28) of the rotation operator, we can write this as

$$\frac{\partial}{\partial\phi}\,\Psi(r, \theta, \phi) = \lim_{\delta\phi\to 0}\left[\frac{1 - P_z(\delta\phi)}{\delta\phi}\right]\Psi(r, \theta, \phi) .$$

The differential operator $\partial/\partial\phi$ may thus be written

$$\frac{\partial}{\partial\phi} = \left[\frac{1 - P_z(\delta\phi)}{\delta\phi}\right],$$

where we have allowed the limit $\delta\phi \to 0$ to be understood. This gives the following form of the angular-momentum operator (2.29)

$$l_z = -\,\mathrm{i}\left[\frac{1 - P_z(\delta\phi)}{\delta\phi}\right]. \tag{2.30}$$

This is equivalent to (2.29) and therefore applies to the orbital angular momentum. We now *require* that (2.30) be valid for *any* angular momentum. If we denote a general angular-momentum operator by j, this condition is

$$j_z = -i\left[\frac{1 - P_z(\delta\phi)}{\delta\phi}\right] \tag{2.31}$$

or

$$\boxed{P_z(\delta\phi) = 1 - i\delta\phi j_z} \ . \tag{2.32}$$

This relation can be used as an alternative definition of the angular-momentum operators. Below, we shall show that this definition leads to the commutator relations (2.5). This implies that the definition (2.32) is more general than the elementary relation (2.29), which is applicable only to integer angular momenta.

In order to show that (2.32) is consistent with the commutation relations (2.5), we consider the operators $P_x(\delta\phi_x)$ and $P_y(\delta\phi_y)$, describing infinitesimal roations about the x and y axis, respectively. These operators do not commute, as we can see by considering the effect of $P_y(\delta\phi_y) P_x(\delta\phi_x)$ and $P_x(\delta\phi_x) P_y(\delta\phi_y)$, respectively, on the functions $\Psi(x, y, z) = x$. In three dimensions, this function is represented by planes perpendicular to the x axis, and the rotation of these planes can be illustrated by means of a unit vector along this axis, as shown in Fig. 2.7. In the first case, the rotation about the x axis is first applied, which has no effect on this vector. The rotation about the y axis moves the vector as indicated in Fig. 2.7a. When the operations are applied in the opposite order, the rotation about the x axis will have a second-order effect. This corresponds to a rotation by the angle $\delta\phi_x\delta\phi_y$ about the z axis. Thus, we get the relation

$$P_x(\delta\phi_x)P_y(\delta\phi_y) = P_z(\delta\phi_x\delta\phi_y)P_y(\delta\phi_y)P_x(\delta\phi_x)$$

(to second order in $\delta\phi_x$, $\delta\phi_y$). Using (2.32), this becomes

$$(1 - i\delta\phi_x j_x)\,(1 - i\delta\phi_y j_y\,) = (1 - i\delta\phi_x\delta\phi_y j_z)\,(1 - i\delta\phi_y j_y)\,(1 - i\delta\phi_x j_x).$$

The term which is independent of $\delta\phi_x$ and $\delta\phi_y$ and those which are linear in these infinitesimals are identical on both sides of the equation. Equating the second-order terms, we obtain

$$-\delta\phi_x\delta\phi_y j_x j_y = -i\delta\phi_x\delta\phi_y j_z - \delta\phi_x\delta\phi_y j_y j_x \ .$$

This leads to the equation

$$[j_x, j_y] = ij_z. \tag{2.33}$$

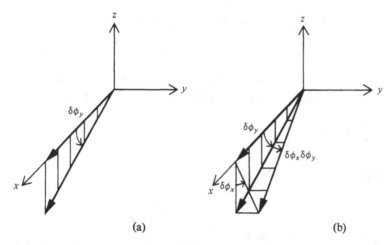

(a)   (b)

Fig. 2.7. (a) Illustrates the effect of $P_y(\delta\phi_y)P_x(\delta\phi_x)$ on a unit vector in the x direction and (b) the effect on the same vector of the rotations in opposite order. The difference corresponds to a rotation by $\delta\phi_x\delta\phi_y$ about the z axis

Thus, the alternative definition (2.32) of angular-momentum operators in terms of infinitesimal rotations leads to the commutator relations (2.5) used previously to define the angular-momentum operators, and it is therefore consistent with them. In particular, the definition (2.32) leads to half-integer angular-momentum quantum numbers.

2.2.2 Transformation of Angular-Momentum States and Spherical Tensor Operators Under Infinitesimal Rotations

We would like now to see how angular-momentum states and spherical tensor operators transform under rotations. According to (2.32) and (2.16), when the rotation operator $P_z(\delta\phi)$ acts on a state $|jm\rangle$, it produces the result

$$P_z(\delta\phi)|jm\rangle = (1 - i\delta\phi m)|jm\rangle . \tag{2.34}$$

Before we discuss how the spherical tensor operators transform, we should recall some of the general transformation properties of operators.

Suppose that there are two sets of states, $\{\Psi_i\}$ and $\{\Phi_i\}$, which are related by means of the operator α

$$\Psi_i = \alpha\Phi_i , \tag{2.35}$$

and we rotate these functions by means of the operator P

$$\Psi'_i = P\Psi_i \quad \text{and} \quad \phi'_i = P\phi_i , \tag{2.36}$$

We can then define a "transformed operator" α' by the equation

$$\Psi'_i = \alpha' \Phi'_i . \tag{2.37}$$

In other words, we require that α' satisfies the same relation between transformed functions as α does for the original ones. Using the inverse of (2.36), we can express the relation (2.35) as

$$P^{-1}\Psi'_i = \alpha P^{-1}\phi'_i$$

or, after operating with P from the left,

$$\Psi'_i = P\alpha P^{-1}\phi'_i .$$

By comparing with (2.37) we see that

$$\alpha' = P\alpha P^{-1} . \tag{2.38}$$

This is called a *similarity transformation*.

We can now see how the tensor operators t^k_q are affected by an infinitesimal rotation of the functions. From (2.32) it follows that

$$[P_z(\delta\phi)]^{-1} = P_z(-\delta\phi) = 1 + i\delta\phi j_z ,$$

which gives the transformed operator

$$(t^k_q)' = P_z(\delta\phi)\, t^k_q P_z(\delta\phi)^{-1} = (1 - i\delta\phi j_z)\, t^k_q (1 + i\delta\phi j_z) .$$

Discarding terms which are second order in the infinitesimal angle $\delta\phi$, this equation becomes

$$(t^k_q)' = t^k_q - i\delta\phi[j_z, t^k_q] , \tag{2.39}$$

and we may use the defining equation of the spherical tensor operators (2.23a) to obtain

$$(t^k_q)' = (1 - i\delta\phi q)\, t^k_q . \tag{2.40}$$

This relation shows how a spherical tensor operator transforms under an infinitesimal rotation.

The point to notice here is that the last equation is completely analogous to the corresponding transformation of the angular-momentum eigenstates (2.34). Since a finite rotation can be built up by means of successive infinitesimal rotations, it follows that the angular-momentum states and the spherical tensor operators transform in the same way under *any* spatial rotation.

Here we have defined spherical tensor operators by means of the commutation relations (2.23) and shown that this implies that they transform under rotation as the angular-momentum states. Alternatively, we can define spherical tensor operators by means of the transformation properties and derive the commutation relations.

....

Problem 2.5. An operator which transforms under a rotation in the same way as the vector **r** (or any other vector) is called a *vector operator*. Show that the gradient operator ∇ satisfies this condition for a rotation of 90° about the z axis, or, in other words, that

$$\nabla'_x = P_z(90°)\,\nabla_x P_z(-90°) = \nabla_y; \quad \nabla'_y = -\nabla_x; \quad \nabla'_z = \nabla_z\,.$$

2.2.3 Transformation of Angular-Momentum States and Spherical Tensor Operators Under Finite Rotations

We would like now to find an explicit expression for the transformation operator $P_z(\phi)$ of a *finite* rotation about the z axis. Suppose that $P_z(\phi)$ is the result of n small rotations $P_z(\delta\phi)$. Then we can write

$$P_z(\phi) = (1 - i\delta\phi j_z)^n$$

or, since $\phi = n\delta\phi$,

$$P_z(\phi) = \left(1 - \frac{i\phi j_z}{n}\right)^n\,.$$

We now let $n \to \infty$ and $\delta\phi \to 0$ in such a way as to hold ϕ constant,

$$P_z(\phi) = \lim_{n\to\infty} \left(1 - \frac{i\phi j_z}{n}\right)^n\,. \tag{2.41}$$

The expression on the right is analogous to one of the representations of the exponential function, and we use it to define an *exponential operator*

$$P_z(\phi) = \exp(-i\phi j_z)\,. \tag{2.42}$$

The exponential may also be expressed as a series expansion.

The operators for rotations about the x and y axes have, of course, a similar form. For instance, the operator, which corresponds to a rotation by an angle θ about the x axis, is

$$P_x(\theta) = \exp(-i\theta j_x)\,. \tag{2.43}$$

When this operator is applied to an angular-momentum eigenfunction, we can express the result

$$P_x(\theta)|\gamma jm\rangle = \sum_{\gamma'j'm'} |\gamma'j'm'\rangle \langle \gamma'j'm'|\exp(-i\theta j_x)|\gamma jm\rangle . \qquad (2.44)$$

Here, we have used the fact that the angular-momentum eigenfunctions form a complete set, which leads to the *completeness relation* (see Eq. B9 in Appendix B)

$$\sum_{\gamma jm} |\gamma jm\rangle \langle \gamma jm| = 1 .$$

Of course, it is here necessary to sum also over the additional quantum numbers represented by γ.

The operator j^2 commutes with j_x and, hence, also with the operator $\exp(-i\theta j_x)$. This can be shown, for instance, by using the analogue of (2.41) for rotations about the x axis or by using the power-series expansion of the exponential function. It then follows from Theorem B. 3 that the matrix in (2.44) is diagonal with respect to j. The matrix elements of j_x—and hence also of $\exp(-i\theta j_x)$—are furthermore independent of the additional quantum numbers γ. We can then simplify (2.44) to

$$P_x(\theta)|\gamma jm\rangle = \sum_{m'} |\gamma jm'\rangle \langle jm'|\exp(-i\theta j_x)|jm\rangle . \qquad (2.45)$$

An arbitrary rotation in three dimensions can be described by means of three rotations about the coordinate axes, using so-called Euler angles [*Goldstein* 1950, p. 107; *Edmonds* 1957, p. 6]. Therefore, a rotation of an angular-momentum eigenfunction can always be expressed as

$$P(\omega)|\gamma jm\rangle = \sum_{m'} |\gamma jm'\rangle D^j_{m'm}(\omega) . \qquad (2.46)$$

$P(\omega)$ represents here the rotation operator for a general rotation ω, and $D^j_{m'm}(\omega)$ is an element of the corresponding rotation matrix $\boldsymbol{D}^j$.

According to (2.40), the spherical tensor operators transform under infinitesimal rotations as do the angular-momentum states. Hence, this must be true also for arbitrary *finite* rotations. The relation for spherical tensor operators analogous to (2.46) is then

$$(t^k_q)' = P(\omega) t^k_q P(\omega)^{-1} = \sum_{q'} t^k_{q'} D^k_{q'q}(\omega) . \qquad (2.47)$$

Using group-theoretical language, (2.46,47) imply that the angular-momentum eigenfunctions and the tensor operators from a basis for a representation of the rotation group. The representation is said to be "*irreducible*", because it is not possible to form any linear combinations of the basis functions which give a representation of lower dimensionality.

2.2.4 The Orbital Angular Momentum. Spherical Harmonics

We would now like to consider in more detail the orbital angular-momentum operator l defined by (2.3). From this definition and the algebraic properties of the gradient operator ∇ one can show that

$$l^2 = -r^2\nabla^2 + \frac{\partial}{\partial r}r^2\frac{\partial}{\partial r} \tag{2.48}$$

(see Problem 2.7). The operator ∇^2 appearing here is the *Laplacian operator*. In Cartesian coordinates it has the form

$$\nabla^2 = \frac{\partial^2}{\partial x^2} + \frac{\partial^2}{\partial y^2} + \frac{\partial^2}{\partial z^2}, \tag{2.49}$$

which in spherical polar coordinates transforms into

$$\nabla^2 = \frac{1}{r^2}\left[\frac{\partial}{\partial r}\left(r^2\frac{\partial}{\partial r}\right) + \frac{1}{\sin\theta}\frac{\partial}{\partial \theta}\left(\sin\theta\frac{\partial}{\partial \theta}\right) + \frac{1}{\sin^2\theta}\frac{\partial^2}{\partial \phi^2}\right] \tag{2.50}$$

(see Problem 2.8). Combining this result with (2.48), we see that

$$l^2 = -\left[\frac{1}{\sin\theta}\frac{\partial}{\partial \theta}\left(\sin\theta\frac{\partial}{\partial \theta}\right) + \frac{1}{\sin^2\theta}\frac{\partial^2}{\partial \phi^2}\right]. \tag{2.51}$$

It follows from (2.48) that the Laplacian operator commutes with the components of the angular-momentum operator. This fact is of fundamental importance for problems with spherical symmetry. It implies that l commutes with the central-field Hamiltonian and hence represents a constant of the motion in such a case, as will be shown in Chap. 6. This is the underlying reason for the importance of angular momentum in atomic physics, and it forms the basis for the atomic shell structure.

The simultaneous eigenfunctions of l^2 and l_z are called *spherical harmonics*, and they form a convenient set of basis functions on the unit sphere. They have the following properties:

$$\begin{cases} l^2 Y_m^l(\theta, \phi) = l(l+1)Y_m^l(\theta, \phi) \\ l_z Y_m^l(\theta, \phi) = mY_m^l(\theta, \phi) \end{cases} \tag{2.52}$$

Here, l is a nonnegative integer and $m = l, l-1, \ldots -l$. Written out more fully, these equations become, using (2.24,51),

$$\begin{cases} \left[\dfrac{1}{\sin\theta}\dfrac{\partial}{\partial\theta}\left(\sin\theta\dfrac{\partial}{\partial\theta}\right) + \dfrac{1}{\sin^2\theta}\dfrac{\partial^2}{\partial\phi^2} + l(l+1)\right] Y_m^l(\theta,\phi) = 0 \\ \left(i\dfrac{\partial}{\partial\phi} + m\right) Y_m^l(\theta,\phi) = 0 . \end{cases} \tag{2.53}$$

The last equation can be solved immediately and shows that the solution must be of the form

$$Y_m^l(\theta,\phi) = f(\theta)\exp(im\phi) . \tag{2.54}$$

To solve the θ equation is considerably more complicated, and interested readers are referred to standard texts in theoretical physics (see, for instance, [*Jackson* 1975, p. 84]). The solutions are normally chosen so that they satisfy the phase condition [*Condon* and *Shortley* 1935]

$$Y_{-m}^l(\theta,\phi) = (-1)^m Y_m^{l*}(\theta,\phi) \tag{2.55}$$

and the orthonormality condition

$$\int_0^{2\pi} d\phi \int_0^{\pi} \sin\theta d\theta \; Y_{m'}^{l'*}(\theta,\phi) \; Y_m^l(\theta,\phi) = \delta(l,l')\delta(m,m') . \tag{2.56}$$

For positive values of m the explicit form of these functions is

$$Y_m^l(\theta,\phi) = (-1)^m \left[\frac{(2l+1)(l-m)!}{4\pi(l+m)!}\right]^{1/2} P_l^m(\cos\theta)\exp(im\phi) , \tag{2.57}$$

where P_l^m is an associated Legendre polynomial.

Some special cases of the general formula (2.57) are sometimes useful. For $m = 0$ and for $\theta = 0$ the associated Legendre polynomial reduces to an ordinary Legendre polynomial

$$P_l^0(\cos\theta) = P_l(\cos\theta) \tag{2.58}$$

$$P_l^m(1) = \delta(m,0) P_l(1) . \tag{2.59}$$

The Legendre polynomial is normalized so that $P_l(1) = 1$. It then follows from these equations and from (2.57) that

$$Y_0^l(\theta,\phi) = \sqrt{\frac{2l+1}{4\pi}} P_l^0(\cos\theta) = \sqrt{\frac{2l+1}{4\pi}} P_l(\cos\theta) \tag{2.60}$$

$$Y_m^l(0,\phi) = \sqrt{\frac{2l+1}{4\pi}} P_l^m(1) = \sqrt{\frac{2l+1}{4\pi}} \delta(m,0) . \tag{2.61}$$

The explicit form of a few spherical harmonics is given in Table 2.1.

Table 2.1. Spherical harmonics

$$l = 0 \quad Y_0^0 = \frac{1}{\sqrt{4\pi}}$$

$$l = 1 \quad \begin{cases} Y_{\pm 1}^1 = \mp \sqrt{\frac{3}{8\pi}} \sin\theta \exp(\pm i\phi) \\[2mm] Y_0^1 = \sqrt{\frac{3}{4\pi}} \cos\theta \end{cases}$$

$$l = 2 \quad \begin{cases} Y_{\pm 2}^2 = \sqrt{\frac{15}{32\pi}} \sin^2\theta \exp(\pm 2i\phi) \\[2mm] Y_{\pm 1}^2 = \mp \sqrt{\frac{15}{8\pi}} \sin\theta \cos\theta \exp(\pm i\phi) \\[2mm] Y_0^2 = \sqrt{\frac{5}{4\pi}} \left(\frac{3}{2} \cos^2\theta - \frac{1}{2} \right) \end{cases}$$

Problem 2.6. Express the spherical harmonics for $l = 1$ and $l = 2$ in Cartesian coordinates.

* Problem 2.7. Verify the relation (2.48).
Hint: Use the vector formulas

$$(a \times b) \cdot c = (b \times c) \cdot a = (c \times a) \cdot b$$
$$a \times (b \times c) = (a \cdot c)b - (a \cdot b)c ,$$

valid for commuting operators, to show that

$$l^2 = -(r \times \nabla) \cdot (r \times \nabla) = -r \cdot \nabla - r^2 \nabla^2 + r \cdot (\nabla \cdot r)\nabla .$$

Problem 2.8 In spherical polar coordinates (see Fig. 2.6) the gradient operator has the form

$$\nabla = \hat{r} \frac{\partial}{\partial r} + \hat{\theta} \frac{1}{r} \frac{\partial}{\partial \theta} + \hat{\phi} \frac{1}{r \sin\theta} \frac{\partial}{\partial \phi} .$$

a) Use this form and the relations

$$\hat{r} \times \hat{r} = 0, \qquad \hat{r} \times \hat{\theta} = \hat{\phi}, \qquad \hat{r} \times \hat{\phi} = -\hat{\theta}$$

to derive (2.24).

* b) Derive the expression (2.50) for the Laplacian operator $\nabla^2 = \nabla \cdot \nabla$.

Hint: Show that the unit vectors in the coordinate directions satisfy the relations

---

* This problem is more difficult and can be left out at the first reading.

$$\frac{\partial \hat{r}}{\partial \theta} = \hat{\theta}; \qquad \frac{\partial \hat{r}}{\partial \phi} = \hat{\phi} \sin \theta; \qquad \frac{\partial \hat{\theta}}{\partial \phi} = \hat{\phi} \cos \theta; \qquad \frac{\partial \hat{\phi}}{\partial \theta} = 0;$$

$$\frac{\partial \hat{\theta}}{\partial \theta} \perp \hat{\theta}; \qquad \frac{\partial \hat{\phi}}{\partial \phi} \perp \hat{\phi}.$$

2.2.5 Example of Rotation of Angular-Momentum Functions

As an example of the techniques which we have developed so far, we consider the following problem. Suppose that a p electron ($l = 1$) has its angular momentum aligned along the z axis ($m = 1$), and we rotate its orbit by 90° about the x axis. The electron will then be described by a wavefunction which is a linear combination of the states $|11\rangle, |10\rangle$, and $|1-1\rangle$. Our problem is to find that linear combination. First we will do this in a straightforward way, and then we will use two other methods which take fuller advantage of the properties of the angular-momentum eigenfunctions.

Method 1. According to (2.43,45), the state which is obtained from $|lm\rangle$ by a rotation by θ about the x axis can be written

$$\exp(-i\theta l_x)|lm\rangle = \sum_{m'} |lm'\rangle \langle lm'|\exp(-i\theta l_x)|lm\rangle. \tag{2.62}$$

So in order to find the desired linear combination, it is necessary to evaluate matrix elements of the kind

$$\langle lm'|\exp(-i\theta l_x)|lm\rangle.$$

For $l = 1$, l_x is represented by the 3×3 matrix (see Problem 2.2)

$$l_x = \frac{1}{\sqrt{2}} \begin{pmatrix} 0 & 1 & 0 \\ 1 & 0 & 1 \\ 0 & 1 & 0 \end{pmatrix},$$

and $\exp(-i\theta l_x)$ is represented by the matrix $\exp(M)$, where

$$M = -\frac{i\theta}{\sqrt{2}} \begin{pmatrix} 0 & 1 & 0 \\ 1 & 0 & 1 \\ 0 & 1 & 0 \end{pmatrix}.$$

It is easy to demonstrate that

$$M^{2n} = (-1)^n \theta^{2n} \begin{pmatrix} \frac{1}{2} & 0 & \frac{1}{2} \\ 0 & 1 & 0 \\ \frac{1}{2} & 0 & \frac{1}{2} \end{pmatrix}$$

and

$$M^{2n-1} = \frac{i}{\sqrt{2}}(-1)^n \theta^{2n-1} \begin{pmatrix} 0 & 1 & 0 \\ 1 & 0 & 1 \\ 0 & 1 & 0 \end{pmatrix},$$

where n is a positive integer. Using the formulas

$$\exp(M) = 1 + \frac{M}{1!} + \frac{M^2}{2!} + \frac{M^3}{3!} + \dots$$

$$\sin \theta = \theta - \frac{\theta^3}{3!} + \frac{\theta^5}{5!} - \frac{\theta^7}{7!} + \dots$$

$$\cos \theta = 1 - \frac{\theta^2}{2!} + \frac{\theta^4}{4!} - \frac{\theta^6}{6!} + \dots,$$

one may show that

$$\exp(-i\theta l_x) = \begin{vmatrix} \dfrac{1+\cos\theta}{2} & \dfrac{-i\sin\theta}{\sqrt{2}} & \dfrac{-(1-\cos\theta)}{2} \\[2mm] \dfrac{-i\sin\theta}{\sqrt{2}} & \cos\theta & \dfrac{-i\sin\theta}{\sqrt{2}} \\[2mm] \dfrac{-(1-\cos\theta)}{2} & \dfrac{-i\sin\theta}{\sqrt{2}} & \dfrac{1+\cos\theta}{2} \end{vmatrix}. \tag{2.63}$$

For the special case $\theta = \pi/2$ the relations, (2.62,63) give

$$\exp\left(-i\frac{\pi}{2}l_x\right)|11\rangle = \sum_m |1m\rangle \langle 1m|\exp\left(-i\frac{\pi}{2}l_x\right)|11\rangle$$

$$= \frac{1}{2}|11\rangle - \frac{i}{\sqrt{2}}|10\rangle - \frac{1}{2}|1-1\rangle. \tag{2.64}$$

Method 2. This particular problem may be solved more simply, if we use explicitly the spherical harmonics for a p function. We introduce the coordinates

$$X = \sqrt{\frac{3}{4\pi}}\frac{x}{r}, \quad Y = \sqrt{\frac{3}{4\pi}}\frac{y}{r}, \quad Z = \sqrt{\frac{3}{4\pi}}\frac{z}{r}$$

and express the spherical harmonics of rank one in terms of them (see Problem 2.6 above)

$$Y_{+1}^1 = -\frac{1}{\sqrt{2}}(X + iY)$$

$$Y_0^1 = Z$$

$$Y_{-1}^1 = \frac{1}{\sqrt{2}} (X - iY).$$

A rotation by $\pi/2$ about the x axis has the effect

$$X \to X, \quad Y \to Z, \quad Z \to -Y,$$

which implics that

$$Y_{+1}^1 \to -\frac{1}{\sqrt{2}} (X + iZ).$$

Making use of the inverse transformation

$$X = -\frac{1}{\sqrt{2}} (Y_{+1}^1 - Y_{-1}^1)$$

$$Y = \frac{i}{\sqrt{2}} (Y_{+1}^1 + Y_{-1}^1)$$

$$Z = Y_0^1,$$

we obtain

$$Y_{+1}^1 \to \frac{1}{2} Y_{+1}^1 - \frac{i}{\sqrt{2}} Y_0^1 - \frac{1}{2} Y_{-1}^1$$

which agrees with (2.64).

Method 3. A rotation by $\pi/2$ about the x axis will transform an angular momentum which is aligned along the z axis to one which is aligned along the negative y axis. Hence the linear combination which we are looking for is just the eigenvector of the matrix

$$l_y = \frac{1}{\sqrt{2}} \begin{pmatrix} 0 & -i & 0 \\ i & 0 & -i \\ 0 & i & 0 \end{pmatrix}$$

corresponding to the eigenvalues -1. We may make use of a general theorem from matrix algebra: the eigenvector of a hermitian matrix M corresponding to the nondegenerate eigenvalue λ is proportional to the minors of any row of matrix $(M - \lambda I)$. We thus find that the rotated function is of the form

$$C (|11\rangle - \sqrt{2} \, i|10\rangle - |1-1\rangle).$$

The constant C can be evaluated by using the normalization condition, and we obtain again (2.64).

Problem 2.9. Find how the spherical harmonics for $l = 1$ transform under a rotation of 45° about the z axis.

Problem 2.10. Use Method 1 above to show that

$$
\exp\left(-i\theta s_x\right) =
\begin{pmatrix}
\cos\dfrac{\theta}{2} & -i\sin\dfrac{\theta}{2} \\
-i\sin\dfrac{\theta}{2} & \cos\dfrac{\theta}{2}
\end{pmatrix}
$$

where s_x is the x component of the spin-angular momentum, represented by the Pauli spin matrix given in (2.22). How is the spin eigenstate α transformed by a rotation θ about the x axis?

Problem 2.11. If a beam of silver atoms, with a total angular momentum of $1/2$, traverses a Stern-Gerlach magnet, it will split up into two components, each of which contains atoms in specific eigenstates with respect to the field orientation. If one of the (polarized) beam components is allowed to traverse a second Stern-Gerlach magnet, rotated an angle θ about the beam direction with respect to the first magnet, it will generally split up in two new components. What is the intensity ratio for the two components?

2.3 Coupling of Angular-Momentum States and Spherical Tensor Operators

Angular-momentum states and spherical tensor operators of the type considered here are entirely adequate to describe the properties of a single spin or a single orbital system. The states of such a system can be expanded in terms of states having a definite angular momentum, and the interaction of a single spin or orbital function with an external field can be expressed in terms of spherical tensor operators. Most real systems, of course, involve several angular momenta. Each electron has a spin and an orbital angular momentum, which are coupled by the spin-orbit interaction, and for many-electron systems the angular momenta involved can give rise to quite complicated coupling schemes.

In this section we consider the coupling of two angular momenta, which leads to the vector-coupling coefficient and the 3-j symbol. The analogy between angular-momentum states and spherical tensor operators will be used to introduce the coupling of spherical tensor operators. In the last section of this chapter, we shall use the transformation properties of the angular-momentum states and spherical tensor operators to prove the important Wigner-Eckart theorem. In the next chapter we shall introduce graphical methods of representing angular momentum, and these methods will be used to extend our treatment to the coupling of three and four angular momenta.

2.3.1 Coupling of States

We consider now a system which is composed of two parts. These two subsystems are described by wave functions $|j_1 m_1\rangle$ and $|j_2 m_2\rangle$, having the property

$$
\begin{aligned}
\boldsymbol{j}^2(k)|j_k m_k\rangle &= j_k(j_k + 1)|j_k m_k\rangle \\
j_z(k)|j_k m_k\rangle &= m_k|j_k m_k\rangle ,
\end{aligned}
\tag{2.65}
$$

where k can be 1 or 2 and, for instance, $\boldsymbol{j}(1)$ is the angular-momentum operator of the first system.

The total angular momentum operator $\boldsymbol{J}$ is the sum of the two angular momenta

$$
\boldsymbol{J} = \boldsymbol{j}(1) + \boldsymbol{j}(2) .
$$

Since $\boldsymbol{j}(1)$ and $\boldsymbol{j}(2)$ refer to two different systems, the components of these operators commute,

$$
[j_i(1), j_k(2)] = 0 .
\tag{2.66}
$$

Then one can easily prove that $\boldsymbol{J}$ satisfies the same commutation relations (2.5) as do $\boldsymbol{j}(1)$ and $\boldsymbol{j}(2)$. So all of the results of Sect. 2.1 apply: $\boldsymbol{J}^2$ has eigenvalues $J(J + 1)$ and for each eigenvalue $J(J + 1)$ there are $2J + 1$ eigenfunctions of J_z having eigenvalues $M = J, J - 1, \ldots, -J$. We shall denote the eigenfunctions of $\boldsymbol{J}^2$ and J_z by $|(j_1 j_2)JM\rangle$. They have the property

$$
\begin{aligned}
\boldsymbol{J}^2|(j_1 j_2)JM\rangle &= J(J + 1)|(j_1 j_2)JM\rangle \\
J_z|(j_1 j_2)JM\rangle &= M|(j_1 j_2)JM\rangle .
\end{aligned}
\tag{2.67}
$$

The product functions

$$
|j_1 m_1, j_2 m_2\rangle = |j_1 m_1\rangle \, |j_2 m_2\rangle
\tag{2.68}
$$

are eigenfunctions of $J_z = j_z(1) + j_z(2)$ corresponding to the eigenvalue $M = m_1 + m_2$; however, these functions are not generally eigenstates of the total angular momentum $\boldsymbol{J}^2$. In order to form eigenstates of $\boldsymbol{J}^2$, it is necessary to take linear combinations of these functions

$$
\boxed{\;|(j_1 j_2)JM\rangle = \sum_{m_1 m_2} |j_1 m_1, j_2 m_2\rangle \langle j_1 m_1, j_2 m_2 | JM\rangle\;}
\tag{2.69}
$$

Taking the scalar product with $|j_1 m_1, j_2 m_2\rangle$, one may readily show that

$$
\langle j_1 m_1, j_2 m_2 | (j_1 j_2)JM\rangle = \langle j_1 m_1, j_2 m_2 | JM\rangle ,
\tag{2.70}
$$

since the functions $|j_1m_1, j_2m_2\rangle$ form an orthonormal set. So the coefficients $\langle j_1m_1, j_2m_2|JM\rangle$ are ordinary scalar products. They vanish unless $m_1 + m_2 = M$, due to the orthogonality of eigenfunctions of $J_z = j_{1z} + j_{2z}$ corresponding to different eigenvalues (Theorem B.2).

The coefficients appearing in (2.69) are known as *Clebsch-Gordan* or *vector-coupling coefficients*. The phase of these coefficients depends upon the phase of the states $|(j_1j_2)JM\rangle$, which is to some extent arbitrary. We note, however, that once the state $|(j_1j_2)JJ\rangle$, with the maximum value of M, is specified, then all of the other states are determined by the phase convention associated with the ladder operators (2.19). In order to specify the phases completely, we require that

the vector-coupling coefficient $\langle j_1m_1, j_2m_2|JM\rangle$ *for which* m_1 *and* M *have their maximum values should be real and nonnegative (positive or zero).*

This phase convention is equivalent with that used by *Condon* and *Shortley* [1935].

The inverse transformation of (2.69) can be written

$$|j_1m_1, j_2m_2\rangle = \sum_{JM} |(j_1j_2)JM\rangle \langle JM|j_1m_1, j_2m_2\rangle, \qquad (2.71)$$

and by taking the scalar product with $|(j_1j_2)JM\rangle$ one may show that the coupling coefficient for this transformation is

$$\langle JM|j_1m_1, j_2m_2\rangle = \langle (j_1j_2)JM|j_1m_1, j_2m_2\rangle = \langle j_1m_1, j_2m_2|JM\rangle^* . \qquad (2.72)$$

With the phase convention we have chosen, the vector-coupling coefficients are real. It then follows from (2.72) that

$$\langle j_1m_1, j_2m_2|JM\rangle = \langle JM|j_1m_1, j_2m_2\rangle, \qquad (2.73)$$

and so the same set of coefficients appears in both transformations.

Before discussing the general properties of the vector-coupling coefficients, we consider the problem of combining the orbital angular momentum of two p electrons ($l_1 = 1$, $l_2 = 1$). The states of this two-electron system are linear combinations of the product functions

$$|1m_1, 1m_2\rangle \quad \text{with} \quad m_1, m_2 = 1, 0, -1.$$

We begin by considering the state $|11, 11\rangle$, for which both m_1 and m_2 have their maximum value. This state has $M_L = m_1 + m_2$ equal to two, and hence must be a linear combination of states for which $L \geq 2$. Furthermore, since two is the largest value M_L can have, there can be no states for which $L > 2$. So we can write

$$|(11)22\rangle = |11, 11\rangle .$$

By operating on this state repeatedly with $L_- = l_-(1) + l_-(2)$, and using (2.19b), we obtain

$$|(11)21\rangle = \frac{1}{\sqrt{2}}(|11, 10\rangle + |10, 11\rangle)$$

$$|(11)20\rangle = \frac{1}{\sqrt{6}}(|11, 1-1\rangle + 2|10, 10\rangle + |1-1, 11\rangle)$$

$$|(11)2-1\rangle = \frac{1}{\sqrt{2}}(|10, 1-1\rangle + |1-1, 10\rangle)$$

$$|(11)2-2\rangle = |1-1, 1-1\rangle .$$

There are two product functions for which $M_L = 1$, namely $|11, 10\rangle$ and $|10, 11\rangle$. As we have seen, one linear combination of these two states corresponds to $|(11)21\rangle$. The other linear combination must correspond to the state $|(11) 11\rangle$, and we write it

$$|(11)11\rangle = a|11, 10\rangle + b|10, 11\rangle .$$

The requirement

$$L_+|(11)11\rangle = 0$$

—or the orthogonality between $|(11)11\rangle$ and $|(11)21\rangle$ —leads directly to the condition $a + b = 0$. In our phase convention, the coefficient is positive for the term where m_1 is maximum. This together with the normalization condition gives

$$|(11)11\rangle = \frac{1}{\sqrt{2}}(|11, 10\rangle - |10, 11\rangle) .$$

By applying L_- successively on this state, we can generate the states $|(11)10\rangle$ and $|(11)1-1\rangle$. The state $|(11)00\rangle$ can then be obtained by requiring that it be orthogonal to the states $|(11)20\rangle$ and $|(11)10\rangle$.

This approach may be readily generalized. A system of two angular momenta, characterized by the quantum numbers j_1 and j_2, may be coupled to a total angular momentum with the quantum number J in the range

$$J = j_1 + j_2, \quad j_1 + j_2 - 1, \ldots |j_1 - j_2| ,$$

and to each value of J correspond $2J + 1$ states with $M = J, J - 1, \ldots, - J$.

We shall now indicate how the general formula for the vector-coupling coefficient can be derived, essentially following *Judd* [1963a]. As in the previous example we begin with the state $|(j_1 j_2)JJ\rangle$ for which M has its maximum value, J. According to (2.69), this state can be written

$$|(j_1j_2)JJ\rangle = \sum_{m_1m_2} |j_1m_1, j_1m_2\rangle \langle j_1m_1, j_2m_2 | JJ\rangle. \tag{2.74}$$

By requiring that

$$J_+|(j_1j_2)JJ\rangle = [j_+(1) + j_+(2)]|(j_1j_2)JJ\rangle = 0,$$

we obtain, using (2.19),

$$\sum_{m_1m_2} \{\langle j_1m_1 - 1, j_2m_2 + 1 | JJ\rangle [j_1(j_1 + 1) - m_1(m_1 - 1)]^{1/2}$$

$$+ \langle j_1m_1, j_2m_2 | JJ\rangle [j_2(j_2 + 1) - m_2(m_2 + 1)]^{1/2}\} |j_1m_1, j_2m_2 + 1\rangle = 0.$$

The states $|j_1m_1, j_2m_2\rangle$ are linearly independent, and so each of the coefficients which appears in this summation must be zero. This leads to the recursion formula

$$\langle j_1m_1 - 1, j_2m_2 + 1 | JJ\rangle = -\left[\frac{j_2(j_2 + 1) - m_2(m_2 + 1)}{j_1(j_1 + 1) - m_1(m_1 - 1)}\right]^{1/2} \langle j_1m_1, j_2m_2 | JJ\rangle$$

$$= -\left[\frac{(j_2 - m_2)(j_2 + m_2 + 1)}{(j_1 - m_1 + 1)(j_1 + m_1)}\right]^{1/2} \langle j_1m_1, j_2m_2 | JJ\rangle, \tag{2.75}$$

where $m_1 + m_2 = J$. This formula allows us to determine all of the coefficients $\langle j_1m_1, j_2m_2 | JJ\rangle$ in terms of $\langle j_1j_1, j_2m_2 | JJ\rangle$. Equations (2.74, 75) serve to define the state $|(j_1j_2)JJ\rangle$, apart from an overall constant factor. This factor can be evaluated by applying the normalization condition

$$\langle (j_1j_2)JJ | (j_1j_2)JJ\rangle = 1$$

and the phase convention discussed previously. The expansion of the other states $|(j_1j_2)JM\rangle$ may then be found by operating successively on (2.74) with $J_- = j_-(1) + j_-(2)$. In this way, the following explicit expression for the vector-coupling coefficients is obtained after some lengthy manipulations:

$$\langle j_1m_1, j_2m_2 | JM\rangle = \delta(m_1 + m_2, M)$$

$$\times \left[\frac{(2J + 1)(j_1 + j_2 - J)!(j_1 - m_1)!(j_2 - m_2)!(J + M)!(J - M)!}{(j_1 + j_2 + J + 1)!(J + j_1 - j_2)!(J + j_2 - j_1)!(j_1 + m_1)!(j_2 + m_2)!}\right]^{1/2}$$

$$\times \sum_r (-1)^{j_1 - m_1 + r} \frac{(j_1 + m_1 + r)!(j_2 + J - m_1 - r)!}{r!(J - M - r)!(j_1 - m_1 - r)!(j_2 - J + m_1 + r)!}. \tag{2.76}$$

This formula was first derived by *Wigner* [1931] using group-theoretical methods. A purely algebraic derivation was first given by *Racah* [1942]. As the reader might judge by the complexity of the final result, the derivation of this formula is quite tedious whatever technique is used. The coupling coefficients for the special cases $j_1 = 1/2$ and $j_1 = 1$ are given in Tables 2.2 and 2.3.

Table 2.2. $\langle \frac{1}{2}m_1, j_2 m_2 | JM \rangle$

J \ m_1	$\frac{1}{2}$	$-\frac{1}{2}$
$j_2 + \frac{1}{2}$	$\sqrt{\dfrac{j_2 + M + \frac{1}{2}}{2j_2 + 1}}$	$\sqrt{\dfrac{j_2 - M + \frac{1}{2}}{2j_2 + 1}}$
$j_2 - \frac{1}{2}$	$\sqrt{\dfrac{j_2 - M + \frac{1}{2}}{2j_2 + 1}}$	$-\sqrt{\dfrac{j_2 + M + \frac{1}{2}}{2j_2 + 1}}$

Table 2.3. $\langle 1\, m_1, j_2 m_2 | JM \rangle$

J \ m_1	1	0	-1
$j_2 + 1$	$\sqrt{\dfrac{(j_2 + M)(j_2 + M + 1)}{(2j_2 + 1)(2j_2 + 2)}}$	$\sqrt{\dfrac{(j_2 - M + 1)(j_2 + M + 1)}{(2j_2 + 1)(j_2 + 1)}}$	$\sqrt{\dfrac{(j_2 - M)(j_2 - M + 1)}{(2j_2 + 1)(2j_2 + 2)}}$
j_2	$\sqrt{\dfrac{(j_2 + M)(j_2 - M + 1)}{2j_2(j_2 + 1)}}$	$-\dfrac{M}{\sqrt{j_2(j_2 + 1)}}$	$-\sqrt{\dfrac{(j_2 - M)(j_2 + M + 1)}{2j_2(j_2 + 1)}}$
$j_2 - 1$	$\sqrt{\dfrac{(j_2 - M)(j_2 - M + 1)}{2j_2(2j_2 + 1)}}$	$-\sqrt{\dfrac{(j_2 - M)(j_2 + M)}{j_2(2j_2 + 1)}}$	$\sqrt{\dfrac{(j_2 + M + 1)(j_2 + M)}{2j_2(2j_2 + 1)}}$

Problem 2.12. According to (2.69) the transformation from the $m_s m_l$ *representation* to the *jm representation* for a single-electron wave function is

$$\Psi(ljm) = \sum_{m_s + m_l = m} \langle \tfrac{1}{2}m_s, lm_l | jm \rangle \phi(lm_s m_l) . \tag{2.77}$$

Using this equation and Table 2.2, show that the explicit form of this transformation is

$$\Psi(ljm) = \sqrt{\frac{l + m + \frac{1}{2}}{2l + 1}}\, \phi(l\tfrac{1}{2}m_l) + \sqrt{\frac{l - m + \frac{1}{2}}{2l + 1}}\, \phi(l -\tfrac{1}{2}m_l)$$

$$\text{for } j = l + \tfrac{1}{2} \tag{2.78}$$

$$\Psi(ljm) = \sqrt{\frac{l - m + \frac{1}{2}}{2l + 1}}\, \phi(l\tfrac{1}{2}m_l) - \sqrt{\frac{l + m + \frac{1}{2}}{2l + 1}}\, \phi(l -\tfrac{1}{2}m_l)$$

$$\text{for } j = l - \tfrac{1}{2} .$$

Problem 2.13. Show that the full transformation matrix between the $m_s m_l$ and *jm* representations for a p electron ($l = 1$) is

jm \\ $m_s m_l$	$\frac{1}{2}\ 1$	$\frac{1}{2}\ 0$	$-\frac{1}{2}\ 1$	$\frac{1}{2}\ -1$	$-\frac{1}{2}\ 0$	$-\frac{1}{2}\ -1$
$\frac{3}{2}\ \frac{3}{2}$	1					
$\frac{3}{2}\ \frac{1}{2}$		$\sqrt{\frac{2}{3}}$	$\sqrt{\frac{1}{3}}$			
$\frac{1}{2}\ \frac{1}{2}$		$\sqrt{\frac{1}{3}}$	$-\sqrt{\frac{2}{3}}$			
$\frac{3}{2}\ -\frac{1}{2}$				$\sqrt{\frac{1}{3}}$	$\sqrt{\frac{2}{3}}$	
$\frac{1}{2}\ -\frac{1}{2}$				$\sqrt{\frac{2}{3}}$	$-\sqrt{\frac{1}{3}}$	
$\frac{3}{2}\ -\frac{3}{2}$						1

We have arranged the rows of the above matrix according to m and the columns according to $m_s + m_l$. The vector-coupling coefficients are zero, if m is not equal to $m_s + m_l$, since the basis functions in both schemes are eigenfunctions of J_z. So the nonzero elements of the transformation matrix appear in blocks—one for each value of m—along the main diagonal.

Although the general formula (2.76) for the vector-coupling coefficients is quite unwieldy for manual work, it can easily be programmed for computers, and several standard routines are now available for calculating these coefficients.

The general formula can also be used to derive some important symmetry properties. For instance, the substitution $r = J - M - t$ leads to the relation

$$\langle j_2 m_2, j_1 m_1 | JM \rangle = (-1)^{j_1 + j_2 - J} \langle j_1 m_1, j_2 m_2 | JM \rangle . \tag{2.79}$$

So the coupling coefficients have a very simple symmetry with respect to the interchange of j_1 and j_2. This is natural, since j_1 and j_2 appear on the same footing as components from which the total angular momentum J is constructed. By contrast the coupling coefficients have rather complicated properties with respect to the interchange of j_1 or j_2 with J. Again using (2.76) it is possible—although not particularly easy—to show that

$$\langle J - M, j_2 m_2 | j_1 - m_1 \rangle = (-1)^{j_2 + m_2} [(2j_1 + 1)/(2J + 1)]^{1/2} \langle j_1 m_1, j_2 m_2 | JM \rangle .$$

Another useful relation which is easier to derive is the following:

$$\langle j_1 - m_1, j_2 - m_2 | J - M \rangle = (-1)^{j_1 + j_2 - J} \langle j_1 m_1, j_2 m_2 | JM \rangle . \tag{2.80}$$

It is obvious from the relations above that the vector-coupling coefficient is not symmetric with respect to all three angular momenta involved. A more symmetric quantity is the *3-j symbol*, introduced by *Wigner*, which is defined by

$$\begin{pmatrix} j_1 & j_2 & j_3 \\ m_1 & m_2 & m_3 \end{pmatrix} = (-1)^{j_1 - j_2 - m_3}(2j_3 + 1)^{-1/2}\langle j_1 m_1, j_2 m_2 | j_3 - m_3 \rangle \quad . \tag{2.81}$$

Using the formula (2.76), also this symbol can be given an exact, algebraic form. It follows from (2.69) that the 3-j symbol vanishes, unless

$$m_1 + m_2 + m_3 = 0 \ . \tag{2.82}$$

Furthermore, the angular momenta must satisfy the *triangular condition*

$$|j_1 - j_2| \leq j_3 \leq j_1 + j_2 \ . \tag{2.83}$$

From (2.79, 81) it follows that an *even* permutation of the columns does not change the value of the 3-j symbol,

$$\begin{pmatrix} j_1 & j_2 & j_3 \\ m_1 m_2 m_3 \end{pmatrix} = \begin{pmatrix} j_2 & j_3 & j_1 \\ m_2 m_3 m_1 \end{pmatrix} = \begin{pmatrix} j_3 & j_1 & j_2 \\ m_3 m_1 m_2 \end{pmatrix} \ , \tag{2.84a}$$

while *odd* permutations introduce the phase factor $(-1)^{j_1 + j_2 + j_3}$,

$$\begin{pmatrix} j_2 & j_1 & j_3 \\ m_2 m_1 m_3 \end{pmatrix} = (-1)^{j_1 + j_2 + j_3} \begin{pmatrix} j_1 & j_2 & j_3 \\ m_1 m_2 m_3 \end{pmatrix} \ . \tag{2.84b}$$

It also follows from (2.80, 81) that

$$\begin{pmatrix} j_1 & j_2 & j_3 \\ -m_1 & -m_2 & -m_3 \end{pmatrix} = (-1)^{j_1 + j_2 + j_3} \begin{pmatrix} j_1 & j_2 & j_3 \\ m_1 m_2 m_3 \end{pmatrix} \ . \tag{2.85}$$

The additional factor $(-1)^{j_1 - j_2 - m_3} (2j_3 + 1)^{-1/2}$, which appears in (2.81), has the effect of removing the asymmetry of the vector-coupling coefficient, so that all angular momenta involved appear on the same footing. Therefore, the 3-j symbol has a much higher degree of symmetry, and it is often more convenient to use in theoretical works.

The condition that the angular-momentum states $|j_1 m_1, j_2 m_2\rangle$ and $|(j_1 j_2) j_3 m_3\rangle$ each form an orthonormal basis set

$$\langle j_1 m_1, j_2 m_2 | j_1 m_1', j_2 m_2' \rangle = \delta(m_1, m_1')\delta(m_2, m_2') \tag{2.86}$$

$$\langle (j_1 j_2) j_3 m_3 | (j_1 j_2) j_3' m_3' \rangle = \delta(j_3, j_3')\delta(m_3, m_3') \tag{2.87}$$

may be used to derive useful relations between the 3-j symbols. Together with (2.69, 71) this leads to

$$\sum_{j_3 m_3} \langle j_1 m_1, j_2 m_2 | j_3 m_3 \rangle \langle j_3 m_3 | j_1 m_1', j_2 m_2' \rangle = \delta(m_1 m_1')\delta(m_2, m_2') \quad \text{and} \tag{2.88}$$

$$\sum_{m_1 m_2} \langle j_3' m_3' | j_1 m_1, j_2 m_2 \rangle \langle j_1 m_1, j_2 m_2 | j_3 m_3 \rangle = \delta(j_3, j_3') \delta(m_3, m_3') \ . \tag{2.89}$$

Using (2.81) it then follows that the 3-j symbols satisfy the equations

$$\sum_{j_3 m_3} (2j_3 + 1) \begin{pmatrix} j_1 & j_2 & j_3 \\ m_1 m_2 m_3 \end{pmatrix} \begin{pmatrix} j_1 & j_2 & j_3 \\ m_1' m_2' m_3 \end{pmatrix} = \delta(m_1, m_1') \delta(m_2, m_2') \text{ and} \tag{2.90}$$

$$\sum_{m_1 m_2} \begin{pmatrix} j_1 & j_2 & j_3 \\ m_1 & m_2 & m_3 \end{pmatrix} \begin{pmatrix} j_1 & j_2 & j_3' \\ m_1 & m_2 & m_3' \end{pmatrix} = (2j_3 + 1)^{-1} \, \delta(j_3, j_3') \, \delta(m_3, m_3') \ . \tag{2.91}$$

Since the line of argument here is based on the orthogonality of the angular-momentum states, the relations (2.90, 91) are often referred to as the *orthogonality relations of the 3-j symbols*.

In Sect. 2.2 [(2.46)] we found that the effect of a rotation operator $P(\omega)$ upon the angular-momentum state $|\gamma j m\rangle$ was to mix the various m values corresponding to the same value of j and γ

$$P(\omega)|\gamma j m\rangle = \sum_{m'} |\gamma j m'\rangle D^j_{m'm}(\omega) \ . \tag{2.92}$$

The states $|\gamma j m\rangle$ form the basis of an irreducible representation of the rotation group. A simultaneous rotation of two functions $|\gamma_1 j_1 m_1\rangle |\gamma_2 j_2 m_2\rangle$ is described by the "direct product" of the corresponding rotation matrices $D^{j_1} \otimes D^{j_2}$,

$$P(\omega)[|\gamma_1 j_1 m_1\rangle |\gamma_2 j_2 m_2\rangle] = \sum_{m_1' m_2'} |\gamma_1 j_1 m_1'\rangle |\gamma_2 j_2 m_2'\rangle D^{j_1}_{m_1' m_1}(\omega) D^{j_2}_{m_2' m_2}(\omega).$$

According to (2.69) these product functions can be combined to form eigenfunctions of a single angular momentum in the combined space,

$$|\gamma_1 \gamma_2 (j_1 j_2) JM \rangle = \sum_{m_1 m_2} |\gamma_1 j_1 m_1 \rangle |\gamma_2 j_2 m_2 \rangle \langle j_1 m_1, j_2 m_2 | JM \rangle \ . \tag{2.93}$$

By definition, these functions transform under rotation according to the rotation matrix D^J. In the language of group theory this is expressed by saying that the direct-product representation can be "reduced" to a sum of irreducible representations formed by the coupled angular-momentum eigenfunctions. A representation of a group is reducible in this sense if by taking linear combinations of the basis functions a smaller set can be formed where the functions transform among themselves under the operations of the group.

Problem 2.14. Use the relation between the vector-coupling coefficient and the 3-j symbol to show that

$$\begin{pmatrix} j & 0 & j \\ -m & 0 & m \end{pmatrix} = (-1)^{j-m}(2j + 1)^{-1/2} \ . \tag{2.94}$$

2.3.2 Coupling of Tensor Operators

Thus far, we have considered coupling of angular-momentum states. The spherical tensor operators can be combined in a similar fashion. In analogy with (2.69), we define a coupled tensor operator X^K, with the components X_Q^K, as follows:

$$X_Q^K = \{t^{k_1}(1)u^{k_2}(2)\}_Q^K = \sum_{q_1q_2} t_{q_1}^{k_1}(1)u_{q_2}^{k_2}(2) \langle k_1q_1, k_2q_2|KQ\rangle . \tag{2.95}$$

Here t^{k_1} and u^{k_2} operate on different parts of the system. Part 1 could be the spin part, for instance, and part 2 the orbital part for a single electron, or, for a two-electron system t^{k_1} may operate on the coordinates of the first electron and u^{k_2} on those of the second. There are many interactions in atomic physics of that kind, for instance, the spin-orbit interaction or the Coulomb interaction between the electrons.

In the expression (2.95) $t_{q_1}^{k_1}$ and $u_{q_2}^{k_2}$ transform upon rotations as the angular-momentum eigenfunctions $|k_1q_1\rangle$ and $|k_2q_2\rangle$, respectively. It then follows that the linear combination (2.95) transforms as the corresponding combination of the angular-momentum eigenfunctions

$$\sum_{q_1q_2} |k_1q_1\rangle |k_2q_2\rangle \langle k_1q_1, k_2q_2|KQ\rangle .$$

According to (2.69), this last expression is equal to the coupled state $|(k_1k_2)KQ\rangle$. Thus, the operator (2.95) transforms upon rotation as the angular-momentum state $|KQ\rangle$, and hence it is a tensor operator of rank K. This operator is called a *tensor operator product* of the tensor operators t^{k_1} and u^{k_2}. The tensor-operator properties of this product can also be shown in a purely algebraic way, using the definitions (2.23).

If the two tensor operators in (2.95) have the same rank, we can combine them to form a tensor of rank zero (a scalar operator),

$$X_0^0 = \{t^k(1)u^k(2)\}_0^0 = \sum_q t_q^k(1)u_{-q}^k(2) \langle kq, k-q|00\rangle .$$

Using the definition of the 3-j symbol (2.81) and the relation (2.94), we get

$$X_0^0 = \sum_q \begin{pmatrix} k & k & 0 \\ q & -q & 0 \end{pmatrix} t_q^k(1)u_{-q}^k(2) = (-1)^k(2k+1)^{-1/2} \sum_q (-1)^q t_q^k(1)u_{-q}^k(2) .$$

Traditionally, the *scalar product* of two tensors is defined by

$$t^k(1)\cdot u^k(2) = \sum_q (-1)^q t_q^k(1)u_{-q}^k(2) . \tag{2.96}$$

Thus

$$\{t^k(1)u^k(2)\}_0^0 = (-1)^k(2k+1)^{-1/2} t^k(1)\cdot u^k(2) . \tag{2.97}$$

2.3.3 A Physical Example: The Coulomb Interaction

As an example of the coupling of tensor operators we consider the Coulomb interaction between two electrons. We represent the electron positions by the coordinate vectors r_1 and r_2, which are separated by an angle ω (Fig. 2.8). In

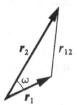

Fig. 2.8

order to expand the Coulomb interaction $(1/r_{12})$ in tensor form, we use the result from classical physics that $1/r_{12}$ can be expanded in partial waves [*Jackson* 1975, p. 100]

$$\frac{1}{r_{12}} = \sum_k \frac{r_<^k}{r_>^{k+1}} P_k(\cos \omega) ,\tag{2.98}$$

and the *addition theorem* for spherical harmonics

$$P_k(\cos \omega) = \frac{4\pi}{2k+1} \sum_q Y_q^k(\theta_1, \phi_1) Y_q^{k*}(\theta_2, \phi_2) ,\tag{2.99}$$

which enables us to write $P_k(\cos \omega)$ in terms of products of spherical harmonics that depend upon the coordinates of each of the electrons separately. Here, $r_<$ is the lesser and $r_>$ the greater of the two radial distances r_1 and r_2. If we define a "C tensor", having components

$$C_q^k = \sqrt{\frac{4\pi}{2k+1}}\, Y_q^k(\theta, \phi) ,\tag{2.100}$$

and use (2.55, 96), we can write (2.99) as

$$P_k(\cos \omega) = C^k(1) \cdot C^k(2) .\tag{2.101}$$

Equation (2.98) then takes the simple form

$$\boxed{\frac{1}{r_{12}} = \sum_k \frac{r_<^k}{r_>^{k+1}} C^k(1) \cdot C^k(2)} .\tag{2.102}$$

The Coulomb interaction is invariant with respect to a simultaneous rotation of the coordinates of the two electrons. For this reason $C^k(1)$ and $C^k(2)$ are coupled together to form a scalar operator.

2.4 The Wigner-Eckart Theorem

So far, we have considered angular-momentum states and spherical tensor operators separately, emphasizing the analogy between their transformation properties. The states and the tensor operators are affected by the rotation operators in the same way, and both transform according to irreducible representations of the rotation group. We shall now consider the simultaneous rotation of a tensor operator and an angular-momentum state. This will be used to show that matrix elements of tensor operators can be factorized into two parts, one of which is independent of the magnetic quantum numbers. This is the important *Wigner-Eckart theorem*.

2.4.1 Proof of the Theorem

We consider a tensor operator acting on an angular-momentum state $t_q^k|\gamma jm\rangle$. A rotation ω transforms this product into

$$
\begin{aligned}
P(\omega)t_q^k|\gamma jm\rangle &= P(\omega)t_q^kP(\omega)^{-1}P(\omega)|\gamma jm\rangle \\
&= \sum_{q'm'} t_{q'}^k D_{q'q}^k(\omega)|\gamma jm'\rangle D_{m'm}^j(\omega) ,
\end{aligned}
\tag{2.103}
$$

using (2.46, 47). This means that the function $t_q^k|\gamma jm\rangle$ transforms according to the direct product $D^k \otimes D^j$, that is, in the same way as the product function $|\gamma_1 kq\rangle|\gamma_2 jm\rangle$. In analogy with (2.93), we can then form linear combination, which transform according to D^J,

$$
\phi(\beta JM) = \sum_{qm} t_q^k|\gamma jm\rangle \langle kq, jm|JM\rangle .
\tag{2.104}
$$

It follows from the transformation properties that these functions are angular-momentum eigenfunctions corresponding to the quantum numbers J and M. They need not, however, be identical to the basis functions $|\gamma JM\rangle$ we have used above. In general, we have

$$
\phi(\beta JM) = \sum_{\gamma''} c(\gamma''J)|\gamma''JM\rangle .
\tag{2.105}
$$

By applying the ladder operators (2.19) to this relation, one can show that the expansion coefficients $c(\gamma''J)$ are independent of the magnetic quantum number M. Obviously, they must also be independent of the quantum numbers q and m, which are summed over, but they do depend on the remaining quantum numbers of the state and, of course, on the tensor t^k.

We would like now to express the function $t_q^k|\gamma jm\rangle$ in terms of the coupled basis functions $|\gamma'JM\rangle$. The coefficient which appears in (2.104) is a vector-coupling coefficient, and the inverse transformation is then obtained in analogy with (2.71)

$$t_q^k |\gamma jm\rangle = \sum_{JM} \phi(\beta JM) \langle JM|kq, jm\rangle .$$

Using (2.105) this last equation becomes

$$t_q^k |\gamma jm\rangle = \sum_{\gamma''JM} c(\gamma''J)|\gamma''JM\rangle \langle JM|kq, jm\rangle .$$

We now take the scalar product with $\langle \gamma'j'm'|$ and use the orthogonality of the functions to obtain

$$\langle \gamma'j'm'|t_q^k|\gamma jm\rangle = c(\gamma'j') \langle j'm'|kq, jm\rangle . \tag{2.106}$$

This shows that the matrix element can be factorized into a vector-coupling coefficient and a factor independent of the magnetic quantum numbers. This is the *Wigner-Eckart theorem* [*Wigner* 1931; *Eckart* 1930]. Interchanging primed and unprimed indices in (2.106) and using the relation (2.81) between the vector-coupling coefficient and the 3-j symbol, we obtain

$$\langle \gamma jm|t_q^k|\gamma'j'm'\rangle = (-1)^{j-m}\begin{pmatrix} j & k & j' \\ -m & q & m' \end{pmatrix} [(-1)^{k-j'+j}(2j+1)^{1/2} c(\gamma j)] .$$

Here, we have also used the symmetry properties of the 3-j symbol and the fact that $(j-m)$ is an integer. We shall denote the quantity in the square bracket as $\langle \gamma j \| t^k \| \gamma'j'\rangle$ and refer to it as the *reduced matrix element*. Then the Wigner-Eckart theorem assumes the following form:

$$\boxed{\langle \gamma jm|t_q^k|\gamma'j'm'\rangle = (-1)^{j-m}\begin{pmatrix} j & k & j' \\ -m & q & m' \end{pmatrix} \langle \gamma j \| t^k \| \gamma'j'\rangle .} \tag{2.107}$$

The reduced matrix element is *independent of m, m′*, and q, which means that the dependence on these quantum numbers is contained entirely in the 3-j symbol and the phase factor.

Problem 2.15. Use the relation (2.94) and the Wigner-Eckart theorem to show that the reduced matrix element of unity (unit tensor operator) is

$$\langle \gamma j \| 1 \| \gamma'j'\rangle = \delta(\gamma, \gamma') \, \delta(j, j') \, (2j+1)^{1/2} . \tag{2.108}$$

Problem 2.16. a) Show that

$$\begin{pmatrix} j & 1 & j \\ -j & 0 & j \end{pmatrix} = \left[\frac{j}{(j+1)(2j+1)}\right]^{1/2} .$$

Hint: Evaluate the corresponding vector-coupling coefficient by means of ladder operators.

b) Use the result above to show that

$$\langle \gamma j \| \boldsymbol{J} \| \gamma' j' \rangle = \delta(\gamma, \gamma')\, \delta(j, j')\, [j(j+1)(2j+1)]^{1/2} \, . \tag{2.109}$$

c) Use the Wigner-Eckart theorem and (2.109) to show that

$$\begin{pmatrix} j & 1 & j \\ -m & 0 & m \end{pmatrix} = (-1)^{j-m}\, [j(j+1)(2j+1)]^{-1/2} m \qquad \text{for} \;\; j \neq 0 \, . \tag{2.110}$$

Problem 2.17. Using the Wigner-Eckart theorem, show that the matrix elements of the components of the angular-momentum operators $\boldsymbol{L}$ and $\boldsymbol{S}$ between states for which $\boldsymbol{S}$ and $\boldsymbol{L}$ are coupled to form $\boldsymbol{J}$, satisfy the equations

$$\langle (SL)JM | L_q | (SL)JM' \rangle = \alpha_L \langle (SL)JM | J_q | (SL)JM' \rangle \tag{2.111a}$$

$$\langle (SL)JM | S_q | (SL)JM' \rangle = \alpha_S \langle (SL)JM | J_q | (SL)JM' \rangle \, , \tag{2.111b}$$

where

$$\alpha_L = \frac{\langle (SL)J \| L \| (SL)J \rangle}{\langle (SL)J \| J \| (SL)J \rangle} \tag{2.112a}$$

and

$$\alpha_S = \frac{\langle (SL)J \| S \| (SL)J \rangle}{\langle (SL)J \| J \| (SL)J \rangle} \, . \tag{2.112b}$$

Show also that the coefficients α_L and α_S satisfy the relation

$$\alpha_L + \alpha_S = 1 \, . \tag{2.113}$$

Equations (2.111) are illustrations of the general fact that a tensor operator of rank 1 (a vector operator) can be replaced by a constant times $\boldsymbol{J}$ itself within the manifold of the states $| \gamma JM \rangle$ with $M = J, J-1, \ldots, -J$. This is a direct consequence of the Wigner-Eckart theorem. Such operators are referred to as "class-T operators" by *Condon* and *Shortley* [1935]. This important property of a vector operator forms the basis for the construction of many "equivalent operators" in atomic physics. As an illustration we shall consider the Zeeman effect.

2.4.2 A Physical Example: the Zeeman Effect

The Zeeman effect is caused by the interaction of the atomic magnetic moment with a weak, external magnetic field. The operator representing this interaction is

$$H_m = -\boldsymbol{\mu} \cdot \boldsymbol{B} = \mu_B B(L_z + g_s S_z), \tag{2.114}$$

where μ is the magnetic moment of the atom, B the magnetic field (directed along the z axis), μ_B is the Bohr magneton, and g_s is the g value of the electron spin. According to Dirac's relativistic theory, g_s is exactly equal to 2. When quantum-electrodynamic effects are taken into account, however, a value of g_s is obtained, which is slightly larger than 2. Accurate calculations yield the value $g_s = 2.002319$, which is also in very good agreement with the experimental result.

Since the magnetic field is assumed to be weak, we are interested in matrix elements of H_m between states $|(SL)JM\rangle$ having definite values of S, L, and J. For states of this kind, we can make the following replacements:

$$L \Rightarrow \alpha_L J \quad \text{and} \quad S \Rightarrow \alpha_S J \tag{2.115}$$

according to the arguments above. Here, α_L and α_S are constant factors, independent of M. Within an SLJ manifold we can then replace the operator H_m by the "equivalent" operator

$$H_m \Rightarrow \mu_B B g_J J_z \,, \tag{2.116}$$

where

$$g_J = \alpha_L + g_s \alpha_S \tag{2.117}$$

is also independent of M. The last constant is known as the *Landé g factor*.

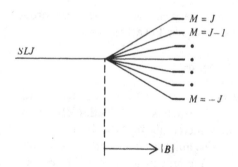

$M = J$
$M = J-1$
SLJ
$M = -J$
$|B|$

Fig. 2.9. Zeeman splitting of an SLJ level

The effective interaction (2.116) splits the states having a definite value of J into $(2J + 1)$ equidistant levels as shown in Fig. 2.9. As long as the magnetic field is sufficiently weak, the separation of these sublevels is proportional to the field strength. In order to calculate the magnitude of the splitting, it is necessary to evaluate α_L and α_S. According to (2.112a,b), these coefficients are related to the reduced matrix elements of L and S. General equations, which will enable us to evaluate these reduced matrix elements, will be given in Sect. 4.3. We evaluate these coefficients here by means of a simple argument.

According to (2.115), the operator

$$L^2 = (J - S)^2 = J^2 + S^2 - 2J \cdot S \tag{2.118}$$

can be replaced by

$$L^2 \Rightarrow J^2 + S^2 - 2\alpha_S J^2 \tag{2.119}$$

when operating within an SLJ manifold. The identity of the diagonal elements leads to

$$L(L + 1) = J(J + 1) + S(S + 1) - 2\alpha_S J(J + 1). \tag{2.120}$$

Equations (2.113, 120) may be solved for α_S and α_L

$$\alpha_S = \frac{\langle(SL)J\|S\|(SL)J\rangle}{\langle(SL)J\|J\|(SL)J\rangle} = \frac{J(J+1) + S(S+1) - L(L+1)}{2J(J+1)} \tag{2.121}$$

and

$$\alpha_L = 1 - \alpha_S = \frac{J(J + 1) - S(S + 1) + L(L + 1)}{2J(J + 1)} \tag{2.122}$$

If we take $g_s = 2$, the expression for the Landé factor (2.117) then becomes

$$g_J = 1 + \frac{J(J + 1) + S(S + 1) - L(L + 1)}{2J(J + 1)}. \tag{2.123}$$

This result, which is usually derived in elementary courses by the nonrigorous vector model, is a direct consequence of the Wigner-Eckart theorem.

2.4.3. Reduced Matrix Elements of the C Tensor

The Winger-Eckart theorem is a fundamental result, which we shall use frequently in the following to evaluate the matrix elements of interaction operators. As can be seen from the Zeeman effect considered above, in order to actually make use of this theorem, it is necessary to be able to evaluate reduced matrix elements. We conclude this section by deriving an expression for the reduced matrix elements of the C tensor (2.100), which occurs often in atomic physics. The Wigner-Eckart theorem (2.107) gives

$$\langle lm | C_q^k | l'm' \rangle = \sqrt{\frac{4\pi}{2k + 1}} \int Y_m^l(\Omega)^* Y_q^k(\Omega) Y_{m'}^{l'}(\Omega) d\Omega \tag{2.124}$$

$$= (-1)^{l-m} \begin{pmatrix} l & k & l' \\ -m & q & m' \end{pmatrix} \langle l\|C^k\|l'\rangle,$$

where Ω represents the angular coordinates θ and ϕ. The integral which occurs here is the coefficient in the expansion

$$Y_q^k(\Omega) Y_{m'}^{l'}(\Omega) = \sum_{lm} a_{lm} Y_m^l(\Omega) .$$

Solving (2.124) for the integral and substituting this into the expansion gives (omitting Ω for brevity)

$$Y_q^k Y_{m'}^{l'} = \sqrt{\frac{2k+1}{4\pi}} \sum_{lm} (-1)^{l-m} \begin{pmatrix} l & k & l' \\ -m & q & m' \end{pmatrix} \langle l \| C^k \| l' \rangle Y_m^l . \qquad (2.125)$$

The orthogonality condition (2.91) can then be used to obtain

$$\sum_{qm'} \begin{pmatrix} l'' & k & l' \\ -m'' & q & m' \end{pmatrix} Y_q^k Y_{m'}^{l'} = \sqrt{\frac{2k+1}{4\pi}} \frac{(-1)^{l''-m''}}{2l''+1} \langle l'' \| C^k \| l' \rangle Y_{m''}^{l''} . \qquad (2.126)$$

This equation is valid for all angles θ and ϕ. According to (2.61), it becomes for $\theta = 0$

$$\begin{pmatrix} l'' & k & l' \\ 0 & 0 & 0 \end{pmatrix} \frac{[2k+1)(2l'+1)]^{1/2}}{4\pi} = (-1)^{l''} \frac{1}{4\pi} \left[\frac{2k+1}{2l''+1} \right]^{1/2} \langle l'' \| C^k \| l' \rangle .$$

Solving for the reduced matrix element, we obtain finally

$$\boxed{\langle l \| C^k \| l' \rangle = (-1)^l [(2l+1)(2l'+1)]^{1/2} \begin{pmatrix} l & k & l' \\ 0 & 0 & 0 \end{pmatrix} .} \qquad (2.127)$$

The 3-j symbol here is zero unless the angular momenta l, k, and l' statisfy the triangular condition (2.83). Furthermore, it follows from (2.85) that *the sum of the angular momenta, $l + k + l'$, must be even.* By applying the Wigner-Eckart theorem (2.107), we can get an expression also for the complete matrix element

$$\langle lm | C_q^k | l'm' \rangle = (-1)^m [2l+1)(2l'+1)]^{1/2} \begin{pmatrix} l & k & l' \\ -m & q & m' \end{pmatrix} \begin{pmatrix} l & k & l' \\ 0 & 0 & 0 \end{pmatrix} . \qquad (2.128)$$

Problem 2.18. Use the relations (2.94,127) to verify the following results:

$$\langle l \| C^0 \| l' \rangle = \delta(l, l')(2l+1)^{1/2} \qquad (2.129)$$

$$\langle l \| C^k \| 0 \rangle = (-1)^l \langle 0 \| C^k \| l \rangle = \delta(l, k) . \qquad (2.130)$$

Problem 2.19. Show the following relation:

$$\{C^{k_1}C^{k_2}\}_Q^K = (-1)^K (2K+1)^{1/2} \begin{pmatrix} k_1 & K & k_2 \\ 0 & 0 & 0 \end{pmatrix} C_Q^K. \tag{2.131}$$

where all the C tensors refer to the same coordinates.
Hint: Use the relation (2.126).

3. Angular-Momentum Graphs

The techniques of dealing with angular momentum developed in the previous chapter are straightforward in principle, but they can become quite complicated when applied to real physical problems. The effort involved in practical applications can be reduced considerably by using graphical methods. Graphical methods of treating angular momenta were first introduced by *Jucys* et al. [1962], who used them to simplify products of 3-j symbols. This technique has been further developed by a number of authors [*Brink* and *Satchler* 1968; *Sandars* 1971, *El Baz* and *Castel* 1972]. In this book we shall use graphical methods extensively, and we shall follow mainly the works of *Brink* and *Satchler* and of *Sandars*.

In the first section of this chapter we shall give all the basic rules which we need to represent angular momenta graphically. These rules will then be used to derive a graphical representation of the vector-coupling coefficient, of coupled states, and of the Winger-Eckart theorem. Rules for treating diagrams with two or more vertices are developed in the second section, and these results are then applied to a particular physical problem: the interaction of a single electron with all electrons of a closed shell. The last two sections of the chapter deal with the problem of coupling three and four angular momenta, using the graphical technique developed thus far. In the next chapter we shall develop the graphical technique further by means of a number of theorems of *Jucys* et al. In the second part of the book we shall find this technique extremely useful in evaluating the spin and angular parts of the Feynman-like diagrams, representing different terms in the many-body perturbation expansion.

3.1 Representation of 3-j Symbols and Vector-Coupling Coefficients

3.1.1 Basic Conventions

In the graphical scheme we shall employ in this book, the Wigner 3-j symbol is represented in the following way:

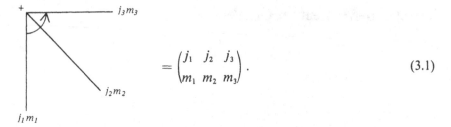

$$= \begin{pmatrix} j_1 & j_2 & j_3 \\ m_1 & m_2 & m_3 \end{pmatrix}. \tag{3.1}$$

The angular momenta correspond to the three lines, which meet at a *vertex*. We have written the value of the angular momentum and its projection near the free end of each line. The arrow and the corresponding sign by the vertex indicate whether the arguments of the 3-*j* symbol are ordered in a counter-clockwise direction (positive sign) or a clockwise direction (negative sign). A change of the cyclic order of the lines around a vertex corresponds to interchanging two of the columns of the 3-*j* symbol. Using (2.84), we then get the relation

$$= (-1)^{j_1+j_2+j_3} \tag{3.2}$$

In the following we shall leave out the arrow at the vertex and denote the order of the angular momenta by the sign only.

In order to incorporate phase factors in our graphical notation, we shall introduce directed lines in the following way:

$$= (-1)^{j_3-m_3} \begin{pmatrix} j_1 & j_2 & j_3 \\ m_1 & m_2 & -m_3 \end{pmatrix} \tag{3.3a}$$

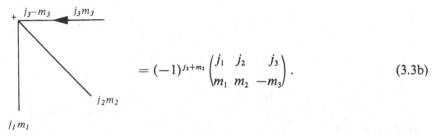

$$= (-1)^{j_3+m_3} \begin{pmatrix} j_1 & j_2 & j_3 \\ m_1 & m_2 & -m_3 \end{pmatrix}. \tag{3.3b}$$

These equations serve as definitions of in- and out-going arrows. Note that the m value changes its sign when passing through an arrow and that the phase factor is $(-1)^{j-m}$, if the arrow points *towards* the end with the quantum number jm.[1]

We shall also need a definition of a single line and adopt the following notation:

$$\underline{jm \qquad j'm'} = \delta(j,j')\,\delta(m,m') \tag{3.4}$$

This implies that a single line without an arrow must have the same quantum numbers at both ends. For numerical purposes it then has the value one. Similarly, for a directed line,

$$\underline{jm \xleftarrow{} j'm'} = (-1)^{j-m}\underline{jm \qquad j'\ -m'} = (-1)^{j-m}\,\delta(j,j')\,\delta(m,-m')$$
$$\underline{jm \xrightarrow{} j'm'} = (-1)^{j'-m'}\underline{jm \qquad j'\ -m'} = (-1)^{j+m}\,\delta(j,j')\,\delta(m,-m'). \tag{3.5}$$

If follows directly from (3.5) that reversing the direction of an arrow introduces a phase factor of $(-1)^{2m} = (-1)^{2j}$,

$$\underline{jm \xleftarrow{} j'm'} = (-1)^{2j}\ \underline{jm \xrightarrow{} j'm'}. \tag{3.6}$$

Furthermore, for a line with two arrows pointing in opposite directions, we get

$$\underline{jm \xleftrightarrow{} j'm'} = (-1)^{j-m}(-1)^{j'-m'}\ \underline{jm \qquad j'm'}$$
$$= (-1)^{2j-2m}\ \underline{jm \qquad j'm'} = \underline{jm \qquad j'm'}. \tag{3.7a}$$

Similarly,

$$\underline{jm \xrightarrow{}\xleftarrow{} j'm'} = \underline{jm \qquad j'm'}. \tag{3.7b}$$

If two arrows point in the same direction, we can reverse the direction of one of them using (3.6), which gives

[1] Note that *Brink* and *Satchler* [1968] as well as *Sandars* [1971] used the opposite phase convention. As we shall see later, our convention will lead to a closer analogy between the angular-momentum diagrams and the Feynman-like diagrams introduced later to represent terms in the perturbation expansion.

$$\xrightarrow[jm]{\quad}\!\!\blacktriangleright\!\!\xrightarrow[\quad]{j'm'} = (-1)^{2J}\ \xrightarrow[jm]{\quad}\!\!\blacktriangleleft\!\!\xrightarrow[\quad]{j'm'} = (-1)^{2J}\ \overline{\ \ jm\qquad\qquad j'm'\ \ } \tag{3.8}$$

$$\xrightarrow[jm]{\quad}\!\!\blacktriangleleft\!\!\blacktriangleleft\!\!\xrightarrow[\quad]{j'm'} = (-1)^{2J}\ \overline{\ \ jm\qquad\quad j'm'\ }\ .$$

We now derive a useful graphical rule which will enable us to add arrows to all three lines of a diagram (3.1). If all arrows at the vertex are directed outwards, we can use (3.3a), (2.82) and (2.85) to get

$$= (-1)^{j_1-m_1+j_2-m_2+j_3-m_3}\begin{pmatrix} j_1 & j_2 & j_3 \\ -m_1 & -m_2 & -m_3 \end{pmatrix}$$

$$= (-1)^{j_1+j_2+j_3}\begin{pmatrix} j_1 & j_2 & j_3 \\ -m_1 & -m_2 & -m_3 \end{pmatrix} = \begin{pmatrix} j_1 & j_2 & j_3 \\ m_1 & m_2 & m_3 \end{pmatrix}.$$

We then have the identity

$$\tag{3.9a}$$

and similary,

$$\tag{3.9b}$$

This means that we can *add arrows pointing either inwards or outwards on all three lines of a vertex* without changing the value of the diagram. In the following we shall leave out the quantum numbers at the vertex, since these can be obtained from those at the free ends by means of (3.5).

As an illustration of the rules given above, we consider the diagram

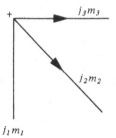

By adding incoming arrows on all lines and using (3.7), we get

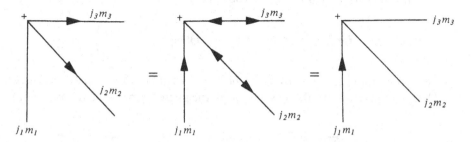

According to (3.2) and (3.6) whenever the orientation of the angular momenta about a vertex is changed or the direction of an arrow is reversed, a phase factor is introduced. In order to simplify the phase factors, which arise in this way, it is often useful to recall the various integer values that can be formed from the angular-momentum quantum numbers. From j and its projection m, it is possible to form the integers $2j$, $2m$, $j - m$, and $j + m$. Also, if j_1, j_2, j_3 satisfy the triangular condition, then $j_1 + j_2 + j_3$ is an integer. For any integer k, one can furthermore use the identities

$$(-1)^k = (-1)^{-k}$$
$$(-1)^{2k} = +1 .$$

Problem 3.1 Show that

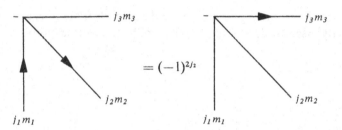

Problem 3.2 If j_1, j_2, j_3 satisfy the triangular condition and m_1, m_2, m_3 are the corresponding projections, show that

a) $(-1)^{j_1-j_2+j_3}(-1)^{j_1+j_2+j_3} = (-1)^{2j_2}$

b) $(-1)^{j_1-j_2+m_3} = (-1)^{j_1-j_2+j_3}(-1)^{j_3-m_3}$

c) $(-1)^{2j_1-j_2+m_3} = (-1)^{j_2-m_3}$.

3.1.2 Representation of the Vector-Coupling Coefficient

We shall now find a graphical representation of the vector-coupling coefficient $\langle j_1 m_1, j_2 m_2 | j_3 m_3 \rangle$. In order to do this we first use the relation between the coupling coefficient and the 3-j symbol (2.81)

$$\langle j_1 m_1, j_2 m_2 | j_3 m_3 \rangle = (-1)^{j_1-j_2+m_3}(2j_3 + 1)^{1/2} \begin{pmatrix} j_1 & j_2 & j_3 \\ m_1 & m_2 & -m_3 \end{pmatrix} .$$

Using the graphical representation of the 3-j symbol (3.1) and the definition (3.3), we can then write the 3-j symbol and the phase factor in the following way:

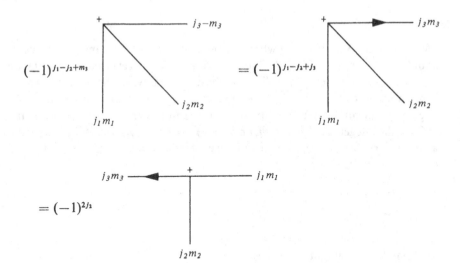

where we have used the fact that $j_3 - m_3$, $2j_2$, and $j_1 + j_2 + j_3$ are integers, as discussed above. By means of (3.9) and (3.6) this can be transformed into

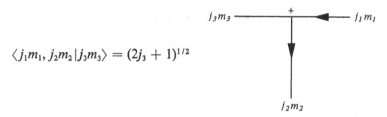

$$\langle j_1 m_1, j_2 m_2 | j_3 m_3 \rangle = (2j_3 + 1)^{1/2}$$

The factor $(2j + 1)$ occurs often in angular-momentum expressions, and particularly in connection with diagrams we shall use the shorthand notation[2]

$$[j] = 2j + 1 . \tag{3.10}$$

We shall also introduce a heavy line in order to incorporate a factor of this kind in the diagrams. This is defined according to

$$\overset{jm}{\rule{2cm}{0.4pt}\blacksquare} = [j]^{1/2} \overset{jm}{\rule{2cm}{0.4pt}} \tag{3.11}$$

Using the notation (3.11) we get the following graphical representation of the vector-coupling coefficient:

$$\langle j_1 m_1, j_2 m_2 | j_3 m_3 \rangle = \qquad\qquad\qquad\qquad . \tag{3.12}$$

This representation of the vector-coupling coefficient can be easily remembered, if one imagines that $j_1 m_1$ is an initial state, which scatters off the heavy line into the final state $j_2 m_2$. This gives the direction of the arrows in agreement with our convention. The heavy line corresponds to the total angular momentum $j_3 m_3$. Note that the angular momenta in the diagram are read in the order $j_3 j_2 j_1$.

Problem 3.3. Use the graphical representation of the vector-coupling coefficient to derive the relation (2.79)

$$\langle j_2 m_2, j_1 m_1 | JM \rangle = (-1)^{j_1 + j_2 - J} \langle j_1 m_1, j_2 m_2 | JM \rangle .$$

---

[2] Generally, we use the following extension of this notation:
$$[a,b, ... / c, ...] = (2a + 1)(2b + 1) ... / (2c + 1) ...$$

Solution:

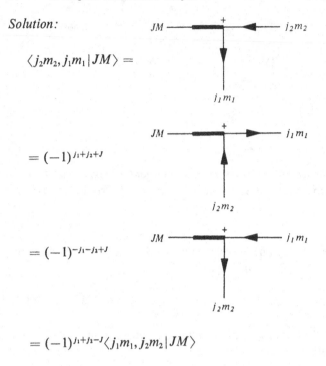

$$\langle j_2 m_2, j_1 m_1 | JM \rangle =$$

$$= (-1)^{j_1+j_2+J}$$

$$= (-1)^{-j_1-j_2+J}$$

$$= (-1)^{j_1+j_2-J} \langle j_1 m_1, j_2 m_2 | JM \rangle$$

3.1.3 Representation of Coupled States

The graphical representation of the vector-coupling coefficient (3.12) can also be used to give a representation of coupled states. The state (2.69)

$$|(j_1 j_2)JM\rangle = \sum_{m_1 m_2} |j_1 m_1, j_2 m_2\rangle \langle j_1 m_1, j_2 m_2 | JM \rangle$$

can be expressed as

$$|(j_1 j_2)JM\rangle = \sum_{m_1 m_2} |j_1 m_1, j_2 m_2\rangle \qquad\qquad\qquad (3.13)$$

We can imagine that the uncoupled states, $|j_1 m_1\rangle$ and $|j_2 m_2\rangle$, are associated with the free ends of the lines to which they correspond. Later, we shall omit these states, but for the time being we write them out explicitly in the interest of clarity. Similarly, the corresponding bra vector is

$$\langle(j_1 j_2)JM| = \sum_{m_1 m_2} \quad JM \quad\quad\quad \langle j_1 m_1, j_2 m_2|. \qquad (3.14)$$

We can also in the same manner represent the uncoupled states by means of coupled ones, using the relation (2.71),

$$|j_1 m_1, j_2 m_2\rangle = \sum_{JM} |(j_1 j_2)JM\rangle \langle JM|j_1 m_1, j_2 m_2\rangle.$$

This gives the graphical representation of the uncoupled ket vector

$$|j_1 m_1, j_2 m_2\rangle = \sum_{JM} |(j_1 j_2)JM\rangle \quad JM \qquad\qquad (3.15)$$

where a coupled state is now associated with the free end to the left. Similarly,

$$\langle j_1 m_1, j_2 m_2| = \sum_{JM} \qquad\qquad JM \ \langle(j_1 j_2)JM|. \qquad (3.16)$$

3.1.4 The Wigner-Eckart Theorem

The Wigner-Eckart theorem (2.107),

$$\langle \gamma j m | t_q^k | \gamma' j' m' \rangle = (-1)^{j-m} \begin{pmatrix} j & k & j' \\ -m & q & m' \end{pmatrix} \langle \gamma j \| t^k \| \gamma' j' \rangle,$$

can also be represented graphically. Using the representation (3.1) of the 3-*j* symbol and the phase convention (3.3), we get

$$\langle \gamma j m | t_q^k | \gamma' j' m' \rangle = \qquad\qquad\qquad \langle \gamma j \| t^k \| \gamma' j' \rangle \qquad (3.17)$$

Note that there is an *outgoing* arrow on the line representing the *final* (bra) state. Due to our above-mentioned phase convention, this is again opposite to the notation of *Brink* and *Satchler* and of *Sandars*.

Equations (3.13–16) give us a graphical representation of the states, and (3.17) represents the matrix elements of a single tensor operator. These equations, which involve diagrams with only one vertex, play an important role in the work which follows. As we shall see, the matrix elements of two-body operators can be formed by joining together the lines of simple diagrams of this kind.

3.1.5 Elimination of a Zero Line

It often happens that a line in a graph corresponds to zero angular momentum. In such a case the graph can be simplified, and we shall illustrate this here for a 3-j symbol. According to (2.94) and (2.84), we have

$$\begin{pmatrix} j_1 & j_2 & 0 \\ m_1 & m_2 & 0 \end{pmatrix} = \delta(j_1, j_2)\delta(m_1, -m_2)(-1)^{j_1-m_1}(2j_1 + 1)^{-1/2},$$

which corresponds to the graphical expression

$$= [j_1]^{-1/2} \qquad\qquad (3.18)$$

using the shorthand notation $[j]$ for $(2j + 1)$ introduced in (3.10). Let us now add incoming arrows on all three lines of the graph to the left and change the direction of one of them. We then get

$$= (-1)^{2j_1}[j_1]^{-1/2} \qquad = [j_1]^{-1/2} \qquad\qquad (3.19)$$

since an arrow on a line with $j = 0$ has no meaning. This gives us the basic rules for removing a $j = 0$ line from a graph.

Problem 3.4. Show the relations

$$\begin{array}{ccc}
j_1 m_1 & & j_1 m_1 \\
\left| \overline{}^{\;00} \right. & = & [j_1]^{-1/2} \\
j_2 m_2 & & j_2 m_2
\end{array} \tag{3.20a}$$

$$\begin{array}{ccc}
j_1 m_1 & & j_1 m_1 \\
\left| \overline{}^{\;00} \right. & = & [j_1]^{-1/2} \\
j_2 m_2 & & j_2 m_2
\end{array} \tag{3.20b}$$

$$\begin{array}{ccc}
j_1 m_1 & & j_1 m_1 \\
\left| \overline{}^{\;00} \right. & = & [j_1]^{-1/2} \\
j_2 m_2 & & j_2 m_2
\end{array} \tag{3.20c}$$

3.2 Diagrams with Two or More Vertices

3.2.1 Summation Rules

In angular-momentum theory one often encounters products of elementary diagrams (3.1) with twice-repeated magnetic quantum numbers. Free lines with the same quantum numbers at the free ends can be joined. For instance,

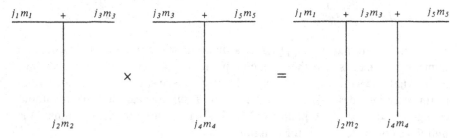

According to (3.1), this diagram corresponds to the product of the 3-j symbols

$$\begin{pmatrix} j_1 & j_2 & j_3 \\ m_1 & m_2 & m_3 \end{pmatrix} \begin{pmatrix} j_3 & j_4 & j_5 \\ m_3 & m_4 & m_5 \end{pmatrix} .$$

Similarly, for a directed line

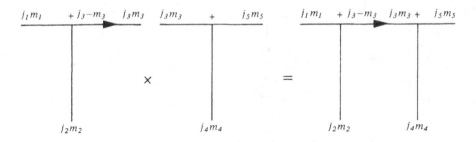

We now introduce the convention that *there is a summation over the m value of an internal line, indicated by leaving out this quantum number from the graph.* Thus,

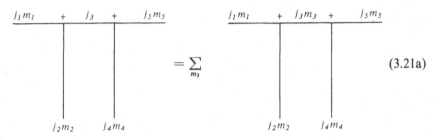

$$(3.21a)$$

and

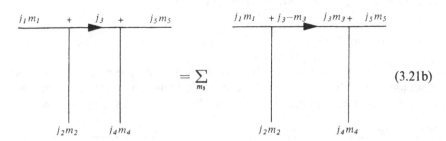

$$(3.21b)$$

So when an internal line appears in a diagram without a magnetic quantum number, this means that the magnetic quantum number is summed over. For an internal line with $j = 0$, the only value m can have is zero. So the summation in this case includes only a single term, and a null line can be broken without changing the diagram. Equations (3.18–20) may then be used to eliminate the null lines completely from the diagram.

3.2.2 Orthogonality Relations

As a first example of the graphical methods, which we have developed so far, we shall write the orthogonality condition (2.87),

$$\langle (j_1 j_2) j_3 m_3 | (j_1 j_2) j_3' m_3' \rangle = \delta(j_3, j_3') \delta(m_3, m_3') \tag{3.22}$$

in graphical form. According to (3.13,14), the states which appear in the scalar product can be written

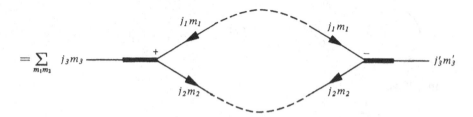

$$\langle (j_1 j_2) j_3 m_3 | = \sum_{m_1 m_2} \; j_3 m_3 \qquad \qquad \langle j_1 m_1, j_2 m_2 | \tag{3.23}$$

$$|(j_1 j_2) j_3' m_3' \rangle = \sum_{m_1' m_2'} |j_1 m_1', j_2 m_2' \rangle \qquad \qquad j_3' m_3' \tag{3.24}$$

and according to (3.4) the delta factors on the right-hand side of (3.22) may be represented by a single line

$$\delta(j_3, j_3') \delta(m_3, m_3') = \overset{j_3 m_3 \qquad \qquad j_3' m_3'}{\rule{3cm}{0.4pt}} .$$

We can now form the scalar product in (3.22) by combining the functions (3.23, 24). Only terms with $m_1 = m_1'$ and $m_2 = m_2'$ will contribute due to the orthogonality of the uncoupled functions. We can then join the corresponding lines

$$\langle (j_1 j_2) j_3 m_3 | (j_1 j_2) j_3' m_3' \rangle$$

$$= \sum_{m_1 m_2} \; j_3 m_3 \qquad \qquad \qquad \qquad \qquad \qquad j_3' m_3'$$

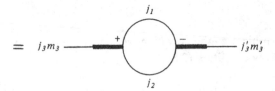

using the summation convention above and the relation (3.7) to eliminate the arrows. The orthogonality relation (3.22) then has the following graphical representation:

$$j_3 m_3 \quad\text{(+)}\quad\text{(−)}\quad j'_3 m'_3 \;=\; \underline{j_3 m_3 \qquad j'_3 m'_3}. \tag{3.25}$$

Using the definition (3.11), this can also be written

$$j_3 m_3 \quad\text{(+)}\quad\text{(−)}\quad j'_3 m'_3 \;=\; [j_3]^{-1}\, \underline{j_3 m_3 \qquad j'_3 m'_3}. \tag{3.26}$$

This is nonzero only if $j'_3 = j_3$ and $m'_3 = m_3$. In that case we can join the free ends of the diagram and sum over m_3, which yields

$$+ \bigoplus_{\,j_3}^{\,j_1\;\; j_2} - \;=\; [j_3]^{-1}\; \bigcirc\; j_3$$

The loop represents simply

$$\bigcirc\, j_3 \;=\; \sum_{m_3} \underline{j_3 m_3 \qquad j_3 m_3} = 2j_3 + 1 = [j_3], \tag{3.27}$$

and, thus,

$$+ \bigoplus_{\,j_3}^{\,j_1\;\; j_2} - \;=\; 1 \,.$$

In order to show explicitly that the angular momenta involved should satisfy the triangular condition, this relation is often written

$$+ \bigcirc_{\substack{j_1 \\ j_2 \\ j_3}}^{} - \; = \Delta(j_1, j_2, j_3) . \tag{3.28}$$

The symbol $\Delta(j_1, j_2, j_3)$ is equal to one if the three angular momenta satisfy the triangular condition (2.83) and zero otherwise.

In a similar way we can use the orthogonality condition (2.86)

$$\langle j_1 m_1, j_2 m_2 | j_1 m'_1, j_2 m'_2 \rangle = \delta(m_1, m'_1)\delta(m_2, m'_2) \tag{3.29}$$

and the graphical representations of the states given in (3.15, 16)

$$\langle j_1 m_1, j_2 m_2 | = \sum_{JM} \quad \begin{array}{c} j_1 m_1 \\ \searrow \\ \nearrow \\ j_2 m_2 \end{array} \; {}^{-} \!\!\!\!\!\!-\!\!\!\!\! JM \; \langle (j_1 j_2) JM | \tag{3.30}$$

$$| j_1 m'_1, j_2 m'_2 \rangle = \sum_{J'M'} |(j_1 j_2)J'M'\rangle \quad J'M' -\!\!\!\!\!\!-\!\!\!\!\!{}^{+} \begin{array}{c} j_1 m'_1 \\ \nearrow \\ \searrow \\ j_2 m'_2 \end{array} \tag{3.31}$$

to obtain the relation

$$\sum_J \; \begin{array}{c} j_1 m_1 \\ \searrow \\ \nearrow \\ j_2 m_2 \end{array} {}^{-}\!\!\!-\!\!\!-\!\!\!{}^J\!\!\!-\!\!\!-\!\!\!{}^{+} \begin{array}{c} j_1 m'_1 \\ \nearrow \\ \searrow \\ j_2 m'_2 \end{array} \; = \; \begin{array}{c} j_1 m_1 -\!\!\!\!-\!\!\!\!- j_1 m'_1 \\[4pt] j_2 m_2 -\!\!\!\!-\!\!\!\!- j_2 m'_2 \end{array} \tag{3.32}$$

Note that there is a factor $[J]^{1/2} = (2J + 1)^{1/2}$ according to (3.11) associated with *each* end of the internal line. By removing the heavy lines and adding arrows to the vertices, we get

$$\sum_J [J] \; \begin{array}{c} j_1 m_1 \\ \diagdown \\ \diagup \\ j_2 m_2 \end{array} {}^{-}\!\!\!-\!\!\!-\!\!\!{}^J\!\!\!-\!\!\!-\!\!\!{}^{+} \begin{array}{c} j_1 m'_1 \\ \diagup \\ \diagdown \\ j_2 m'_2 \end{array} \; = \; \begin{array}{c} j_1 m_1 -\!\!\!\!-\!\!\!\!- j_1 m'_1 \\[4pt] j_2 m_2 -\!\!\!\!-\!\!\!\!- j_2 m'_2 \end{array} \tag{3.33}$$

The relations (3.26, 33) are quite useful in reducing angular-momentum graphs, and we shall use them frequently in the following. Equation (3.26), for instance, makes it possible to remove all simple loops from a line. By means of this rule, we get, for instance,

$$j_1 m_1 \underline{} 00 \quad = \quad \delta(j_1, 0).$$

We can also remove the zero line using (3.20),

$$= [j_2]^{-1/2} \quad j_1 m_1$$

Thus,

$$= [j_2]^{1/2} \quad j_1 m_1 \underline{} 00 \qquad (3.34)$$

which provides us with a rule for removing a loop at the end of a line. Note that *if a single line is connected to a loop, it must have $j = 0$.*

Problem 3.5. Verify the following evaluations:

a)

$$j_1' m_1' \quad = \delta(j_1, j_1')\delta(m_1, m_1')[j_1]^{-1} \quad (3.35)$$

b)

$$j_3 \quad = \delta(j_1, j_1')\delta(m_1, m_1')\delta(j_2, 0) \, [j_3/j_1]^{1/2} . \quad (3.36)$$

Problem 3.6. Evaluate the following diagrams:

a)   b)

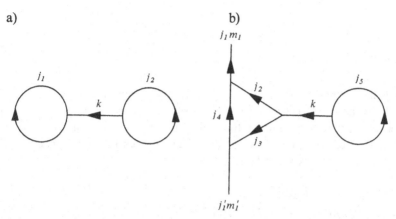

All angular-momentum quantum numbers are assumed to be integers and the sum of these quantum numbers at each vertex is assumed to be even.

3.3 A Physical Example: The Coulomb Interaction

3.3.1 Representation of a Single Matrix Element

As a second example of graphical methods, we now evaluate matrix elements of the Coulomb interaction between two electrons separated by a distance r_{12}. According to (2.102) and (2.97), the Coulomb interaction can be written

$$\frac{1}{r_{12}} = \sum_k (-1)^k [k]^{1/2} \frac{r_<^k}{r_>^{k+1}} \{C^k(1)C^k(2)\}_0^0 , \qquad (3.37)$$

where $C^k(1)$ and $C^k(2)$ are functions of the coordinates of the first and second electron, respectively, and $r_<$ is the lesser, $r_>$ the greater of the two radial distances from the origin, r_1 and r_2. The Coulomb operator is thus of the general form

$$g_{12} = \gamma(r_1, r_2) \{(t^{k_1}(1)u^{k_2}(2)\}_Q^K , \qquad (3.38)$$

where $\gamma(r_1, r_2)$ is a function of the two radii r_1 and r_2, and $t^{k_1}(1)$ and $u^{k_2}(2)$ are two spin-independent tensor operators, coupled together to form a tensor operator of rank K in the combined orbital space.

Before we consider the Coulomb interaction, we shall evaluate the general matrix element

$$\langle ab | \gamma(r_1, r_2) \{t^{k_1}(1)u^{k_2}(2)\}_Q^K | cd \rangle , \qquad (3.39)$$

where a, b, c, d represent uncoupled single-electron states

$$|a\rangle = |n_a l_a m_s^a m_l^a\rangle \,,$$

and

$$|nlm_s m_l\rangle = \frac{1}{r} P_{nl}(r) Y_{m_l}^l(\theta, \phi) \chi_{m_s} \,.$$

Y_m^l is a spherical harmonic defined in (2.57) and χ_{m_s} a spin function [see (2.21)].
The radial function $P_{nl}(r)$ is discussed in Sect. 5.1. The matrix element (3.39)
may be calculated by expanding the operator $\{t^{k_1}(1) u^{k_2}(2)\}^K$ in terms of the
single-electron tensor operators using (2.95) and evaluating the matrix elements
of each operator independently,

$$\langle ab | \gamma(r_1, r_2) \{t^{k_1}(1) u^{k_2}(2)\}_Q^K | cd\rangle$$
$$= \iint P_a(r_1) P_b(r_2) \gamma(r_1, r_2) P_c(r_1) P_d(r_2) dr_1 dr_2$$
$$\times \left[\sum_{q_1, q_2} \langle a | t_{q_1}^{k_1} | c\rangle \langle k_1 q_1, k_2 q_2 | KQ\rangle \langle b | u_{q_2}^{k_2} | d\rangle \right] . \tag{3.40}$$

The symbols $\langle a | t_{q_1}^{k_1} | c\rangle$ and $\langle b | u_{q_2}^{k_2} | d\rangle$ appearing here are shorthand notations
for the integrals over the angular coordinates and sums over the spin coordinates
of the corresponding single-particle matrix elements. Assuming for simplicity
that the operators are spin independent, we can use the Wigner-Eckart theorem
(3.17) to represent these in the following way:

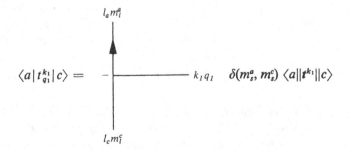

The vector-coupling coefficient $\langle k_1 q_1, k_2 q_2 | KQ \rangle$ can be represented by (3.12), and we can then evaluate the spin-angular part of the matrix element of (3.40), enclosed in the square brackets, by joining the lines of the corresponding graphs together in the following way:

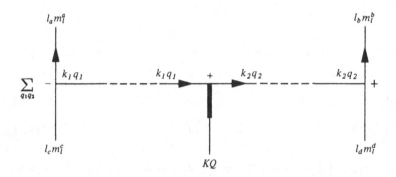

We represent the summation over the magnetic quantum numbers as before by omitting them from the graph. The matrix element (3.40) then becomes

$$\langle ab | \gamma(r_1, r_2) \{ t^{k_1}(1) u^{k_2}(2) \}_Q^K | cd \rangle = \delta(m_s^a, m_s^c) \delta(m_s^b, m_s^d)$$
$$\times \langle a \| t^{k_1} \| c \rangle \langle b \| u^{k_2} \| d \rangle \iint P_a(r_1) P_b(r_2) \gamma(r_1, r_2) P_c(r_1) P_d(r_2) \, dr_1 dr_2 \qquad (3.41)$$

times the angular-momentum diagram

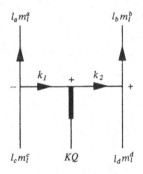

The rank of the Coulomb operator (3.37) is zero. So the matrix elements of the Coulomb interaction may be obtained by setting $K = 0$ in the angular-momentum diagram above. Using the graphical identity (3.19), we then obtain

$$\langle ab | r_{12}^{-1} | cd \rangle = \delta(m_s^a, m_s^c) \delta(m_s^b, m_s^d) \sum_k X^k(ab, cd) \qquad (3.42)$$

times the diagram

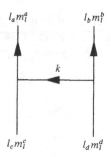

where

$$X^k(ab, cd) = X(k, l_a l_b l_c l_d)\, R^k(ab, cd) \quad \text{and} \tag{3.43}$$

$$X(k, l_a l_b l_c l_d) = (-1)^k \langle l_a \| C^k \| l_c \rangle \langle l_b \| C^k \| l_d \rangle \tag{3.44a}$$

$$R^k(ab, cd) = \iint P_a(r_1) P_b(r_2)\, \frac{r_<^k}{r_>^{k+1}}\, P_c(r_1) P_d(r_2)\, dr_1 dr_2 \;. \tag{3.44b}$$

The reduced matrix element $\langle l_a \| C^k \| l_c \rangle$ is given in (2.127). It is nonzero only if the sum of the angular momenta $l_a + k + l_c$ is even. Similarly, $l_b + k + l_d$ must be even, and for this reason the diagram (3.42) is independent of the order of the angular momenta l_a, k, l_c and l_b, k, l_d. So we have omitted the sign by the two vertices. Since the Coulomb interaction is independent of spin, it is necessary that $m_s^a = m_s^c$ and $m_s^b = m_s^d$. The two-electron radial integrals R^k (ab, cd), which are characteristic of the Coulomb interaction, are called *Slater integrals*.

We shall sometimes find it useful to represent also the spin part of the Coulomb matrix element (3.42) graphically. Using (3.4), we then get

$$\langle ab | r_{12}^{-1} | cd \rangle = \sum_k X(k, l_a l_b l_c l_d) R^k(ab, cd) \tag{3.45}$$

times the diagrams

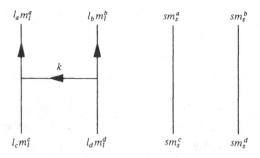

The spin diagram is analogous to the orbital one apart from the fact that the k line is missing and there are no arrows on the outgoing lines. The Coulomb interaction is independent of spin, and the two spin states pass through the interaction unaffected. We shall see later that the analogy between the orbital and spin diagrams is preserved also in more complex situations.

3.3.2 Summation Over Filled Shells

In order to actually make use of the graphical representation of the Coulomb matrix elements (3.42) and to show its utility, we evaluate something which we will need in a later chapter. According to (5.21), the interaction of a single electron in the state $|a\rangle$ with all of the electrons $|b\rangle$ in a filled shell B (with the same nl quantum numbers) is given by

$$e^0(a, B) = \sum_{m_s^b m_l^b} [\langle ab|r_{12}^{-1}|ab\rangle - \langle ab|r_{12}^{-1}|ba\rangle] . \tag{3.46}$$

The first matrix element in the summation is called the *direct* term, since the ordering of the quantum numbers is the same on both sides, and the second matrix element is called the *exchange* term. The Slater integrals associated with these two special cases are usually denoted

$$\begin{cases} F^k(a, b) = R^k(ab, ab) \\ G^k(a, b) = R^k(ab, ba) . \end{cases} \tag{3.47}$$

Using the graphical representation of the Coulomb matrix elements (3.42), the direct term in the summation (3.46) may be written

$$\sum_{m_s^b m_l^b} \langle ab|r_{12}^{-1}|ab\rangle = \sum_{m_s^b m_l^b} \sum_k X(k, l_a l_b l_a l_b)\, F^k(a, b)$$

Joining the b lines together and performing the summation over m_l^b graphically gives

$$\sum_{m_s^b} \sum_k X(k, l_a l_b l_a l_b)\, F^k(a, b)$$

Since all of these factors are independent of spin, the summation over m_s^b just introduces a factor of two, and we may use (3.36) to obtain

$$\sum_{m_s^b m_l^b} \langle ab| r_{12}^{-1} |ab \rangle = 2 \, X(0, l_a l_b l_a l_b) F^0(a, b) \, [l_b/l_a]^{1/2} \, . \tag{3.48}$$

Similarly, for the exchange term in (3.46)

$$- \sum_{m_s^b m_l^b} \langle ab| r_{12}^{-1} |ba \rangle$$

$$= - \sum_{m_s^b m_l^b} \delta(m_s^a, m_s^b) \sum_k X(k, l_a l_b l_b l_a) \, G^k(a, b)$$

Since the Coulomb interaction is independent of spin, the electron a interacts in this way only with electrons with the same m_s quantum number. Carrying out the summations over m_s^b and m_l^b, we get,

$$- \sum_k X(k, l_a l_b l_b l_a) \, G^k(a, b) \tag{3.49}$$

$$= - [l_a]^{-1} \sum_k X(k, l_a l_b l_b l_a) \, G^k(a, b) \, ,$$

where we have used (3.35) in the last step.

Alternatively, we could in the example above have started from (3.45). The spin sum would then in the direct case give rise to the diagram

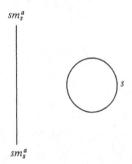

which according to (3.4, 27) has the value $2s + 1 = 2$. Similarly, in the exchange case the spin diagram becomes

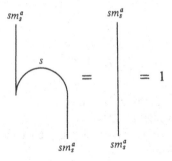

Thus, the spin summation in the direct case gives a factor of two and in the exchange case a factor of one.

Collecting together the direct and exchange contributions (3.48, 49), and using (2.127) for the reduced matrix elements, we obtain finally

$$e^0(a, B) = (4l_b + 2)\left[F^0(a, b) - \frac{1}{2}\sum_k \begin{pmatrix} l_a & k & l_b \\ 0 & 0 & 0 \end{pmatrix}^2 G^k(a, b)\right]. \qquad (3.50)$$

This equation gives the interaction of a single electron a with all of the electrons in a filled shell B. It applies whether or not electron a belongs to the shell B.

3.4 Coupling of Three Angular Momenta. The 6-j Symbol

In atomic physics it is often necessary to deal with systems for which there are more than two angular momenta. For instance, a configuration of two electrons has two orbital angular momenta and two spins. As a first step towards the many-electron problem, we consider the coupling of three angular momenta j_1, j_2, j_3 to give a resultant J.

3.4.1 The 6-j Symbol

There are, in general, two distinct ways of coupling j_1, j_2, j_3 to form the resultant J. We may first couple j_1 and j_2 together to form an intermediate angular momentum J_{12} and then couple this to j_3 to form J

$$|(j_1 j_2)J_{12}, j_3, JM\rangle . \qquad (3.51)$$

Alternatively, we may first couple j_2 and j_3 to form J_{23} and then j_1 to J_{23} to form J

$$|j_1, (j_2 j_3)J_{23}, JM\rangle \ . \tag{3.52}$$

The states (3.51) and (3.52) both constitute complete sets, and they are thus related by a unitary transformation,

$$|j_1, (j_2 j_3)J_{23}, JM\rangle$$
$$= \sum_{J_{12}} |(j_1 j_2)J_{12}, j_3, JM\rangle \langle (j_1 j_2)J_{12}, j_3, JM | j_1, (j_2 j_3)J_{23}, JM\rangle \ . \tag{3.53}$$

By operating on this relation with the ladder operators (2.19), it is easy to show that the transformation coefficients are independent of M.

We shall now use the graphical technique to evaluate these coefficients. Using (3.13, 14), the coupled states can be written

$$\langle (j_1 j_2)J_{12}, j_3, JM | \ = \tag{3.54a}$$

$$|j_1, (j_2 j_3)J_{23}, JM\rangle = \tag{3.54b}$$

leaving out the uncoupled states for simplicity. The scalar product may then be formed by joining the corresponding lines together

$$\langle (j_1 j_2)J_{12}, j_3, JM | j_1, (j_2 j_3) J_{23}, JM\rangle$$

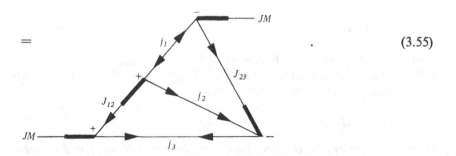

$$= \tag{3.55}$$

This is the transformation coefficient between the two sets of states (3.51) and (3.52), which recouples the angular momenta according to (3.53). Since it is independent of M, we can replace the two factors $(2J + 1)^{1/2}$ associated with the JM lines by a summation over M. So the free ends can be joined, and we may use the graphical identities (3.7,8) to remove the double arrows. In this way we obtain the following representation of the recoupling coefficient:

$$\langle (j_1 j_2)J_{12}, j_3, JM | j_1, (j_2 j_3)J_{23}, JM \rangle$$

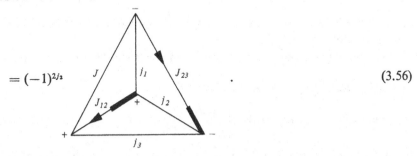

$$= (-1)^{2j_2} \qquad \qquad (3.56)$$

By adding outgoing arrows to the lower left-hand vertex and changing two of the vertex signs, one may readily transform this graph into the more symmetric form (see Problem 3.7)

$$\langle (j_1 j_2)J_{12}, j_3, JM | j_1, (j_2 j_3)J_{23}, JM \rangle$$

$$= (-1)^{j_1+j_2+j_3+J} [J_{12}, J_{23}]^{1/2} \qquad \qquad (3.57)$$

The diagram which appears in (3.57) has a high degree of symmetry, and it is generally denoted

$$\begin{Bmatrix} j_1 & J_{12} & j_2 \\ j_3 & J_{23} & J \end{Bmatrix} = \qquad \qquad (3.58)$$

Since it involves six angular momenta, it is called a *6-j symbol,* a notation which was introduced by *Wigner.*

The 6-j symbol is nonvanishing only if the angular momenta meeting at each vertex satisfy the triangular condition (2.83). These conditions can be illustrated in the following way:

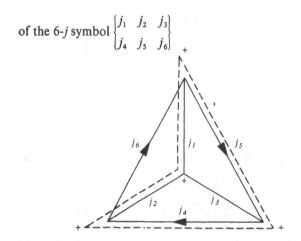

$$\left\{ \begin{matrix} O-O-O \end{matrix} \right\} \left\{ \begin{matrix} O \\ O-O \end{matrix} \right\} \left\{ \begin{matrix} O \\ O \quad O \end{matrix} \right\} \left\{ \begin{matrix} O \\ O-O \end{matrix} \right\}. \tag{3.59}$$

From (3.58) it is evident that the 6-j symbol is invariant with respect to any permutation of the columns.

Using the 6-j symbol, the transformation coefficient (3.57) becomes

$$\langle (j_1 j_2) J_{12}, j_3, J | j_1, (j_2 j_3) J_{23}, J \rangle$$

$$= (-1)^{j_1+j_2+j_3+J} [J_{12}, J_{23}]^{1/2} \begin{Bmatrix} j_1 & J_{12} & j_2 \\ j_3 & J_{23} & J \end{Bmatrix}. \tag{3.60}$$

Since it is independent of M, we have left M out of the recoupling coefficient.

Problem 3.7. Use the phase rules discussed in connection with Problem 3.2 to verify the phase of the coupling coefficient (3.57).

3.4.2 Equivalent Forms of the 6-j Symbol. The Hamilton Line

In dealing with a complex angular-momentum diagram, it is often useful to draw a line called a *Hamilton line* of the diagram, which has the property that it *passes through each vertex once* [*Jucys* et al. 1962]. This means that it cannot pass along any line of the diagram more than once. After a Hamilton line is drawn, the graph may be pulled apart so that this line lies along the edges of a polygon.

As an example of the use of a Hamilton line, we consider the diagram (3.58)

of the 6-j symbol $\begin{Bmatrix} j_1 & j_2 & j_3 \\ j_4 & j_5 & j_6 \end{Bmatrix}$

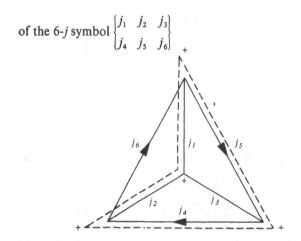

The dotted line is a Hamilton line of the graph. Rewriting the diagram so that the Hamilton line appears on the periphery, we obtain

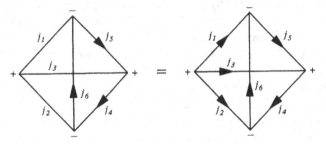

Note that the signs of two vertices are changed in this process. In the second graph above we have added arrows to all the lines of the vertex on the left, using the graphical identity (3.9a). These two diagrams provide equivalent representation of the 6-*j* symbol.

In Fig. 3.1 we have collected together the different diagramatic representations of the 6-*j* symbol.

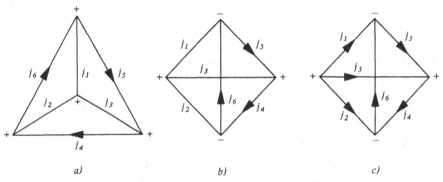

a)   b)   c)

Fig. 3.1 a-c. Equivalent representations of the 6-*j* symbol $\begin{Bmatrix} j_1 & j_2 & j_3 \\ j_4 & j_5 & j_6 \end{Bmatrix}$

Each of these diagrams can be written out explicitly, of course, in terms of 3-*j* symbols. For instance, if we begin at the left-hand corner of diagram (c), go around the edge of the diagram in a clockwise direction, and write down explicitly the 3-*j* symbols and also supply a phase for each arrow, we obtain the following expression:

$$
\begin{Bmatrix} j_1 & j_2 & j_3 \\ j_4 & j_5 & j_6 \end{Bmatrix} = \sum_{\text{all } m} (-1)^{j_1-m_1+j_2-m_2+j_3-m_3+j_4-m_4+j_5-m_5+j_6-m_6}
$$

$$
\times \begin{pmatrix} j_1 & j_2 & j_3 \\ -m_1 & -m_2 & -m_3 \end{pmatrix} \begin{pmatrix} j_1 & j_5 & j_6 \\ m_1 & -m_5 & m_6 \end{pmatrix} \begin{pmatrix} j_2 & j_6 & j_4 \\ m_2 & -m_6 & m_4 \end{pmatrix} \begin{pmatrix} j_3 & j_4 & j_5 \\ m_3 & -m_4 & m_5 \end{pmatrix}. \quad (3.61)
$$

In all cases the m value represents the value at the end of line towards which the arrow points.

The 6-j symbol can be evaluated by means of (3.61) and the general formula for the 3-j symbol (2.76, 81). As shown by *Racah* [1942], however, it is possible to reduce this expression considerably and give it a form which is not more complicated than that of the 3-j symbol (see, for instance, [*Brink* and *Satchler* 1968]). The coefficient used by *Racah*, the so-called *Racah coefficient*, differs from the 6-j symbol only in the phase. The 6-j symbol is more symmetric, however, and is nowadays used almost exclusively.

Simplified formulas can be derived for specific 6-j symbols, where one of the j values is small. A few formulas of this kind are given in Appendix C. Further expressions of this kind are given in the book of *Edmonds* [1957].

Efficient computer programs are available for evaluating 3-j and 6-j symbols in a routine way, and extensive tabulations are also available, for instance, that of *Rotenberg* et al. [1959]. A small part of this table is reproduced in Appendix D.

Problem 3.8. Using the graphical representation of the 6-j symbol, given in Fig. 3.1(c), show that the symbol is invariant under an interchange of the upper and lower arguments in two of its columns, such as

$$\begin{Bmatrix} j_1 & j_2 & j_3 \\ j_4 & j_5 & j_6 \end{Bmatrix} = \begin{Bmatrix} j_4 & j_5 & j_3 \\ j_1 & j_2 & j_6 \end{Bmatrix} .$$

Problem 3.9. Verify the following evaluation:

$$= (-1)^{j_1+j_2+j_3+j_4} \begin{Bmatrix} j_1 & j_4 & j_5 \\ j_3 & j_2 & j_6 \end{Bmatrix} .$$

Problem 3.10. a) Derive the following orthogonality relation for the coupling coefficients (3.53)

$$\sum_{J_{12}} \langle j_1, (j_2 j_3)J_{23}, J | (j_1 j_2)J_{12}, j_3, J \rangle \times \langle (j_1 j_2)J_{12}, j_3, J | j_1, (j_2 j_3)J'_{23}, J \rangle$$

$$= \delta(J_{23}, J'_{23}) .$$

b) Use this relation to derive the corresponding relation for 6-j symbols

$$\sum_f [f] \begin{Bmatrix} a & b & c \\ d & e & f \end{Bmatrix} \begin{Bmatrix} a & b & c' \\ d & e & f \end{Bmatrix} = \delta(c, c') [c]^{-1} . \tag{3.62}$$

Problem 3.11. Use the phase rule (2.79) and the definition (3.60) to verify the following relation:

$$\langle (j_1, j_2) J_{12}, j_3, J \,|\, (j_1 j_3) J_{13}, j_2, J \rangle$$
$$= (-1)^{j_1 - J_{12} - J_{13} + J} \langle (j_2 j_1) J_{12}, j_3, J \,|\, j_2 (j_1 j_3) J_{13}, J \rangle$$
$$= (-1)^{j_2 + j_3 + J_{12} + J_{13}} [J_{12}, J_{13}]^{1/2} \begin{Bmatrix} j_2 & J_{12} & j_1 \\ j_3 & J_{13} & J \end{Bmatrix}$$

Problem 3.12. Derive the following sum rule for 6-j symbols:

$$\sum_g (-1)^{e+f+g} [g] \begin{Bmatrix} a & d & g \\ b & c & f \end{Bmatrix} \begin{Bmatrix} a & d & g \\ c & b & e \end{Bmatrix} = \begin{Bmatrix} a & b & e \\ d & c & f \end{Bmatrix}. \tag{3.63}$$

Hint: Start from the coupling coefficient

$$\langle (j_1 j_2) J_{12}, j_3, J \,|\, j_1, (j_2 j_3) J_{23}, J \rangle$$

and insert a complete set of intermediate states where j_1 and j_3 are coupled to J_{13}. Use the result of the previous problem.

3.4.3 A Physical Example: The *ls* Configuration

As a simple example of the coupling of three angular momenta, we consider an *ls* configuration for which one electron has an orbital angular momentum $l > 0$ and the other electron has $l = 0$. This configuration has three angular momenta associated with it: l_1, s_1, and s_2, and the two ways of coupling them together are

$$|(l_1 s_1) j_1, s_2, J \rangle \qquad \text{and} \qquad |l_1, (s_1 s_2) S, J \rangle \, .$$

These two coupling schemes are represented by the following vector diagrams:

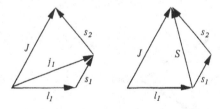

Which of these coupling schemes provides the best description of the system depends upon the nature of the interaction between the two electrons. l_1 and s_1 are coupled together by the spin-orbit interaction, and as we shall see the two spins are coupled by the exchange part of the Coulomb interaction. We shall return to this problem in Chap. 7. At this point we only give the recoupling coefficient between the two basis sets, using (3.60),

$$\langle (l_1 s_1) j_1, s_2, J | l_1, (s_1 s_2) S, J \rangle = (-1)^{l_1 + J + 1} [j_1, S]^{1/2} \begin{Bmatrix} l_1 & j_1 & s_1 \\ s_2 & S & J \end{Bmatrix}.$$

Here s_1 and s_2 both have, of course, the value $1/2$.

3.5 Coupling of Four Angular Momenta. The 9-j Symbol

The recoupling coefficient involving four angular momenta has an important place in atomic physics, because it may be used to transform states from LS to j-j coupling. In LS coupling the orbital angular momenta are first coupled to a total orbital angular momentum L and the spins to a total spin S, while in j-j coupling the orbital and spin angular momenta of each electron are first coupled together.

For a two-electron system, the coefficients for a transformation between these two coupling schemes can be denoted

$$\langle (s_1 s_2) S, (l_1 l_2) L, J | (s_1 l_1) j_1, (s_2 l_2) j_2, J \rangle . \tag{3.64}$$

In analogy with (3.54), the coupled states may be written

$$\langle (s_1 s_2) S, (l_1 l_2) L, JM | \;\; = \;\; JM \tag{3.65}$$

$$| (s_1 l_1) j_1, (s_2 l_2) j_2, JM \rangle = \tag{3.66}$$

and their scalar product is

$$\langle(s_1 s_2)S, (l_1 l_2)L, J | (s_1 l_1)j_1, (s_2 l_2)j_2, J\rangle$$

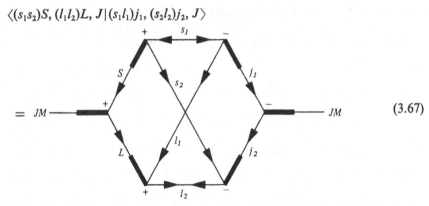

$$(3.67)$$

As is the case with three angular momenta, this recoupling coefficient is independent of M. So we may remove the two factors $(2J+1)^{1/2}$ and join the free ends together. After reversing the directions of the arrows on the s_1 and l_2 lines, we obtain the diagram

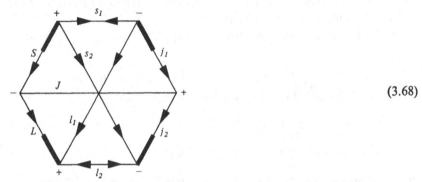

$$(3.68)$$

Since all of the arrows now go out of the upper vertices and into the lower vertices, they may be removed entirely from the diagram, using (3.9). We may also use (3.11) to remove the solid lines and change all of the vertex signs to plus by means of (3.2). In this way we obtain a more symmetric diagram, and (3.67) may be written

$$\langle(s_1 s_2)S, (l_1 l_2)L, J | (s_1 l_1)j_1, (s_2 l_2)j_2, J\rangle$$

$$= (-1)^R [j_1, j_2, S, L]^{1/2}$$

$$(3.69)$$

where R is the sum of the nine angular momenta that appear in the diagram. This diagram has a high degree of symmetry, and again we shall use a special symbol to denote it. It is called a *9-j symbol* and is defined by

$$
\begin{Bmatrix} j_{11} & j_{12} & j_{13} \\ j_{21} & j_{22} & j_{23} \\ j_{31} & j_{32} & j_{33} \end{Bmatrix} =
$$

(3.70)

Sometimes it is convenient to have a representation of the 9-j symbol for which there is one arrow on each of the lines. A diagram of this kind can be obtained by adding arrows on every other vertex.

The 9-j symbol can also be written out explicitly in terms of 3-j symbols,

$$
\begin{Bmatrix} j_{11} & j_{12} & j_{13} \\ j_{21} & j_{22} & j_{23} \\ j_{31} & j_{32} & j_{33} \end{Bmatrix} = \sum_{\text{All } m\text{'s}} \begin{pmatrix} j_{11} & j_{12} & j_{13} \\ m_{11} & m_{12} & m_{13} \end{pmatrix} \begin{pmatrix} j_{21} & j_{22} & j_{23} \\ m_{21} & m_{22} & m_{23} \end{pmatrix} \begin{pmatrix} j_{31} & j_{32} & j_{33} \\ m_{31} & m_{32} & m_{33} \end{pmatrix}
$$

$$
\times \begin{pmatrix} j_{11} & j_{21} & j_{31} \\ m_{11} & m_{21} & m_{31} \end{pmatrix} \begin{pmatrix} j_{12} & j_{22} & j_{32} \\ m_{12} & m_{22} & m_{32} \end{pmatrix} \begin{pmatrix} j_{13} & j_{23} & j_{33} \\ m_{13} & m_{23} & m_{33} \end{pmatrix}.
$$

(3.71)

The symmetry properties of the 9-j symbol are quite apparent from this form (see, for example, Problem 3.13). It follows from (3.71)—as well as from the graph (3.70)—that the angular momenta in each row and in each column must satisfy the triangular condition (2.83).

Using the definition of the 9-j symbol (3.70) and the symmetry properties given in Problem 3.13, the transformation coefficient (3.69) between LS and j-j coupling for a two-electron system can be written

$$
\langle (s_1 s_2) S, (l_1 l_2) L, J | (s_1 l_1) j_1, (s_2 l_2) j_2, J \rangle = [j_1, j_2, S, L]^{1/2} \begin{Bmatrix} s_1 & s_2 & S \\ l_1 & l_2 & L \\ j_1 & j_2 & J \end{Bmatrix}.
$$

(3.72)

Problem 3.13. Using graphical methods, show that the 9-j symbol has the following symmetry properties: An odd permutation of the rows or columns multiplies the symbol by $(-1)^R$, where R is the sum of the nine angular momenta. An even permutation produces no change of phase and neither does a transposition, which interchanges rows and columns.

Problem 3.14. Using the rules discussed in Sect. 3.1 for removing a zero line from an angular-momentum graph, derive the following relations:

$$\begin{Bmatrix} j_1 & j_2 & j_3 \\ l_1 & l_2 & 0 \end{Bmatrix} = (-1)^{j_1+j_2+j_3}[j_1,j_2]^{-1/2}\, \delta(j_1, l_2)\delta(j_2, l_1) \tag{3.73}$$

$$\begin{Bmatrix} j_1 & j_2 & j_3 \\ l_1 & l_2 & j_3' \\ k & k' & 0 \end{Bmatrix} = \delta(j_3, j_3')\delta(k, k')\,(-1)^{j_2+j_3+l_1+k}[j_3, k]^{-1/2} \begin{Bmatrix} j_1 & j_2 & j_3 \\ l_2 & l_1 & k \end{Bmatrix}. \tag{3.74}$$

Problem 3.15. Derive an orthogonality relation for the 9-j symbols corresponding to the relation (3.62) for the 6-j symbols.

4. Further Developments of Angular-Momentum Graphs. Applications to Physical Problems

In this chapter we shall develop the graphical methods of representing angular momentum further by means of a number of theorems, first given by *Jucys, Levinson* and *Vanagas* (JLV) [1962]. We shall then use these theorems to derive a general expression for the matrix elements of a tensor-operator product between coupled states in a purely graphical way. Finally, we shall apply these results to an important physical problem—the Coulomb interaction of a two-electron system—in some detail. This example clearly demonstrates the usefulness of the graphical technique in dealing with angular-momentum problems of certain complexity. In a later chapter (Chap. 8)—after having studied the physical properties of atoms more systematically—we shall extend the technique further and apply it to general many-electron systems.

4.1 The Theorems of Jucys, Levinson and Vanagas

4.1.1 The Basic Theorem

The "basic" theorem, which forms the foundation of the technique we shall develop here, concerns diagrams that are composed of two parts which are joined together by a single line. In order for the theorem to apply it is necessary that one of the parts be *closed*. This means that it has no free lines. The diagram is then of the general form

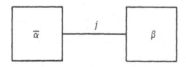

where $\bar{\alpha}$ represents a closed part. The basic JLV theorem states that the angular momentum of the connecting line j must be zero, provided that $\bar{\alpha}$ has one arrow on each internal line. Such a diagram is said to be in *normal form*. The proof of this important theorem, which depends on the relation between angular-momentum and rotation operators, will be omitted here. Interested readers are referred to the original works of *Jucys* et al. [1962] or to the article of *Sandars* [1971].

We may obtain a more convenient form of the basic theorem by showing explicitly the lines to which the line j is attached. Since j must be zero, we can then remove the line, using (3.19, 20). In this way we obtain the result

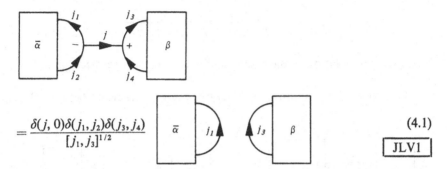

$$= \frac{\delta(j, 0)\delta(j_1, j_2)\delta(j_3, j_4)}{[j_1, j_3]^{1/2}}$$

(4.1)

JLV1

(The arrow on the j line has no significance and can be left out.) We shall refer to this theorem as JLV1, since it applies to diagrams with a closed part which is connected to the rest of the diagram by a single line. Such a diagram is said to be "separable on a single line". By means of this theorem it is straightforward to derive rules for dealing with diagrams that are separable on more than one line, as we shall see below.

If there are no lines at all on the β part of the diagram, JLV1 reduces to

$$\qquad jm \quad = \frac{\delta(j, 0)\delta(m, 0)\delta(j_1, j_2)}{[j_1]^{1/2}} \qquad j_1 \quad .$$

(4.2)

This equation can also be obtained directly from the basic theorem by removing the zero line.

4.1.2 Diagrams Separable on Two Lines

The tool which we shall use to derive the more general theorems from JLV1 is the orthogonality relation (3.33)

$$\sum_{j_3} [j_3] \qquad = \qquad$$

(4.3)

By adding arrows at each vertex, this can be written

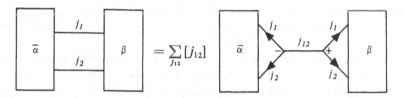

$$\sum_{j_3} [j_3] \quad = \quad \text{(4.4)}$$

This last representation of the orthogonality relation will be used quite often in the following.

We are now in a position to consider diagrams separable on *two* lines. Using (4.4), we obtain

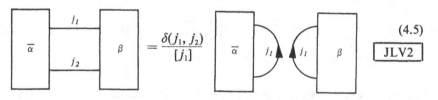

In order to ensure that $\bar{\alpha}$ is still in normal form, after adding the arrows on the j_1 and j_2 lines, we must regard any original arrows on the j_1 and j_2 lines as belonging to the β part. After reversing the arrows on the j_2 lines, we can then apply JLV1, which gives

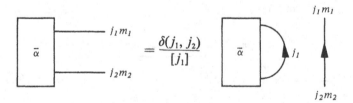

(4.5)

JLV2

This theorem will be called JLV2. The direction of both arrows in this last diagram may, of course, be changed. It is only important that they point in the same direction.

If the block β is completely empty, this theorem can be written

If there is an arrow on any of the free lines, it must be separated from $\bar{\alpha}$ and included in the β part before the theorem can be applied. It will then appear on the line to the far right, as illustrated below,

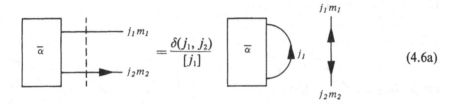

$$\begin{array}{c} \boxed{\bar{\alpha}} \quad \begin{array}{l} j_1 m_1 \\ j_2 m_2 \end{array} \end{array} = \frac{\delta(j_1, j_2)}{[j_1]} \quad \boxed{\bar{\alpha}} \quad j_1 \qquad \qquad \begin{array}{c} j_1 m_1 \\ \\ j_2 m_2 \end{array} \tag{4.6a}$$

This result can also be obtained more directly by applying the orthogonality relation above (see Problem 4.1). The arrows on the line to the right can be eliminated and the relation expressed as

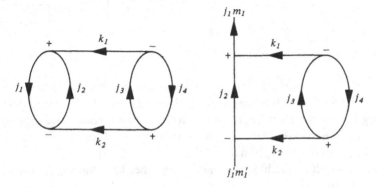

$$\begin{array}{c} j_2 m_2 \xleftarrow{} \boxed{\bar{\alpha}} \xrightarrow{j_1 m_1} = \delta(j_1, j_2)\, \delta(m_1, m_2) \\ \times\ j_1 \left(\boxed{\bar{\alpha}} \right) \end{array} \tag{4.6b}$$

where we have used heavy lines to eliminate the factor $[j_1]^{-1}$. This is a very useful relation, which we shall apply frequently in the following.

One important conclusion can be drawn directly: if an angular-momentum diagram in normal form (with exactly one arrow on each internal line) has only two free lines, these must represent the same angular-momentum quantum numbers. Another conclusion is that such a diagram is independent of the magnetic quantum number. We shall see that this type of diagram appears when forming scalar products or matrix elements of scalar operators [see, for instance, (3.55) and (3.67)].

Problem 4.1. Use the orthogonality relation (4.4) together with JLV1 to verify the relation (4.6a).

Problem 4.2.
a) Evaluate the following diagrams:

$$\begin{array}{cc}
\begin{array}{c} + \quad k_1 \quad - \\ j_1 \quad j_2 \quad j_3 \quad j_4 \\ - \quad k_2 \quad + \end{array}
&
\begin{array}{c} j_1 m_1 \\ + \quad k_1 \quad - \\ j_2 \quad j_3 \quad j_4 \\ - \quad k_2 \quad + \\ j_1' m_1' \end{array}
\end{array}$$

assuming all angular momenta to be integers.

b) Verify the following evaluation:

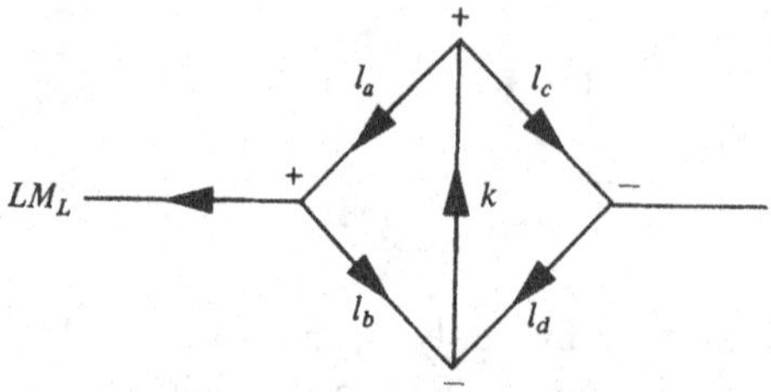

$$LM_L = (-1)^{l_b+l_c+L+k}\,[L]^{-1}\begin{Bmatrix} l_a & l_b & L \\ l_d & l_c & k \end{Bmatrix}$$

4.1.3 Diagrams Separable on Three Lines

For a diagram which is separable on three lines, we use (4.4) to write

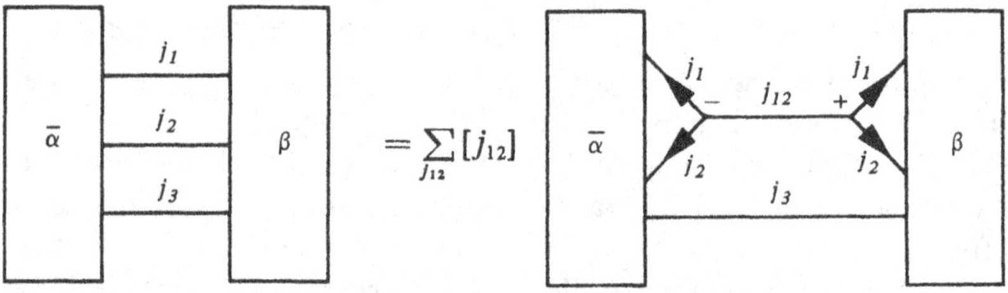

Again we assume that $\bar{\alpha}$ is normal and that any original arrows on j_1, j_2, and j_3 are included in β. The theorem JLV2 may then be applied to obtain

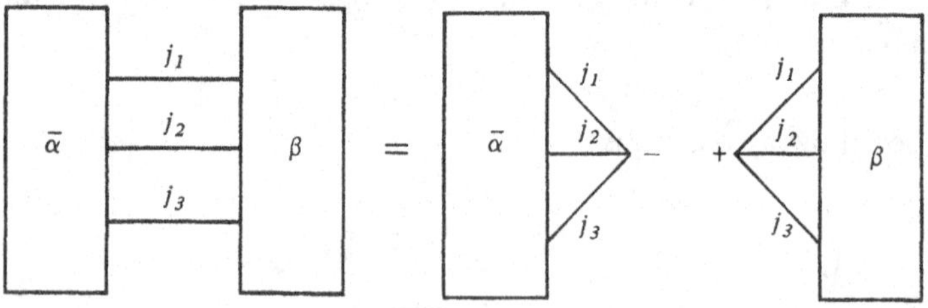

(4.7)

JLV3

after removing the arrows at the vertices. We shall refer to this theorem as JLV3. It says that a diagram, which is separable on three lines, can be factorized into two parts by joining the three lines together to form two vertices. The signs at the vertices are such that the three angular momenta at each vertex are read in the same order; with the signs given above this order is j_3, j_2, j_1.

For the case in which the block β contains no other lines than j_1, j_2, and j_3, this theorem can be written

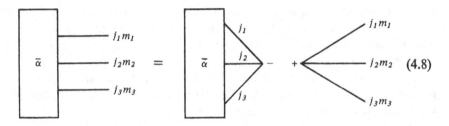

Any arrow on the free lines of the diagram on the left-hand side will reappear on the corresponding free lines of the diagram to the far right. This last equation is essentially the *Wigner-Eckart theorem* (3.17). It says that a graph, which is closed apart from three free lines, may be replaced by a graph for which the three lines are joined at a point and the graph of a 3-j symbol. The closed graph represents the angular part of the reduced matrix element.

4.1.4 Diagrams Separable on Four Lines

A diagram, which is separable on four lines, can be transformed in the following way, using (4.4) and JLV2,

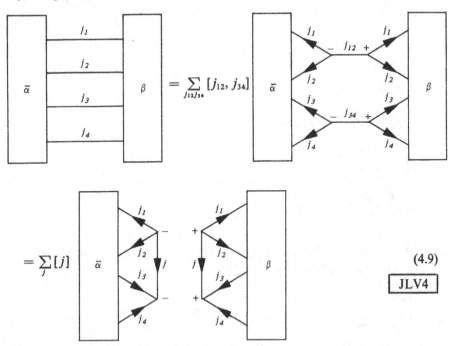

As before, we have assumed that $\bar{\alpha}$ is normal and that all the original arrows on $j_1, j_2, j_3,$ and j_4 are incorporated in β. This theorem, which we shall call JLV4, shows that a diagram, which is separable on four lines, can be factorized, but

only at the expense of a sum over the additional angular momentum j. All of the lines diverge from one vertex in each block and converge to the other vertex. The signs at the vertices show that the order of corresponding angular momenta about each vertex is the same.

Similar theorems can be derived for diagrams which are separable on more than four lines (see, for instance, [*El Baz* and *Castel* 1972]). We shall not use these theorems, however, in our applications.

4.2 Some Applications of the JLV Theorems

We now consider the following diagram as an example of the use of the JLV theorems:

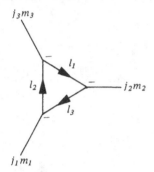

Using (4.8) this graph may be written

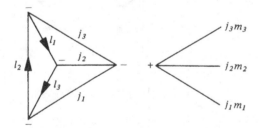

By moving the vertex formed by $j_1 j_2 j_3$ of the closed diagram inside the triangle of $l_1 l_2 l_3$, all vertex signs will be changed,

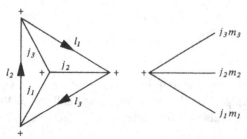

According to (3.58), the closed diagram corresponds to a 6-j symbol, and hence the diagram above can be written

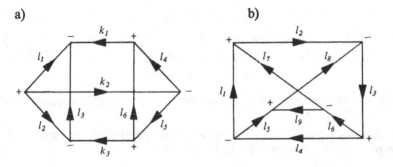

$$\begin{Bmatrix} j_1 & j_2 & j_3 \\ l_1 & l_2 & l_3 \end{Bmatrix}$$ (4.10)

This identity will allow us to remove a triangle from a graph.

Problem 4.3. Verify the following relation:

(4.11)

$$= \sum_k [k] (-1)^{k+l_2-l_4} \begin{Bmatrix} j_1 & j_4 & k \\ l_4 & l_2 & l_1 \end{Bmatrix} \begin{Bmatrix} j_2 & j_3 & k \\ l_4 & l_2 & l_3 \end{Bmatrix}$$

Problem 4.4 Evaluate the following diagrams, assuming all angular momenta to be integers:

a)

b)

4.3 Matrix Elements of Tensor-Operator Products Between Coupled States

In the last two sections we have developed a number of general theorems concerning angular-momentum diagrams, and applied these theorems to a few simple problems. In this section we shall derive formulas for the matrix elements of tensor operators between coupled states. This will provide further examples of the use of the graphical theorems. Also formulas of this kind may be applied to a wide range of atomic problems, as we shall see later.

4.3.1 The General Formula

We consider now a tensor-operator product of the form (2.95)

$$X_Q^K = \{t^{k_1}(1)u^{k_2}(2)\}_Q^K = \sum_{q_1 q_2} t_{q_1}^{k_1}(1)u_{q_2}^{k_2}(2) \langle k_1 q_1, k_2 q_2 | KQ \rangle \tag{4.12}$$

and form its matrix elements with the coupled states, defined in (2.69),

$$\langle (\gamma_1 j_1 \gamma_2 j_2)JM | X_Q^K | (\gamma_1' j_1' \gamma_2' j_2')J'M' \rangle . \tag{4.13}$$

Here, we have included the additional quantum numbers γ_1 and γ_2, needed to specify the states completely. The operator (4.12) as well as the states in (4.13) is made up of two parts. The operator $t^{k_1}(1)$ acts on the first part of the system, $u^{k_2}(2)$ acts on the second, and j_1, j_1', for instance, are quantum numbers of the first part of the system. As in Chap. 2, we assume that the systems 1 and 2 are *distinguishable*. In the next section we shall apply the results to indistinguishable systems, in which case properly antisymmetric functions must be used.

According to (3.13, 14), the coupled states above may be written

$$\langle (\gamma_1 j_1 \gamma_2 j_2)JM | = \sum_{m_1 m_2} \quad \text{[diagram]} \quad \langle \gamma_1 j_1 m_1, \gamma_2 j_2 m_2 | \tag{4.14}$$

$$|(\gamma_1' j_1' \gamma_2' j_2')J'M'\rangle = \sum_{m_1' m_2'} |\gamma_1' j_1' m_1', \gamma_2' j_2' m_2'\rangle \quad \text{[diagram]} \tag{4.15}$$

The additional quantum numbers do not appear in the vector-coupling coefficients.

Using the completeness relation (B.9) and the Wigner-Eckart theorem (3.17), the single-particle tensor operators may be written

$$t_{q_1}^{k_1}(1) = \sum_{\gamma_1 j_1 m_1 \gamma_1' j_1' m_1'} |\gamma_1 j_1 m_1\rangle \langle \gamma_1 j_1 m_1 | t_{q_1}^{k_1} |\gamma_1' j_1' m_1'\rangle \langle \gamma_1' j_1' m_1'|$$

$$= \sum_{\gamma_1 j_1 m_1 \gamma_1' j_1' m_1'} |\gamma_1 j_1 m_1\rangle \; j_1 m_1 \!\!\!\overset{+}{\underset{k_1 q_1}{\longleftarrow}}\!\!\! j_1' m_1' \; \langle \gamma_1' j_1' m_1' | \langle \gamma_1 j_1 \| t^{k_1} \| \gamma_1' j_1'\rangle \qquad (4.16)$$

$$u_{q_2}^{k_2}(2) = \sum_{\gamma_2 j_2 m_2 \gamma_2' j_2' m_2'} |\gamma_2 j_2 m_2\rangle \; j_2 m_2 \!\!\!\overset{+}{\underset{k_2 q_2}{\longleftarrow}}\!\!\! j_2' m_2' \; \langle \gamma_2' j_2' m_2' | \langle \gamma_2 j_2 \| u^{k_2} \| \gamma_2' j_2'\rangle .$$

$$(4.17)$$

The vector-coupling coefficient which couples the two operators in (4.12) together has the graphical representation (3.12)

$$\langle k_1 q_1, k_2 q_2 | KQ \rangle = \quad KQ \!\!\!\overset{+}{\underset{k_2 q_2}{\longleftarrow}}\!\!\! k_1 q_1 \qquad (4.18)$$

As in the previous applications in Sects. 3.2.2 and 3.3.1, we can now form the matrix element (4.13) by joining corresponding lines of the graphical pieces together in the following way:

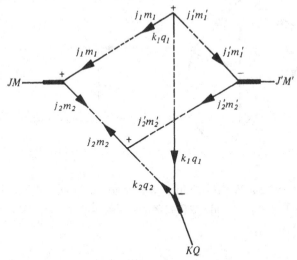

Due to the orthogonality of the states associated with the free ends, the angular-momentum quantum numbers of the lines joined together by dotted lines must be the same. According to (4.16,17), there is a reduced matrix element $\langle \gamma_1 j_1 \| t^{k_1} \| \gamma_1' j_1' \rangle$ associated with the first interaction vertex and $\langle \gamma_2 j_2 \| u^{k_2} \| \gamma_2' j_2' \rangle$ associated with the second. By performing the summations over the magnetic quantum numbers in the usual way, we obtain the graph

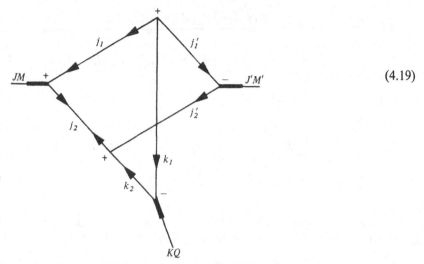

(4.19)

The matrix element (4.13) is equal to this diagram times the reduced matrix elements $\langle \gamma_1 j_1 \| t^{k_1} \| \gamma_1' j_1' \rangle$ and $\langle \gamma_2 j_2 \| u^{k_2} \| \gamma_2' j_2' \rangle$.

Diagram (4.19) may be brought to normal form by removing the heavy lines and adding outgoing arrows to the leftmost vertex. Equation (4.8) may then be used to obtain the following expression:

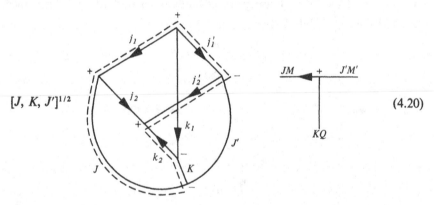

$[J, K, J']^{1/2}$

(4.20)

The dotted line shown here is a Hamilton line, which goes through all of the vertices once (see Sect. 3.4.2). Rewriting the diagram so that this line appears on the periphery, we obtain

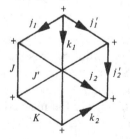

We may remove the arrows from this diagram and identify it as a 9-j symbol (3.70). Collecting together these results and discarding the open diagram

which appears in the Wigner-Eckart theorem (3.17), we obtain the following expression for the reduced matrix element:

$$\langle(\gamma_1 j_1 \gamma_2 j_2)J \| \{t^{k_1}(1)u^{k_2}(2)\}^K \|(\gamma_1' j_1' \gamma_2' j_2')J'\rangle$$

$$= \langle\gamma_1 j_1 \| t^{k_1} \| \gamma_1' j_1'\rangle \langle\gamma_2 j_2 \| u^{k_2} \| \gamma_2' j_2'\rangle [J, K, J']^{1/2} \begin{Bmatrix} j_1 & j_1' & k_1 \\ j_2 & j_2' & k_2 \\ J & J' & K \end{Bmatrix}. \qquad (4.21)$$

4.3.2 Special Cases

Equation (4.21) is particularly valuable, since three important special cases can be obtained from it by setting each of the tensor ranks K, k_1 and k_2 equal to zero.

We consider first the case for which K is zero. According to (2.97), we then have essentially a scalar product. Using the Wigner-Eckart theorem (2.107) and the relations (2.94) and (3.74), we get

$$\langle(\gamma_1 j_1 \gamma_2 j_2)JM \,|\, t^k(1)\cdot u^k(2) \,|\, (\gamma_1' j_1' \gamma_2' j_2')J'M'\rangle$$

$$= (-1)^{k+J-M}[k]^{1/2} \begin{pmatrix} J & 0 & J' \\ -M & 0 & M' \end{pmatrix} \langle(\gamma_1 j_1 \gamma_2 j_2)J \| \{t^k(1)u^k(2)\} \, {}^0_0 \|(\gamma_1' j_1' \gamma_2' j_2')J'\rangle$$

$$= (-1)^{j_1'+j_2+J} \delta(J, J')\delta(M, M') \begin{Bmatrix} j_1 & j_1' & k \\ j_2' & j_2 & J \end{Bmatrix} \langle\gamma_1 j_1 \| t^k \| \gamma_1' j_1'\rangle\langle\gamma_2 j_2 \| u^k \| \gamma_2' j_2'\rangle. \qquad (4.22)$$

Next, we set k_2 equal to zero. The operator acting on part 2 is then a scalar, and we can without loss of generality assume it to be unity. We are thus left with an operator $t^k(1)$, *acting only on part 1* of the system. Equation (4.21) gives in this case, again using (3.74),

$$\langle(\gamma_1 j_1 \gamma_2 j_2)J \| t^k \|(\gamma_1' j_1' \gamma_2' j_2')J'\rangle = \delta(\gamma_2, \gamma_2')\delta(j_2, j_2')(-1)^{j_1+j_2+J'+k}[J, J']^{1/2}[j_2]^{-1/2}$$

$$\times \begin{Bmatrix} J & k & J' \\ j_1' & j_2 & j_1 \end{Bmatrix} \langle\gamma_1 j_1 \| t^k \| \gamma_1' j_1'\rangle \langle\gamma_2 j_2 \| 1 \| \gamma_2 j_2\rangle .$$

According to (2.108), the reduced matrix element of 1 is $[j_2]^{1/2}$, and the expression reduces to

$$\langle(\gamma_1 j_1 \gamma_2 j_2)J\|t^k\|(\gamma_1' j_1' \gamma_2' j_2')J'\rangle$$

$$= \delta(\gamma_2, \gamma_2')\delta(j_2, j_2')(-1)^{j_1+j_2+J'+k}[J, J']^{1/2}\begin{Bmatrix} J & k & J' \\ j_1' & j_2 & j_1 \end{Bmatrix}\langle\gamma_1 j_1\|t^k\|\gamma_1' j_1'\rangle . \quad (4.23a)$$

Analogously, we get for an operator $\boldsymbol{u}^{(k)}$ *acting only on part 2,*

$$\langle(\gamma_1 j_1 \gamma_2 j_2)J\|\boldsymbol{u}^k\|(\gamma_1' j_1' \gamma_2' j_2')J'\rangle$$

$$= \delta(\gamma_1, \gamma_1')\delta(j_1, j_1')(-1)^{j_1+j_2'+J+k}[J, J']^{1/2}\begin{Bmatrix} J & k & J' \\ j_2' & j_1 & j_2 \end{Bmatrix}\langle\gamma_2 j_2\|\boldsymbol{u}^k\|\gamma_2' j_2'\rangle . \quad (4.23b)$$

....

Problem 4.5. Derive the relation (4.23a) directly using graphical methods.
Solution. We consider first the complete matrix element

$$\langle(\gamma_1 j_1 \gamma_2 j_2)JM \,|\, t^k_q \,|\, (\gamma_1' j_1' \gamma_2' j_2')J'M'\rangle .$$

Using the graphical representations of the coupled functions and the Wigner-Eckart theorem (3.17), we find, as in the more general case above, that the matrix element becomes

$$\delta(\gamma_2, \gamma_2') \langle\gamma_1 j_1\|t^k\|\gamma_1' j_1'\rangle \qquad \text{}$$

By adding outgoing arrows to the left vertex, the diagram becomes

$$\delta(j_2, j_2') [J, J']^{1/2} \qquad \text{}$$

After reversing the direction of the arrow of the j_1' line and changing two vertex signs, we can apply (4.10), which yields

$$\delta(j_2, j_2') [J, J']^{1/2}(-1)^{j_1+j_2+J'+k}\begin{Bmatrix} J & k & J' \\ j_1' & j_2 & j_1 \end{Bmatrix} \qquad \text{} .$$

By removing the diagram of the 3-*j* symbol and including the additional factors, we obtain the desired reduced matrix element (4.23a).

Problem 4.6. Use the graphical technique to verify the phase of (4.23b).

Problem 4.7. Show that the reduced matrix element of S in the coupled $(SL)J$ scheme is

$$\langle (SL)J\|S\|(SL)J\rangle = (-1)^{S+L+J+1}(2J+1)[S(S+1)(2S+1)]^{1/2}\begin{Bmatrix} J & 1 & J \\ S & L & S \end{Bmatrix}.$$

Use the expression for the 6-*j* symbol in Appendix C to recover the well-known expression for the Landé *g*-factor in *LS* coupling given by (2.123).

Problem 4.8. Derive graphically the formula

$$\langle \gamma J\| \{t^{k_1}u^{k_2}\}^{K}\|\gamma'J'\rangle = [K]^{1/2}\sum_{J''}(-1)^{J+J'+K}\begin{Bmatrix} k_1 & k_2 & K \\ J' & J & J'' \end{Bmatrix}$$

$$\times \sum_{\gamma''}\langle \gamma J\|t^{k_1}\|\gamma''J''\rangle\,\langle \gamma''J''\|u^{k_2}\|\gamma'J'\rangle\,, \tag{4.24}$$

where the reduced matrix elements are expressed in the *same* scheme.
Hint: Start from the identity

$$\langle \gamma JM\,|\,\{t^{k_1}u^{k_2}\}^{K}_{Q}\,|\,\gamma'J'M'\rangle$$
$$= \sum_{q_1q_2\gamma''J''M''}\langle k_1q_1, k_2q_2\,|\,KQ\rangle\,\langle \gamma JM\,|\,t^{k_1}_{q_1}\,|\,\gamma''J''M''\rangle\,\langle \gamma''J''M''\,|\,u^{k_2}_{q_2}\,|\,\gamma'J'M'\rangle\,.$$

4.4 The Coulomb Interaction for Two-Electron Systems in *LS* Coupling

We consider now the Coulomb interaction of a two-electron system in some detail. This is an important problem in its own right, and it also provides other useful examples of the graphical technique.

In Sect. 3.3 we evaluated the matrix elements of the Coulomb interaction (2.102)

$$\frac{1}{r_{12}} = \sum_{k}\frac{r^{k}_{<}}{r^{k+1}_{>}}\,C^{k}(1)\cdot C^{k}(2)$$

for uncoupled, direct-product states. We are now in a position to extend this treatment to coupled states. We shall first consider a system where the spin and the orbital angular momenta of the two electrons are coupled separately (*LS* coupling) and then a system where the spin and the orbital angular momenta of each electron are first coupled (*j-j* coupling).

4.4.1 The Basic Formula

The matrix elements of the Coulomb interaction for states where the orbital angular momenta of the two electrons are coupled to a specific L and the spins to a specific S can be evaluated using (4.22). Only the diagonal matrix elements, for which SM_SLM_L are the same for both states, are nonzero. These diagonal elements are

$$\langle (n_a l_a)_1 (n_b l_b)_2, SM_SLM_L | r_{12}^{-1} | (n_c l_c)_1 (n_d l_d)_2, SM_SLM_L \rangle$$
$$= \sum_k R^k(ab, cd)(-1)^{l_c+l_b+L} \begin{Bmatrix} l_a & l_b & L \\ l_d & l_c & k \end{Bmatrix} \langle l_a \| C^k \| l_c \rangle \langle l_b \| C^k \| l_d \rangle . \qquad (4.25)$$

Here, the subscripts 1 and 2 refer to the individual electrons in the same manner as in Sect. 3.3, and the two-electron radial integrals $R^k(ab, cd)$ are the Slater integrals, defined by (3.44b).

From a physical point of view, it is clear that the Coulomb interaction $1/r_{12}$ is invariant with respect to a simultaneous rotation of the coordinates of the two electrons. This fact may be used to show that the Coulomb operator commutes with L and S. It follows then as an elementary result of quantum mechanics that the Coulomb matrix elements should be diagonal with respect to SM_SLM_L (Theorem B. 4) and that they should be independent of the magnetic quantum numbers M_S and M_L which serve to define the orientation of the two electrons (Theorem B. 6).

As a further illustration, we also derive (4.25) graphically. Since the spin and the orbital angular momenta of the electrons are coupled separately, we can represent the spin-orbital part of the coupled state by means of *two* vector-coupling coefficients

$$\langle (l_a l_b)SM_SLM_L | = \langle (s_a s_b)SM_S | \langle (l_a l_b)LM_L |$$

where we have left out the uncoupled states. The other state can be represented in a similar way. According to (3.45), the Coulomb interaction is represented by

By combining the orbital parts together, we get the diagram

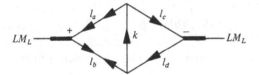

and similarly the spin diagram is

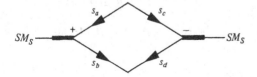

The orbital diagram above has already been evaluated (Problem 4.2b),

$$
\begin{array}{c}
\text{(diagram)}
\end{array}
= (-1)^{l_b + l_c + L + k}
\begin{Bmatrix}
l_a & l_b & L \\
l_d & l_c & k
\end{Bmatrix}.
$$

The spin diagram above is easily found to be unity, after eliminating the arrows and applying (3.25). This is quite obvious, since the spin part of the matrix element is simply a scalar product of the two spin functions.

Combining the results above with the phase, the radial integral, and the reduced matrix elements of the Coulomb interaction in (3.45), we recover (4.25).

4.4.2 Antisymmetric Wave Functions

The wave functions in (4.25) are *not* antisymmetric. In order to use this formula for real systems, made up of indistinguishable particles, we must, of course, use properly antisymmetrized functions.

We consider first a system of two electrons in *LS* coupling. The antisymmetric wave function of two electrons with quantum numbers nl and $n'l'$ is

$$
| \{nl\, n'l'\}\, SM_S LM_L \rangle
$$
$$
= F[| (nl)_1 (n'l')_2,\, SM_S LM_L \rangle - | (nl)_2 (n'l')_1,\, SM_S LM_L \rangle], \tag{4.26}
$$

where F is a normalization factor. The subscripts 1 and 2 refer, as before, to the individual electrons. *The curly brackets, introduced here, will be used in the following to denote an antisymmetric combination.* Using (2.79), we may reverse the order of the angular momenta in the second term on the right side. Note that the spins are also coupled in the same order, although we have left out the spin quantum numbers for the individual electrons from the expression. Thus, we get two phase factors, $(-1)^{l+l''+L}$ and $(-1)^{S+1}$, by reversing the order. Equation (4.26) can then be rewritten as

$$| \{nl\,n'l'\}\,SM_SLM_L\rangle \qquad\qquad (4.27)$$
$$= F[|(nl)_1(n'l')_2,\,SM_SLM_L\rangle + (-1)^{l+l''+L+S}|(n'l')_1(nl)_2,\,SM_SLM_L\rangle]\,.$$

If $nl = n'l'$, the electrons are said to be *equivalent*. The right-hand side of (4.27) then reduces to

$$F[1 + (-1)^{S+L}]|(nl)_1(nl)_2,\,SM_SLM_L\rangle\,.$$

This will vanish, if $S + L$ is odd, and hence *only functions where $S + L$ is even correspond to physical states.* The normalization factor F is then equal to $1/2$, and the normalized, antisymmetric function becomes

$$| \{nl^2\}\,SM_SLM_L\rangle = |(nl)_1(nl)_2,\,SM_SLM_L\rangle \qquad S + L\text{ even} \qquad (4.28)$$

Thus, ordinary vector coupling leads in this case automatically to an antisymmetric function, if $S + L$ is even.

If, on the other hand, the two electrons are not equivalent, then the two states in (4.27) are orthogonal, and the normalization factor F is $1/\sqrt{2}$. Then the normalized, antisymmetric function is

$$| \{nl\,n'l'\}\,SM_SLM_L\rangle$$
$$= \frac{1}{\sqrt{2}}[|(nl)_1(n'l')_2,\,SM_SLM_L\rangle - |(nl)_2(n'l')_1,\,SM_SLM_L\rangle] \qquad (4.29)$$
$$= \frac{1}{\sqrt{2}}[|(nl)_1(n'l')_2,\,SM_SLM_L\rangle + (-1)^{l+l''+L+S}|(n'l')_1(nl)_2,\,SM_SLM_L\rangle]\,.$$

4.4.3 Two Equivalent Electrons in LS Coupling

We can now use the wave function (4.28) to obtain the Coulomb interaction for a state of two equivalent electrons in LS coupling. Using the basic formula (4.25) and the expression (2.127) for the reduced matrix elements of the C tensor, we obtain (leaving out the M_SM_L quantum numbers)

$$\langle \{nl^2\}\,SL\,|\,1/r_{12}\,|\,\{nl^2\}\,SL\rangle = \sum_k F^k(nl,\,nl)(-1)^L \begin{Bmatrix} l & l & L \\ l & l & k \end{Bmatrix} \langle l\|C^k\|l\rangle^2$$

$$= \sum_k F^k(nl,\,nl)(-1)^L(2l+1)^2 \begin{Bmatrix} l & l & L \\ l & l & k \end{Bmatrix} \begin{pmatrix} l & k & l \\ 0 & 0 & 0 \end{pmatrix}^2\,. \qquad (4.30)$$

The F^k integrals are defined by (3.47).

As an example we consider two p electrons ($l = 1$), which give rise to three *LS* terms, namely ^1S ($S = 0, L = 0$), ^1D ($S = 0, L = 2$), and ^3P ($S = 1, L = 1$). The diagonal elements can be obtained from (4.30) using the tables of 3-j and 6-j symbols in Appendix D. The result is

$$
\begin{cases}
\langle np^2\,{}^1S | 1/r_{12} | np^2\,{}^1S \rangle = F^0 + \frac{2}{5}F^2 \\[2mm]
\langle np^2\,{}^1D | 1/r_{12} | np^2\,{}^1D \rangle = F^0 + \frac{1}{25}F^2 \\[2mm]
\langle np^2\,{}^3P | 1/r_{12} | np^2\,{}^3P \rangle = F^0 - \frac{1}{5}F^2 \,.
\end{cases}
\tag{4.31}
$$

From these equations we can obtain the splitting of the *LS* terms which is due to the Coulomb interaction between the electrons. This splitting is illustrated in Fig. 4.1.

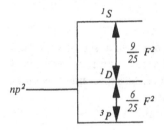

Fig. 4.1. Electrostatic splitting of an np^2 configuration

Although the absolute energies depend upon the radial integrals, the ratio of the upper interval to the lower interval is independent of F^0 and F^2,

$$
\frac{E({}^1S) - E({}^1D)}{E({}^1D) - E({}^3P)} = \frac{\frac{9}{25}F^2}{\frac{6}{25}F^2} = \frac{3}{2} \,.
$$

The experimental value of this ratio is 1.501 for the $4p^2$ configuration of Ge and 1.391 for the $5p^2$ configuration of Sn. The departures from the value 1.50 are due to the breakdown of *LS* coupling and to the admixture of other configurations.

4.4.4 Two Nonequivalent Electrons in *LS* Coupling

For a system with two nonequivalent electrons (different n or l) the antisymmetric function is given by (4.29).

$$
| \{nl\,n'l'\}\,SM_SLM_L \rangle = \frac{1}{\sqrt{2}} [\,|(nl)_1(n'l')_2,\,SM_SLM_L \rangle
$$
$$
- |(nl)_2(n'l')_1,\,SM_SLM_L \rangle\,].
$$

The expectation value of the Coulomb interaction for this state consists of two direct terms for which the ordering of the quantum numbers is the same on both sides and two exchange terms for which the order is reversed. As we shall see in Chap. 5, the symmetry of the Coulomb interaction with respect to an interchange of the two electrons may be used to show that the two direct terms are equal. This is also true of the two exchange terms. The Coulomb matrix elements may then be written as the difference of a direct and an exchange matrix element

$$\langle \{nl\, n'l'\}\, SM_S LM_L | r_{12}^{-1} | \{nl\, n'l'\}\, SM_S LM_L \rangle$$
$$= \langle (nl)_1 (n'l')_2,\, SM_S LM_L | r_{12}^{-1} | (nl)_1 (n'l')_2,\, SM_S LM_L \rangle$$
$$- \langle (nl)_1 (n'l')_2,\, SM_S LM_L | r_{12}^{-1} | (nl)_2 (n'l')_1,\, SM_S LM \rangle .$$

As in (4.27) we may reverse the order of the angular momenta on the right side of the second matrix element to obtain

$$\langle \{nl\, n'l'\}\, SL | r_{12}^{-1} | \{nl\, n'l'\}\, SL \rangle = \langle (nl\, n'l')SL | r_{12}^{-1} | (nl\, n'l')SL \rangle$$
$$+ (-1)^{l+l''+S+L} \langle (nl\, n'l')SL | r_{12}^{-1} | (n'l'\, nl)\, SL \rangle .$$

We have used here a simplified notation in which the first set of quantum numbers within ordinary parentheses represents electron 1 and the second set electron 2. Antisymmetric combinations will, as before, be represented by curly brackets. As before, we leave out the $M_S M_L$ values.

We can now apply the basic formula (4.25) to obtain

$$\langle \{nl\, n'l'\}\, SL | r_{12}^{-1} | \{nl\, n'l'\}\, SL \rangle$$
$$= (-1)^{l+l''+L} \sum_k F^k(nl,\, n'l') \begin{Bmatrix} l & l' & L \\ l' & l & k \end{Bmatrix} \langle l \| C^k \| l \rangle \langle l' \| C^k \| l' \rangle$$
$$+ (-1)^{l+l''+S} \sum_k G^k(nl,\, n'l') \begin{Bmatrix} l & l' & L \\ l & l' & k \end{Bmatrix} \langle l \| C^k \| l' \rangle \langle l' \| C^k \| l \rangle . \qquad (4.32)$$

For an ls configuration ($l' = 0$, $L = l$), this reduces to the simple expression (see Problem 4.9)

$$\langle \{nl\, n's\}\, SL | r_{12}^{-1} | \{nl\, n's\}\, SL \rangle = F^0(nl,\, n's) + (-1)^S \frac{G^l(nl,\, n's)}{2l + 1} . \qquad (4.33)$$

This configuration has two LS terms: a *triplet* term with $S = 1$ (^3l) and a *singlet* term with $S = 0$ (^1l). According to (4.33) these two terms are split by an amount $2\, G^l(nl,\, n's)/(2l + 1)$ by the exchange part of the Coulomb interaction. Thus, this splitting of the two spin states is a consequence of the antisymmetry of the functions.

Problem 4.9. Use the relation (3.73) and the reduced matrix elements (2.127) to derive the formula (4.33).

4.5 The Coulomb Interaction for Two-Electron Systems in j-j Coupling

Next we shall consider the evaluation of the Coulomb interaction for two-electron systems in j-j coupling. This problem has been treated without using angular-momentum diagrams by *Judd* [1962]. If the electrons belong to the same nlj shell, then, in analogy with the LS case, the wave function can be shown to be

$$| \{(nlj)^2\} JM \rangle = | (nlj\,nlj)JM \rangle \,,$$

where we have used the same notation as in the LS case. The diagonal element of the Coulomb interaction then becomes simply

$$\langle \{(nlj)^2\} JM | r_{12}^{-1} | \{(nlj)^2\} JM \rangle = \langle (nlj\,nlj)JM | r_{12}^{-1} | (nlj\,nlj)JM \rangle \,. \tag{4.34}$$

The antisymmetric wave function for electrons belonging to *different* nlj states is

$$| \{nlj, n'l'j'\} JM \rangle$$
$$= \frac{1}{\sqrt{2}} [| (nlj)_1(n'l'j')_2, JM \rangle - | (nlj)_2(n'l'j')_1 JM \rangle]$$
$$= \frac{1}{\sqrt{2}} [| (nlj\,n'l'j')JM \rangle - (-1)^{j+j'-J} | (n'l'j'\,nlj)JM \rangle] \,.$$

In analogy with the LS case, we then obtain

$$\langle \{nlj\,n'l'j'\} JM | r_{12}^{-1} | \{nlj\,n'l'j'\} JM \rangle$$
$$= \langle (nlj\,n'l'j')JM | r_{12}^{-1} | (nlj\,n'l'j')JM \rangle$$
$$- (-1)^{j+j'-J} \langle (nlj\,n'l'j')JM | r_{12}^{-1} | (n'l'j'\,nlj)JM \rangle \,. \tag{4.35}$$

As an example of evaluating a matrix element in j-j coupling, we consider in detail the second term (the exchange term) in the formula above. The coupled functions can be represented graphically in the way shown in Sect. 3.1.3,

$$\langle (nlj\,n'l'j')JM | = \quad JM$$

and similarly for the other state in the matrix element. Using (3.42), we get

$$\langle (nlj\, n'l'j')JM\, |\, r_{12}^{-1}\, |\, (n'l'j'\, nlj)JM \rangle = \sum_k X(k,\, ll'l'l)G^k(nl,\, n'l')\qquad(4.36)$$

times the angular-momentum graph

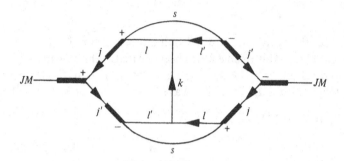

By adding arrows to the three vertices at the left, this diagram may be cast into normal form with one and only one arrow on each internal line. After removing the heavy lines and applying (4.6), the diagram becomes

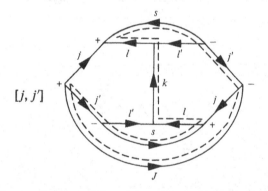

Again, the dotted line drawn in the diagram is a Hamilton line. Rewriting the diagram so that this line appears on the periphery, we obtain

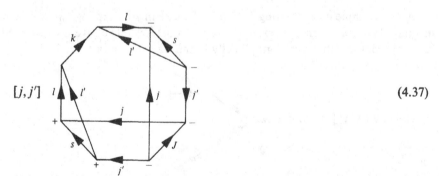

(4.37)

This graph may be evaluated using JLV3. We leave this as an exercise (Problem 4.10).

Collecting all the results together, we obtain finally

$$\langle (nlj\ n'l'j')JM\,|\,r_{12}^{-1}\,|\,(n'l'j'\ nlj)JM\rangle$$

$$= (-1)^{j+j'-J}\sum_{k}[l,l',j,j']\begin{pmatrix} l & k & l' \\ 0 & 0 & 0 \end{pmatrix}^2 \begin{cases} k & j' & j \\ s & l & l' \end{cases} \begin{cases} k & j' & j \\ J & j' & j \end{cases} G^k(nl,\,n'l')\,.$$

$$(4.38)$$

This equation gives the matrix elements of the exchange interaction in *j-j* coupling. The direct term can be evaluated in a similar way. For the special case of an *ls* configuration, $l' = 0,\ j' = 1/2$, the entire interaction reduces to

$$\langle \{nlj\ n's\}\,JM\,|\,r_{12}^{-1}\,|\,\{nlj\ n's\}\,JM\rangle$$

$$= F^0(nl,\,n's) - \frac{(2j+1)}{(2l+1)}\begin{cases} s & l & j \\ s & J & j \end{cases} G^l(nl,\,n's)\,.$$

$$(4.39)$$

This leads to the same energy level splitting as the formula given by *Judd*[1962].

....

Problem 4.10. Show that the angular-momentum diagram appearing in (4.37) is equal to

$$(-1)^{j+j'-J}\begin{cases} k & j' & j \\ s & l & l' \end{cases}^2 \begin{cases} k & j' & j \\ J & j' & j \end{cases}.$$

Solution: The diagram may be split into two parts by cutting the lines k, j', and j, using JLV3,

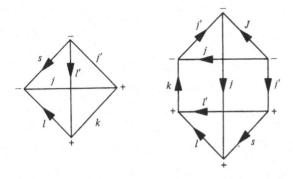

The first part is easily found to be equal to

$$-(-1)^{s+l+j}\begin{cases} k & j' & j \\ s & l & l' \end{cases}.$$

The second part can be cut into two parts by applying JLV3 a second time, yielding

$$- (-1)^{s+l+J} (-1)^{j+j'-J} \begin{Bmatrix} k & j & j' \\ s & l' & l \end{Bmatrix} \begin{Bmatrix} k & j' & j \\ J & j' & j \end{Bmatrix}.$$

The final result is obtained by combining these two expressions together.

Problem 4.11. Verify the direct term in (4.39).

5. The Independent-Particle Model

In Chaps. 2–4 we have developed the theory of angular momentum and, in particular, its graphical representation. As illustrations we have also considered some physical problems, like the Coulomb interaction. We are now in a position to study atomic properties in a more systematic way. In the present chapter we shall investigate the Hamiltonian for many-electron atoms and the independent-particle model in which the electrons are assumed to move independently of each other in an average field due to the nucleus and the other electrons. The independent-particle model and the variational principle then lead to the (unrestricted) Hartree-Fock equations. In the next chapter we shall make the additional assumption that the average potential is spherically symmetric. This is the central-field model, which together with the variation principle leads to the restricted Hartree-Fock equations. The central-field model forms the basis of the atomic shell structure and the chemical regularity of the elements.

5.1 The Magnetic Interactions

For an N-electron atom with a nuclear charge Z the nonrelativistic Hamiltonian may be written (in atomic units, see Appendix A)

$$H = -\frac{1}{2} \sum_{i=1}^{N} \nabla_i^2 - \sum_{i=1}^{N} \frac{Z}{r_i} + \sum_{i<j}^{N} \frac{1}{r_{ij}} + V_{\text{mag}} . \tag{5.1}$$

In this expression, r_i is the distance of electron i from the nucleus, and r_{ij} its distance from electron j. The third summation represents the Coulomb repulsion among the electrons, and the fourth term the interaction of the spin of the electrons with the magnetic fields produced by their spin and by their orbital motion (the so-called magnetic interactions).

We have already studied the Coulomb interaction in some detail, and we shall now discuss briefly the magnetic interactions.

The electric field at the position of electron i due to the nucleus and the other electrons is

$$E_i = Z \frac{r_i}{r_i^3} - \sum_{j}^{j \neq i} \frac{r_{ij}}{r_{ij}^3} .$$

Here, r_i and r_j are the position vectors for electron i and j, respectively, and $r_{ij} = |r_{ij}| = |r_i - r_j|$ is the interelectronic distance. The motion of electron i in this electric field gives rise to a magnetic field in the rest frame of the electron which interacts with its spin. A quantum-mechanical treatment based on the relativistic Dirac equation then leads to the interaction (see, for instance, [*Schiff* 1968, Chap. 12] or [*Messiah* 1961, Chap. 20])

$$V_{so} = \frac{\alpha^2}{2} \sum_i (E_i \times p_i) \cdot s_i \tag{5.2}$$

$$= \frac{\alpha^2}{2} \sum_i \left[\frac{Z}{r_i^3} r_i \times p_i - \sum_j^{j \neq i} \frac{r_{ij}}{r_{ij}^3} \times p_i \right] \cdot s_i ,$$

where α is the fine-structure constant. This expression can also be derived in a semiclassical way, using classical relativistic theory, as first shown by *Thomas* [1926].

The first term in (5.2) can also be written

$$\frac{\alpha^2}{2} \sum_i \frac{Z}{r_i^3} l_i \cdot s_i \tag{5.3}$$

and this represents the interaction of the electron spin with the magnetic field caused by the orbital motion in the nuclear Coulomb field. The second term in (5.2) represents the corresponding effect due to the orbital motion in the Coulomb field of the remaining electrons. Both terms describe the *spin-(own-)orbit interaction*, that is, the interaction between the spin and the orbital motions of the same electron.

There is also an interaction between the spin motion of one electron and the orbital motion of the other electrons, the so-called *spin-other-orbit interaction*. This interaction can be derived in a semiclassical way as follows.

The electric current due to the motion of electron i is

$$j_i(r) = v_i \rho_i(r) = - e v_i \, \delta(r - r_i) ,$$

where v_i is the velocity of the electron and $\rho_i(r) = -e \, \delta(r - r_i)$ is the charge density. According to the Biot-Savart law [*Jackson* 1975] this current produces at the position of electron j the magnetic field

$$B_{ij} = \frac{\mu_0}{4\pi} \int j_i \times \frac{r_{ij}}{r_{ij}^3} dr = - \frac{\mu_0}{4\pi} e \, v_i \times \frac{r_{ij}}{r_{ij}^3} ,$$

dr being the volume element. In atomic units we have $\mu_0/4\pi = \alpha^2$, $e = 1$ and $v_i = p_i$, so this field becomes

$$B_{ij} = - \alpha^2 p_i \times \frac{r_{ij}}{r_{ij}^3} .$$

The field interacts with the spin of electron j, which—after summing over i and j—yields the following interaction:

$$V_{soo} = \alpha^2 \sum_{ij}^{i \neq j} \left(\boldsymbol{p}_i \times \frac{\boldsymbol{r}_{ij}}{r_{ij}^3} \right) \cdot \boldsymbol{s}_j \,.$$

This represents an interaction between the spin of one electron (j) and the orbital motion of another electron (i), and therefore the effect is called the spin-other-orbit interaction. Combining this with the spin-(own-)orbit interaction (5.2), we get the following "generalized spin-orbit interaction":

$$V_{so} + V_{soo} = \frac{\alpha^2}{2} \sum_i \frac{Z}{r_i^3} \boldsymbol{l}_i \cdot \boldsymbol{s}_i - \frac{\alpha^2}{2} \sum_{ij}^{i \neq j} \left(\frac{\boldsymbol{r}_{ij}}{r_{ij}^3} \times \boldsymbol{p}_i \right) \cdot (\boldsymbol{s}_i + 2\boldsymbol{s}_j) \,. \tag{5.4}$$

In addition to these interactions, there are other magnetic interactions in a many-electron system which can be interpreted as *spin-spin* and *orbit-orbit interactions*. The orbit-orbit interaction is usually very weak and can often be neglected. The orbit-orbit interaction does not cause any *splitting* of the different J states. For these reasons, we shall not consider these interactions further here.

The Dirac equation leads to a correct description of the spin-orbit interaction for a single-electron system. Treating the magnetic interaction for many-electron systems, on the other hand, is a *quantum-electrodynamical* (QED) problem, which is very difficult to handle in a rigorous way. The problem was first treated by *Breit* [1929, 1930, 1932], who derived the lowest-order corrections (second order in α) to the electron-electron interaction for a two-electron system, known as the *Breit interaction*. A more rigorous treatment, based on the modern form of QED, was later given by *Brown* and *Ravenhall* (1951). Combined with the Pauli approximation, where the Dirac equation is treated to the same order, the Breit interaction leads to a Hamiltonian which contains all the magnetic interactions discussed above (see, for instance, [*Bethe* and *Salpeter* 1957]).

Instead of using the Pauli approximation, it is nowdays common to combine the Breit interaction with the exact Dirac equation. In this way, the spin-(own-)orbit interaction is treated to all orders, while the spin-other-orbit and other magnetic interactions are treated only in second order. Although fundamental objections can be raised against such a procedure [*Brown* and *Ravenhall* 1951; *Armstrong* and *Feneuille* 1974; *Armstrong* 1978; *Sucher* 1980], it has been found to work quite well in practical applications on ordinary atomic systems (Sect. 14.7).

Within the central-field approximation—to be discussed in the next chapter—it is possible to make important simplifications in the treatment of the magnetic interactions. It can be shown that not only the one-body nuclear part (5.3) but also most of the two-body electronic part can be represented by means of an "effective" one-body interaction

$$V_{so}^{eff} = \sum_i \zeta_i \boldsymbol{l}_i \cdot \boldsymbol{s}_i \,. \tag{5.5}$$

Here, the summation runs over the electrons in the open shells only. The parameter ζ_i is called the spin-orbit parameter, but it should be observed that it can be defined to contain most of the spin-other-orbit interaction as well.

In a local, central potential u—neglecting the spin-other-orbit interaction—one finds that the spin-orbit parameter introduced above simply reduces to the well-known expression

$$\zeta = \frac{\alpha^2}{2} \left\langle \frac{1}{r} \frac{du}{dr} \right\rangle, \tag{5.6}$$

where the symbol $\langle \ \rangle$ represents the expectation value in the open shell.

The more general spin-orbit interaction (5.4) has been analysed in the central-field approximation by *Blume* and *Watson* [1962, 1963] and by *Blume* et al. [1964], who derived an expression for the spin-orbit parameter (5.5) in the Hartree-Fock scheme for atoms with a single open shell. A more complete study of the spin-orbit interaction has recently been made by *Huang* and *Starace* [1978], using graphical methods of angular-momentum theory.

In this book we shall not be particularly concerned with relativistic effects, and for that reason we shall not discuss the magnetic interactions any further. Interested readers are referred to the works mentioned above and to references found therein.

In our analysis of atomic energy levels in Chap. 7, we shall use the form (5.5) of the spin-orbit interaction and the expressions of *Blume* and *Watson* in order to compare theory and experiments. In Chap. 14, we shall also briefly discuss the effect of higher-order electrostatic perturbations on the spin-orbit interaction, using the effective-operator approach of many-body perturbation theory. In this way it can be shown that also many higher order effects—beyond the central-field model—can be represented by means of an effective spin-orbit operator of the type (5.5).

5.2 Determinantal Wave Functions

Even if we neglect the magnetic interactions completely, the Hamiltonian (5.1) is quite complex for any atom having more than a few electrons. As a further approximation it is then customary to assume that each electron moves independently of the other electrons in an *average* field caused by the nucleus and the electrons. This assumption leads to the *independent-particle* model, which provides an approximate description of the atom and also serves as the starting point of more accurate calculations. The independent-particle model may be put on a mathematical footing by separating the full Hamiltonian (5.1) in the following way:

$$H = H_0 + V_{es}, \tag{5.7}$$

omitting the magnetic interactions. Here,

$$H_0 = \sum_{i=1}^{N} h_0(i) \tag{5.8}$$

is a sum of one-electron operators

$$h_0(i) = -\frac{1}{2} \nabla_i^2 - \frac{Z}{r_i} + u(\mathbf{r}_i)$$

and

$$V_{es} = -\sum_i u_i(\mathbf{r}_i) + \sum_{i>j} \frac{1}{r_{ij}}.$$

The approximate Hamiltonian H_0 represents an average interaction with the nucleus and the other electrons. V_{es} causes departures from this single-particle description, and it will be treated as a perturbation. In order for V_{es} to be reasonably small, the average potential u must contain most of the effect of the Coulomb repulsion among the electrons. The Coulomb repulsion itself is usually too large to be treated by perturbation theory.

The simplest wave function which describes a system of N electrons moving in an average potential, is the product function

$$\Psi = \varphi_a(1)\varphi_b(2) \dots \varphi_n(N). \tag{5.9}$$

Here $a, b, \dots$ stand for the set of quantum numbers, necessary to specify a single-electron state, and $1, 2, \dots$ stand for the space and spin coordinates of electrons $1, 2, \dots$ If the single-electron functions satisfy the eigenvalue equations

$$\left[-\frac{1}{2} \nabla_1^2 - \frac{Z}{r_1} + u(\mathbf{r}_1) \right] \varphi_a(1) = \varepsilon_a \varphi_a(1), \tag{5.10}$$

then Ψ will be an eigenfunction of the Hamiltonian

$$H_0 = \sum_{i=1}^{N} \left[-\frac{1}{2} \nabla_i^2 - \frac{Z}{r_i} + u(\mathbf{r}_i) \right], \tag{5.11}$$

corresponding to the eigenvalue

$$E_0 = \sum_{i=1}^{N} \varepsilon_i. \tag{5.12}$$

Any permutation of the electrons among the single-particle functions also leads to an eigenfunction of H_0. For instance, the function $\varphi_a(2)\varphi_b(1)\varphi_c(3) \dots$ $\varphi_n(N)$, for which the coordinates of the first and second electrons have been interchanged, is also an eigenfunction of H_0 with the same eigenvalue. There will usually be many eigenfunctions of H_0 which can be formed in this way. In order to satisfy the Pauli exclusion principle, however, the wave functions of a many-electron system must be *antisymmetric* with respect to the interchange of any two of the electrons. We can form a wave function having this property by antisymmetrizing the product function (5.9) in the following way:

$$\Phi = \sqrt{(1/N!)}\, \Sigma_P (-1)^p\, P\varphi_a(1)\varphi_b(2) \dots \varphi_n(N).$$

The sum extends over all permutations P of the electrons among the states defined by the quantum numbers $a, b, \dots n$, and p is the parity of the permutation; $(-1)^p$ is $+1$ if the permutation is even and -1 if it is odd. There are $N!$ possible permutations among the N electrons, and the factor $\sqrt{(1/N!)}$ ensures that the state is normalized. Since each of the terms in the sum is an eigenfunction of H_0 corresponding to the eigenvalue E_0, it is clear that Φ will also have this property. The state Φ can be written in the form of a determinant

$$\Phi = \sqrt{(1/N!)} \begin{vmatrix} \varphi_a(1)\varphi_a(2) \dots \varphi_a(N) \\ \varphi_b(1)\varphi_b(2) \dots \varphi_b(N) \\ \dots\dots\dots\dots\dots\dots \\ \varphi_n(1)\varphi_n(2) \dots \varphi_n(N) \end{vmatrix} \tag{5.13}$$

and it is called a determinantal product state or *Slater determinant*. An interchange of any two of the electrons amounts to interchanging two of the columns of the determinant and this leads to a sign change. If any two of the single-particle functions are the same, two rows of the determinant become identical, and the determinant will vanish. This shows that the exclusion principle is satisfied.

In order to construct a determinantal state Φ it is necessary to know which single-particle states are occupied and how these states are ordered within the determinant. We may provide this information by writing simply

$$|\Phi\rangle = |\{a\, b \dots n\}\rangle,$$

where we have made use of the curly brackets introduced in the previous chapter to indicate that the function is antisymmetric. The properties of these states follow from (5.13), for example,

$$|\{a\, b\, c \dots n\}\rangle = - |\{b\, a\, c \dots n\}\rangle,$$

5.3 Matrix Elements Between Slater Determinants

In order to calculate the energy levels of an atom, it will be necessary to evaluate matrix elements of the Hamiltonian between antisymmetric states. The Hamiltonian (5.1) includes one-electron operators of the type Z/r_i, which act on the coordinates of one electron, and two-electron operators of the kind $1/r_{ij}$. We will therefore need matrix elements of one- and two-electron operators between determinants of orthonormal functions.

5.3.1 Matrix Elements of Single-Particle Operators

We consider first a general single-electron operator, which may be written

$$F = \sum_{i=1}^{N} f(i) , \qquad (5.14)$$

where $f(i)$ acts only on the coordinates of the ith electron. For simplicity, we restrict ourselves to a two-electron system for which

$$F = f(1) + f(2) .$$

The diagonal matrix elements of F for the antisymmetric wave function $|\{ab\}\rangle$ are

$$\langle \{ab\} | F | \{ab\} \rangle = \frac{1}{2} \iint [\varphi_a(1)\varphi_b(2) - \varphi_a(2)\varphi_b(1)]^*$$
$$\times [f(1) + f(2)] [\varphi_a(1)\varphi_b(2) - \varphi_a(2)\varphi_b(1)] \, dr_1 \, dr_2 , \qquad (5.15)$$

where dr_1 and dr_2, denote the volume elements, and the integrations also represent summations over the spin coordinates. Cross terms of the kind

$$\iint \varphi_a^*(1)\varphi_b^*(2)f(1)\varphi_a(2)\varphi_b(1) \, dr_1 \, dr_2$$

are obviously zero, since $f(1)$ operates only on the first wave function and $\varphi_b(2)$ and $\varphi_a(2)$ are assumed to be orthogonal. Furthermore, by interchanging the coordinates of the first and second electron, one may easily see that

$$\iint \varphi_a^*(1)\varphi_b^*(2)f(1)\varphi_a(1)\varphi_b(2) \, dr_1 \, dr_2 = \iint \varphi_a^*(2)\varphi_b^*(1)f(2)\varphi_a(2)\varphi_b(1) \, dr_1 \, dr_2 .$$

So (5.15) reduces to

$$\langle \{ab\} | F | \{ab\} \rangle = \iint \varphi_a^*(1)\varphi_b^*(2) [f(1) + f(2)] \varphi_a(1)\varphi_b(2) \, dr_1 \, dr_2$$
$$= \langle a|f|a \rangle + \langle b|f|b \rangle . \qquad (5.16)$$

This is exactly what one would have obtained had one used ordinary product functions $\varphi_a(1)\varphi_b(2)$, rather than determinats $|\{ab\}\rangle$.

The nondiagonal matrix element between two determinantal states, which differ by a single state, can be shown in a similar way to be

$$\langle \{ab\} | F | \{ac\} \rangle = \langle b|f|c \rangle , \qquad (5.17)$$

and, finally, if both states are different, we get

$$\langle \{ab\} | F | \{cd\} \rangle = 0 . \tag{5.18}$$

5.3.2 Matrix Elements of Two-Particle Operators

A two-electron operator can be written generally

$$G = \sum_{i<j} g(i,j) ,$$

where $g(i,j)$ acts on the ith and jth electrons, and the summation includes each *pair* of electrons. For a two-electron system this is simply

$$G = g(1, 2).$$

The diagonal matrix element of G is then

$$\langle \{ab\} | G | \{ab\} \rangle = \frac{1}{2} \iint [\varphi_a(1)\varphi_b(2) - \varphi_a(2)\varphi_b(1)]^* g(1, 2)$$

$$\times [\varphi_a(1)\varphi_b(2) - \varphi_a(2)\varphi_b(1)] \, dr_1 \, dr_2 = \frac{1}{2} \iint [\varphi_a^*(1)\varphi_b^*(2)g(1, 2) \, \varphi_a(1)\varphi_b(2)$$

$$- \varphi_a^*(1)\varphi_b^*(2)g(1,2)\varphi_a(2)\varphi_b(1) - \varphi_a^*(2)\varphi_b^*(1)g(1, 2)\varphi_a(1)\varphi_b(2)$$

$$+ \varphi_a^*(2)\varphi_b^*(1)g(1, 2)\varphi_a(2)\varphi_b(1)] \, dr_1 \, dr_2 .$$

Since the two-electron interaction $g(1,2)$ is symmetric with respect to an interchange of the coordinates of the two electrons $(1 \leftrightarrow 2)$, the first and fourth terms in this expansion are equal, and similarly the second and third terms are equal. So the matrix element may be written simply

$$\langle \{ab\} | G | \{ab\} \rangle = \langle ab | g | ab \rangle - \langle ba | g | ab \rangle . \tag{5.19}$$

The symbols on the right represent here matrix elements with ordinary product functions. We shall call the first matrix element the *direct* term and the second matrix element the *exchange* term. The exchange matrix element would not occur if one used product functions $\varphi_a(1)\varphi_b(2)$ rather than proper antisymmetric wave functions.

5.3.3 A New Notation

The results obtained above may be generalized to N-electron systems. For this purpose we shall use a special notation. We will allow Greek letters to stand for ordered sets of quantum numbers representing Slater determinants. So, for instance, if we let α correspond to the quantum numbers $a,b,\ldots n$, the determinantal state $| \{ab \ldots n\} \rangle$ can be written simply as $|\alpha\rangle$. Single-particle functions, which appear in the determinant, will be called *occupied* orbitals and the remaining functions of the set will be called *excited* or *virtual* orbitals. We shall use the notation $|\alpha_a^r\rangle$ to denote a determinant for which an occupied orbital a in α is replaced by the virtual orbital r. Similarly, double substitutions for which two electrons (here a and b) are excited from the sea of occupied orbitals will be written $|\alpha_{ab}^{rs}\rangle$.

Using this notation the formulas for the matrix elements of one- and two-particle operators between determinantal states of a many-particle system can be generalized in the following way.

1) For diagonal matrix elements

$$\langle\alpha|F|\alpha\rangle = \sum_{a}^{occ} \langle a|f|a\rangle \tag{5.20}$$

$$\langle\alpha|G|\alpha\rangle = \sum_{a<b}^{occ} (\langle ab|g|ab\rangle - \langle ba|g|ab\rangle), \tag{5.21}$$

where the sums run over orbitals a and b that are occupied in $|\alpha\rangle$. The summation (5.21) includes each pair of orbitals a and b once. This equation can also be written

$$\langle\alpha|G|\alpha\rangle = \frac{1}{2} \sum_{ab}^{occ} (\langle ab|g|ab\rangle - \langle ba|g|ab\rangle), \tag{5.22}$$

where a and b run independently.

2) For elements between states which differ by the quantum numbers of a single orbital

$$\langle\alpha_a^r|F|\alpha\rangle = \langle r|f|a\rangle \tag{5.23}$$

$$\langle\alpha_a^r|G|\alpha\rangle = \sum_{b}^{occ} (\langle rb|g|ab\rangle - \langle br|g|ab\rangle). \tag{5.24}$$

3) For elements between states which differ by the quantum numbers of two orbitals

$$\langle\alpha_{ab}^{rs}|F|\alpha\rangle = 0 \tag{5.25}$$

$$\langle\alpha_{ab}^{rs}|G|\alpha\rangle = \langle rs|g|ab\rangle - \langle sr|g|ab\rangle. \tag{5.26}$$

All matrix elements of F and G between states for which more than two quantum numbers are different vanish.

5.3.4 Feynman Diagrams

The matrix elements that appear in the above equations may be represented by simple diagrams, which are related to the Feynman diagrams introduced in field theory [*Feynman* 1949]. In the later part of this book we shall use similar diagrams to describe the departures from the independent-particle model. Here we use the diagrams only as a simple pictorial way of describing these matrix elements. The diagrams which correspond to (5.23,24,26) are shown in Fig. 5.1.

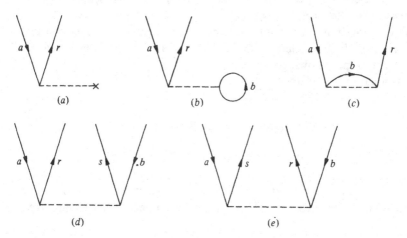

Fig. 5.1a-e. Diagrams for one- and two-body interactions. (a) corresponds to (5.23), (b,c) to (5.24), and (d, e) to (5.26)

Diagram (*a*) corresponds to the matrix element $\langle \alpha_a^r | F | \alpha \rangle$ in (5.23) for which a single electron is excited from the occupied orbital a to a virtual orbital r by a one-body interaction. Diagrams (*b*) and (*c*) correspond to the direct and exchange matrix elements in (5.24). The internal line represents a summation over occupied orbitals. Each vertex of a diagram corresponds to one variable of integration. So in the diagram for the direct matrix element the line for the occupied state b is associated with a single variable, while in the exchange diagram it is associated with both integration variables. Diagrams (*d*) and (*e*) correspond to the two matrix elements on the right-hand side of (5.26). They represent double excitations.

5.4 The Hartree-Fock Equations

Equations (5.20–26) may be used to evaluate the matrix elements of the atomic Hamiltonian (5.1). Neglecting the magnetic interactions, the expectation value of the total energy for a state represented by a Slater determinant $|\alpha\rangle$ is

$$\langle E \rangle = \langle \alpha | H | \alpha \rangle = \left\langle \alpha \left| \sum_{i=1}^{N} \left(-\frac{1}{2} \nabla_i^2 - \frac{Z}{r_i} \right) + \sum_{i<j}^{N} \frac{1}{r_{ij}} \right| \alpha \right\rangle. \tag{5.27}$$

According to the variational principle, the "best" determinant for the ground state can be determined by minimizing this expectation value. A necessary condition is then that the expection value be stationary with respect to small changes in the form of the occupied orbitals, and this condition will be used to derive the Hartree-Fock (HF) equations as follows.

Small changes in the occupied orbitals (a) can be expressed by means of small admixtures of virtual orbitals (r)

$$|a\rangle \rightarrow |a\rangle + \eta|r\rangle\,, \tag{5.28}$$

where η is a small, real number. This leads to an admixture of single substitutions $|\alpha_a^r\rangle$ into $|a\rangle$

$$|a\rangle \rightarrow |a\rangle + \eta|\alpha_a^r\rangle\,, \tag{5.29}$$

and a corresponding change in the expectation value of the energy

$$\langle E\rangle \rightarrow \langle E\rangle + \eta(\langle \alpha_a^r|H|a\rangle + \langle a|H|\alpha_a^r\rangle)\,,$$

neglecting terms quadratic in η. Since H is a hermitian operator, the two matrix elements above are the complex conjugates of each other. With the conventions used here, the elements are real and hence equal. The energy will then be stationary if

$$\langle \alpha_a^r|H|a\rangle = 0\,. \tag{5.30}$$

This condition, called *Brillouin's theorem* [*Brillouin* 1933, 1934], implies that the Hamiltonian H has no matrix elements between $|a\rangle$ and states obtained from $|a\rangle$ by a single substitution. As we shall see later, this implies that there will be no first-order mixing of such states.

Using (5.23,24), the Hartree-Fock condition (5.30) may be written out explicitly in terms of one- and two-particle matrix elements,

$$\left\langle r\left| -\frac{1}{2}\nabla^2 - \frac{Z}{r}\right|a\right\rangle + \sum_b^{occ}(\langle rb|r_{12}^{-1}|ab\rangle - \langle br|r_{12}^{-1}|ab\rangle) = 0\,. \tag{5.31}$$

In order to write (5.31) more simply, we define a Hartree-Fock operator (h_{HF}) and potential (u_{HF}) by the equations

$$h_{HF} = -\frac{1}{2}\nabla^2 - \frac{Z}{r} + u_{HF} \tag{5.32}$$

$$\langle i|u_{HF}|j\rangle = \sum_b^{occ}(\langle ib|r_{12}^{-1}|jb\rangle - \langle bi|r_{12}^{-1}|jb\rangle)\,, \tag{5.33}$$

where the sum b runs over all of the orbitals occupied in the determinant $|a\rangle$. Then the condition (5.31) to be satisfied becomes simply

$$\langle r|h_{\mathrm{HF}}|a\rangle = 0\,,$$
(5.34)

where a is an occupied and r a virtual orbital. Using the completeness relation (B.9), this leads to the equation

$$h_{\mathrm{HF}}|a\rangle = \sum_i |i\rangle\,\langle i|h_{\mathrm{HF}}|a\rangle = \sum_b^{\mathrm{occ}} |b\rangle\,\langle b|h_{\mathrm{HF}}|a\rangle\,,$$

where i runs over all orbitals and b over occupied ones. Thus, when acting on an occupied orbital the Hartree-Fock operator produces only occupied orbitals.

It follows directly from the symmetry of the Coulomb interaction that

$$\langle a|h_{\mathrm{HF}}|b\rangle = \langle b|h_{\mathrm{HF}}|a\rangle\,,$$

which means that the HF operator is hermitian. Furthermore, it can be shown that this operator is invariant for a unitary transformation. Therefore, we can find a new set of orbitals, where h_{HF} is diagonal,

$$h_{\mathrm{HF}}|a'\rangle = \varepsilon'_a|a'\rangle\,.$$
(5.35)

This is the normal form of the general *Hartree-Fock equation*.

Using (5.32) the HF equation (5.35) can be written out explicitly (leaving out the primes)

$$\left(-\frac{1}{2}\boldsymbol{\nabla}^2 - \frac{Z}{r} + u_{\mathrm{HF}}\right)|a\rangle = \varepsilon_a|a\rangle\,.$$
(5.36)

Each term here can be given a simple physical interpretation. The first term represents the kinetic energy of electron a and Z/r its attraction to the nucleus. The potential u_{HF} represents the average Coulomb interaction of electron a with the other electrons in the atom, *including* the exchange interaction. The interpretation of the eigenvalue ε_a will be discussed in the next section.

Using the Feynman-like diagrams introduced above, we can represent the HF potential (5.33), in the way shown in Fig. 5.2.

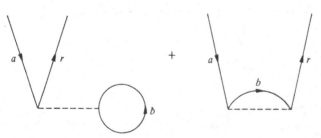

Fig. 5.2. Diagrammatic representation of the Hartree-Fock potential

5.5 Koopmans' Theorem

According to (5.27) and (5.20–22) the expectation value of the total energy of an atomic state represented by the Slater determinant $|a\rangle$ is

$$\langle E_{\text{atom}}\rangle = \sum_b^{\text{occ}} \left\langle b\left|-\frac{1}{2}\nabla^2 - \frac{Z}{r}\right|b\right\rangle + \frac{1}{2}\sum_{bc}^{\text{occ}}(\langle bc|r_{12}^{-1}|bc\rangle - \langle cb|r_{12}^{-1}|bc\rangle)\,.$$

Let us now remove one electron—occupying orbital a, say—from this system without changing any of the other orbitals. The energy of the new system, $\langle E_{\text{ion}}\rangle$, is given by the same expression, except for the fact that all elements involving orbital a are removed. The energy difference is then

$$\langle E_{\text{atom}}\rangle - \langle E_{\text{ion}}\rangle = \left\langle a\left|-\frac{1}{2}\nabla^2 - \frac{Z}{r}\right|a\right\rangle$$
$$+ \sum_b^{\text{occ}}(\langle ab|r_{12}^{-1}|ab\rangle - \langle ba|r_{12}^{-1}|ab\rangle)\,.$$

But from (5.36) and (5.32,33) we find that this is exactly equal to

$$\langle a|h_{\text{HF}}|a\rangle = \varepsilon_a\,.$$

Thus, we see that the eigenvalue of the Hartree-Fock operator is equal to the negative of the work required to remove the electron from the system, or the negative of the *binding energy*. This is called *Koopmans' theorem* [*Koopmans* 1933].

It should be observed that it is assumed above that all the remaining orbitals of the system were unaffected by the removal of one electron. Physically, the system is expected to readjust its orbitals to the new situation. Koopmans' theorem is valid only if such a readjustment or "*relaxation*" can be neglected. For real systems it is usually found that Koopmans' theorem gives a good approximation to the binding energy, although the relaxation effect is normally quite significant for the more tightly bound electrons. The relaxation is a form of many-body effect, and it can be treated by means of many-body perturbation theory as we shall indicate in later chapters.

In the derivation of the HF equation above, the only restriction on the wave function is that it be represented by a single Slater determinant. On the other hand, there is no restriction on the single-electron orbitals that form the determinant. This equation is therefore often called the "*unrestricted*" *Hartree-Fock* equation. As we shall see in the next chapter, it is often convenient to restrict the orbitals to have the form of those of an electron moving in a central field. This leads to the "*restricted*" *Hartree-Fock* equation, which will be discussed in Chap. 7.

6. The Central-Field Model

In the previous chapter we considered the independent-particle model in which the electrons are assumed to move independently of each other in an average field due to the nucleus and the other electrons. This assumption reduces the many-electron problem to a number of independent single-electron problems, and it leads to the general unrestricted Hartree-Fock model. For an arbitrary potential the single-particle functions can have quite a complicated form, however, and the problem is considerably simplified if we assume that the potential is spherically symmetric. The single-electron orbitals can then be written as a product of a radial function, a spherical harmonic and a spin function. This is the well-known central-field approximation.

We shall begin the present chapter by separating the single-particle equation for a central field. Then we shall briefly discuss the concepts of electron configuration and the building-up principle of the elements. All states within a configuration are degenerate in the central-fileld model, but the residual noncentral part of the Coulomb interaction among the electrons leads to a splitting of the configuration into LS terms. We shall consider this splitting in detail for an np^2 configuration, utilizing the results obtained in Chap. 4. A more complete treatment of the term splitting will be given in Chap. 8. The present chapter ends with a derivation of the average energy of a configuration in the general case. Such expressions are useful in many cases, for instance, as a starting point for the restricted Hartree-Fock procedure, treated in the next chapter. Hartree-Fock orbitals of this kind will form the basis for the many-body perturbation treatment in the second part of this book.

6.1 Separation of the Single-Electron Equation for a Central Field

We return now to the Hamiltonian (5.7)

$$H = H_0 + V_{es} \tag{6.1}$$

and assume that the potential $u(\mathbf{r}_i)$ is spherically symmetric. This means that it depends only on r_i and not on the angular coordinates. We then have

$$H_0 = \sum_{i=1}^{N} h_0(i) = \sum_{i=1}^{N} \left[-\frac{1}{2} \nabla_i^2 - \frac{Z}{r_i} + u(r_i) \right] \tag{6.2}$$

$$V_{es} = -\sum_i u(r_i) + \sum_{i>j} \frac{1}{r_{ij}} \ . \tag{6.3}$$

We shall first consider the single-electron equation (5.10) in this case,

$$\left[-\frac{1}{2} \nabla_1^2 - \frac{Z}{r_1} + u(r_1) \right] \varphi_a(1) = \varepsilon_a \varphi_a(1) \ . \tag{6.4}$$

From the solutions of this equation—for a particular central fileld, $u(r_1)$—we can then construct the determinantal wave functions (5.13) of the system.

Again, we use the result from quantum mechanics that there exists a complete set of simultaneous eigenfunctions for a set of mutually commuting observables (Theorem B.5). From what has been said above and from the properties of angular-momentum operators, discussed in Chap. 2, we know that the operators h_0, l^2, l_z, s^2 and s_z are mutually commuting. Thus, we can choose the eigenfunctions of h_0 so that these are simultaneously eigenfunctions of l^2, l_z, s^2 and s_z.

The eigenfunctions of l^2 and l_z are the spherical harmonics, discussed in Sect. 2.2.4, and the eigenfunctions of s^2 and s_z are the spin functions (2.21). We can then write the solutions of (6.4) in the form

$$\varphi = R(r) Y^l_{m_l}(\theta, \phi) \chi_{m_s} \ ,$$

where χ_{m_s} (with $m_s = \pm 1/2$) represents the spin functions α and β. The only unknown part here is the radial function $R(r)$, and this is determined by the requirement that φ be a solution of (6.4). We shall now find the equation for this radial part of the wave function.

It is convenient at this point to introduce another radial function, $P(r) = rR(r)$. The single-particle wave functions then become

$$\varphi = \frac{P(r)}{r} Y^l_{m_l}(\theta, \phi) \chi_{m_s} \tag{6.5}$$

and the effect of the Laplacian (2.50) upon the radial part of the wave function is

$$\frac{1}{r^2} \frac{\partial}{\partial r} r^2 \frac{\partial}{\partial r} R(r) = \frac{1}{r} \frac{d^2}{dr^2} P(r) \ .$$

Combining this equation with the relation (2.48), we see that the Laplacian acting on the function (6.5) gives

$$\nabla^2 \varphi = \frac{1}{r} \left[\frac{d^2}{dr^2} - \frac{l(l+1)}{r^2} \right] P(r) Y^l_{m_l} \chi_{m_s} \ . \tag{6.6}$$

Substituting this into (6.4) and cancelling the factor $Y^l_{m_l} \chi_{m_s}/r$ from both sides, we obtain the *radial equation*

$$\left[-\frac{1}{2}\frac{d^2}{dr^2} + \frac{l(l+1)}{2r^2} - \frac{Z}{r} + u(r) \right] P(r) = \varepsilon P(r) . \tag{6.7}$$

For the wave function φ to be finite at the origin, it is necessary that $P(r) \to 0$ as $r \to 0$. Furthermore, φ must be normalizable, which means that the integral

$$\int |\varphi|^2 \, dv = \int_0^\infty P(r)^2 dr$$

should be finite. To fulfill this condition, it is necessary that $P(r) \to 0$ as $r \to \infty$.

The values of ε for which there exist solutions of the radial equation, satisfying the subsidiary conditions, are the possible energy eigenvalues of an electron moving in the central potential $-Z/r + u(r)$. The solutions may be distinguished by the value of l to which they correspond and by the number of nodes of the radial function. In atomic physics it is customary to denote the orbital angular momentum with the spectroscopic notation

$$l = 0 \ 1 \ 2 \ 3 \ 4 \ 5 \ 6 \ 7 \ 8 \dots$$
$$s \ p \ d \ f \ g \ h \ i \ k \ l \dots , \tag{6.8}$$

and to use a *principal quantum number n*, defined by

$$n = l + \nu + 1 , \tag{6.9}$$

where ν is the number of nodes of the radial function.

It should be observed that the eigenvalue ε depends only on the quantum numbers n and l and not on m_l or m_s for any central field. This is obvious from a physical point of view, since the energy of an electron cannot depend on the *orientation* of its angular momenta in a potential field that is invariant with respect to rotations.

It is well known that the energy of hydrogen-like systems depends only on the principal quantum number n (neglecting relativistic and other small effects). This is due to a special property of the Coulomb field. In any other central field, the energy depends also on the orbital angular-momentum quantum number l.

6.2 The Electron Configuration and the "Building-Up" Principle

6.2.1 The Meaning of a Configuration

In the central-field approximation, the wave function for a many-electron atom can be represented by means of a Slater determinant (5.13), formed of orbitals of the type (6.5). Such a function is an eigenfunction of the central-field Hamiltonian (6.2) with the eigenvalue according to (5.12) equal to

$$E_0 = \sum_i \varepsilon_i \, , \tag{6.10}$$

or the sum of the eigenvalues of the orbitals appearing in the determinant.

As shown in the previous section, the eigenvalue of a single electron in a central field depends only on n and l. The orbitals with the same values of n and l are said to form a *shell*. Therefore, the eigenvalue of the determinantal product depends only on the number of electrons in each shell. This specification is called the *electron configuration*, and it is usually denoted by means of the spectroscopic symbols given above. For instance, $1s^2 2s^2 2p^3$ means that there are two electrons in each of the $1s$ and $2s$ shells and three electrons in the $2p$ shell. The energy (6.10) of the configuration can then be written

$$E_0 = \sum_{nl} q_{nl} \, \varepsilon_{nl} \, , \tag{6.11}$$

where q_{nl} is the occupation number of the shell nl, and the sum runs over all shells. Observe that *in the central-field approximation all states belonging to the same configuration are degenerate.*

6.2.2 The "Building-Up" Principle

According to the exclusion principle there can be only one electron in each orbital, characterized by the set of quantum numbers nlm_sm_l. Therefore, there can be at most $2(2l + 1)$ electrons in a particular shell nl. A shell which has all orbitals occupied is said to be *filled* or *closed*. A partially filled shell is said to be *open*.

Since the energy in the central-field approximation is according to (6.11) given by the occupation numbers of each shell, we would expect to get the lowest energy for a particular atom by successively filling the lowest-lying electron shells. The electron configuration of the ground state (ground configuration) would then consist of a number of closed shells and at most one open shell. This leads to the historically important building-up (or "Aufbau") principle, which was suggested by *Bohr* to explain the periodic table of the elements already in 1922 [*Bohr* 1922]. It was only after the discovery of the exclusion principle [*Pauli* 1926], though, that this idea was fully understood.

Of course, as electrons are being added, the average central field changes, so the picture is somewhat more complicated than indicated here. Furthermore, the electron-electron interaction also causes departures from the central-field description. These facts lead to some irregularities in the filling of the electron shells, as successively heavier elements are being formed. As a consequence, there can be two open shells in the ground configuration. This is the case for some of the *transition elements,* where an nd shell is being filled, and for some *rare-earth elements*, which have a partially filled 4*f* shell. Competing shells may lie so close in these cases that the departures from the central-field model can be sufficient to change the order in which the shells are filled.

6.3 Russell-Saunders Coupling

We have seen that the central-field model leads to highly degenerate energy levels, and we shall now investigate how this degeneracy is removed by the noncentral part of the Coulomb interaction among the electrons. In order to do so, we apply first-order perturbation theory. In Chap. 9 we shall treat the effect of perturbations more systematically. For the moment we only need the well-known fact that the first-order energy contribution for a degenerate system is given by the eigenvalues of the perturbation matrix *within* the degenerate level and that the corresponding eigenvectors yield the correct zeroth-order functions [*Merzbacher* 1961, p.385].

As an example we consider a configuration with two equivalent *p* electrons np^2 (two electrons with $l = 1$ and with the same principal quantum number *n*). This configuration contains $\binom{6}{2} = 15$ determinantal states. In order to determine the first-order splitting, we could, of course, set up the 15×15 electrostatic matrix and diagonalize it. The effort involved can be considerably reduced, however, by considering the *LS* symmetry of the problem.

The components of the total angular momenta *S* and *L* commute with the central-field Hamiltonian H_0, and we can choose eigenfunctions of H_0 that are simultaneous eigenfunctions also of S^2, L^2, S_z, and L_z (Theorem B.5). It is easy to show that the electrostatic perturabtion V_{es} also commutes with these operators (Problem 6.1), so the perturbation matrix will be diagonal with respect to the quantum numbers S, L, M_S and M_L (Theorem B.4). Therefore, instead of diagonalizing V_{es} for the entire configuration in a single, big step, we can consider one S, L, M_S, M_L combination at a time. For a two-electron configuration (or two electrons outside closed shells), one can show that there is never more than one state with a particular combination of these four quantum numbers, which therefore specify the state uniquely within the configuration. This means that the SLM_SM_L functions make the electrostatic matrix entirely diagonal, and the first-order energy contribution is given directly by the expectation value

$$\Delta E = \langle \gamma SLM_SM_L | V_{es} | \gamma SLM_SM_L \rangle , \tag{6.12}$$

where γ represents the configuration. This energy contribution should be added to the zeroth-order energy (6.11) to give the total energy in this approximation.

It follows from the symmetry of the system that the expectation value (6.12) cannot depend on the magnetic quantum numbers M_S and M_L, which define the orientation of *S* and *L*. This follows also formally from the fact that V_{es} commutes with *all* components of *S* and *L* (Theorem B.6). Consequently, the electrostatic interaction splits the configuration into a number of *LS terms*, each of which is characterized by the eigenvalues of S^2 and L^2, or the quantum numbers S and L. *This is called LS coupling or Russell-Saunders coupling.*

Problem 6.1. Show that the electrostatic interaction commutes with the total orbital angular-momentum operator for a two-electron system.
Hint: Show that

$$r_{12}^{-1} = \{r_1^2 + r_2^2 - 2r_1 r_2[\cos\theta_1\cos\theta_2 + \sin\theta_1\sin\theta_2\cos(\phi_1 - \phi_2)]\}^{-1/2},$$

where $(r_i\theta_i\phi_i)$ are the spherical polar coordinates of the electrons and r_{12} the interelectronic distance. Use this result to show that r_{12}^{-1} commutes with the orbital angular-momentum operator

$$L_z = l_{1z} + l_{2z} = -i\left(\frac{\partial}{\partial\phi_1} + \frac{\partial}{\partial\phi_2}\right).$$

6.4 Angular-Momentum Properties of Determinantal States

Before we consider the problem of the term splitting of a configuration, we shall look into the angular-momentum properties of the determinantal states from which our wave functions will be constructed.

If the configuration consists entirely of closed shells, there is only one possible determinantal function. For example, if we have an np shell with six p electrons, the determinant must be

$$| \{np1^+, np0^+, np-1^+, np1^-, np0^-, np-1^-\}\rangle,$$

where, as before, curly brackets are used to denote an antisymmetric product. Here $np1^+$, for instance, represents an np orbital with $m_l = 1$ and $m_s = +1/2$. Apart from rearranging the rows and columns, which does not yield any new physical state, there is no other determinant that is nonzero for this particular system.

If one or several shells are open, there are normally many determinants belonging to the same configuration. We first consider a two-electron system, where we may denote the determinantal states by

$$| \{nlm_s m_l, n'l'm'_s m'_l\}\rangle.$$

Operating on this state with $L_+ = l_{1+} + l_{2+}$ and using (2.19), we obtain

$$L_+| \{nlm_s m_l, n'l'm'_s m'_l\}\rangle$$
$$= d_+| \{nlm_s m_l + 1, n'l'm'_s m'_l\}\rangle + d'_+| \{nlm_s m_l, n'l'm'_s m'_l + 1\}\rangle, \tag{6.13}$$

where

$$d_+ = [l(l + 1) - m_l(m_l + 1)]^{1/2}$$
$$d'_+ = [l'(l' + 1) - m'_l(m'_l + 1)]^{1/2} .$$

We also find that

$$L_z | \{nlm_sm_l, n'l'm'_sm'_l\} \rangle = (m_l + m'_l) | \{nlm_sm_l, n'l'm'_sm'_l\} \rangle , \qquad (6.14)$$

which implies that the determinantal state is an eigenstate of $L_z = l_{1z} + l_{2z}$ with the eigenvalue $m_l + m'_l$.

Equation (6.13) may readily be generalized to a many-electron system,

$$L_+ | \{a_1a_2 \dots a_N\} \rangle = d^1_+ | \{a'_1a_2 \dots a_N\} \rangle$$
$$+ d^2_+ | \{a_1a'_2 \dots a_N\} \rangle + \dots + d^N_+ | \{a_1a_2 \dots a'_N\} \rangle , \qquad (6.15)$$

where a_i and a'_i stand for the quantum numbers n_i, l_i, m^i_s, m^i_l and n_i, l_i, m^i_s, $m^i_l + 1$, respectively, and the coefficients d^i_+ are defined as before,

$$d^i_+ = [l_i(l_i + 1) - m^i_l(m^i_l + 1)]^{1/2} .$$

The generalization of (6.14) is

$$L_z | \{a_1a_2 \dots a_N\} \rangle = (m_1 + m_2 + \dots + m_N) | \{a_1a_2 \dots a_N\} \rangle . \qquad (6.16)$$

We now consider a determinant for which the quantum numbers a_1, a_2, ..., a_r represent electrons in closed shells and a_{r+1}, ..., a_N electrons in open shells. Because of the Pauli exclusion principle, it is not possible to raise the m_l value of the states belonging to filled shells. Then (6.15) becomes

$$L_+ | \{a_1 \dots a_r a_{r+1} \dots a_N\} \rangle$$
$$= d^{r+1}_+ | \{a_1 \dots a_r a'_{r+1} \dots a_N\} \rangle + \dots + d^N_+ | \{a_1 \dots a_r a_{r+1} \dots a'_N\} \rangle . \qquad (6.17)$$

Similarly, since the m_l values of the electrons in each filled shell sum to zero, we have

$$L_z | \{a_1 \dots a_r a_{r+1} \dots a_N\} \rangle = (m_{r+1} + \dots + m_N) | \{a_1 \dots a_r a_{r+1} \dots a_N\} \rangle . \quad (6.18)$$

Apparently the determinant $| \{a_1 \dots a_r a_{r+1} \dots a_N\} \rangle$ transforms under L in the same way as the state $| \{a_{r+1} \dots a_N\} \rangle$, which involves only open-shell electrons. This is, of course, a consequence of the fact that a closed shell of electrons is spherically symmetric. Suppose that we had expressed a particular LS state of an open-shell system as a linear combination of determinantal states. Then by adding to each of the determinants the quantum numbers corresponding to a number of filled shells, we may form a many-electron state, which is also an eigenstate of L^2. The same is true, of course, of S^2 and J^2. So in order to form a state which has a definite angular-momentum symmetry, it is only necessary to consider the open-shell electrons.

6.5 *LS* Terms of a Given Configuration

As an example of forming linear combinations of determinants, which have definite *LS* symmetry, we consider a configuration with two electrons in an np shell (and any number of closed shells). There are, as mentioned, 15 determinants of this configuration, and we want to find the linear combinations that are eigenfunctions of S^2, L^2, S_z, and L_z,

$$|np^2, SLM_SM_L\rangle . \tag{6.19}$$

We know from (6.18) that the determinantal states are eigenfunctions of L_z with the eigenvalue equal to the sum of the m_l values of the open-shell electrons, and similarly for S_z. A particular function (6.19), which is an eigenfunction of S_z and L_z with the eigenvalues M_S and M_L, respectively, can, of course, contain only determinants with the same eigenvalues of these operators. The other determinants are orthogonal and cannot appear in the expansion of this state (Theorem B.2). For that reason it is convenient to arrange the determinants according to the eigenvalues of S_z and L_z, as in Table 6.1.

Table 6.1. Determinants of the np^2 configuration

M_L \ M_S	1	0	−1
2		$\{1^+1^-\}$	
1	$\{1^+0^+\}$	$\{1^+0^-\}$ $\{0^+1^-\}$	$\{1^-0^-\}$
0	$\{1^+ -1^+\}$	$\{1^+ -1^-\}$ $\{0^+0^-\}$ $\{-1^+1^-\}$	$\{1^- -1^-\}$
−1	$\{-1^+0^+\}$	$\{-1^+0^-\}$ $\{0^+ -1^-\}$	$\{-1^-0^-\}$
−2		$\{-1^+ -1^-\}$	

In order to determine the possible values of S and L, we shall utilize the fact that if a state $|\gamma SLM_SM_L\rangle$ exists, then all states with $|M_S| \leq S$ and $|M_L| \leq L$ (and the same values of S, L, γ) must also exist. This can easily be seen by applying the ladder operators (2.19). Thus, there can be no ^3D state ($S = 1$, $L = 2$) in an np^2 configuration, since the box with $M_S = 1$, $M_L = 2$ is empty. Consequently, the state $| \{1^+1^-\}\rangle$ with $M_S = 0$, $M_L = 2$ must belong to a ^1D state ($S = 0$, $L = 2$). Since there is only one determinant in this box, we must have

$$|np^2, {}^1D(M_S = 0, \ M_L = 2)\rangle = |\{1^+1^-\}\rangle, \tag{6.20}$$

apart from an undetermined phase factor, which we have here set equal to unity.

The other states within the ^1D term can now be obtained by applying the step-down operators. By operating with $L_- = l_{1-} + l_{2-}$ on (6.20) and using (2.19), we obtain

$$\sqrt{2 \cdot 3 - 2 \cdot 1}\,|np^2, {}^1D(0, 1)\rangle = \sqrt{1 \cdot 2 - 1 \cdot 0}\,|\{0^+1^-\}\rangle + \sqrt{1 \cdot 2 - 1 \cdot 0}\,|\{1^+0^-\}\rangle$$

or

$$|np^2, {}^1D(0, 1)\rangle = \frac{1}{\sqrt{2}}[|\{0^+1^-\}\rangle + |\{1^+0^-\}\rangle]. \tag{6.21}$$

This function is a linear combination of the determinants in the box $M_S = 0$, $M_L = 1$. Since there are two determinants in that box, there must be two independent LS functions with these values of M_S and M_L. The remaining one must derive from a ^3P state, since the box $M_S = 1$, $M_L = 1$ is occupied. We have

$$|np^2, {}^3P(1, 1)\rangle = |\{1^+0^+\}\rangle, \tag{6.22}$$

and by applying $S_- = s_{1-} + s_{2-}$ we obtain

$$|np^2, {}^3P(0, 1)\rangle = \frac{1}{\sqrt{2}}(|\{1^-0^+\}\rangle + |\{1^+0^-\}\rangle)$$

$$= \frac{1}{\sqrt{2}}(-|\{0^+1^-\}\rangle + |\{1^+0^-\}\rangle), \tag{6.23}$$

using the antisymmetry of the determinants. This is the remaining combination for $M_S = 0$, $M_L = 1$, which, of course, could have been obtained directly from (6.21), utilizing the orthonormality of the LS functions.

By applying the step-down operators on the ^1D and ^3P functions above, we see that we can account for all the states within the configuration, *except* one state in the box $M_S = M_L = 0$. Evidently, this final state must be a ^1S state. By generating the $^1D(0,0)$ and $^3P(0,0)$ states and forming a third orthogonal combination, it is easy to show that

$$|np^2, {}^1S(0,0)\rangle = \frac{1}{\sqrt{3}}(|\{1^+ -1^-\}\rangle - |\{0^+0^-\}\rangle + |\{-1^+1^-\}\rangle). \tag{6.24}$$

We have then found that the LS terms within an np^2 configuration are ^1S, ^1D and ^3P. Obviously, it is not necessary to generate the explicit form of the LS functions, as we have done here, in order to determine the possible terms. These can be found quite easily just by inspection of the arrangement of the determinants like that in Table 6.1.

Problem 6.2. Show that an np^3 configuration gives rise to the LS terms ^2D, ^2P and ^4S.

6.6 Term Energies

We have now seen that the electrostatic perturbation V_{es} splits up a configuration into a number of LS terms. In this section, we shall discuss how the first-order energy of such terms can be evaluated for systems with one or two electrons in the open shell(s) and apply the results to the np^2 configuration. More general systems will be treated in Chap. 8.

As mentioned earlier, there is never more than one LS term of a particular kind for atoms with one or two electrons (outside closed shells), and, hence, the LS functions are the correct zeroth-order functions, making the electrostatic perturbation diagonal within the configuration. The first-order energy contribution is then given by the expectation value (6.12). By adding to this the zeroth-order energy E_0 (which is the expectation value of H_0), we can write the energy to first order as

$$E = \langle \gamma S L M_S M_L | H | \gamma S L M_S M_L \rangle \,, \tag{6.25}$$

where

$$H = H_0 + V_{es} = \sum_{i=1}^{N} \left(-\frac{1}{2} \nabla_i^2 - \frac{Z}{r_i} \right) + \sum_{i<j} r_{ij}^{-1} \tag{6.26}$$

is the Hamiltonian of the system.

By forming the proper LS functions, as illustrated in the previous section for an np^2 configuration, we can, in principle, evaluate the term energies, using the general formulas (5.20–26) for the matrix elements between determinantal states; however, apart from the very simplest systems, this is a very tedious procedure. Instead, we shall take advantage of the angular-momentum theory, which we have already developed in the previous chapters.

6.6.1 The Single-Particle Operator

The Hamiltonian (6.26) contains the single-particle operator

$$f = -\frac{1}{2} \nabla^2 - \frac{Z}{r} \tag{6.27}$$

and the two-particle operator

$$g_{12} = r_{12}^{-1} \,. \tag{6.28}$$

Since the operator f is a tensor operator of rank zero, the single-particle matrix elements

$$I(nl) = \left\langle nlm_s m_l \left| -\frac{1}{2}\nabla^2 - \frac{Z}{r} \right| nlm_s m_l \right\rangle$$

are independent of the quantum numbers m_s and m_l (Theorem B.6). Using (6.6), this matrix element can be written

$$I(nl) = \int_0^\infty P_{nl}(r)\left[-\frac{1}{2}\frac{d^2}{dr^2} + \frac{l(l+1)}{2r^2} - \frac{Z}{r} \right] P_{nl}(r)\,dr \,. \tag{6.29}$$

The first term in this integral corresponds to the radial part and the second term to the angular part of the kinetic energy of the nl electron, and the last term corresponds to the attraction of the electron to the nucleus. The diagonal elements of the operator

$$F = \sum_i \left(-\frac{1}{2}\nabla_i^2 - \frac{Z}{r_i} \right) \tag{6.30}$$

for a determinantal state may then be obtained using (5.20),

$$\langle F \rangle = \sum_a^{occ} \left\langle a \left| -\frac{1}{2}\nabla^2 - \frac{Z}{r} \right| a \right\rangle = \sum_{nl} q_{nl} I(nl) \,, \tag{6.31}$$

where q_{nl} is the occupation number of the nl shell. The integral $I(nl)$ is independent of m_s and m_l, and therefore the diagonal element (6.31) is the same for all determinants within a configuration.

6.6.2 The Two-Particle Operator

According to (5.22), the expectation value of the two-particle operator

$$G = \sum_{i<j} r_{ij}^{-1}$$

is

$$\langle G \rangle = \frac{1}{2}\sum_{ab}^{occ} \left(\langle ab | r_{12}^{-1} | ab \rangle - \langle ba | r_{12}^{-1} | ab \rangle \right) \,,$$

where a and b run independently and the factor of $1/2$ compensates for the fact that each electron pair is counted twice. This sum can be split into three parts for which

1) both electrons (a, b) belong to closed shells,
2) one electron is in a closed shell and one electron in an open shell,
3) both electrons belong to open shells.

In the first two cases we can use the formula (3.50) for the interaction between an electron in orbital a and *all* the electrons of a closed shell B,

$$e^0(a, B) = (4l_b + 2)\left[F^0(a, b) - \frac{1}{2}\sum_k \begin{pmatrix} l_a & k & l_b \\ 0 & 0 & 0 \end{pmatrix}^2 G^k(a, b)\right], \qquad (6.32)$$

which holds whether or not a belongs to the shell B. The contribution 1) above can then be written

$$\langle G \rangle_1 = \frac{1}{2}\sum_A q_a e^0(a, A) + \sum_{A>B} q_a e^0(a, B), \qquad (6.33)$$

where a belongs to the shell A and b to the shell B.

Obviously, for closed shells we have

$$q_a e^0(a, B) = q_b e^0(b, A), \qquad (6.34)$$

since both expressions represent the total interaction between the two shells. Equation (6.33) can then be rewritten as

$$\langle G \rangle_1 = \frac{1}{2}\sum_{AB} q_a e^0(a, B). \qquad (6.35)$$

Similarly, we get for the contribution 2), involving one electron (a) in an open shell and one electron in a closed shell (B),

$$\langle G \rangle_2 = \sum_{AB} q_a e^0(a, B). \qquad (6.36)$$

Since $e^0(a, B)$ is independent of the $m_s m_l$ quantum numbers of the orbital a, it follows that the contributions (6.33) and (6.36) are the same for all determinants (and any other state) within the configuration. The third contribution, on the other hand, involving two open shells, depends on the state considered, and it is responsible for the splitting of the LS terms of a configuration. As an example of evaluating the term energies we consider below the ground configuration of the neutral carbon atom, $1s^2\,2s^2\,2p^2$.

6.6.3 Example: Term Energies of the $1s^2 2s^2 2p^2$ Configuration

For the configuration $1s^2\,2s^2\,2p^2$ the energy contribution due to the single-particle operator (6.30) is according to (6.31)

$$\langle F \rangle = 2I(1s) + 2I(2s) + 2I(2p). $$

The Coulomb interaction within the closed shells, $1s^2$ and $2s^2$, gives according to (6.33) the contribution

$$\langle G \rangle_1 = e^0(1s, 1s^2) + e^0(2s, 2s^2) + 2e^0(2s, 1s^2), $$

where we use the notation $1s^2$ and $2s^2$ to indicate that a summation over the closed shell is involved. Similarly, we get from (6.36) the energy contribution due to the interaction between the open $2p$ shell and the closed shells,

$$\langle G \rangle_2 = 2e^0(2p, 1s^2) + 2e^0(2p, 2s^2) \, .$$

We have then accounted for all the contributions to the first-order energy (6.25), except the Coulomb interaction within the open shell. Thus,

$$E = E^0 + \langle 2p^2, SL | r_{12}^{-1} | 2p^2, SL \rangle \, , \tag{6.37}$$

where E^0 is given by the three expressions above,

$$E^0 = \langle F \rangle + \langle G \rangle_1 + \langle G \rangle_2 = 2I(1s) + 2I(2s) + 2I(2p) + e^0(1s, 1s^2)$$
$$+ e^0(2s, 2s^2) + 2e^0(2s, 1s^2) + 2e^0(2p, 1s^2) + 2e^0(2p, 2s^2) \, . \tag{6.38}$$

It should be observed that E^0 is different from the zeroth-order energy E_0, given by (6.11), and, in fact, contains also a substantial part of the first-order energy contribution (6.12).

The matrix element appearing in (6.37) has already been evaluated in Chap.4. Using (4.31) we find that the first-order term energies for the $1s^2 \, 2s^2 \, 2p^2$ configuration are

$$\begin{cases} E(^1S) = E^0 + F^0(2p, 2p) + \dfrac{10}{25} F^2(2p, 2p) \\[2mm] E(^1D) = E^0 + F^0(2p, 2p) + \dfrac{1}{25} F^2(2p, 2p) \\[2mm] E(^3P) = E^0 + F^0(2p, 2p) - \dfrac{5}{25} F^2(2p, 2p) \, . \end{cases} \tag{6.39}$$

Problem 6.3. Use the formula (6.32) and the table of 3-j symbols in Appendix D to show that the constant term E^0 given by (6.38) is equal to

$$E^0 = 2I(1s) + 2I(2s) + 2I(2p) + F^0(1s, 1s) + F^0(2s, 2s)$$
$$+ 4F^0(2s, 1s) - 2G^0(2s, 1s) + 4F^0(2p, 1s) - \frac{2}{3} G^1(2p, 1s) \tag{6.40}$$
$$+ 4F^0(2p, 2s) - \frac{2}{3} G^1(2p, 2s) \, .$$

6.6.4 A General Energy Expression

We have seen that the first-order energy is a linear combination of the single-particle integrals (6.29) and the Slater integrals F^k and G^k. In the next chapter we shall derive the restricted Hartree-Fock equations by requiring that the energy be stationary with respect to small changes in the form of the radial wave functions. For this purpose it is convenient to use the following general expression for the energy:

$$E = \sum_a q_a I_a + \frac{1}{2} \sum_{abk} q_a q_b [c(abk)F^k(a, b) + d(abk)G^k(a, b)], \tag{6.41}$$

where a and b run over all nl shells. This expression is sufficiently general to include the cases we have treated previously, and it also includes (as a special case) the average energy considered in the next section.

If one of the orbitals, b say, belongs to a closed shell, then

$$\sum_k q_b [c(abk)F^k(a, b) + d(abk)G^k(a, b)] \tag{6.42}$$

represents the interaction between one electron in the state a and all electrons of the closed shell B, which is identical to the quantity $e^0(a,B)$ given by (6.32). This gives us

$$\begin{cases} c(abk) = \delta(k, 0) \\ d(abk) = -\frac{1}{2} \begin{pmatrix} l_a & k & l_b \\ 0 & 0 & 0 \end{pmatrix}^2 \end{cases} \tag{6.43}$$

which may be used whenever one of the interacting shells is closed.

If both shells are open, the coefficients will depend on the state considered. For the $1s^2 2s^2 2p^2$ configuration we can, for instance, make the following identifications between (6.39) and (6.41) for the three terms:

$$\begin{cases} c(2p, 2p, 0) = 1/2 \\ c(2p, 2p, 2) = a/50 \\ d(2p, 2p, k) = 0, \end{cases} \tag{6.44}$$

where $a = 10$, 1, and -5 for the ^1S, ^1D, and ^3P states, respectively. Since the direct and exchange integrals are identical here, we need not use the d coefficients.

In the following section we shall derive general expressions for the *average* first-order energy of a configuration, where each level is given a weight proportional to the degeneracy of the level. For the configuration considered above, a weighted average of the energy levels produces

$$E_{av} = \frac{1}{15}[1 \cdot E(^1S) + 5 \cdot E(^1D) + 9 \cdot E(^3P)] = E^0 + F^0(2p, 2p)$$

$$- \frac{2}{25} F^2(2p, 2p). \tag{6.45}$$

The expression in brackets is the sum of all diagonal matrix elements of the

Hamiltonian within the configuration or the *trace* of the corresponding matrix. Since the trace of a matrix is invariant with respect to similiaty transformations, the average energy is independent of the choice of basis functions. This property of the average energy is necessary, of course, to make the concept useful.

Equation (6.39) for the term energies of the $1s^2 2s^2 2p^2$ configuration and (6.45) for the average energy can now be summarized in the following way:

$$E(1s^2 2s^2 2p^2) = E^0 + F^0(2p, 2p) + \frac{a}{25} F^2(2p, 2p) ,$$

where $a = 10, 1, -5$ for the ^1S, ^1D and ^3P states and $a = -2$ for the weighted average.

6.7 The Average Energy of a Configuration

The idea of the configuration average was discussed early by *Shortley* [1936] and has been treated in detail by *Slater* [1960, Chap. 14; 1968]. The formalism can easily be extended to include also the average of several configurations, which is of particular interest in relativistic calculations [*Lindgren* and *Rosén* 1974]. Since we are mainly concerned with nonrelativistic problems in this book, we shall restrict ourselves here to the simpler case of a single configuration.

6.7.1 Derivation of the General Formula

We start by considering the average interaction between an electron in a state a and the electrons in an open shell B. We choose a p shell with two electrons again as an example. First we assume that a does not belong to B. There are $\binom{6}{2} = 15$ possible ways of putting two electrons into a p shell, as illustrated in Fig. 6.1. The average interaction energy between the electron a and the open shell B is obtained by summing this interaction over all states of the configuration and dividing by the number of states. We then see that the numerator contains in this case the interaction between a and all the electrons of the *closed* shell B exactly five times, and, hence, the average becomes

$$\langle e(a, B) \rangle = \frac{5}{15} e^0(a, B) . \tag{6.46}$$

The factor five is the number of possible ways of placing the remaining electron in the available positions, or $\binom{5}{1}$. We can then easily generalize this result to

$$\langle e(a, B) \rangle = \left[\binom{q_B - 1}{q_b - 1} \Big/ \binom{q_B}{q_b} \right] e^0(a, B) , \tag{6.47}$$

	1^+	0^+	-1^+	1^-	0^-	-1^-
$\lvert a\rangle$	x	x				
	x		x			
	x			x		
	x				x	
	x					x
		x	x			
		x		x		
		x			x	
		x				x
			x	x		
			x		x	
			x			x
				x	x	
			,	x		x
					x	x

Fig. 6.1. Illustration of the average interaction between an electron (a) and the electrons of an open shell, when the electron does not belong to that shell

where q_b is the occupation number of shell B and $q_B = 4l_b + 2$ is equal to the number of states in the shell (the occupation number of the closed shell). Using the explicit form of the binomial coefficients $\binom{n}{m} = n!/m!(n-m)!$, this reduces to

$$\langle e(a, B)\rangle = \frac{q_b}{q_B} e^0(a, B), \qquad a \notin B . \tag{6.48}$$

Next, we consider the average interaction between one electron of an open shell and all the other electrons of the *same* shell. As an example for this case we choose a p shell with three electrons. If we let a be the 1^+ orbital, say, there are $\binom{5}{2} = 10$ possible ways of placing the remaining two electrons, as shown in Fig. 6.2. The sum of the interactions for all states is now $\binom{4}{1} = 4$ times the interaction between a and the closed shell, and the average becomes

$$\langle e(a, A)\rangle = \frac{4}{10} e^0(a, A) . \tag{6.49}$$

Generalizing, as before, we get

$$\langle e(a, A)\rangle = \left[\binom{q_A - 2}{q_a - 2} \Big/ \binom{q_A - 1}{q_a - 1} \right] e^0(a, A)$$

or

1^+	0^+	-1^+	1^-	0^-	-1^-
	x	x			
	x		x		
	x			x	
$\lvert a\rangle$	x				x
		x	x		
		x		x	
		x			x
			x	x	
			x		x
				x	x

Fig. 6.2. Illustration of the average interaction between an electron (a) and an open shell when the electron belongs to the same shell

$$\langle e(a, A)\rangle = \frac{q_a - 1}{q_A - 1} e^0(a, A), \qquad a \in A. \tag{6.50}$$

We now see that in both cases the average interaction is independent of the $m_s m_l$ quantum numbers of the orbital a. Therefore, we can express the average interaction between open shells in the same way as the interaction between closed shells (6.35),

$$\langle G\rangle_3 = \frac{1}{2} \sum_{AB} q_a \langle e(a, B)\rangle. \tag{6.51}$$

When closed shells are considered, the average interaction $\langle e(a, B)\rangle$ becomes equal to $e^0(a, B)$. Therefore, we can write the average of the entire first-order energy as

$$E_{av} = \sum_a q_a I_a + \frac{1}{2} \sum_{AB} q_a \langle e(a, B)\rangle. \tag{6.52}$$

This formula can be compared with the general expression (6.41). Using (6.32, 48, 50), we see that the coefficients become in this case

$$\begin{cases} c(abk) = Q\delta(k, 0) \\ d(abk) = -\frac{Q}{2} \begin{pmatrix} l_a & k & l_b \\ 0 & 0 & 0 \end{pmatrix}^2, \end{cases} \tag{6.53}$$

where

$$\begin{cases} Q = 1 & \text{for } a \neq b \\ Q = \dfrac{q_A(q_a - 1)}{q_a(q_A - 1)} & \text{for } a = b. \end{cases}$$

6.7.2 Example: Average of the $1s^22s^22p^2$ Configuration

As an example of the configuration average, we consider again the carbon atom with the configuration $1s^22s^22p^2$. The contributions from the single-particle operator F and from the Coulomb interactions, where at least one closed shell is involved, is as before equal to E^0 which is given by (6.38). For the open $2p$ shell we have according to (6.53)

$$\begin{cases} c(2p, 2p, 0) = \dfrac{3}{5} \\[2mm] d(2p, 2p, 0) = -\dfrac{1}{10} \\[2mm] d(2p, 2p, 2) = -\dfrac{1}{25}. \end{cases} \tag{6.54}$$

Since the direct and exchange integrals within a shell are identical, we can, as before, eliminate the d coefficients from the general expression for the energy (6.41), which gives

$$\begin{cases} c(2p, 2p, 0) = 1/2 \\[2mm] c(2p, 2p, 2) = -1/25. \end{cases} \tag{6.55}$$

So the contribution to (6.41) due to the interaction of the $2p$ electrons is

$$F^0(2p, 2p) - \frac{2}{25} F^2(2p, 2p),$$

and the total expression for the average energy is

$$E(1s^22s^22p^2)_{av} = E^0 + F^0(2p, 2p) - \frac{2}{25} F^2(2p, 2p), \tag{6.56}$$

which agrees with (6.45). The expression (6.56) was calculated using determinantal states, while (6.45) was evaluated in the LS scheme. This demonstrates the fact mentioned before that the average energy is independent of the choice of basis functions.

7. The Hartree-Fock Model

In the present chapter we shall optimize the description within the central-field model by choosing the radial part of the orbitals so that the first-order energy is minimized. This energy could either be that of a particular (LS) term or of the average of a configuration, as discussed at the end of the previous chapter. This process leads to the ordinary Hartree-Fock (HF) model, and it is often called the "restricted" HF procedure, in order to distinguish it from the more general or "unrestricted" HF method, discussed in Chap. 5.

There is now a vast literature concerning the HF procedure, and numerous calculations have been performed. We shall by no means try to give a complete survey of this field. The HF model provides, however, a qualitative understanding of some of the general features of atoms, and it will be used as a starting point for the more accurate calculations we shall consider in the second part of the book. For that reason we shall discuss this model in some detail in the present chapter. We shall start by deriving the HF equations in the standard way. This will be done using a general expression for the energy which includes the term energy and the configuration average as special cases. Rules will be given for writing down the HF equations directly from the energy of the system. The interpretation of the eigenvalue (Koopmans' theorem) is somewhat more complicated in the restricted case than in the unrestricted one and requires a special investigation. Furthermore, we shall give a number of examples of HF calculations and try to give some intuitive idea of the magnitude of the different interactions represented by the Slater integrals and the spin-orbit coupling constants.

7.1 Radial Equations for the Restricted Hartree-Fock Procedure

In order to derive the radial equations for the restricted HF procedure, we shall—as in Chap. 5—use the Ritz variational method. We require that the expectation value E of the total energy be stationary,

$$\delta E = 0, \tag{7.1}$$

for small changes in the *radial* part of the orbitals. This means that the orbitals are restricted to be of the form (6.5). Furthermore, the orbitals are assumed to be orthonormal,

$$\int\limits_0^\infty P_{nl}(r)P_{n'l}(r)dr = \delta(n, n') \tag{7.2}$$

and to satisfy the same boundary condition as in the single-electron case (see Sect. 6.1).

It is an elementary result of the calculus of variations [*Messiah* 1961, p. 776; *Slater* 1960, Chap. 17] that satisfying (7.1) with the subsidiary condition (7.2) is equivalent to satisfying the equation

$$\delta[E - \sum_a q_a \lambda_{aa} N_{aa} - \sum_{a\neq b} \delta(l_a, l_b) q_a q_b \lambda_{ab} N_{ab}] = 0 \tag{7.3}$$

for *all* changes, $\delta P_a(r)$, in the radial functions, satisfying the boundary conditions. The summations here run over all nl shells, and N_{ab} is used as a short-hand notation for the overlap integral (7.2). The parameters λ_{ab} are the *Lagrange multipliers*, and they have the effect of preserving the orthonormality. If $l_a \neq l_b$, then the angular parts are orthogonal, and no Lagrange multipliers are needed in the radial equation.

It follows from (6.41) and (7.3) that the expression to be varied depends upon the integrals I_a, $F^k(a, b)$, $G^k(a, b)$ and the overlap integral N_{ab}. We consider first the variation of these integrals separately. Using (6.29) we obtain

$$\delta I_a = \int\limits_0^\infty \delta P_a(r)\left[-\frac{1}{2}\frac{d^2}{dr^2} + \frac{l_a(l_a+1)}{2r^2} - \frac{Z}{r} \right] P_a(r)dr$$
$$+ \int\limits_0^\infty P_a\left[-\frac{1}{2}\frac{d^2}{dr^2} + \frac{l_a(l_a+1)}{2r^2} - \frac{Z}{r} \right]\delta P_a(r)\, dr\;.$$

Integrating by parts and using the fact that $P_a(r)$ and $\delta P_a(r)$ vanish at the origin and at infinity, one may readily show that

$$\int\limits_0^\infty P_a(r)\frac{d^2}{dr^2}\delta P_a(r)dr = \int\limits_0^\infty \delta P_a(r)\frac{d^2}{dr^2}P_a(r)dr\;.$$

So the two integrals that appear in the expression for δI_a above are equal, giving

$$\delta I_a = 2\int\limits_0^\infty \delta P_a(r)\left[-\frac{1}{2}\frac{d^2}{dr^2} + \frac{l_a(l_a+1)}{2r^2} - \frac{Z}{r} \right]P_a(r)\, dr\;. \tag{7.4}$$

In order to study the properties of the two-electron Slater integrals, it is convenient to introduce a set of functions of r, introduced by *Hartree* [1957]. These are defined by

$$Y_k(ab, r) = r\int\limits_0^\infty \frac{r_<^k}{r_>^{k+1}} P_a(r')P_b(r')dr'$$

$$= \frac{1}{r^k} \int_0^r r'^k P_a(r') P_b(r') \, dr' + r^{k+1} \int_r^\infty \frac{1}{r'^{k+1}} P_a(r') P_b(r') \, dr' , \qquad (7.5)$$

where $r_<$ is the lesser and $r_>$ the greater of the two radial distances r and r'. Using these functions and the definitions (3.44b) and (3.47), the F^k and G^k integrals can be written

$$F^k(a, b) = \int_0^\infty P_a^2(r) \frac{1}{r} Y_k(bb, r) \, dr = \int_0^\infty P_b^2(r) \frac{1}{r} Y_k(aa, r) \, dr \qquad (7.6a)$$

$$G^k(a, b) = \int_0^\infty P_a(r) P_b(r) \frac{1}{r} Y_k(ab, r) \, dr. \qquad (7.6b)$$

We now consider the variation of the $F^k(a, b)$ and $G^k(a, b)$ integrals with respect to variations of the orbital P_a. All of the other orbitals will be held constant. In the case $a \neq b$, the function $Y_k(bb, r)$ does not vary with a change of P_a. So from (7.6a) we get

$$\delta F^k(a, b) = 2 \int_0^\infty \delta P_a(r) P_a(r) \frac{1}{r} Y_k(bb, r) \, dr \qquad (a \neq b). \qquad (7.7)$$

For the case $a = b$, however, there will be an additional term, due to the change of the second factor,

$$\delta F^k(a, a) = 2 \int_0^\infty \delta P_a(r) P_a(r) \frac{1}{r} Y_k(aa, r) \, dr + \int_0^\infty P_a^2(r) \delta \left[\frac{1}{r} Y_k(aa, r) \right] dr .$$

Using the definition of the Y_k integral (7.5), these two terms can easily be shown to be equal, which gives

$$\delta F^k(a, a) = 4 \int_0^\infty \delta P_a(r) P_a(r) \frac{1}{r} Y_k(aa, r) \, dr .$$

Combined with (7.7), this gives

$$\delta F^k(a, b) = 2(1 + \delta_{ab}) \int_0^\infty \delta P_a(r) P_a(r) \frac{1}{r} Y_k(bb, r) \, dr . \qquad (7.8)$$

Similarly, the variations of the exchange integral and of the overlap integral are

$$\delta G^k(a, b) = 2(1 + \delta_{ab}) \int_0^\infty \delta P_a(r) P_b(r) \frac{1}{r} Y_k(ab, r) \, dr \qquad (7.9)$$

$$\delta N_{ab} = (1 + \delta_{ab}) \int_0^\infty \delta P_a(r) P_b(r) \, dr . \qquad (7.10)$$

We now require that the condition (7.3) be satisfied for small changes in the orbital $P_a(r)$, while the remaining orbitals are held constant. From the general energy expression (6.41) we then get the following variation in the energy:

$$\delta E = q_a \delta I_a + \frac{q_a}{2} \sum_{bk}^{b \neq a} q_b [c(abk) \delta F^k(a, b) + d(abk) \delta G^k(a, b)]$$

$$+ \frac{q_a}{2} \sum_{bk}^{b \neq a} q_b [c(bak) \delta F^k(b, a) + d(bak) \delta G^k(b, a)]$$

$$+ \frac{1}{2} q_a^2 \sum_k [c(aak) \delta F^k(a, a) + d(aak) \delta G^k(a, a)]$$

and similarly for the remaining terms in (7.3). With the expressions for the variations of the integrals given above we then get the following equation:

$$2q_a \int_0^\infty \delta P_a(r) \left\{ \left[-\frac{1}{2} \frac{d^2}{dr^2} + \frac{l_a(l_a + 1)}{2r^2} - \frac{Z}{r} \right] P_a(r) \right.$$

$$+ \sum_{bk} q_b \left[c(abk) \frac{1}{r} Y_k(bb, r) P_a(r) + d(abk) \frac{1}{r} Y_k(ab, r) P_b(r) \right] \qquad (7.11)$$

$$\left. - \lambda_{aa} P_a(r) - \frac{1}{2} \sum_b^{b \neq a} \delta(l_a, l_b) q_b \lambda_{ab} P_b(r) - \frac{1}{2} \sum_b^{b \neq a} \delta(l_a, l_b) q_b \lambda_{ba} P_b(r) \right\} dr = 0 \,.$$

This integral will vanish for arbitrary $\delta P_a(r)$, only if the radial functions satisfy the equation

$$\left[-\frac{1}{2} \frac{d^2}{dr^2} + \frac{l_a(l_a + 1)}{2r^2} - \frac{Z}{r} \right] P_a(r)$$

$$+ \sum_{bk} q_b \left[c(abk) \frac{1}{r} Y_k(bb, r) P_a(r) + d(abk) \frac{1}{r} Y_k(ab, r) P_b(r) \right]$$

$$= \lambda_{aa} P_a(r) + \frac{1}{2} \sum_b^{b \neq a} \delta(l_a, l_b) q_b (\lambda_{ab} + \lambda_{ba}) P_b(r) \,.$$

In this equation λ_{aa} serves as a single-electron eigenvalue. If we denote the eigenvalue by $\varepsilon_a (= \lambda_{aa})$ and introduce the notation

$$\varepsilon_{ab} = \frac{1}{2} \delta(l_a, l_b)(\lambda_{ab} + \lambda_{ba}) \,, \qquad (7.12)$$

the equation above becomes

$$\left[-\frac{1}{2} \frac{d^2}{dr^2} + \frac{l_a(l_a + 1)}{2r^2} - \frac{Z}{r} \right] P_a(r)$$

$$+ \sum_{bk} q_b \left[c(abk) \frac{1}{r} Y_k(bb, r) P_a(r) + d(abk) \frac{1}{r} Y_k(ab, r) P_b(r) \right] \qquad (7.13)$$

$$= \varepsilon_a P_a(r) + \sum_b^{b \neq a} q_b \varepsilon_{ab} P_b(r) \,.$$

This is the *restricted Hartree-Fock equation* for the radial part of the orbitals in shell a.

The HF equation (7.13) contains the same parameters as the energy expression (6.41), which was minimized. Thus, once the energy of the system is given, we can immediately write down the corresponding HF equations by identifying the coefficients of the Slater integrals. Alternatively, we can use directly the fact that there is a one-to-one correspondence between the terms in the expression for the energy (6.41) and the left-hand side of the HF equation (7.13). Then the HF equation for a particular orbital can be obtained immediately by making the following replacements in the energy terms which depend on that orbital,

$$
\begin{cases}
I_a \rightarrow \dfrac{1}{q_a}\left[-\dfrac{1}{2}\dfrac{d^2}{dr^2} + \dfrac{l_a(l_a+1)}{2r^2} - \dfrac{Z}{r}\right]P_a(r) \\[2ex]
F^k(a, b) \rightarrow \dfrac{1+\delta(a, b)}{q_a}\dfrac{1}{r}\,Y_k(bb, r)P_a(r) \\[2ex]
G^k(a, b) \rightarrow \dfrac{1+\delta(a, b)}{q_a}\dfrac{1}{r}\,Y_k(ab, r)P_b(r)\,.
\end{cases}
\tag{7.14}
$$

As can be seen from (6.29) and (7.6), these replacements are obtained essentially by breaking an integration associated with the orbital $P_a(r)$, and they can be verified by reconsidering the derivation of the HF equations above. This approach is often quite convenient, as we shall demonstrate below.

Each term on the left-hand side of the HF equation (7.13) can be given a simple physical interpretation. The first two terms represent the radial and angular parts of the kinetic energy of electron a and the third term represents its attraction to the nucleus. The terms in the sum on the left-hand side represent the electrostatic interaction of electron a with all the other electrons in the atom. Here, the contribution of each shell appears separately. The term

$$
c(abk)\,\frac{1}{r}\,Y_k(bb, r)\,P_a(r)
$$

corresponds to the direct part of the interaction and

$$
d(abk)\,\frac{1}{r}\,Y_k(ab, r)\,P_b(r)
$$

to the exchange part.

The interpretation of the eigenvalue ε_a requires a special investigation and will be discussed in the next section. The parameters ε_{ab} are related to the off-diagonal Lagrange multipliers by (7.12), and they are nonzero only for $l_a = l_b$. It can furthermore be shown that they vanish if the shells a and b are both closed. Usually, these multipliers have only minor effects on the solutions even for open-shell systems, and we shall not consider them further here.

There is one HF equation (7.13) for each occupied shell a. These equations are *coupled* and must therefore be solved iteratively. An initial estimate of the radial functions is made and used to evaluate the Y_k functions. The equations are then solved again, and this procedure is continued until self-consistency is reached. Then the stationary condition (7.1) is fulfilled with the restrictions made in this case. The problem of convergence and numerical accuracy is a very delicate one, which we shall not be concerned with here. Interested readers are referred to extensive treatments on self-consistent procedures available in the literature (see, for instance, [*Hartree* 1957; *Slater* 1960; *Froese-Fischer* 1977]).

* 7.2 Koopmans' Theorem in Restricted Hartree-Fock

As in the unrestricted case the eigenvalue ε in the HF equation (7.13) is related to the binding energy of the electron (Koopmans' theorem). Due to the restrictions on the orbitals in the present case, however, this theorem will have a somewhat different formulation here.

We begin by multiplying the HF equation (7.13) by $P_a(r)$ and integrating. Using (6.29) and (7.6), this gives

$$\varepsilon_a = I_a + \sum_{bk} q_b [c(abk) F^k(a, b) + d(abk) G^k(a, b)] . \tag{7.15}$$

This quantity is the energy associated with an electron in shell a according to (6.41). In general, however, it does not represent the binding energy of that electron. The reason is that the final configuration—after the removal of one electron from a particular shell—has generally a term splitting, and the work required to remove the electron depends on the final (as well as on the initial) state.

In the configurational average procedure, the expression (7.15) does represent the *average*, though, of the electron binding energies, as we can see in the following way. We first consider the average interaction *between* two shells, A and B, which is, according to (6.48),

$$\langle e(A, B) \rangle = q_a \langle e(a, B) \rangle = \frac{q_a q_b}{q_B} e^0(a, B), \qquad (a \in A \neq B). \tag{7.16}$$

If we remove one electron from the shell A, then the average becomes

$$\langle e(A, B) \rangle_{-1} = (q_a - 1) \langle e(a, B) \rangle = \frac{(q_a - 1) q_b}{q_B} e^0(a, B) ,$$

where q_a is the original occupation number. The difference in the average interaction is, using (6.32),

---

* This section can be omitted in the first reading.

$$\Delta = \langle e(a, B) \rangle = \frac{q_b}{q_B} e^0(a, B) = q_b \left[F^0(a, b) - \frac{1}{2} \sum_k \begin{pmatrix} l_a & k & l_b \\ 0 & 0 & 0 \end{pmatrix}^2 G^k(a, b) \right]. \tag{7.17}$$

But according to (6.53) this is just equal to the expression

$$\sum_k q_b [c(abk)F^k(a, b) + d(abk)G^k(a, b)]$$

appearing in (7.15) for the configurational average.

Similarly, for the average interaction *within* a shell A we find that

$$\langle e(A, A) \rangle = \frac{1}{2} q_a \langle e(a, A) \rangle = \frac{q_a(q_a - 1)}{2(q_A - 1)} e^0(a, A) \tag{7.18}$$

and

$$\langle e(A, A) \rangle_{-1} = \frac{(q_a - 1)(q_a - 2)}{2(q_A - 1)} e^0(a, A) .$$

The difference is here

$$\Delta = \frac{q_a - 1}{q_A - 1} e^0(a, A)$$

$$= \frac{q_a - 1}{q_A - 1} q_A \left[F^0(a, a) - \frac{1}{2} \sum_k \begin{pmatrix} l_a & k & l_a \\ 0 & 0 & 0 \end{pmatrix}^2 G^k(a, a) \right] \tag{7.19}$$

$$= \sum_k q_a [c^k(aak)F^k(a, a) + d(aak)G^k(a, a)] .$$

Thus, we have seen that Koopmans' theorem is valid in the configuration-average HF in the sense that the eigenvalue is equal to the negative of the *average* binding energy or the difference in *average* energy for the initial and final configurations.

7.3 The Hartree-Fock Potential

In order to simplify the HF equation (7.13), we introduce the HF potential $u_{HF}(r)$, defined by

$$u_{HF}(r)P_a(r) = \sum_{bk} q_b \left[c(abk) \frac{1}{r} Y_k(bb, r)P_a(r) \right.$$

$$\left. + d(abk) \frac{1}{r} Y_k(ab, r)P_b(r) \right]. \tag{7.20}$$

This potential is nonlocal. By this we mean that its effect on an orbital at a certain radius depends also—via the Y_k functions—on the orbitals at other radii. By means of this potential the HF equation can be written

$$\left[-\frac{1}{2}\frac{d^2}{dr^2} + \frac{l_a(l_a+1)}{2r^2} - \frac{Z}{r} + u_{HF}(r)\right]P_a(r) = \varepsilon_a P_a(r), \tag{7.21}$$

leaving out the off-diagonal Lagrange multipliers.

It is sometimes convenient to define separate direct and exchange functions as follows:

$$\begin{cases} J_b(r)P_a(r) = \sum_k c(abk)\frac{1}{r}Y_k(bb,r)P_a(r) \\ K_b(r)P_a(r) = -2\sum_k d(abk)\frac{1}{r}Y_k(ab,r)P_b(r). \end{cases} \tag{7.22}$$

Then the HF potential becomes

$$u_{HF}(r) = \sum_b q_b\left[J_b(r) - \frac{1}{2}K_b(r)\right] \tag{7.23}$$

In the case of closed shells, we have according to (6.43)

$$\begin{cases} c(abk) = \delta(k,0) \\ d(abk) = -\frac{1}{2}\begin{pmatrix} l_a & k & l_b \\ 0 & 0 & 0 \end{pmatrix}^2, \end{cases} \tag{7.24}$$

which gives

$$\begin{vmatrix} J_b(r)P_a(r) = \frac{1}{r}Y_0(bb,r)P_a(r) \\ K_b(r)P_a(r) = \sum_k \begin{pmatrix} l_a & k & l_b \\ 0 & 0 & 0 \end{pmatrix}^2 \frac{1}{r}Y_k(ab,r)P_b(r). \end{vmatrix} \tag{7.25}$$

This agrees with the usual definition of these functions, and (7.22) is a natural extension of this definition.

For closed shells we have for $k = 0$ according to (7.24) and (2.94),

$$d(aa0) = -(4l_a + 2)^{-1}.$$

Thus, in this case the exchange contribution to (7.20)

$$q_a d(aa0)\frac{1}{r}Y_0(aa,r)P_a(r) = -\frac{1}{r}Y_0(aa,r)P_a(r)$$

cancels exactly one unit of the direct contributions. This is the so-called *self-*

interaction. The direct part of the HF potential contains the interaction with *all* electrons of the atom, and the exchange part compensates for the fact that an electron does not interact with itself. It is not difficult to show formally that this kind of cancellation occurs also for open-shell systems, which is obvious from a physical point of view.

We would like now to consider the behavior of the potential $u_{HF}(r)$ as r becomes large. According to (7.5), for values outside the charge distribution described by $P_b(r)$, we have

$$\frac{1}{r} Y_0(bb, r) \rightarrow \frac{1}{r} \int_0^\infty P_b(r')P_b(r')dr' = \frac{1}{r} .$$

So outside the atom the direct part of the potential $\sum(4l_b + 2)J_b(r)$ becomes simply

$$(1/r) \sum_b (4l_b + 2) = N/r,$$

where N is the number of electrons in the atom. But the self-interaction has the effect of reducing this potential by one unit of r^{-1}. It can be shown that the remaining parts of the exchange potential fall off more rapidly, and, hence, at large distances the HF potential has the limiting form

$$u_{HF} \rightarrow \frac{N - 1}{r} .$$

This means that the nucleus is screened by $(N - 1)$ electrons. This is, of course, what we would expect physically at large distances for an electron which is a part of the system.

7.4 Examples of Hartree-Fock Equations

7.4.1 A Closed-Shell System: $1s^2 2s^2$

As a first example of the restricted Hartree-Fock equations we consider the ground state of the beryllium atom $1s^2 2s^2$ ^1S, which is a closed-shell system. In this case the HF potential (7.23) becomes

$$u_{HF} = (2J_{1s} - K_{1s}) + (2J_{2s} - K_{2s}) .$$

As we have seen, the self-interaction K_{ns} cancels one of the J_{ns} functions in the ns equation

$$(J_{1s} - K_{1s})P_{1s} = 0$$

$$(J_{2s} - K_{2s})P_{2s} = 0$$

as follows directly from (7.25). Taking advantage of this fact, we obtain the following HF equations for the $1s$ and $2s$ orbitals:

$$\left[-\frac{1}{2}\frac{d^2}{dr^2} - \frac{Z}{r} + J_{1s} + (2J_{2s} - K_{2s})\right]P_{1s} = \varepsilon_{1s}P_{1s}$$

$$\left[-\frac{1}{2}\frac{d^2}{dr^2} - \frac{Z}{r} + (2J_{1s} - K_{1s}) + J_{2s}\right]P_{2s} = \varepsilon_{2s}P_{2s}\ .$$

The form of these two equations is quite natural, if one recalls that the two electrons in each shell are in opposite spin states. Each $1s$ electron interacts directly with the other $1s$ electron (but not with itself); it interacts directly with two $2s$ electrons and by the exchange interaction with the $2s$ electron which is in the same spin state.

Using (7.25) we can express the HF equations in terms of the Y_k functions as follows:

$$\begin{cases}\left[-\frac{1}{2}\frac{d^2}{dr^2} - \frac{Z}{r} + \frac{1}{r}\,Y_0(1s1s, r) + \frac{2}{r}\,Y_0(2s2s, r)\right]P_{1s}(r) \\[2mm] \qquad - \frac{1}{r}\,Y_0(1s2s, r)P_{2s}(r) = \varepsilon_{1s}P_{1s}(r) \\[3mm] \left[-\frac{1}{2}\frac{d^2}{dr^2} - \frac{Z}{r} + \frac{2}{r}\,Y_0(1s1s, r) + \frac{1}{r}\,Y_0(2s2s, r)\right]P_{2s}(r) \\[2mm] \qquad - \frac{1}{r}\,Y_0(2s1s, r)P_{1s}(r) = \varepsilon_{2s}P_{2s}(r)\ .\end{cases} \qquad (7.26)$$

As an illustration of the HF formalism for open-shell systems, we return again to the ground configuration of the carbon atom $1s^2\,2s^2\,2p^2$. The c and d coefficients of the HF equation (7.13) are given by (6.43), apart from those concerning the interaction *within* the open shell, which are given by (6.44)

$$\begin{cases}c(2p, 2p, 0) = 1/2 \\ c(2p, 2p, 2) = a/50\ ,\end{cases}$$

where $a = 10, 1, -5$, and -2 for the ^1S, ^1D, ^3P and the configuration average, respectively. This gives us the following HF equations:

$$1s: \left[-\frac{1}{2}\frac{d^2}{dr^2} - \frac{Z}{r} + \frac{1}{r}\,Y_0(1s1s, r) + \frac{2}{r}\,Y_0(2s2s, r)\right.$$

$$\left. + \frac{2}{r}\,Y_0(2p2p, r)\right]P_{1s}(r) - \frac{1}{r}\,Y_0(1s2s, r)P_{2s}(r) \qquad (7.27a)$$

$$- \frac{1}{3}\frac{1}{r}\,Y_1(1s2p, r)\,P_{2p}(r) = \varepsilon_{1s}P_{1s}(r)$$

$$2s: \left[-\frac{1}{2}\frac{d^2}{dr^2} - \frac{Z}{r} + \frac{2}{r} Y_0(1s1s, r) + \frac{1}{r} Y_0(2s2s, r) \right.$$

$$\left. + \frac{2}{r} Y_0(2p2p, r) \right] P_{2s}(r) - \frac{1}{r} Y_0(2s1s, r) P_{1s}(r) \tag{7.27b}$$

$$- \frac{1}{3}\frac{1}{r} Y_1(2s2p, r) P_{2p}(r) = \varepsilon_{2s} P_{2s}(r)$$

$$2p: \left[-\frac{1}{2}\frac{d^2}{dr^2} + \frac{1}{r^2} - \frac{Z}{r} + \frac{2}{r} Y_0(1s1s, r) + \frac{2}{r} Y_0(2s2s, r) \right.$$

$$\left. + \frac{1}{r} Y_0(2p2p, r) + \frac{a}{25}\frac{1}{r} Y_2(2p2p, r) \right] P_{2p}(r) \tag{7.27c}$$

$$- \frac{1}{3}\frac{1}{r} Y_1(2p1s, r) P_{1s}(r) - \frac{1}{3}\frac{1}{r} Y_1(2p2s, r) P_{2s}(r) = \varepsilon_{2p} P_{2p}(r).$$

Note that the equations for the closed shells do not contain the energy parameter a. The self-consistent solutions, though, will depend indirectly on that parameter via the $2p$ orbital.

Alternatively, we could have obtained the equations above directly from the energy expression (6.39, 40), using the replacements (7.14), without explicitly identifying the c and d coefficients. For instance, in order to obtain the $1s$ equation we first discard all energy terms which do not depend on the $1s$ orbital, leaving

$$E_{1s} = 2I(1s) + F^0(1s, 1s) + 4F^0(2s, 1s) - 2G^0(2s, 1s)$$

$$+ 4F^0(2p, 1s) - \frac{2}{3}G^1(2p, 1s). \tag{7.28}$$

With $q_a = 2$ and $l_a = 0$ we then make the following replacements:

$$2I(1s) \rightarrow \left[-\frac{1}{2}\frac{d^2}{dr^2} - \frac{Z}{r} \right] P_{1s}(r)$$

$$F^0(1s, 1s) \rightarrow \frac{1}{r} Y_0(1s1s, r) P_{1s}(r) \tag{7.29}$$

$$4F^0(2s, 1s) \rightarrow \frac{2}{r} Y_0(2s2s, r) P_{1s}(r)$$

$$- 2G^0(2s, 1s) \rightarrow -\frac{1}{r} Y_0(2s1s, r) P_{2s}(r)$$

and so on, which yields (7.27a). We can retrieve the other HF equations for this configuration in a similar way.

7.5 Examples of Hartree-Fock Calculations

7.5.1 The Carbon Atom

In Fig. 7.1 we show the radial HF functions for the carbon atom, using the configuration-average procedure [$a = -2$ in (7.27)]. The $1s$ function has a single maximum, which occurs quite close to the nucleus. The $2s$ function has two maxima, the first of which is slightly inside that of the $1s$. The $2p$ function again has a single maximum, and its shape is quite similar to that of $2s$ outside the node. In Fig. 7.2 we compare the $2p$ functions for the configuration average and for the ^1S state ($a = 10$).

In Table 7.1 we give the total energies, orbital eigenvalues and the expectation values of r of the $1s$, $2s$ and $2p$ orbitals for the ^3P, ^1D and ^1S states and

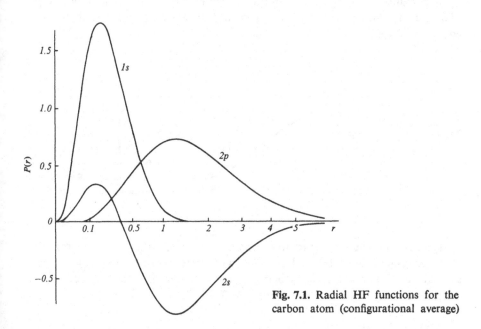

Fig. 7.1. Radial HF functions for the carbon atom (configurational average)

Table 7.1. Total energy, orbital eigenvalues and expectation values of r in HF calculations for the $1s^22s^22p^2$ configuration of carbon

	$E(^3P)$		E_{av}		$E(^1D)$		$E(^1S)$	
E	-37.6886		-37.6602		-37.6313		-37.5495	
nl	ε_{nl}	$\langle r \rangle_{nl}$	ε_{nl}	$\langle r \rangle_{nl}$	ε_{nl}	$\langle r \rangle_{nl}$	ε_{nl}	$\langle r \rangle_{nl}$
$1s$	11.326	0.268	11.338	0.268	11.352	0.268	11.392	0.268
$2s$	0.706	1.589	0.712	1.586	0.719	1.582	0.740	1.571
$2p$	0.433	1.715	0.407	1.743	0.381	1.772	0.310	1.871

for the configuration average. The average energy falls between that of the ^1D and ^3P states, which explains the order of the columns in the table. The energy increases as one goes from left to right in the table. From the values of $\langle r \rangle_{2p}$ and from Fig. 7.2 it is apparent that the $2p$ function is more diffuse for the higher lying energy states. This reduces the screening of the $1s$ and $2s$ functions. As a result, these orbitals become more tightly bound as the total energy increases, and they are drawn in toward the nucleus. The $2p$ orbital lies far outside the $1s$ and it provides essentially a shell of charge or a Faraday cage inside which the $1s$ electron moves. From classical physics, we know that the force on a charged particle inside a charged shell of spherical symmetry is zero and that the potential energy inside the sphere is constant, equal to its value at the center. As the $2p$ function becomes more diffuse, the effective radius of this shell of charge around the $1s$ orbital increases. This does not influence very much the form of the $1s$ function, but it does alter the potential energy of the $1s$ and its energy eigenvalue. Obviously, the eigenvalue of the $2p$ orbital is more sensitive to the energy than are the others, since the energy parameter appears explicitly in the HF equation of the $2p$.

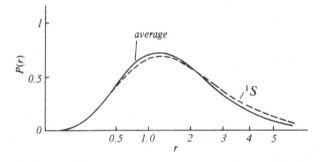

Fig. 7.2. Radial HF $2p$ functions for the carbon atom in the ^1S state and in the configurational average

It would now be interesting to compare the term splitting obtained from the three term-dependent HF calculations with that obtained in configuration average using first-order perturbation theory. In the former case we get the splitting directly from the total energies in Table 7.1,

$$E(^1S) - E(^1D) = 0.0818$$
$$E(^1D) - E(^3P) = 0.0573.$$

In the latter case the splittings are obtained from the first-order expressions (6.39). They depend only upon the Slater integral $F^2(2p, 2p)$, which in the configuration-average HF has the value

$$F^2(2p, 2p) = 0.2387.$$

This gives

$$E(^1S) - E(^1D) = \frac{9}{25} F^2(2p, 2p) = 0.0859$$

$$E(^1D) - E(^3P) = \frac{6}{25} F^2(2p, 2p) = 0.0573 ,$$

which agrees quite well with the values above.

The difference between the term-dependent and the configuration-average results can be regarded as higher-order contributions in a perturbation expansion based on the latter. In most cases, however, this difference represents only a small fraction of the higher-order contributions. For that reason it is in many cases more convenient to base the treatment on a Hartree-Fock calculation of the configuration-average type, where a single set of orbitals is used for the entire configuration. This is the procedure we shall normally follow in this book.

7.5.2 The Size of the Atom

We would like now to give further examples of Hartree-Fock calculations carried out on various elements of the periodic table. Our purpose is to give some intuitive idea of the physical properties of atoms, such as their size and the magnitude of their interactions. The Hartree-Fock method is a useful tool for understanding in a qualitative way some of these general features.

From a physical and a chemical point of view a natural definition of the size of an atom is the expectation value of r of the outermost electrons. In Table 7.2 we show this value for the elements of the second row of the periodic table, obtained by means of the configuration-average HF.

Table 7.2. Expectation values of r for the second-row elements

	Li	Be	B	C	N	O	F
$1s$	0.573	0.415	0.326	0.268	0.228	0.199	0.176
$2s$	3.874	2.649	1.977	1.586	1.326	1.141	1.001
$2p$	—	—	2.205	1.743	1.447	1.239	1.085

As one goes from one neutral atom to the next in the row, the nuclear charge increases by one, and one electron is added to the charge cloud. The screening effect of the additional, outer electrons is quite small, though, so the atoms contract in size as the $2p$ shell is filled. From the table it is clear that the fluorine atom is much smaller than the lithium atom. The situation is the same in other p shells. For instance, the expectation value $\langle r \rangle_{3p}$ for aluminum, which has only a single $3p$ electron, is 3.434, while for chlorine, which has five $3p$ electrons, it is 1.842. For the transition elements as well as the rare earths and the actinides, on the other hand, where d and f shells are being filled, the effect is less dramatic. The reason for this is that these shells lie deeper inside the atom—particularly the f shells—and hence shield the outer electrons more effectively.

In Table 7.3 we give the expectation values of r of the rare-gas elements from helium to xenon. The xenon atom has 54 electrons and yet it is only between two and three times the size of helium. Xenon is considerably smaller than the hydrogen ion H^-, which has only two electrons.

Table 7.3. Expectation values of r for noble gas atoms

nl	He $\langle r \rangle_{nl}$	Ne $\langle r \rangle_{nl}$	A $\langle r \rangle_{nl}$	Kr $\langle r \rangle_{nl}$	Xe $\langle r \rangle_{nl}$
1s	0.93	0.16	0.09	0.04	0.03
2s		0.89	0.41	0.19	0.12
2p		0.96	0.38	0.16	0.10
3s			1.42	0.54	0.32
3p			1.66	0.54	0.31
3d				0.55	0.28
4s				1.63	0.75
4p				1.95	0.78
4d					0.87
5s					1.98
5p					2.34

7.5.3 The Magnitude of the Spin-Orbit Coupling Constant

We consider now the results of some calculations of the spin-orbit-coupling constant and some other atomic quantities. According to (5.3–6) the nuclear part of the spin-orbit-coupling constant is

$$\zeta = \frac{\alpha^2}{2} Z \langle r^{-3} \rangle . \tag{7.30}$$

The magnetic dipole and electric quadrupole hyperfine interactions also depend upon $\langle r^{-3} \rangle$, so this is an important physical quantity. In Table 7.4 we show the expectation values of r and r^{-3} of the 2p electron for the second-row elements. The Slater integrals involving the 2p electrons and the spin-orbit constant are also shown. As mentioned before, the 2p orbital is drawn in by the increasing nuclear charge. As a result, the expectation value $\langle r \rangle$ decreases and $\langle r^{-3} \rangle$ increases with increasing nuclear charge. The value of $\langle r^{-3} \rangle$ increases by nearly a factor of ten in going from boron to fluorine. This is a much larger relative change than for the expectation value of r. The large increase in $\langle r^{-3} \rangle$ can be understood by considering the form of this expectation value

$$\langle r^{-3} \rangle_{nl} = \int_0^\infty \frac{1}{r^3} P_{nl}^2(r) \, dr .$$

Table 7.4. Hartree-Fock expectation values and interaction constants for the second-row elements. The spin-orbit constant is given in cm^{-1} and the remaining values in atomic units. One atomic energy unit (1 Hartree) corresponds to 219475 cm^{-1}

	B(5) 2p	C(6) $2p^2$	N(7) $2p^3$	O(8) $2p^4$	F(9) $2p^5$
$\langle r \rangle$	2.205	1.743	1.447	1.239	1.085
$\langle r^{-3} \rangle$	0.776	1.662	3.021	4.949	7.545
$F^0(2p,2p)$	—	0.531	0.642	0.751	0.860
$F^2(2p, 2p)$	—	0.239	0.287	0.334	0.381
ζ	10.12	31.36	73.66	147.80	266.51
Z_t	2.23	3.23	4.17	5.11	6.05

The factor of r^{-3} weights very strongly the part of the wave function near the nucleus, and it is just this part of the function for which the nuclear attraction is the strongest and which is drawn in most as Z increases. In Fig. 7.3 we illustrate this effect by means of the Hartree-Fock $2p$ orbitals for boron and fluorine. The latter orbital is much larger close to the nucleus, which explains the large difference in $\langle r^{-3} \rangle$.

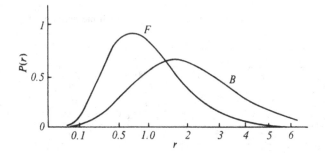

Fig. 7.3. Radial HF $2p$ functions for boron (B) and fluorine (F)

As we have discussed in Sect. 5.1, the contributions to the spin-orbit constant ζ include a large nuclear term proportional to $\langle r^{-3} \rangle$ and inter-electronic terms due to the spin-(own)-orbit and spin-other-orbit interactions, which magnetically shield the electron from the nucleus. The relative importance of this shielding effect decreases as the electron wave function is drawn in toward the nucleus. This explains why the spin-orbit constants shown in Table 7.4 increase more than $\langle r^{-3} \rangle$.

In order to describe the shielding of the nucleus by the two-body magnetic interactions, we express the spin-orbit constant ζ in analogy with the nuclear part (7.30) as

$$\zeta = \frac{\alpha^2}{2} Z_t \langle r^{-3} \rangle \, ,$$

where Z_t is an "effective" nuclear charge. In Table 7.4 we have given the corresponding values of Z_t for the elements boron to fluorine. We see that this parameter increases by almost exactly one unit in going from one element to the next in the series. If we define the screening parameter σ by

$$Z_i = Z - \sigma \,,$$

we find that σ is close to 3 for all elements of this period. In Table 7.5 we give $\langle r^{-3} \rangle$, ζ and σ for the $2p$, $3p$ and $4p$ shell elements. The value of σ varies slowly from 2.77 for boron to 3.41 for bromine. As we shall see in the next section, the first maximum of d and f orbitals lie outside the first maxima of p functions. This is due to the fact that they have a larger value of the angular momentum. It is just the inner part of the wave function which contributes most to the nuclear part of ζ. Thus, as far as the spin-orbit interaction is concerned, d functions generally lie outside p functions and the screening constant for the d shell is considerably larger. In Table 7.6 we show $\langle r^{-3} \rangle$ and ζ for the $3d$ and $4d$ transition elements. The screening constant σ varies from 11.8 for scandium to 12.5 for silver. The $4f$ electrons of the rare earths have a larger angular momentum than the d electrons of the transition elements, and as a consequence the screening is even more

Table 7.5. Hartree-Fock values of $\langle r^{-3} \rangle$, ζ and σ for the $2p$, $3p$ and $4p$ shell elements (from [*Froese-Fischer* 1977])

		np	np^2	np^3	np^4	np^5
$2p$		B(5)	C(6)	N(7)	O(8)	F(9)
	$\langle r^{-3} \rangle$	0.776	1.662	3.021	4.949	7.545
	ζ	10.12	31.36	73.66	147.80	266.51
	σ	2.77	2.77	2.83	2.89	2.96
$3p$		A(13)	Si(14)	P(15)	S(16)	Cl(17)
	$\langle r^{-3} \rangle$	1.088	2.028	3.267	4.836	6.769·
	ζ	62.38	128.57	226.53	363.78	548.64
	σ	3.19	3.15	3.13	3.13	3.13
$4p$		Ga(31)	Ge(32)	As(33)	Se(34)	Br(35)
	$\langle r^{-3} \rangle$	2.891	4.733	6.854	9.272	11.999
	ζ	463.92	788.85	1183.61	1656.31	2214.70
	σ	3.54	3.48	3.45	3.43	3.41

Table 7.6 Hartree-Fock values of $\langle r^{-3} \rangle$ and ζ for the $3d$ and $4d$ shells (from [*Froese-Fischer* 1977])

		nd	nd^2	nd^3	nd^4	nd^5	nd^6	nd^7	nd^8	nd^9
$3d$		Sc(21)	Ti(22)	V(23)	Cr(24)	Mn(25)	Fe(26)	Co(27)	Ni(28)	Cu(29)
	$\langle r^{-3} \rangle$	1.429	1.975	2.589	3.281	4.060	4.931	5.900	6.972	8.154
	ζ	77.21	119.96	172.73	237.98	317.54	413.30	527.23	661.39	817.95
$4d$		Y(39)	Zr(40)	Nb(41)	Mo(42)	Tc(43)	Ru(44)	Rh(45)	Pd(46)	Ag(47)
	$\langle r^{-3} \rangle$	1.712	2.397	3.122	3.900	4.739	5.644	6.619	7.667	8.791
	ζ	260.76	381.51	516.73	669.73	842.67	1037.52	1256.12	1500.36	1772.09

pronounced for these elements. The screening constant is close to 31 for the entire lanthanide series.

It is interesting to note that the screening parameters obtained here for p and d electrons are quite close to those derived semiempirically long ago by *Barnes* and *Smith* [1954], namely $\sigma = 4$ for p electrons and $\sigma = 10$ for d electrons.

We conclude here our discussion of the spin-orbit interaction by emphasising that the screening associated with the magnetic interactions can be quite different from ordinary electrostatic shielding. The $5d$ function of Ce^{2+} shown in Fig. 7.4, for instance, lies outside the $4f$ function. The $5d$ is shielded more electrostatically from the nucleus than the $4f$. However, since it has two maxima near the nucleus, which contribute to the spin-orbit constant, the magnetic shielding of the $5d$ is less than for the $4f$ and ζ_{5d} is larger than ζ_{4f}.

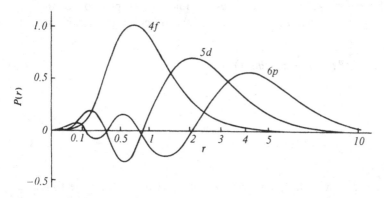

Fig. 7.4. Radial $4f$ and $5d$ functions for the configuration $4f5d$ and the $6p$ function for the configuration $4f\,6p$ of Ce^{2+}

7.5.4 The Ce^{2+} Ion

As two further examples of Hartree-Fock calculations, we consider the $4f6p$ and $4f5d$ configurations of the Ce^{2+} ion. This will provide illustrations of the wave functions of a heavy system, and we shall use these examples in the next section when we consider the properties of the two-electron Slater integrals. Figure 7.4 shows the Hartree-Fock $4f$ and $5d$ radial functions for the $4f5d$ configuration of Ce^{2+} and the $6p$ function for the $4f6p$ configuration. The Hartree-Fock $4f$ function is nearly the same for these configurations. The $6p$ electron has less angular momentum than the $5d$, and hence the centrifugal term, $l(l+1)/2r^2$, in (7.21) is smaller. So the first maximum of the $6p$ function occurs inside the first maximum of the $5d$. The maximum of the $4f$ function occurs outside the first maximum of the $6p$ as well as the $5d$.

In the next section we shall consider the Coulomb interaction in some detail. We shall find that the spatial extent of a radial function—the region in space over which the function has an appreciable value—plays an important role in determining the magnitude of the Coulomb interaction integrals between that function and other functions in the atom. The spatial extent of the $6p$ function is much larger than that of the $5d$ and $4f$ functions. For instance, the expectation value of r is 4.52 for the $6p$ functions, while it is 1.97 and 0.97 for the $5d$ and $4f$, respectively. We shall find that the $6p$ electron interacts weakly with the $4f$, while the interaction of the $4f$ and the $5d$ is rather large.

7.6 Properties of the Two-Electron Slater Integrals

In this section, we shall study some of the general properties of the Slater integrals $F^k(nl, n'l')$ and $G^k(nl, n'l')$. These integrals determine the strength of the Coulomb interaction, which is mainly responsible for the position of the LS terms of a configuration. According to (7.5, 6a), the direct integral can be written

$$F^k(nl, n'l') = \int_0^\infty P^2_{n'l'}(r) \frac{1}{r} Y_k(nlnl, r) \, dr , \tag{7.31}$$

where

$$\frac{1}{r} Y_k(nlnl, r) = \int_0^\infty \frac{r^k_<}{r^{k+1}_>} P^2_{nl}(r') \, dr' \tag{7.32}$$

and $r_<$ is the lesser and $r_>$ the greater of r, r'. In order to interpret the function $(1/r) \, Y_k(nlnl, r)$ physically, we consider the potential $u(r)$ due to the charge distribution of the nl electron

$$|\varphi_{nlm}(r')|^2 = \frac{P^2_{nl}(r')}{r'^2} |Y^l_m(\theta', \phi')|^2 , \tag{7.33}$$

The potential due to a general charge density $\rho(r')$ may be written

$$u(r) = \int \frac{\rho(r')}{|r - r'|} dr' , \tag{7.34}$$

and we may use (2.98), (2.99) and (2.55) to write this in terms of the coordinates r and r' separately,

$$u(r) = 4\pi \sum_{kq} \frac{(-1)^q}{2k + 1} \int Y^k_{-q}(\theta', \phi') \frac{r^k_<}{r^{k+1}_>} \rho(r') \, dr' \, Y^k_q(\theta, \phi) . \tag{7.35}$$

For the charge distribution (7.33) this becomes

$$u(r) = 4\pi \sum_{kq} \frac{(-1)^q}{2k+1} \langle lm| Y^k_{-q} | lm \rangle \frac{1}{r} Y_k(nlnl, r) Y^k_q(\theta, \phi) \qquad (7.36)$$

and we see that $(1/r) Y_k(nlnl, r)$ is just the radial part of the kth term in the multipole expansion of the potential due to $|\varphi_{nlm}|^2$. Outside the charge distribution we have the relation

$$\frac{1}{r} Y_k(nlnl, r) = \frac{1}{r^{k+1}} \int r'^k P^2_{nl}(r') \, dr' \qquad (7.37)$$

and (7.36) assumes the usual form for the multipole expansion of a charge distribution [*Jackson* 1975, p 136]

$$u(r) = \sum_{kq} \frac{4\pi}{2k+1} a_{kq} \frac{Y^k_q(\theta, \phi)}{r^{k+1}} . \qquad (7.38)$$

The coefficients a_{kq} may be evaluated, using (2.100), (7.36), and (7.37), to be

$$a_{kq} = (-1)^q \sqrt{\frac{2k+1}{4\pi}} \langle lm| C^k_{-q} | lm \rangle \langle r^k \rangle_{nl} . \qquad (7.39)$$

We now consider the spatial form of the function $(1/r) Y_0 (nlnl, r)$. This function is associated with the monopole term in the expansion of the potential of $|\varphi_{nlm}|^2$. According to (7.32), it is

$$\frac{1}{r} Y_0(nlnl, r) = \int_0^\infty \frac{1}{r_>} P^2_{nl}(r') \, dr' . \qquad (7.40)$$

We may identify the factor $P^2_{nl} (r')dr'$ in the integral as the amount of charge of the wave function φ_{nlm} in a spherical shell of radius r'

$$P^2_{nl}(r')dr' = \int |\varphi_{nlm}(r')|^2 r'^2 d\Omega' ,$$

where the integration is performed over the angular coordinates. So the integrand in (7.40) represents the potential at a point r of the charge distributed uniformly in a spherical shell of radius r', and the integral is the potential at r due to a spherical average of the entire charge distribution of the wave function φ_{nlm}. At small values of r (7.40) becomes simply

$$\frac{1}{r} Y_0(nlnl, r) = \int_0^\infty \frac{1}{r'} P^2_{nl}(r')dr' = \langle r^{-1} \rangle_{nl} .$$

If r is near the origin, the value of $r_>$ for each spherical shell of charge is equal to the radius of the shell. However, as the point r moves away from the origin, it passes through spherical shells of charge, and for each shell through which r has passed, the value of $r_>$ is replaced by r which is larger. Thus, as r passes through the charge distribution, the integral (7.40) decreases. Outside the charge distribution it becomes

$$\frac{1}{r} Y_0(nlnl, r) = \frac{1}{r} \int_0^\infty P_{nl}^2(r')dr' = \frac{1}{r} .$$

The form of the function (7.40) is shown in a schematic way in Fig. 7.5. The spatial extent of the function $P_{nl}(r)$ is located between the two vertical dotted lines in the figure. Also shown in the figure are two wave functions $P_{n'l'}(r)$ (A and B) which have a very small overlap with $P_{nl}(r)$. If $P_{n'l'}(r)$ has the form A, then it follows from (7.31) that $F^0(nl, n'l')$ will be close to $\langle r^{-1} \rangle_{nl}$. On the other hand, if $P_{n'l'}(r)$ has the form B, then it will be close to $\langle r^{-1} \rangle_{n'l'}$. Any spatial overlap of the functions $P_{nl}(r)$ and $P_{n'l'}(r)$ in each case will lower the value of $F^0(nl, n'l')$ below the corresponding value of $\langle r^{-1} \rangle$. These results may be summarized by writing

$$F^0(nl, n'l') \leqq \min (\langle r^{-1} \rangle_{nl}, \langle r^{-1} \rangle_{n'l'}) . \tag{7.41}$$

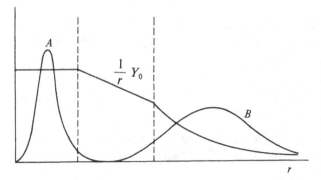

Fig. 7.5. Schematic form of the function $(1/r)Y_0(nlnl, r)$. The function $P_{nl}(r)$ is assumed to be located between the dotted lines. The meaning of the functions A and B is discussed in the text

The equality holds if there is no overlap of the functions $P_{nl}(r)$ and $P_{n'l'}(r)$ or—what amounts to the same thing—if $\langle r^{-1} \rangle_{nl}$ and $\langle r^{-1} \rangle_{n'l'}$ are quite different. In Table 7.7 expectation values of r^{-1} and the $F^0(nl, 4f)$ integrals are given for the orbitals of the two configurations of Ce^{2+} discussed above. All values except those in the last line are taken from the configuration $4f5d$.

For the $1s$, $2s$ and $2p$ functions $F^0(nl, 4f)$ is very nearly equal to $\langle r^{-1} \rangle_{4f}$. The value of $F^0(nl, 4f)$ decreases as n becomes larger due to the increasing overlap of the nl and $4f$ wave functions. Finally, for the $5d$ and $6p$ functions, $F^0(nl, 4f)$ is fairly well represented by $\langle r^{-1} \rangle_{nl}$.

Table 7.7. Expectation values of $1/r$ and $F^0(nl\,4f)$ integrals for the configurations $4f5d$ and $4f6p$ of Ce^{2+}

n	$\langle r^{-1} \rangle_{nl}$	$F^0(nl, 4f)$
$1s$	57.46	1.299
$2s$	13.29	1.299
$2p$	13.28	1.299
$3s$	4.953	1.278
$3p$	4.875	1.279
$3d$	4.745	1.284
$4s$	2.078	1.099
$4p$	1.981	1.087
$4d$	1.769	1.055
$4f$	1.299	0.941
$5s$	0.808	0.655
$5p$	0.722	0.604
$5d$	0.504	0.454
$6p$	0.270	0.256

The argument which led to (7.41) may be generalized for arbitrary values of k to obtain

$$F^k(nl, n'l') \leq \min \left(\langle r^k \rangle_{nl} \langle 1/r^{k+1} \rangle_{n'l'},\ \langle r^k \rangle_{n'l'} \langle 1/r^{k+1} \rangle_{nl} \right). \tag{7.42}$$

For values of k which are equal to or greater than 2, (7.42) provides a useful upper bound to the Slater integrals only when the two functions are well separated in space. The reason for this is that the functions r^k and $1/r^{k+1}$ tend to weight the radial functions heavily in the region where they overlap. In order to use (7.42) to estimate the value of $F^2(2p, 4f)$, for instance, it is necessary to evaluate $\langle r^2 \rangle_{2p} \langle r^{-3} \rangle_{4f}$. The function r^2 weights the tail of the $2p$ function, while r^{-3} weights the part of the $4f$ function closest to the nucleus.

In order to estimate a direct integral $F^k(nl, n'l')$, it is often best to use (7.41) to get some idea of the value of F^0 $(nl, n'l')$ and then to estimate the ratio F^k $(nl, n'l')/F^0$ $(nl, n'l')$. According to (7.31, 32), the integral F^k $(nl, n'l')$ may be written in the form

$$F^k(nl, n'l') = \int_0^\infty \int_0^\infty \left(\frac{r_<}{r_>} \right)^k \frac{1}{r_>} P_{nl}^2(r_1)\, P_{n'l'}^2(r_2)\, dr_1 dr_2 .$$

Since all of the contributions to this integral are positive, and since $(r_</r_>)$ is always less than unity, the direct integrals F^k $(nl, n'l')$ will decrease as k increases. How rapidly they decrease will depend on the amount of separation of the centers of gravity of the two functions. The ratio F^2 $(np, 4f)/F^0$ $(np, 4f)$ and the expectation values $\langle r \rangle_{np}$ are given for all of the p functions which are occupied in the $4f5d$ configuration of Ce^{2+} in Table 7.8. The expectation value $\langle r \rangle_{4f}$ is 0.483. Of all the p functions, the spatial properties of the $4p$ most closely resemble that of the $4f$, and the integrals F^k $(4p, 4f)$ decrease most slowly as k increases.

Table 7.8. The expectation values of r and the ratios $F^2(np, 4f)/F^0(np, 4f)$ for the $4f5d$ configuration of Ce^{2+}

n	$\langle r \rangle_{np}$	$F^2(np, 4f)/F^0(np, 4f)$
2	0.095	0.040
3	0.282	0.241
4	0.687	0.482
5	1.783	0.354

The exchange integrals, $G^k(nl, n'l')$, are more difficult to characterize than are the direct integrals. However, it follows immediately from the definition of these integrals that they vanish when the two functions $P_{nl}(r)$ and $P_{n'l'}(r)$ are separated in space. This is reasonable, since, if the electrons are localized in different regions of space, they may be regarded as distinguishable particles and adequately described by ordinary product functions without exchange.

7.7 Coupling Schemes for Two-Electron Systems

In the previous sections, we have considered the magnitude of the spin-orbit and the Coulomb interactions in some detail. We would like now to discuss generally the different coupling schemes which may be used to describe a configuration of two electrons. As we shall see, the coupling scheme which best describes the system and the general appearance of the energy levels depends upon the relative magnitude of the Coulomb and spin-orbit interactions. Our discussion here will follow mainly the work of *Cowan* and *Andrew* [1965].

There are four angular momenta associated with a two-electron system

$$s_1, s_2, l_1, l_2,$$

and we begin by considering the interactions involving pairs of these angular momenta. The simplest interaction of this kind is the spin-orbit interaction

$$H_{so} = \zeta_1 s_1 \cdot l_1 + \zeta_2 s_2 \cdot l_2 . \tag{7.43}$$

The one-electron spin-orbit operator $\zeta s \cdot l$ can be written

$$\zeta s \cdot l = \frac{\zeta}{2} (j^2 - s^2 - l^2) ,$$

and its effect upon the coupled states $|(sl)j\rangle$ is

$$\zeta|(sl)j\rangle = \frac{\zeta}{2}[j(j+1) - s(s+1) - l(l+1)], \tag{7.44}$$

Using this relation, it is easy to show that the spin-orbit operator (7.43) will split the energy levels of each of the electrons in the way shown in Fig. 7.6. So the angular momentum s_i and l_i interact in such a way that s_i and l_i are good quantum numbers and the energy associated with the resultant angular momenta $j_i = l_i \pm 1/2$ are different. This situation is described by saying that the spin-orbit interaction couples s_i and l_i. If the spin-orbit constant ζ_i were zero, then the two j_i levels would be degenerate.

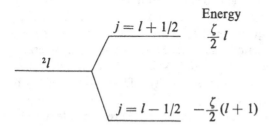

Fig. 7.6. Spin-orbit splitting of the single-electron energy levels

In Sects. 4.4 and 4.5 we derived general expressions for the Coulomb matrix elements in the LS and j–j schemes. Using (4.32) and (2.127), the matrix elements for two nonequivalent electrons can be written

$$\langle \{nln'l'\} SL | r_{12}^{-1} | \{nln'l'\} SL \rangle = \sum_k [f_k F^k(nl, n'l') + g_k G^k(nl, n'l')], \tag{7.45}$$

where

$$f_k = (-1)^L[l, l'] \begin{pmatrix} l & k & l \\ 0 & 0 & 0 \end{pmatrix} \begin{pmatrix} l' & k & l' \\ 0 & 0 & 0 \end{pmatrix} \begin{Bmatrix} l & l' & L \\ l' & l & k \end{Bmatrix} \tag{7.46}$$

and

$$g_k = (-1)^S[l, l'] \begin{pmatrix} l & k & l' \\ 0 & 0 & 0 \end{pmatrix}^2 \begin{Bmatrix} l & l' & L \\ l & l' & k \end{Bmatrix}. \tag{7.47}$$

According to (7.45 46), the direct matrix element depends only upon the principal quantum numbers nl, $n'l'$ and the total orbital angular momentum L. The direct part of the Coulomb interaction then couples the two orbital angular momenta l and l' together to form L. The exchange matrix elements, which are given by (7.45, 47), depend also upon the total spin quantum number S and so the exchange interaction couples the two spins.

We may summarize the way in which the various angular momenta are coupled as follows:

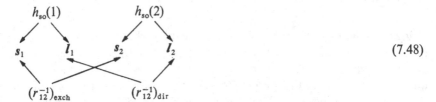

$$(7.48)$$

The spin-orbit interaction couples the spin and the orbital angular momenta of each electron. The direct part of the Coulomb interaction couples the two orbits, and the exchange interaction couples the two spins. From (7.47), it is clear that the exchange interaction also couples the two orbital angular momenta l_1 and l_2; however, this effect is usually quite small compared to the direct coupling. The effect of other magnetic interactions, such as the spin-other-orbit interaction, is normally at least an order of magnitude smaller than the spin-orbit interaction, and the simple description (7.48) is usually sufficient to understand the general appearance of the energy levels.

For the low-Z elements the electrostatic interactions are predominate, which means that $(r_{12}^{-1})_{\text{dir}}, (r_{12}^{-1})_{\text{exch}} > h_{so}(1), h_{so}(2)$. In that case the coupling of the four angular momenta is usually best described by the LS coupling

$$|(ss')S, (ll')L, J\rangle . \qquad (7.49)$$

For the heavy elements the spin-orbit interaction is comparable to or sometimes larger than the electrostatic interactions

$$h_{so}(1), h_{so}(2) \gtrsim (r_{12}^{-1})_{\text{dir}}, (r_{12}^{-1})_{\text{exch}} \qquad (7.50)$$

and the coupling approaches the j-j limit. The states in this coupling scheme are of the form

$$|(sl)j, (s'l')j', J\rangle . \qquad (7.51)$$

Both the LS and j-j coupling schemes are symmetric. The angular momenta of each electron are coupled together in the same way. These coupling schemes are usually best for the ground and low-lying configurations. Here, the wave functions of the two electrons have about the same spatial extent and the interactions of the electrons are about the same.

When one of the outer electrons is more highly excited, and particularly when it has a high value of l, the configuration is highly asymmetric in the two electrons, and the symmetric couplings (7.49,51) are usually no longer good approximations. The highly excited electron, which we give the quantum numbers $n's'l'$, usually has a very small spin-orbit interaction. The wave functions of the two electrons are localized in different regions of space, so the exchange interaction will also be small. Under these circumstances the coupling of l' and s' and of s and s' will be small, and the states of the atom will be of the general from

$$|(sll')\ K,\ s',\ J'\rangle \tag{7.52}$$

in which the angular momenta s, l, and l' are coupled together in some way to form an intermediate angular momentum K. Whenever the $(s'l')$ and (ss') interactions are small compared with the (ll') and (sl) couplings, the states will be of the form (7.52), and K will be a good quantum number. Under these circumstances the energy levels of the atom will occur in pairs with the J values of the levels of each pair differing by unity.

How the angular momenta (sll') in the present scheme are coupled together to form K depends upon the relative magnitudes of the direct part of the Coulomb interaction $(r_{12}^{-1})_{\text{dir}}$ and the spin-orbit interaction h_{so} of the sl electron. These two interactions couple (ll') and (sl), respectively. If the direct interaction is largest, then a third type of coupling is appropriate

$$|((ll')\ L,\ s)\ K,\ s',\ J\rangle . \tag{7.53}$$

Following *Cowan* and *Andrew*, we shall call this LK coupling, since in this case both L and K are fairly good quantum numbers.

For highly excited states, however, the direct interaction may become smaller than the spin-orbit interaction of the inner valence electron. We will then have another kind of coupling, called jK coupling,

$$|((sl)j,\ l')\ K,\ s',\ J\rangle . \tag{7.54}$$

In each case the angular momenta used to denote the coupling scheme are good quantum numbers to the extent that the interactions which are responsible for the coupling dominate over the other interactions of the two-electron system. In the LK coupling scheme L and K are approximately good quantum numbers, and for jK coupling j and K are good quantum numbers.

We conclude this section with two examples. The first example, which concerns the pf configuration, is taken from the article by *Cowan* and *Andrew* [1965]. In Fig. 7.7, the energy levels of this configuration are shown for values of the interaction integrals that correspond to LS and LK coupling. The LS energies can be calculated using (7.45–47), and the energy matrix in the LK scheme can then be obtained by recoupling the spins to form S. For both sets of energy levels shown, the simplifying assumptions $G^4 = \zeta_f = 0$ are made. The direct integral F^2 separates the energy levels into three levels characterized by the L quantum numbers D, F and G. At the left-hand side of the figure the G^2 integral splits each of these L levels into a singlet and a degenerate triplet. For small values of ζ_p,

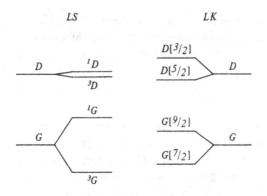

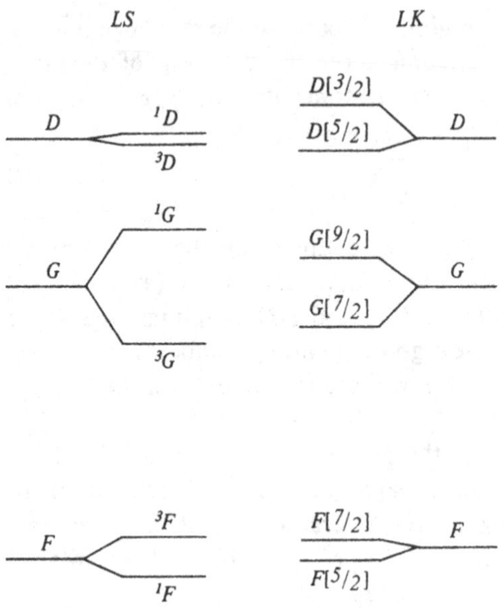

Fig. 7.7. Typical energy-level structure of a pf configuration in LS and in LK coupling

the spin-orbit interaction will remove the remaining degeneracy to produce the singlet and triplet structure characteristic of the LS coupling scheme. Similarly, at the right-hand side of the figure, ζ_p separates each L level into two doubly degenerate sublevels with $K = L \pm 1/2$. Nonzero values of G^2 will remove this final degeneracy to produce a pair level structure characteristic of the LK or of any pair-coupling scheme.

As a final example we consider again the ls configuration. According to (4.33), the matrix elements of the Coulomb interaction in the LS scheme can be written

$$\langle \{nln's\}\, SL \,|\, r_{12}^{-1} \,|\, \{nln's\}\, SL \rangle = F^0(nl, n's) + (-1)^s \frac{G^l(nl, n's)}{2l + 1} . \tag{7.55}$$

This configuration has two LS terms: ^3l and ^1l. The splitting of these two LS states by the exchange part of the Coulomb interaction is shown in Fig. 7.8.

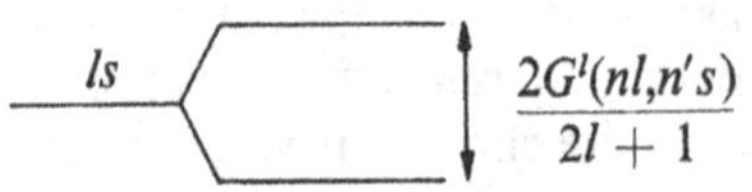

$$ls \left\{ \begin{array}{c} \\ \\ \end{array} \right. \quad \updownarrow \ \frac{2G^l(nl,n's)}{2l + 1}$$

Fig. 7.8. Splitting of the ls state in LS coupling

The states of ls may also be described in j-j coupling, the states in this scheme being of the form $|\, \{^2l_j,\, {}^2s_{1/2}\}\, J \rangle$. The spin-orbit interaction will split the two j states of the l electron by an amount $\zeta_l(l + 1/2)$, and it will not affect the s electron. Consequently, the states $|\, \{^2l_j,\, {}^2s_{1/2}\}\, J \rangle$ for which $j =$

$l + 1/2$ and $j = l - 1/2$, respectively will be separated by the amount $\zeta_l(l + 1/2)$ independent of the value of J, as shown in Fig. 7.9.

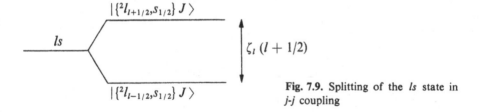

Fig. 7.9. Splitting of the ls state in j-j coupling

We consider as an example the $5d6s$ configuration of Ce^{2+}. The Hartree-Fock values of the relevant integrals are

$$G^2(5d, 6s) = 12734 \text{ cm}^{-1}$$
$$\zeta_{5d} = \qquad 963 \text{ cm}^{-1}.$$

Using the results illustrated in Figs. 7.8 and 7.9, we readily find that the Coulomb splitting of the two LS terms and the spin-orbit splitting of the two j-j states are 5094 cm^{-1} and 2408 cm^{-1}, respectively. We expect then that the $5d6s$ configuration of Ce^{2+} will be fairly well described by LS coupling and that the center of gravity of the ^3D levels will lie about 5000 cm^{-1} below the ^1D level. The $5d6s$ levels have been found experimentally to be 0, 676, 2216, and 7089 cm^{-1} with respect to the lowest level of the configuration. If we identify the lowest three levels to be the $^3D_1$, $^3D_2$, and $^3D_3$, respectively, then a weighted average of these levels is 5839 cm^{-1} below the $^1D_2$ level. Similarly, the Hartree-Fock integrals for the $4f6s$ configuration of Ce^{2+} are

$$G^3(4f, 6s) = 2449 \text{ cm}^{-1}$$
$$\zeta_{4f} = \qquad 775 \text{ cm}^{-1}.$$

In this case, the Coulomb splitting of the ^3F and ^1F states is 700 cm^{-1} and the spin-orbit splitting of the j-j states is 2713 cm^{-1}, and the configuration is best described then by j-j coupling. The splitting of the two $\{^2l_j\, ^2s_{1/2}\}\, J$ levels due to the Coulomb interaction is given by (4.39). Using Hartree-Fock values of the interaction integrals, we obtain the energy levels shown in the left-hand column of Fig. 7.10.

It is apparent from these results that the agreement between theory and experiment could be improved by reducing the value of the integrals $G^3(4f6s)$ and ζ_{4f}. The radial functions which were used to obtain these integrals were produced in a single-configuration Hartree-Fock calculation. This procedure assumes that

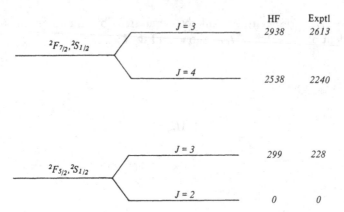

Fig. 7.10. Energy levels of the $4f6s$ configuration of Ce^{2+}. The values are given in cm^{-1}

the configuration $4f6s$ is pure. In reality there are other configurations which lie above the $4f6s$ configuration, interact with it, and compress its structure. Effects of this kind will be treated extensively in the second part of this book, particularly in Chap. 13.

8. Many-Electron Wave Functions

In the last two chapters we have applied the central-field model mainly to systems with two electrons in open shell(s). The extension to more general systems provides no fundamental difficulty. Using the standard technique of *Condon* and *Shortley* [1935], described in Chap. 5, one can set up the matrix of the Hamiltonian with Slater determinants as basis functions. The zeroth-order wave function and the first-order energy splitting are then obtained from the eigenvectors and the eigenvalues of this matrix. This procedure, which is still widely used, is for all but the very simplest cases quite tedious, however. The technique of *fractional parentage*, developed by *Racah* [1943, 1949], represents an enormous improvement. This technique has been described in several textbooks and monographs [*Slater* 1960; *Judd* 1963a, 1967; *Wybourne* 1965a; *Judd* and *Elliott* 1970].

We have found earlier that it is very convenient to represent the angular-momentum properties of two-electron systems in a graphical way and to evaluate the matrix elements using diagrammatic rules. We shall in the present chapter extend this technique to more general systems by means of a graphical representation of the fractional-parentage expansion. Our presentation is along the lines of the review article by *Briggs* [1971], although our angular-momentum representation is more developed.

8.1 Graphical Representation of the Fractional-Parentage Expansion

8.1.1 The $(nl)^3$ Configuration

As an illustration of the idea of fractional parentage, we consider the simple case of three equivalent electrons, discussed in Racah's original paper [*Racah* 1943; see also *Judd* 1963a, p. 171].

We know from Sect. 4.4.2. that two equivalent electrons in LS coupling can only form states for which $S + L$ is even, in order to satisfy the exclusion principle. Let us now take one of these antisymmetric states $(S'L')$ and couple to it a third electron to form a final state (SL), using the ordinary vector-coupling rules,

$$|nl^2(S'L')nl, SL\rangle = \sum_{M'_S m_s M'_L m_l} |nl^2, S'M'_S L'M'_L\rangle |nlm_s m_l\rangle$$

$$\times \langle S'M'_S sm_s | SM_S\rangle \langle L'M'_L lm_l | LM_L\rangle .$$

(8.1)

This state will normally *not* be antisymmetric with respect to an exchange of electrons 1 and 3 or 2 and 3. To see this more clearly, we can make a transformation to a scheme where electrons 2 and 3 are first coupled, as discussed in Sect. 3.4.1,

$$|nl^2(S'L')nl, SL\rangle = \sum_{S''L''} |nl, nl^2(S''L''), SL\rangle \tag{8.2}$$

$$\times \langle s, (ss)S'', S|(ss)S', s, S\rangle \langle l, (ll)L'', L|(ll)L', l, L\rangle .$$

The summation must here normally run over all possible values of S'' and L'', fulfilling the triangular conditions. But the function $|nl,nl^2(S''L''), SL\rangle$ is antisymmetric with respect to electrons 2 and 3 only if $S'' + L''$ *is even*. Thus, it is *not* possible to form a completely antisymmetric function in this case, by starting from a particular $S'L'$ combination for two of the electrons. Instead, we have to form a *linear combination* of functions (8.2) for different values of $S'L'$, so that the entire function becomes antisymmetric,

$$|nl^3, SL\rangle = \sum_{S'L'} c(S'L')|nl^2(S'L')nl, SL\rangle . \tag{8.3}$$

The condition here is that the terms where $S'' + L''$ is odd should disappear when the transformation (8.2) is applied,

$$\sum_{S'L'} c(S'L')\langle s, (ss)S'', S|(ss)S', s, S\rangle\langle l, (ll)L'', L|(ll)L', l, L\rangle = 0$$

$$\text{for} \quad S'' + L'' \text{ odd} . \tag{8.4}$$

The function will then be antisymmetric with respect to electrons 2 and 3. Since it is already antisymmetric with respect to 1 and 2, it will then also be antisymmetric with respect to an exchange $1 \leftrightarrow 3$. In order to see this, we can generate this permutation in the following way:

$$1\,2\,3 \to 1\,3\,2 \to 3\,1\,2 \to 3\,2\,1.$$

Each step involves a sign change, and hence also the exchange $1 \to 3$ changes the sign of the wave function.

The coefficients $c(S'L')$ in (8.3) are called the *coefficients of fractional parentage* (*cfp*).

Problem 8.1. Determine the fractional-parentage coefficients for the ^4S, ^2D and ^2P terms of the np^3 configuration.

Solution: The possible parent states are here the ^1S, ^1D and ^3P states of the np^2 configuration. It is obvious that if the first two spins are coupled to a singlet ($S' = 0$), it is not possible to reach a quartet state ($S = 3/2$) by means of the third electron. Thus, the ^4S state has only one parent, namely ^3P. By similar arguments for the orbital angular momentum, it is found that the ^2D state has the ^1D and ^3P parents, while the ^2P state can have any one of the np^2 states as a parent.

Equation (8.4) becomes in the present case

$$\sum_{S'L'} c(S'L') \left\langle \frac{1}{2}, \left(\frac{1}{2}\,\frac{1}{2}\right) S'', S \middle| \left(\frac{1}{2}\,\frac{1}{2}\right) S', \frac{1}{2}, S \right\rangle \langle 1, (11)L'', L | (11)L', 1, L \rangle = 0$$

or, using (3.60),

$$\sum_{S'L'} c(S'L')(-1)^{3/2+S+3+L}[S', S'', L', L'']^{1/2} \begin{Bmatrix} 1/2 & S' & 1/2 \\ 1/2 & S'' & S \end{Bmatrix} \begin{Bmatrix} 1 & L' & 1 \\ 1 & L'' & L \end{Bmatrix} = 0 .$$

For the ^2D state we then get the following equations for $S'' = 0$, $L'' = 1$:

$$c(^1D)(-1)^7[1 \cdot 1 \cdot 5 \cdot 3]^{1/2} \begin{Bmatrix} 1/2 & 0 & 1/2 \\ 1/2 & 0 & 1/2 \end{Bmatrix} \begin{Bmatrix} 1 & 2 & 1 \\ 1 & 1 & 2 \end{Bmatrix}$$

$$+ c(^3P)(-1)^7[3 \cdot 1 \cdot 3 \cdot 3]^{1/2} \begin{Bmatrix} 1/2 & 1 & 1/2 \\ 1/2 & 0 & 1/2 \end{Bmatrix} \begin{Bmatrix} 1 & 1 & 1 \\ 1 & 1 & 2 \end{Bmatrix} = 0$$

or

$$-\frac{\sqrt{3}}{4} c(^1D) - \frac{\sqrt{3}}{4} c(^3P) = 0 ,$$

which together with the normalization condition leads to

$$c(^1D) = - c(^3P) = \pm 1/\sqrt{2} .$$

The other odd combination, $S'' = 1$ and $L'' = 2$, leads, of course, to the same result. The phase cannot be determined by this procedure, but it turns out that the lower sign is in agreement with our phase conventions. Thus, the $|np^3, {}^2D\rangle$ state can be expanded in terms of parent states in the following way:

$$|np^3, {}^2D\rangle = \frac{1}{\sqrt{2}} |np^2(^3P)np, {}^2D\rangle - \frac{1}{\sqrt{2}} |np^2(^1D)np, {}^2D\rangle . \tag{8.5}$$

In a similar way one would find [Racah 1943; Judd 1963a]

$$|np^3, {}^2P\rangle = - \frac{1}{\sqrt{2}} |np^2(^3P)np, {}^2P\rangle - \sqrt{\frac{5}{18}} |np^2(^1D)np, {}^2P\rangle$$
$$+ \sqrt{\frac{2}{9}} |np^2(^1S)np, {}^2P\rangle . \tag{8.6}$$

It is, of course, possible to determine these coefficients also by expanding the states in determinantal wave functions [Slater 1960, Vol. II, p. 108], but this is a very tedious procedure, which is not tractable for more complicated systems.

8.1.2 Classification of States

It follows from our discussion of vector coupling in Chap. 2 that the coupling of two electrons—or two holes—is in LS coupling completely defined by the $SM_S LM_L$ quantum numbers. We have also seen that this is the case for an np^3 configuration. For configurations involving more than two electrons (or holes) with $l \geq 2$, on the other hand, it is in general not sufficient to specify these four quantum numbers in order to define the state. For an nd^3 configuration, for instance, there are two ^2D terms, and some additional information is required to distinguish them.

   Racah [1943] has shown that the terms of any nd^N configuration can be distinguished by means of a *seniority quantum number* (v). This is related to a two-particle operator Q, which can be diagonalized together with S^2, S_z, L^2 and L_z. It can then be shown that each term of the nd^N configuration with a non-zero value of Q is directly connected (via the fractional-parentage coefficients) to a term of the same kind of the nl^{N-2} configuration [*Judd* 1967, p. 41]. This chain terminates when Q vanishes, and each member of the series is given a seniority quantum number equal to the number of electrons of the last member of the chain. A state within an nd^N configuration is then completely specified by

$$|nd^N v S M_S L M_L\rangle .$$

For instance, of the two ^2D states of the configuration nd^3, one is connected in this way to the ^2D state of the nd configuration and hence has the seniority 1. The other state appears for the first time in the nd^3 configuration and has the seniority 3. One can say that the seniority is the number of electrons in the configuration for which the state occurs for the first time, when the shell is being filled.

   For systems of equivalent electrons with $l \geq 3$ the seniority is in general not sufficient to distinguish between terms of the same kind. For the case of f electrons ($l = 3$), *Racah* [1949] has shown that the states can be classified in a very elegant way by introducing continuous groups. This approach provides a single, unified way of classifying atomic states. The ordinary quantum numbers S, M_S, L, M_L can be regarded as labels of irreducible representations of the three-dimensional rotation group R_3. Similarly, the representation labels of the higher groups, which *Racah* introduced, contain R_3 as a subgroup and provide the additional quantum numbers that are necessary to specify the state completely. *Racah* was able to classify the states of the nd^N and nf^N configurations in this way.

   It is outside the scope of this book to consider the group-theoretical aspects any further, and interested readers are referred to the books by *Judd* [1963a] and *Judd* and *Elliott* [1970] and to the original papers of *Racah*.

8.1.3. The Expansion of the nl^N State

We shall now generalize the previous treatement and expand an arbitrary nl^N state in terms of its parents as follows:

$$|nl^N\gamma SL\rangle = \sum_{\bar{\gamma}\bar{S}\bar{L}} |nl^{N-1}(\bar{\gamma}\bar{S}\bar{L})nl, SL\rangle(l^{N-1}\bar{\gamma}\bar{S}\bar{L}, l|\} l^N\gamma SL) . \tag{8.7}$$

The coefficient $(l^{N-1}\bar{\gamma}\bar{S}\bar{L}, l|\} l^N \gamma SL)$ is the *coefficient of fractional parentage* (*cfp*). γ and $\bar{\gamma}$ represent here additional quantum numbers that may be needed to specify the states completely. If the atomic states are classified using group theory, then the *cfp* coefficients may be interpreted simply as coupling coefficients of the higher continuous groups. Equation (8.7) transforms states which are coupled together at the R_3 level to states which correspond to definite representations of the higher groups. The $M_S M_L$ values are omitted in (8.7), since it is obvious that the *cfp* are independent of these quantum numbers.

We shall sometimes find it convenient to use the following abbreviations:

$$\Omega = \gamma SL(M_S L_L) \tag{8.8}$$
$$\bar{\Omega} = \bar{\gamma}\bar{S}\bar{L}(\bar{M}_S\bar{M}_L) .$$

Equation (8.7) then reads

$$|nl^N\Omega\rangle = \sum_{\bar{\Omega}} |nl^{N-1}(\bar{\Omega})nl, \Omega\rangle(l^{N-1}\bar{\Omega}, l|\} l^N\Omega) . \tag{8.9}$$

With these notations the corresponding complex conjugate wave function (bra vector) has the expansion

$$\langle nl^N\Omega| = \sum_{\bar{\Omega}} (l^N\Omega\{|l^{N-1}\bar{\Omega}, l)\langle nl^{N-1}(\bar{\Omega})nl, \Omega| , \tag{8.10}$$

where obviously

$$(l^N\Omega\{|l^{N-1}\bar{\Omega}, l) = (l^{N-1}\bar{\Omega}, l|\} l^N\Omega)^* . \tag{8.11}$$

There exist nowadays efficient computer programs for evaluating the *cfp*, and also extensive tabulations are available. The tables of *Nielson* and *Koster* [1963] contains all coefficients for the p^N, d^N and f^N configurations, and shorter tables are found in standard books [*Slater* 1960, Appendix 27; *Judd* 1963a, Appendix 2]. Therefore, we can in our treatment here assume the *cfp* to be known quantities.

8.1.4 Graphical Representation of the Fractional-Parentage Coefficient

In the state

$$|nl^{N-1}(\bar{\Omega})nl, \Omega\rangle = |nl^{N-1}(\bar{\gamma}\bar{S}\bar{L})nl, SM_SLM_L\rangle ,$$

appearing in the expansion (8.7, 9), there is an ordinary vector coupling (which disregards the antisymmetry requirement) of the parent state

$$|nl^{N-1}\bar{\Omega}\rangle = |nl^{N-1}\bar{\gamma}\bar{S}\bar{M}_S\bar{L}\bar{M}_L\rangle$$

and the single-electron state $|nlm_sm_l\rangle$. This coupled state can also be expressed in terms of *uncoupled* states by means of vector-coupling coefficients (2.69)

$$|nl^{N-1}(\bar{\Omega})nl, \Omega\rangle = \sum_{\bar{M}_S m_s \bar{M}_L m_l} |nl^{N-1} \bar{\gamma}\bar{S}\bar{M}_S\bar{L}\bar{M}_L\rangle |nlm_sm_l\rangle \tag{8.12}$$
$$\times \langle \bar{S}\bar{M}_S sm_s | SM_S\rangle \langle \bar{L}\bar{M}_L lm_l | LM_L\rangle .$$

The antisymmetric many-electron state (8.7, 9) will then be

$$|nl^N\gamma SM_SLM_L\rangle = \sum_{\substack{\bar{\gamma}\bar{S}\bar{L} \\ \bar{M}_S m_s \bar{M}_L m_l}} (l^{N-1} \bar{\gamma}\bar{S}\bar{L}, l|\} l^N\gamma SL) \tag{8.13}$$
$$\times |nl^{N-1} \bar{\gamma}\bar{S}\bar{M}_S\bar{L}\bar{M}_L\rangle |nlm_sm_l\rangle \langle \bar{S}\bar{M}_S sm_s | SM_S\rangle \langle \bar{L}\bar{M}_L lm_l | LM_L\rangle ,$$

and we shall represent this graphically by

$$|nl^N\gamma SM_SLM_L\rangle = \sum_{fp} \qquad\qquad\qquad \tag{8.14}$$

in analogy with the representation of ordinary vector-coupled states (3.13). We have here omitted the uncoupled states associated with the free ends to the left of the symbol and assumed that a summation is performed over the magnetic quantum numbers of these states. The summation over the parent states is indicated by "$\sum_{fp}$". The black square represents the *cfp*, and the diagram also includes the two vector-coupling coefficients in the expansion (8.13). Whenever this symbol occurs, it can then be replaced by

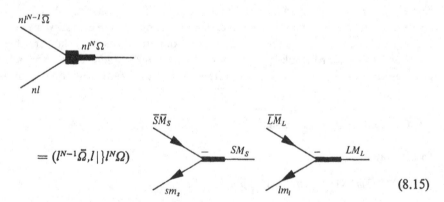

$$= (l^{N-1}\bar{\Omega}, l|\} l^N\Omega) \qquad\qquad\qquad \tag{8.15}$$

In the graphical representation the summation over the magnetic quantum numbers will be automatically accomplished by joining the lines of the vector-coupling diagrams to the rest of the graph.

8.2 Matrix Elements of a Single-Particle Operator

8.2.1 General

We shall now use the idea of fractional parentage to evaluate matrix elements of different kinds of operators. We first consider a single-particle operator of the type (5.14),

$$F = \sum_{i=1}^{N} f_i , \tag{8.16}$$

where f_i operates only on electron i. Each term in this operator must give an identical contribution to a matrix element

$$\langle nl^N \Omega | F | nl^N \Omega' \rangle \tag{8.17}$$

between properly antisymmetrized functions. Thus, we can pick one term and multiply the result by the number of electrons in the system,

$$\langle nl^N \Omega | F | nl^N \Omega' \rangle = N \langle nl^N \Omega | f_N | nl^N \Omega' \rangle . \tag{8.18}$$

By means of the *cfp* we can express the wave functions in terms of vector couplings between the $(N-1)$ system and the Nth particle. Since the operator f_N only operates on the latter, we can apply the results of Sect. 4.3.2. We can also easily do this directly in a graphical way. The initial and final states in (8.18) contain the following graphs:

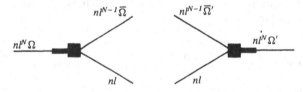

The parent states are unaffected by the interaction f_N and they must therefore be the same. We can then join the corresponding lines, and in this way also indicate the summation over the magnetic quantum numbers, in analogy with the pure angular-momentum graphs. The interaction f_N acts on the single-particle line, and we can then write the element (8.18) symbolically as

$$\langle nl^N \Omega | F | nl^N \Omega' \rangle = N \sum_{\bar{\Omega}} \tag{8.19}$$

The diagram can be evaluated using the relation (8.15). We get in this way one diagram for the angular part and one for the spin part—in addition to the *cfp* and the reduced single-particle element of f. We shall now consider this kind of evaluation in some detail.

8.2.2 Orbital Operator

Let us first assume that f is a tensor operator that affects only the *orbital* part of the wave function

$$F = T^k = \sum_i t_i^k .$$

Then the matrix element (8.19) becomes

$$\langle nl^N\Omega | T^k | nl^N\Omega' \rangle$$

$$= N\langle nl \| t^k \| nl \rangle \sum_{\bar{\Omega}} cfp$$

$$(8.20)$$

where "*cfp*" stands for the two fractional-parentage coefficients. The spin diagram represents the orthonormality condition for the spin functions (as described in Sect. 3.2.2) and yields simply the delta factor $\delta(S, S')$. By adding arrows at one of the vertices and changing two of the vertex signs, the orbital diagram becomes by means of (4.10)

$$= (-1)^{L+\bar{L}+l+k}[L, L']^{1/2}\begin{Bmatrix} L & k & L' \\ l & \bar{L} & l \end{Bmatrix}$$

By removing the diagram of the 3-j symbol, which appears also in the Wigner-Eckart theorem (3.17), we get the reduced matrix element

$$\langle nl^N\Omega \| T^k \| nl^N\Omega' \rangle = N\langle nl \| t^k \| nl \rangle \delta(S, S')[L, L']^{1/2}$$

$$\times \sum_{\bar{\Omega}} (-1)^{L+\bar{L}+l+k} (l^N\Omega \{| l^{N-1}\bar{\Omega}, l) (l^{N-1}\bar{\Omega}, l| \} l^N\Omega') \begin{Bmatrix} L & k & L' \\ l & \bar{L} & l \end{Bmatrix}. \quad (8.21)$$

Of course, we could have obtained that result directly without graphs, using (8.7) and (4.23b). However, we shall find the graphical technique quite convenient in dealing with more complicated systems.

8.2.3 Double-Tensor Operators

It is straightforward to extend the treatment above to a single-particle operator, acting on the spin as well as the orbital part of the wave function. This represents an important type of operator, including for instance the ordinary spin-orbit interaction and operators appearing in the magnetic hyperfine interaction. These interactions will be further treated in Chap. 14. For the moment we shall only derive some general formulas.

We consider a *double-tensor operator*

$$W^{\kappa k} = \sum_{i=1}^{N} w_i^{\kappa k} , \tag{8.22}$$

and for simplicity we restrict the single-particle operators to be of the form

$$w_i^{\kappa k} = s_i^{\kappa} t_i^k, \qquad \kappa = 0, 1 .$$

s^1 is here the ordinary spin operator and s^0 is the identity operator. t^k is a tensor operator of rank k, operating only on the orbital part of the wave function. We shall later couple the two parts together to a tensor-operator product of the type discussed in Sect. 2.3.2 and 4.3.

The representation (8.19) is valid also for the double-tensor operator (8.22), if we replace the interaction f by the operator $w^{\kappa k}$. The main difference is now that there will be an interaction also on the spin diagram in (8.20), which then becomes

In analogy with (8.21) we then get

$$\langle nl^N \Omega \| W^{\kappa k} \| nl^N \Omega' \rangle = N \langle s \| s^{\kappa} \| s \rangle \langle nl \| t^k \| nl \rangle$$

$$\times [L, L', S, S']^{1/2} \sum_{\bar{\Omega}} (-1)^f (l^N \Omega \{| l^{N-1} \bar{\Omega}, l) (l^{N-1} \bar{\Omega}, l |\} l^N \Omega') \tag{8.23}$$

$$\times \begin{Bmatrix} S & \kappa & S' \\ s & \bar{S} & s \end{Bmatrix} \begin{Bmatrix} L & k & L' \\ l & \bar{L} & l \end{Bmatrix}$$

with $f = S + \bar{S} + 1/2 + \kappa + L + \bar{L} + l + k$. We have here adopted the natural extension of the Wigner-Eckart theorem to double-tensor operators [*Judd* 1963a, p. 154],

$$\langle nl^N\gamma SM_SLM_L|\, W^{\kappa k}_{\mu q}\,|nl^N\gamma'S'M'_SL'M'_L\rangle$$

$$= \langle nl^N\gamma SL\|\, W^{\kappa k}\|nl^N\gamma'S'L'\rangle \quad \overset{SM_S \quad + \quad S'M'_S}{\underset{\kappa\mu}{\rule{0pt}{0pt}}} \quad \overset{LM_L \quad + \quad L'M'_L}{\underset{kq}{\rule{0pt}{0pt}}} \tag{8.24}$$

A word of warning should be given here concerning double-tensor operators of the type (8.22). For $\kappa = 0$ this operator is identical to the operator T^k considered before. However, the reduced matrix element (8.23) differs from (8.21) by a factor of $(2S+1)^{1/2}$. The matrix elements themselves must, of course, be identical, and the difference in the reduced elements is compensated by the extra 3-j symbol for the spin part in the Wigner-Eckart theorem for double-tensor operators.

We now couple the operators (8.22) to an operator of rank K in the combined space, as in (2.95),

$$W^{(\kappa k)\,K} = \sum_{i=1}^{N} w_i^{(\kappa k)K} = \sum_{i=1}^{N} \{s_i^\kappa t_i^k\}^K. \tag{8.25}$$

The reduced element of this operator can then be obtained in exactly the same way as for the two-electron case, discussed in Sect. 4.3.1,

$$\langle nl^N\,\gamma SLJ\|\, W^{(\kappa k)\,K}\|nl^N\,\gamma'S'L'J'\rangle$$

$$= [J, J', K]^{1/2}\begin{Bmatrix} S & S' & \kappa \\ L & L' & k \\ J & J' & K \end{Bmatrix}\langle nl^N\gamma SL\|\, W^{\kappa k}\|nl^N\gamma'S'L'\rangle. \tag{8.26}$$

8.2.4 Standard Unit-Tensor Operators

We have seen above that the matrix elements of a single-particle operator between many-electron states can be factorized into a reduced single-particle matrix element and a factor that depends on the states involved and the *tensor structure* of the interaction. It is then very convenient to define some *unit-tensor operators*, whose matrix elements can be evaluated once and for all.

For a single-particle operator acting only on the orbital part we use the definition

$$U = \sum_{i=1}^{N} u_i \tag{8.27}$$

$$\langle nl\|u\|n'l'\rangle = \delta(n, n')\delta(l, l').$$

This agrees with the definition used by *Racah* [1942], *Nielson* and *Koster* [1963], *Wybourne* [1965a, p. 32] and others, but it should be observed that other basic tensors are also commonly used [*Judd* 1963a, p. 101]. The reduced matrix elements of this operator are obtained from (8.21) by removing the single-particle element. For an arbitrary operator of this type we then get simply

$$\langle nl^N\gamma SL\| T^k\|nl^N\gamma'S'L'\rangle = \langle nl\|t^k\|nl\rangle\langle nl^N\gamma SL\|U^k\|nl^N\gamma'S'L'\rangle \ . \tag{8.28}$$

It is also convenient to introduce the unit double-tensor operator [*Racah* 1942]

$$V^{1k} = \sum_{i=1}^{N} v_i^{1k} = \sum_{i=1}^{N} (su^k)_i \ , \tag{8.29}$$

where the u^k tensor operator is the same as in (8.27). The matrix elements of this operator are obtained from (8.23) by replacing the reduced elements by

$$\langle s\|s\|s\rangle = \sqrt{3/2} \ , \tag{8.30}$$

according to (2.109) and (8.27). For an arbitrary double-tensor operator with $\kappa = 1$ we then get from (8.23)

$$\begin{aligned} &\langle nl^N \gamma SL\| W^{1k}\|nl^N \gamma'S'L'\rangle \\ &= \langle nl\|t^k\|nl\rangle\langle nl^N \gamma SL\| V^{1k}\|nl^N \gamma'S'L'\rangle \ . \end{aligned} \tag{8.31}$$

(Note that the reduced element of s ($=\sqrt{3/2}$) is included in the V^{1k} operator!) This gives in combination with (8.26)

$$\begin{aligned} &\langle nl^N \gamma SLJ\| W^{(1k)\,K}\|nl^N \gamma'S'L'J'\rangle \\ &= \langle nl\|t^k\|nl\rangle [J, J', K]^{1/2} \begin{Bmatrix} S & S' & 1 \\ L & L' & k \\ J & J' & K \end{Bmatrix} \langle nl^N \gamma SL\| V^{1k}\|nl^N \gamma'S'L'\rangle \ . \end{aligned} \tag{8.32}$$

This formula is directly applicable to a number of important interactions in atomic physics which involve the electron spin.

The reduced matrix elements of the unit-tensor operators U^2, V^{11} and V^{12} have been tabulated by *Nielson* and *Koster* [1963] for all p^N, d^N and f^N configurations. Tables for the p^N and d^N configurations are also found in the paper of *Racah* [1942] and in the book of *Slater* [1960, Appendix 26].

8.2.5 A Physical Example: The Spin-Orbit Interaction

As an illustration of the technique described above, we shall consider the spin-orbit interaction. In the case of a single open shell nl^N, this interaction can be approximately represented by the equivalent operator (5.5)

$$V_{so}^{eff} = \zeta_{nl} \sum_{i=1}^{N} s_i \cdot l_i \ , \tag{8.33}$$

where the summation runs over the electrons of the open shell. According to (2.97) we have

$$s \cdot l = -\sqrt{3} \, \{sl\}_0^0 \ , \tag{8.34}$$

so this operator is of the double-tensor type (8.25) with $K = 0$. Using

$$\langle nl \| t^k \| nl \rangle = \langle nl \| l \| nl \rangle = [l(l + 1)(2l + 1)]^{1/2} , \qquad (8.35)$$

given by (2.109), we obtain from (8.32) in analogy with (4.22)

$$\langle nl^N \gamma SLJM \,|\, V_{so}^{eff} \,|\, nl^N \gamma' S'L'J'M' \rangle = (-1)^{S'+L+J} \, \delta(J, J') \delta(M, M')$$
$$\times \, \zeta_{nl} [l(l + 1)(2l + 1)]^{1/2} \begin{Bmatrix} S & S' & 1 \\ L' & L & J \end{Bmatrix} \langle nl^N \gamma SL \| V^{11} \| nl^N \gamma' S'L' \rangle . \qquad (8.36)$$

8.3 Matrix Elements of a Two-Particle Operator

In his 1943 paper (cited above) *Racah* showed that the technique of fractional parentage could be extended also to two-particle operators

$$G = \sum_{i<j} g_{ij} \qquad (8.37)$$

by expanding the N-particle state in terms of "grandparents" in the $(N-2)$-particle system. Such a procedure, however, is too complicated to be of general interest. In a later paper, *Racah* [1949] showed that the treatment could be greatly simplified by considering the relation between the interactions within the N-particle and the $(N-1)$-particle systems. This made it possible to generate the matrix elements of a two-particle operator using ordinary *cfp* only.

In order to illustrate Racah's method, we shall first expand the state in terms of grandparent states,

$$|nl^N \Omega \rangle = \sum_{\bar{\Omega} \tilde{\Omega}} |nl^{N-2}(\bar{\Omega}) nl^2(\tilde{\Omega}), \Omega \rangle (l^{N-2}\bar{\Omega}, l^2 \tilde{\Omega} |\} l^N \Omega) . \qquad (8.38)$$

Here, $\bar{\Omega}$ represents the quantum numbers of the nl^{N-2} system and $\tilde{\Omega}$ those of the nl^2 system. This expansion can be represented graphically as

$$(8.39)$$

which is a straightforward generalization of the representation (8.14) of the ordinary *cfp* expansion.

Since the wave functions are completely antisymmetric, all electron pairs will make the same contribution to the matrix element. We can therefore pick one pair and multiply the result by the number of pairs,

$$\langle nl^N\Omega|G|nl^N\Omega\rangle = \frac{1}{2}N(N-1)\langle nl^N\Omega|g_{N,N-1}|nl^N\Omega\rangle \ . \tag{8.40}$$

For the Coulomb interaction of the electrons, the right-hand side can be represented graphically in analogy with (8.19) as

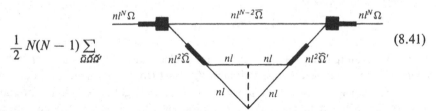

$$\frac{1}{2}N(N-1)\sum_{\bar{\Omega}\bar{\Omega}\bar{\Omega}'} \tag{8.41}$$

using the expansion (8.39). If the double *cfp* are known [*Dolan* 1970], it is straightforward—although usually quite tedious—to evaluate this expression.

As an illustration let us consider again the nl^3 configuration. From the equations above we then get

$$\langle nl^3\gamma SL|V_{es}|nl^3\gamma SL\rangle$$

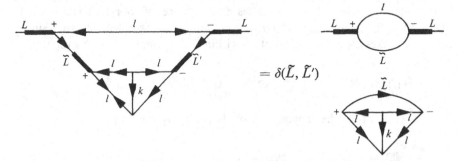

$$= 3\sum_{\bar{S}\bar{L}\bar{S}'\bar{L}'} \tag{8.42}$$

The relevant orbital angular-momentum diagram is

using JLV2 in Chap. 4. Note that in order to apply this theorem, it is necessary to transform the lower part of the diagram to normal form with exactly one arrow on each internal line, which can be accomplished by adding arrows at some of the vertices. The upper diagram on the right-hand side yields unity, while the lower diagram represents the orbital part of the matrix element of the Coulomb interaction for a two-electron system,

$$\langle nl^2 \tilde{S}\tilde{L} | V_{es} | nl^2 \tilde{S}\tilde{L} \rangle , \tag{8.43}$$

as discussed in Sect. 4.4.1. The corresponding spin diagram is equal to one. Equation (8.42) then reduces to

$$\begin{aligned} &\langle nl^3 \gamma SL | V_{es} | nl^3 \gamma SL \rangle \\ &= 3 \sum_{\tilde{S}\tilde{L}} |(l, l^2 \tilde{S}\tilde{L} |\} l^3 \gamma SL)|^2 \langle nl^2 \tilde{S}\tilde{L} | V_{es} | nl^2 \tilde{S}\tilde{L} \rangle . \end{aligned} \tag{8.44}$$

Since the grandparent state is here a single-electron state, it follows that the coefficients are identical to the ordinary *cfp* apart from sign due to the coupling order. Thus, if we know these *cfp*, we can immediately get the Coulomb interaction of the nl^3 system from the interaction of the nl^2 system, which can be obtained in an elementary way (see Sect. 4.4).

For an np^3 configuration we obtain the following expectation values of the Coulomb interaction (also equal to the first-order energy contribution in a perturbation expansion) by means of the *cfp* in (8.5, 6) and the results for the np^2 configuration given in Sect. 4.4.3:

$$\begin{cases} E(np^3 \, {}^4S) = 3E(np^2 \, {}^3P) = 3F^0 - \dfrac{3}{5} F^2 \\[2mm] E(np^3 \, {}^2D) = 3 \left[\dfrac{1}{2} E(np^2 \, {}^3P) + \dfrac{1}{2} E(np^2 \, {}^1D) \right] = 3F^0 - \dfrac{6}{25} F^2 \\[2mm] E(np^3 \, {}^2P) = 3 \left[\dfrac{1}{2} E(np^2 \, {}^3P) + \dfrac{5}{18} E(np^2 \, {}^1D) + \dfrac{2}{9} E(np^2 \, {}^1S) \right] = 3F^0 . \end{cases} \tag{8.45}$$

The procedure used above can easily be generalized. Instead of splitting the system into a two-particle and an $(N-2)$-particle system, however, we can separate out a single particle and consider the interaction within the $(N-1)$ system. (For a three-particle system this makes no difference, of course.) The N system has $N(N-1)/2$ pairs and the $(N-1)$ system $(N-1)(N-2)/2$ pairs. Since each pair makes the same contribution within each system, we get the relation

$$\langle nl^N \Omega | \sum_{i<j}^{N} g_{ij} | nl^N \Omega \rangle = \frac{N}{N-2} \langle nl^N \Omega | \sum_{i<j}^{N-1} g_{ij} | nl^N \Omega \rangle . \tag{8.46}$$

The last element can be represented symbolically by the graph

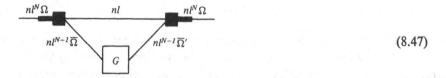

$$\tag{8.47}$$

If G is a scalar, we can show in the same way as before that this is precisely the interaction within the $(N-1)$ system—times the *cfp*. Thus,

$$\langle nl^N\Omega\,|\,G\,|\,nl^N\Omega\rangle$$

$$= \frac{N}{N-2}\sum_{\bar{\Omega}\bar{\Omega}'}(l^N\Omega\,\{|\,l^{N-1}\bar{\Omega},\,l)\,(l^{N-1}\bar{\Omega}',\,l|\}\,l^N\Omega)\,\langle nl^{N-1}\bar{\Omega}\,|\,G\,|\,nl^{N-1}\bar{\Omega}'\rangle\ . \tag{8.48}$$

Obviously, only terms with $\bar{S}' = \bar{S}$ and $\bar{L}' = \bar{L}$ contribute here. By means of this relation we can evidently generate the Coulomb energies for all nl^N states by means of the *cfp*. These energies are tabulated for all p^N, d^N and f^N states by *Nielson* and *Koster* [1963].

8.4 More Than One Open Shell

So far, we have only considered atoms with a single open shell. The methods we have used can be extended to more general systems, although the practical difficulties usually increase drastically. This problem has been treated in some detail by *Wybourne* [1965a]. Here, we shall restrict ourselves to a few examples, the purpose being mainly to illustrate how the graphical technique can be applied to this type of problem.

8.4.1 Transition Probability Between the nl^N and $nl^{N-1}\,n'l'$ Configurations

Let us as a first example consider the matrix element

$$\langle 3p^5(^2P),\,nd,\,^1P\,|\,\sum_{i=1}^{6}z_i\,|\,3p^6\,^1S\rangle\ . \tag{8.49}$$

The operator is here the dipole operator, and this matrix element occurs in calculating the probability for an optical transition in argon, where an electron in the $3p$ shell is excited to an nd state [*Kelly* 1976]. Both states appearing here must, of course, be antisymmetric. Suppose now that it is the sixth electron which makes the transition $3p - nd$; then the matrix element would be

$$\langle 3p^5(^2P),\,nd_6,\,^1P\,|\,z_6\,|\,3p^6\,^1S\rangle\ , \tag{8.50}$$

since only $i = 6$ would contribute. On the left side we have an ordinary vector coupling (2.69), which implies that this state is not antisymmetric with respect to an exchange of the sixth electron with any one of the other five. The $3p^5$ part is, of course, antisymmetric and contains 5! permutations and a normalization factor of $(5!)^{-1/2}$. In order to make it completely antisymmetric, we have to make 6 times more permutations and divide the function by $\sqrt{6}$. But each permutation must for symmetry reasons yield the same contribution to the matrix element. As a result, the element (8.50) must be multiplied by $\sqrt{6}$ to give the desired result,

$$\langle 3p^5(^2P), nd, {}^1P \mid \sum_{i=1}^{6} z_i \mid 3p^6 \, {}^1S \rangle$$
$$= \sqrt{6} \, \langle 3p^5(^2P), nd_6, {}^1P \mid z_6 \mid 3p^6 \, {}^1S \rangle \, .$$

(8.51)

The state to the left in the matrix element (the bra state) can be represented graphically as

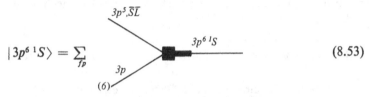

$$\langle 3p^5(^2P), nd_6, {}^1P \mid \; =$$

(8.52)

This involves two vector-coupling coefficients, in analogy with the representation (8.15), and this is here indicated by the heavy line. Since no *cfp* is involved, the black square is omitted. The state to the right (the ket state) has the representation (8.14)

$$\mid 3p^6 \, {}^1S \rangle = \sum_{fp}$$

(8.53)

We can now combine the two functions (8.52, 53) to get the following representation of the element (8.51):

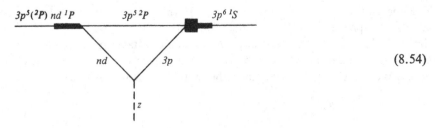

(8.54)

The state $3p^6 \, {}^1S$ has obviously only one parent, namely $3p^5 \, {}^2P$, so the *cfp* must here be equal to unity. Since the interaction is spin independent the spin diagram will be the usual orthogonality diagram. The orbital part is

$$\langle nd \| r \| 3p \rangle$$

(8.55)

By eliminating the zero line and the loop (3.20, 25), the diagram becomes 1/3. Including the factor of $\sqrt{6}$, due to the non-antisymmetry, the final result becomes

$$\langle 3p^5(^2P)nd, {}^1P|\sum_i z_i|3p^6\ {}^1S\rangle \tag{8.56}$$

$$= \sqrt{\frac{2}{3}}\langle nd\|r\|3p\rangle = \frac{2}{\sqrt{3}}\int P_{nd}(r)rP_{3p}(r)dr.$$

In the last step we have used the identity $r = rC^1$ and (2.127) to evaluate the reduced matrix element of the C tensor.

....

Problem 8.2. Verify the result (8.56).

8.4.2 Coulomb Interaction Between the Configurations nl^N and $nl^{N-1}\,n'l'$

As a second example, we consider the following matrix element of the Coulomb interaction:

$$\langle 2p^2(S'L')3p, SL|\sum_{i<j} r_{ij}^{-1}|2p^3, SL\rangle . \tag{8.57}$$

Let us here fix the 3p orbital to electron 3, giving

$$\langle 2p^2(S'L')3p_3, SL|r_{13}^{-1} + r_{23}^{-1}|2p^3, SL\rangle , \tag{8.58}$$

since only the part of the interaction which involves electron 3 contributes. In order to make the state to the left antisymmetric, we have to make three permutations and divide by $\sqrt{3}$. Since each permutation gives the same contribution to the matrix element (8.57), we get the desired result by multiplying the element (8.58) by $\sqrt{3}$. Furthermore, r_{13}^{-1} and r_{23}^{-1} must give the same contribution. Thus, we get

$$\langle 2p^2(S'L')\,3p, SL|\sum_{i<j} r_{ij}^{-1}|2p^3, SL\rangle$$
$$= 2\sqrt{3}\,\langle 2p^2(S'L')3p_3, SL|r_{13}^{-1}|2p^3, SL\rangle . \tag{8.59}$$

The states involved can here be represented graphically as

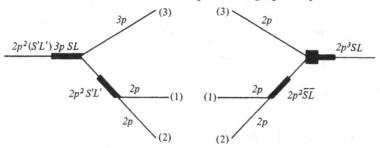

This leads to the following representation of the matrix element on the right-hand side of (8.59):

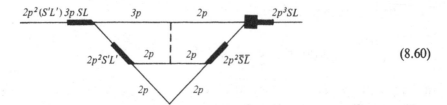

$$(8.60)$$

The associated orbital diagram is

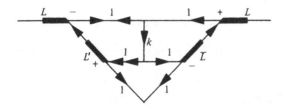

which after transforming to normal form and applying JLV3 becomes

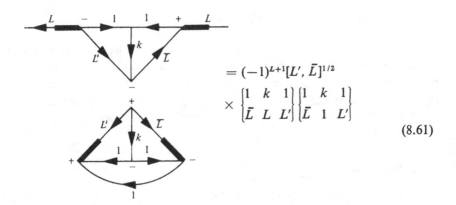

$$= (-1)^{L+1}[L', \bar{L}]^{1/2}$$
$$\times \begin{Bmatrix} 1 & k & 1 \\ \bar{L} & L & L' \end{Bmatrix} \begin{Bmatrix} 1 & k & 1 \\ \bar{L} & 1 & L' \end{Bmatrix}$$

$$(8.61)$$

The corresponding spin diagram is

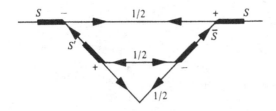

which after applying JLV2 becomes

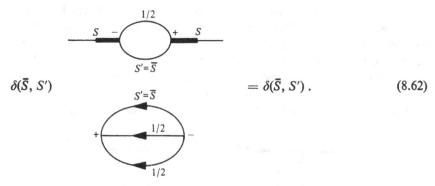

$$\delta(\bar{S}, S') = \delta(\bar{S}, S') . \tag{8.62}$$

This is an expected result, since the Coulomb interaction is independent of spin.

The final result is obtained by collecting together the diagrams (8.61, 62), the numerical factor in (8.59), the *cfp* and the expression for the Coulomb interaction (3.42)

$$\langle 2p^2(S'L')\,3p, SL| \sum r_{ij}^{-1} |2p^3, SL\rangle$$
$$= 2\sqrt{3} \sum_{\bar{L}} (p^2(S'\bar{L}), p|\}p^3SL)\,[L', \bar{L}]^{1/2}\,(-1)^{L+1} \tag{8.63}$$
$$\times \sum_k \begin{Bmatrix} 1 & k & 1 \\ \bar{L} & L & L' \end{Bmatrix} \begin{Bmatrix} 1 & k & 1 \\ \bar{L} & 1 & L' \end{Bmatrix} \langle 1\|C^k\|1\rangle^2\, R^k(2p2p, 2p3p) .$$

8.4.3 Transition Probability Between the Configurations $nl^N\,n'l'$ and $nl^{N-1}\,n'l'^2$

As a final example, we consider the following matrix element of the dipole operator:

$$\langle nl^N(\Omega_1)n'l', \Omega| \sum z_i |nl^{N-1}(\Omega_1')n'l'^2(\Omega_2), \Omega'\rangle . \tag{8.64}$$

This appears, for instance, in calculating the transition probability between configurations of the type $4f^N5d \rightarrow 4f^{N-1}5d^2$ for the rare-earth elements. This matrix element has been evaluated by *Wybourne* [1965a, p. 88], and we shall here make the corresponding evaluation using graphical methods. This fairly complicated example demonstrates very clearly the power of the graphical technique.

Let us first fix the $n'l'$ orbital on the left to the $(N+1)$th electron and the $n'l'$ orbitals to the right to the Nth and the $(N+1)$th electrons. Only the term $i = N$ of the operator will then contribute,

$$\langle nl^N(\Omega_1)n'l'_{N+1}, \Omega| z_N |nl^{N-1}(\Omega_1')n'l'^2(\Omega_2)_{N,N+1}, \Omega'\rangle . \tag{8.65}$$

In order to make the states antisymmetric, we have to make $(N+1)$ more permutations on the left and divide by $(N+1)^{1/2}$, and $N(N+1)/2$ permutations to the right and divide by $[N(N+1)/2]^{1/2}$. The operator shall connect one of the nl orbitals to the left with one of the $n'l'$ orbitals to the right. For each permutation on the right there are two permutations on the left which contribute, namely when $(N - 1)$ of the "spectator" electrons in the nl shell are ordered in the same way as to the right and the remaining two electrons are ordered in any one of the two possible ways. This means that we get $N(N + 1)$ identical contributions. Including the normalization factors above, we find that we should multiply the matrix element (8.65) by $(2N)^{1/2}$ to get the desired result.

The matrix element (8.65) can be represented by the following diagram:

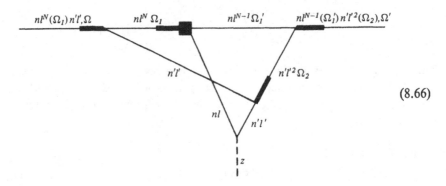

$$(8.66)$$

It is left as an exercise for the reader (Problem 8.3) to show that the *orbital* part of this diagram becomes

$$[L, L_1, L', L_2]^{1/2} \begin{Bmatrix} l' & L_2 & l'' \\ L & L' & 1 \\ L_1 & L_1' & l \end{Bmatrix}$$

and the *spin* part

$$\delta(S, S')(-1)^{S+S_1'+1}[S_1, S_2]^{1/2} \begin{Bmatrix} S_1' & 1/2 & S_1 \\ 1/2 & S & S_2 \end{Bmatrix}.$$

The final result is then obtained by including the *cfp*, the reduced matrix element of the dipole operator, and the factor $(2N)^{1/2}$ given above.

Problem 8.3. Derive the orbital and spin parts given in the text of the diagram (8.66).

Part II

Perturbation Theory and the Treatment of Atomic Many-Body Effects

9. Perturbation Theory

9.1 Basic Problem

The basic problem in nonrelativistic atomic theory is to find an approximate solution of the time-independent Schrödinger equation

$$H\Psi = E\Psi \tag{9.1}$$

for different kinds of atomic systems. In the first part of the book we have built up an approximate atomic model, based on the central-field concept. In that model the electrons move independently in a central potential, and the single-electron wave functions can be separated into radial, angular and spin parts. The theory of angular momentum, which was studied extensively in Chaps. 2–4 then plays a fundamental role. The radial part of the wave functions can be determined by requiring that the state of the atom corresponds to a single configuration for which there is a certain number of electrons in each shell and that the expectation value of the energy is stationary with respect to small changes in the form of the radial functions. This procedure leads to the restricted Hartree-Fock model, which has been quite successful in atomic calculations. The Hartree-Fock energy of the ground state of the carbon atom, for instance, is -37.689 H[1], compared to the "true" nonrelativistic energy estimated to be -37.839 H. While the difference between these two values is relatively small, it is comparable in magnitude to the binding energy of the outer electrons, which are mainly responsible for the optical and chemical properties of the atom. Therefore, in order to be able to understand such properties even qualitatively, it is usually necessary to go beyond the Hartree-Fock model and to treat the atom as a true many-body system. We shall be mainly concerned with this problem in this second part of the book.

As we have mentioned, the central-field model depends upon two assumptions,

  a) *that the electrons move independently of each other in the average field of the nucleus and the remaining electrons, and*
  b) *that this average field is spherically symmetric.*

If restriction b) is relaxed, we get the more general independent-particle model,

---

[1] H stands for "Hartree", which is the energy unit in the atomic unit system used in this book (see Appendix A).

treated in Chap. 5. There is then no restriction on the spin-orbitals, and there is, in principle, no shell structure. Therefore, in going from the central-field model to the independent-particle model, the spin-orbitals are distorted and the filled shells are broken up. This effect is usually referred to as the *polarization* of the atomic core. Minimizing the energy then leads to the *unrestricted Hartree-Fock* model. If also restriction a) above is relaxed, it implies that we take into account the instantaneous motion of the electrons or the *correlation* between the electrons in their mutual Coulomb field. This represents true many-body effects.

Many-body calculations which go beyond the Hartree-Fock model can be performed in essentially two ways, using either a *variational* or a *perturbation* procedure. The most frequently used form of variational procedure is the so-called *configuration-interaction* (CI) approach, which has been applied to a large number of atomic and molecular systems (for reviews, see [*Schaefer* 1972, 1977]). As the starting point of such calculations, the Hartree-Fock wave functions are usually expanded in terms of a set of analytic functions. The number of basis functions is then extended to form an essentially complete set, and a many-body wave function is constructed using this basis set. The parameters in the basis functions and the mixing coefficients are then determined by minimizing the expectation value of the energy. The success of such calculations depends critically on the choice of the basis set, and a great amount of effort has been devoted to optimizing this set within a given size.

Another approach, which is based on the variational principle, is the *multi-configuration Hartree-Fock* (MCHF) procedure, which is a natural extension of the Hartree-Fock scheme discussed earlier. Here, a number of configurations are chosen and the mixing coefficients—as well as the radial parts of the spin-orbitals—are varied. This method does not depend upon a choice of basis set, and it is in principle a complete, straightforward means of carrying out a many-body calculation. In practice, however, a full MCHF calculation would require variation of a large number of small interrelated quantities, and it is usually necessary to divide the interacting configurations into subgroups that are treated separately. The MCHF procedure has been applied to various atomic systems, in particular by *Froese-Fischer* and *Saxena* [1974].

In the *perturbation* approach the Hamiltonian of the system (H) is split into two parts, a *model Hamiltonian* (H_0) and a *perturbation* (V)

$$H = H_0 + V. \tag{9.2}$$

The model Hamiltonian should be a reasonable approximation to the full Hamiltonian and it should be simple enough to manage. For atoms, the natural choice of H_0 is some central-field approximation, such as the Hartree-Fock model, and it is, in fact, the presence of a natural first approximation which makes perturbation theory such a promising method for atoms.

The most well-known perturbation schemes are the *Rayleigh-Schrödinger* and the *Brillouin-Wigner* expansions. In a straightforward approach, these schemes are quite hard to employ beyond, say, third order. During the last two decades, however, several new schemes have been developed which make it possible to treat the perturbation theory of many-body systems in a more general fashion. Many of these new ideas originate from developments in field theory [*Feynman* 1949; *Dyson* 1949; *Wick* 1950], and the field-theoretical approach was first applied to many-body systems by *Hugenholtz* [1957] and *Goldstone* [1957]. This formalism has also led to the development of the time-dependent Green's-function technique, which nowadays is widely used in many-body theory, particularly for extended systems like the electron gas and infinite nuclear matter (see, for instance, [*March* et al. 1967] and [*Fetter* and *Waleĉka* 1971]). Another important development has taken place in particle scattering theory [*Lippman* and *Schwinger* 1950; *Watson* 1953; *Francis* and *Watson* 1953]. This formalism has been further developed in nuclear theory in order to overcome the problem of the strongly repulsive core .[*Brueckner* and *Levinson* 1955; *Bethe* 1956].

It was first conjectured by *Brueckner* [1955] that so-called unlinked terms should disappear from the Rayleigh-Schrödinger expansion, since they do not have the correct linear dependence on the number of particles in the system. This cancellation was shown formally by *Goldstone* [1957] using a graphical representation similar to that introduced by *Feynman* in field theory. This is the linked-diagram theorem, which has been frequently used in atomic and nuclear calculations. The first atomic application was made by *Kelly* [1963] on the beryllium atom.

An important feature of modern perturbation theory is the so-called *partitioning technique* [*Feshbach* 1958a,b, 1962; *Löwdin* 1965, 1966]. In this procedure, the functional space for the wave functions is separated into two parts, a *model space* and an *orthogonal space*. The basic idea here is to find "effective operators" which act only within the limited model space but which generate the same result as do the original operators acting on the entire functional space. An effective-operator formalism suitable for atomic systems has been developed mainly by *Judd* [1967, 1969], *Wybourne* [1965a] and *Sandars* [1969], following early ideas of *Racah* [1952]. An essentially independent development has taken place in nuclear physics (for reviews, see [*Barrett* and *Kirson* 1973] or [*Ellis* and *Osnes* 1977]).

Instead of carrying out the perturbation calculations order by order, it is often advantageous to arrange the terms of the expansion according to the number of particles excited. This means that we distinguish between single-particle effects, two-particle effects and so on. In the former case we study how a single particle-hole excitation introduced into the system "propagates" as the perturbation expansion is carried to higher orders. Such effects can relatively easily be taken into account to all orders, and they lead to an improved single-particle model, which is physically appealing. The inclusion of certain two-particle effects to all orders of perturbation theory leads similarly to an "exact" pair equation. Such an equation describes two electrons moving in an average potential due to the other electrons and interacting among themselves with the full Coulomb force. A two-particle approach of this kind was first developed in nuclear physics [*Bethe* and *Goldstone* 1957] and has been extensively applied to atomic and molecular systems (see, for instance, [*Kutzelnigg* 1977] for a review). For light and medium-heavy elements the entire correlation energy can generally be represented by independent pairs to within about 10%.

After this brief sketch of some different approaches to the many-body theory of finite systems, we shall outline how this problem will be treated in this book.

In the present chapter we shall first consider the Brillouin-Wigner formalism for nondegenerate systems, such as closed-shell atoms. This scheme contains the *exact* energy of the state explicitly and therefore is best suited for problems where only a *single energy level* is involved. The Green's-function formalism is closely related to the Brillouin-Wigner perturbation theory and this analogy will be briefly outlined.

In the second part of the present chapter we shall consider the Rayleigh-Schrödinger formalism, which will form the basis for our development. This scheme has the advantage that the exact energy does not appear explicitly. Therefore, it can be applied to a *group of states* simultaneously in order to investigate various kinds of energy level splittings. For that reason, we shall treat the Rayleigh-Schrödinger formalism in its most general form. We shall see that this leads naturally to the concept of (energy-independent) effective operators, which are very useful in describing the effect of different physical interactions.

As a first application of perturbation theory, we shall in the next chapter determine the first-order contributions to the wave function and the energy for a closed-shell system. A graphical representation, using so-called Goldstone diagrams, will be introduced. In the following chapter, a more general graphical procedure is developed, based on the formalism of second quantization and Wick's theorem. In the order-by-order approach, this procedure leads to the linked-diagram theorem, which is demonstrated for closed-shell systems in Chap. 12 and for general open-shell systems in Chap. 13. In Chap. 14 the idea of effective operators is applied to the hyperfine interaction. In the last chapter the graphical procedure is developed further to derive one-, two-, ... particle equations, where single, double, ... excitations are taken into account to all orders, and finally to the so-called coupled-cluster approach for open-shell systems.

9.2   Nondegenerate Brillouin-Wigner Perturbation Theory [2]

9.2.1   Basic Concepts

We shall begin our formal treatment of perturbation theory by defining some basic concepts. First of all, we need a zeroth-order or model Hamiltonian which

---

[2] The treatment can easily be extended to the degenerate case, but we restrict ourselves here to the non-degenerate case mainly for pedagogical reasons. Degenerate perturbation theory will be treated in its full generality in connection with the Rayleigh-Schrödinger perturbation theory to be discussed in Sect. 9.4.

is a reasonable approximation to the full Hamiltonian and which is easy to handle. For atoms a natural choice is the central-field Hamiltonian (6.2)

$$H_0 = \sum_{i=1}^{N} \left[-\frac{1}{2} \nabla_i^2 - \frac{Z}{r_i} + u(r_i) \right]. \tag{9.3}$$

The perturbation is then the noncentral part of the electrostatic interaction, and it may also include magnetic interactions between the electrons (V_{mag}) and other interactions, such as the interactions with external fields (V_{ext}) or the hyperfine interaction (h),

$$V = \sum_{i<j} \frac{1}{r_{ij}} - \sum_{i=1}^{N} u(r_i) + V_{mag} + V_{ext} + h . \tag{9.4}$$

We shall leave the definition of H_0 open for the moment, and assume only that it is a hermitian operator having a complete set of eigenfunctions,

$$H_0 \phi^\alpha = E_0^\alpha \phi^\alpha . \tag{9.5}$$

In the present section, we shall assume that the eigenvalues of H_0 are *non-degenerate*, in which case the eigenfunctions are automatically orthogonal (see Appendix B)

$$\langle \phi^\alpha | \phi^\beta \rangle = \delta(\alpha, \beta) . \tag{9.6}$$

We assume now that one of the eigenfunctions of H_0, ϕ^α, is a reasonable approximation of the exact wave function Ψ for the state we are investigating. This zeroth-order approximation is called the *reference function* or the *model function*,

$$\Psi_0 = \phi^\alpha . \tag{9.7}$$

It is then the purpose of perturbation theory to find a scheme for generating a sequence of successive improvements to this zeroth-order approximation.

We introduce a *projection operator* (see Appendix B) associated with the model function (9.7)

$$P = |\Psi_0\rangle\langle\Psi_0| = |\phi^\alpha\rangle\langle\phi^\alpha| \tag{9.8a}$$

and a corresponding projection operator for the complementary part of the functional space, which we call the *orthogonal space* or the *Q space*,

$$Q = \sum_{\beta \neq \alpha} |\phi^\beta\rangle\langle\phi^\beta| . \tag{9.8b}$$

The P operator projects out of any function, representing the system we are considering, the part that is proportional to the model function, and the Q operator projects out the part that is orthogonal to this function.

Since H_0 is hermitian, it follows from the completeness relation (B.9) that its eigenfunctions form a complete set. This implies that

$$P + Q = 1 . \tag{9.9}$$

It is easy to derive a number of important relations for the projection operators, such as

$$
\begin{aligned}
P^\dagger &= P; & Q^\dagger &= Q \\
PP &= P; & QQ &= Q \\
PQ &= QP = 0 \\
[P, H_0] &= [Q, H_0] = 0 ,
\end{aligned}
\tag{9.10}
$$

the proof of which is left as an exercise for the reader. Note, in particular, that the projection operators, which are formed from eigenfunctions of the model Hamiltonian H_0, commute with this operator.

If we operate with P on the exact wave function Ψ, we get

$$P\Psi = |\Psi_0\rangle\langle\Psi_0|\Psi\rangle .$$

We now *assume* that the scalar product appearing here is equal to unity,

$$\langle\Psi_0|\Psi\rangle = 1 , \tag{9.11}$$

which gives

$$\boxed{P\Psi = \Psi_0} . \tag{9.12}$$

Furthermore, we assume that the model function Ψ_0 is normalized to unity, which implies generally that the exact wave function Ψ is not. This procedure, called *intermediate normalization*, will be employed in the following.

Using (9.9) and (9.12), the exact wave function can be expressed as

$$\Psi = (P + Q)\Psi = \Psi_0 + Q\Psi . \tag{9.13}$$

As mentioned above, the model function Ψ_0 represents the zeroth-order approximation and the remaining part, $Q\Psi$, can be regarded as a "correction". If Ψ_0 is generated in an independent-particle model, such as the Hartree-Fock, then $Q\Psi$ is often referred to as the *correlation function* [Sinanoğlu 1964, 1969].

9.2.2 A Resolvent Expansion of the Wave Function

We assume now that the model function Ψ_0 is known, and we want to find a series expansion for the remaining part, $Q\Psi$, of the exact wave function (9.13). In order to accomplish this we return to the Schrödinger equation (9.1), which we write in the form

$$(E - H_0)\Psi = V\Psi,$$

using the separation (9.2) of the Hamiltonian. Operating from the left with Q and using the fact that Q commutes with H_0, we get

$$(E - H_0)Q\Psi = QV\Psi. \qquad (9.14)$$

This is an *inhomogeneous differential equation*, and it can be solved for $Q\Psi$ by introducing a *resolvent* T_E [*Messiah* 1961, p. 712], defined by

$$T_E(E - H_0) = Q; \qquad T_EQ = QT_E = T_E. \qquad (9.15)$$

This operator, which is often written

$$T_E = \frac{Q}{E - H_0}, \qquad (9.16)$$

is the inverse of $(E - H_0)$ in the Q space, and it gives zero when operating on the model function.

The effect of operating with the resolvent on the eigenfunctions of H_0 can be obtained directly from (9.15)

$$T_E(E - H_0)|\phi^\beta\rangle = T_E(E - E_0^\beta)|\phi^\beta\rangle = Q|\phi^\beta\rangle$$

$$T_E|\phi^\beta\rangle = \frac{Q}{E - E_0^\beta}|\phi^\beta\rangle.$$

This shows that T_E has the same eigenfunctions as H_0 and that the corresponding eigenvalues are $(E - E_0^\beta)^{-1}$ for $\beta \neq \alpha$ and zero for $\beta = \alpha$. Using the completeness condition for the states $\{\phi^\beta\}$, we then obtain the following *spectral resolution* for the resolvent:

$$T_E = \sum_\beta T_E|\phi^\beta\rangle\langle\phi^\beta| = \sum_{\beta \neq \alpha} \frac{|\phi^\beta\rangle\langle\phi^\beta|}{E - E_0^\beta}. \qquad (9.17)$$

By operating with T_E from the left on (9.14) and using (9.15), we get furthermore

$$Q\Psi = T_EV\Psi. \qquad (9.18)$$

This expression for the part of the wave function in the Q space may then be substituted into (9.13) to obtain

$$\Psi = \Psi_0 + T_E V \Psi \, , \tag{9.19}$$

which is a very useful relation. It is *exact*, and it may be used to generate a series expansion for the wave function. Replacing Ψ on the right by the entire right-hand side, we get

$$\Psi = \Psi_0 + T_E V \Psi_0 + T_E V T_E V \Psi \, . \tag{9.20}$$

By continuing this process, we obtain the infinite series

$$\Psi = (1 + T_E V + T_E V T_E V + T_E V T_E V T_E V + \cdots) \Psi_0 \tag{9.21}$$

or using the form (9.16) of the resolvent

$$\Psi = \left(1 + \frac{Q}{E - H_0} V + \frac{Q}{E - H_0} V \frac{Q}{E - H_0} V + \cdots\right) \Psi_0 \, . \tag{9.22}$$

This is the *Brillouin-Wigner perturbation expansion*. Using the spectral resolution (9.17) of the resolvent, we find that the first-order term becomes

$$\Psi^{(1)} = \sum_{\beta \neq \alpha} \frac{|\phi^\beta\rangle \langle \phi^\beta | V | \Psi_0\rangle}{E - E_0^\beta} \tag{9.23}$$

which is a familiar result. Similarly, the second-order term becomes

$$\Psi^{(2)} = \sum_{\beta, \gamma \neq \alpha} \frac{|\phi^\beta\rangle \langle \phi^\beta | V | \phi^\gamma\rangle \langle \phi^\gamma | V | \Psi_0\rangle}{(E - E_0^\beta)(E - E_0^\gamma)} \tag{9.24}$$

and so on.

9.2.3 The Wave Operator

The quantity within brackets in (9.21, 22),

$$\begin{aligned}
\Omega_E &= 1 + T_E V + T_E V T_E V + \cdots \\
&= 1 + \frac{Q}{E - H_0} V + \frac{Q}{E - H_0} V \frac{Q}{E - H_0} V + \cdots \, ,
\end{aligned} \tag{9.25}$$

defines an operator with the property

$$\Psi = \Omega_E \Psi_0 \, . \tag{9.26}$$

In other words, this operator generates the full wave function when operating on the model function, which is given by (9.12),

$$\Psi_0 = P\Psi .\qquad(9.27)$$

Such an operator is called a *model operator* or *wave operator* [*Møller* 1945, 1946; *Eden* and *Frances* 1955; *Löwdin* 1962, 1965]. It should be noted that in the form given here this operator depends explicitly on the exact energy of the state considered. In treating the Rayleigh-Schrödinger perturbation theory in Sect. 9.4 we shall derive a form of the wave operator which does not have such an explicit energy dependence.

The wave operator (9.26) and the projection operator (9.27) are illustrated in a simple way in Fig. 9.1.

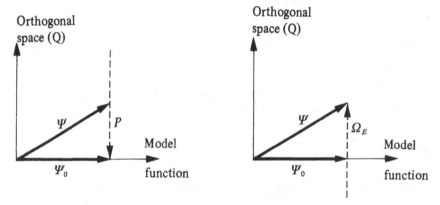

Fig. 9.1. Simple illustration of the wave operator (Ω_E) and the projection operator (P). P projects out of any function the component along the model function Ψ_0, and the wave operator generates the exact wave function by operating on Ψ_0

We can also derive an implicit equation for the wave operator. From (9.19, 26) we get

$$\Omega_E \Psi_0 = \Psi_0 + T_E V \Omega_E \Psi_0 ,\qquad(9.28)$$

which is consistent with

$$\Omega_E = 1 + T_E V \Omega_E ,\qquad(9.29)$$

This relation can be used to generate the expansion (9.25), but it is valid also when such an expansion does not converge. .

9.2.4 Determination of the Energy

As mentioned, the wave operator introduced above depends on the exact energy of the state, which we do not know *a priori*. Therefore, before we can use this operator, we have to determine the energy—to the desired degree of accuracy.

In the intermediate normalization (9.11) we get immediately from the Schrödinger equation (9.1)

$$E = \langle \Psi_0 | H | \Psi \rangle \tag{9.30}$$

or, using the partitioning (9.2)

$$E = \langle \Psi_0 | H_0 | \Psi \rangle + \langle \Psi_0 | V | \Psi \rangle . \tag{9.31}$$

By operating with H_0 to the left, the first term reduces to

$$\langle \Psi_0 | H_0 | \Psi \rangle = E_0 \langle \Psi_0 | \Psi \rangle = E_0 , \tag{9.32}$$

where E_0 is the eigenvalue of the model function

$$H_0 \Psi_0 = E_0 \Psi_0 . \tag{9.33}$$

This represents the zeroth-order energy. Using the definition of the wave operator (9.26), we then get from (9.31)

$$E = E_0 + \langle \Psi_0 | V \Omega_E | \Psi_0 \rangle . \tag{9.34}$$

The operator in the last term,

$$W_E = V \Omega_E , \tag{9.35}$$

can be considered as an "*effective interaction*". It has the same effect on the model function as has the perturbation V on the exact wave function,

$$W_E \Psi_0 = V \Psi , \tag{9.36}$$

as can be seen directly from the relation (9.26). The exact energy can then be written

$$E = E_0 + \langle \Psi_0 | W_E | \Psi_0 \rangle . \tag{9.37}$$

We can use the expansion (9.25) of the wave operator also to get an expansion of the effective interaction (9.35)

$$W_E = V + VT_E V + VT_E VT_E V + \cdots . \tag{9.38}$$

With (9.37) this leads to the energy expansion

$$E = E_0 + \langle \Psi_0 | V + VT_E V + VT_E VT_E V + \cdots | \Psi_0 \rangle \tag{9.39}$$

or, using the explicit form (9.16) of the resolvent operator,

$$E = E_0 + \langle \Psi_0| V + V\frac{Q}{E - H_0}V + V\frac{Q}{E - H_0}V\frac{Q}{E - H_0}V + \cdots |\Psi_0\rangle \ .$$

(9.40)

Equation (9.40) is the *Brillouin-Wigner energy expansion*. The effective interaction (9.38) appearing in the last term is equivalent to the *transition matrix* or *T matrix*, used in scattering theory. The expansion (9.38) is then known as the *Born series* [*Francis* and *Watson* 1953; *Messiah* 1961, p. 835; *Roman* 1965, p. 302]. In the Brueckner theory of nuclear matter the effective interaction is usually referred to as the *reaction matrix* or the *K matrix* [*Brueckner* and *Levinson* 1955; *Thouless* 1972, p. 76; *Brown* 1967, Chap. 11].

The energy expansion (9.40) can be written in the form

$$E = E_0 + E^{(1)} + E^{(2)} + E^{(3)} + \cdots ,$$

(9.41)

where $E^{(n)}$ is the general nth order energy contribution, containing n interactions of the perturbation V,

$$E^{(n)} = \langle \Psi_0| V\left(\frac{Q}{E - H_0}V\right)^{n-1}|\Psi_0\rangle \ .$$

(9.42)

This gives, in particular, the following well-known expressions for the low-order terms:

$$E^{(1)} = \langle \Psi_0| V |\Psi_0\rangle$$

(9.43a)

$$E^{(2)} = \sum_{\beta \neq \alpha} \frac{\langle \Psi_0| V |\phi^\beta\rangle\langle\phi^\beta| V |\Psi_0\rangle}{E - E_0^\beta}$$

(9.44b)

$$E^{(3)} = \sum_{\beta, \gamma \neq \alpha} \frac{\langle \Psi_0| V |\phi^\beta\rangle\langle\phi^\beta| V |\phi^\gamma\rangle\langle\phi^\gamma| V |\Psi_0\rangle}{(E - E_0^\beta)(E - E_0^\gamma)} ,$$

(9.43c)

etc., using the spectral resolution (9.17) of the resolvent. We note, in particular, that the sum of the zeroth- and first-order energy contributions is

$$E_0 + E^{(1)} = \langle \Psi_0| H_0 + V |\Psi_0\rangle = \langle \Psi_0| H |\Psi_0\rangle = \langle E\rangle ,$$

(9.44)

which is the expectation value of the energy in the state represented by the model function. This is the quantity which is minimized in the Hartree-Fock procedure. It is then known as the Hartree-Fock energy, and the remaining part of the energy as the *correlation energy*.

Since we do not know the exact energy E a priori, we cannot evaluate the energy contributions order by order. Instead, we have a self-consistency problem to solve. For that reason we may replace the energy in the denominators by a parameter ε and define a function

$$F(\varepsilon) = E_0 + E^{(1)}(\varepsilon) + E^{(2)}(\varepsilon) + \cdots . \tag{9.45}$$

The true energy E is then given by the solution of the equation

$$F(\varepsilon) = \varepsilon . \tag{9.46}$$

This equation can be solved iteratively by means of successive approximations, using, for instance, the Newton-Raphson procedure [*Arfken* 1970, p. 265; *Löwdin* 1966].

9.2.5 The Feshbach Operator

It is also possible to find an explicit "effective operator", which yields the exact energy when operating on the model function, as shown by *Feshbach* and others [*Feshbach* 1958a, b, 1962; *Löwdin* 1962, 1966; *Messiah* 1961, p. 995]. For this purpose we write the Schrödinger equation in the form

$$H(P + Q)\Psi = E\Psi$$

and operate from the left with P and Q, respectively. We then get

$$H_{PP}\Psi_P + H_{PQ}\Psi_Q = E\Psi_P$$

$$H_{QP}\Psi_P + H_{QQ}\Psi_Q = E\Psi_Q ,$$

using the shorthand notations

$$H_{PP} = PHP, \qquad H_{PQ} = PHQ, \ldots$$

and

$$\Psi_P = P\Psi, \ \Psi_Q = Q\Psi .$$

These equations can also be written in matrix form,

$$\begin{pmatrix} H_{PP} & H_{PQ} \\ H_{QP} & H_{QQ} \end{pmatrix} \begin{pmatrix} \Psi_P \\ \Psi_Q \end{pmatrix} = E \begin{pmatrix} \Psi_P \\ \Psi_Q \end{pmatrix} . \tag{9.47}$$

We can now eliminate Ψ_Q and obtain an equation for Ψ_P

$$\Psi_Q = (E - H_{QQ})^{-1} H_{QP} \Psi_P$$

$$[H_{PP} + H_{PQ}(E - H_{QQ})^{-1} H_{QP}] \Psi_P = E\Psi_P . \tag{9.48}$$

The operator in the last equation can be regarded as an "effective Hamiltonian", since it yields the exact energy when operating on the model function. Again, this operator depends explicitly upon the energy. In connection with the Rayleigh-Schrödinger perturbation theory in Sect. 9.4 we shall consider an equivalent operator without such an energy dependence. The operator in (9.48) is known in nuclear physics as the *Feshbach operator*.

The Brillouin-Wigner form of perturbation theory is formally very simple. As we have seen, however, the operators depend on the exact energy of the state considered. This requires a self-consistency procedure, and limits the application to one energy level at a time. There are also other more fundamental difficulties with the Brillouin-Wigner theory, particularly for extended systems, as we shall demonstrate later. The Rayleigh-Schrödinger perturbation theory, to be described in Sect. 9.4, does not have these shortcomings, and it is therefore a more suitable basis for many-body calculations than the Brillouin-Wigner form of the theory.

9.3 The Green's Function

9.3.1 The Green's-Function Operator or the Propagator

In the previous section we have developed the Brillouin-Wigner perturbation theory, using the resolvent-operator technique. Before continuing with the energy-independent Rayleigh-Schrödinger formalism we consider a closely related approach, namely the time-independent Green's-function technique.[3]

Consider the *inhomogeneous* equation

$$(E - H)\Psi(x) = \phi(x), \tag{9.49}$$

where H is the Hamiltonian of the system and x represents all the coordinates. For $\phi(x) \equiv 0$, this equation becomes identical to the (homogeneous) Schrödinger equation. It is tempting to try to solve the inhomogeneous equation (9.49) by means of the inverse of the operator on the left-hand side, $(E - H)^{-1}$, but this operator is singular and undefined whenever E equals an eigenvalue of H. This can be seen from the spectral resolution (9.17) of the operator. The eigenvalues of $(E - H)^{-1}$ are the inverses of the eigenvalues of $(E - H)$, and they approach $+$ or $-\infty$, when E approaches an eigenvalue of H from above or below.

We can define a general resolvent operator [*Kato* 1949; *van Hove* 1955, 1956; *Hugenholtz* 1957; *Roman* 1965, p. 392] by

---

[3] In this book we are concerned only with stationary problems, and for that reason we do not consider the time-dependent form of the Green's-function technique, which is nowadays frequently used in many-body theory. Interested readers are referred to existing books on the subject, for instance [*Fetter* and *Waleĉka* 1971].

$$R(z) = (z - H)^{-1} , \tag{9.50}$$

where z is a complex number. This operator has *poles* on the real axis corresponding to the *discrete* eigenvalues of H and a *cut* corresponding to the *continuum*. It is analytic at all other points of the complex plane. We shall not be particularly concerned here with the mathematical properties of the resolvent operator. Interested readers are referred to the book by *Kato* [1966] for further details.

A special form of the resolvent above is

$$G^{+}(\varepsilon) = (\varepsilon - H + i\eta)^{-1} , \tag{9.51}$$

where ε and η are real numbers. This is well behaved for all $\eta \neq 0$. In the following we shall assume that η *is a small, positive number*. By definition, $G^{+}(\varepsilon)$ satisfies the equation

$$(\varepsilon - H + i\eta)G^{+}(\varepsilon) = 1 .$$

This operator is sometimes called the *true Green's-function operator* or the *true propagator* [*Roman* 1965, p. 382]. By means of this operator the solution of the equation

$$(\varepsilon - H + i\eta)\Psi'(x) = \phi(x)$$

can be written

$$\Psi'(x) = G^{+}(\varepsilon)\phi(x) = (\varepsilon - H + i\eta)^{-1}\phi(x) .$$

As η approaches zero for $\varepsilon = E$ this function approaches the solution of (9.49),

$$\Psi(x) = \lim_{\eta \to 0}(E - H + i\eta)^{-1}\phi(x) . \tag{9.52}$$

In order to make contact with the resolvent T_E (9.15, 16) used above, we introduce the "zeroth-order" Green's-function operator or propagator

$$G_0^{+}(\varepsilon) = (\varepsilon - H_0 + i\eta)^{-1} , \tag{9.53}$$

which is obtained from $G^{+}(\varepsilon)$ by replacing the full Hamiltonian H by the model Hamiltonian H_0. By operating on the identity

$$(\varepsilon - H_0 + i\eta) = (\varepsilon - H + i\eta) + V$$

with $(\varepsilon - H_0 + i\eta)^{-1}$ from the left and with $(\varepsilon - H + i\eta)^{-1}$ from the right, we get

$$(\varepsilon - H + i\eta)^{-1} = (\varepsilon - H_0 + i\eta)^{-1} + (\varepsilon - H_0 + i\eta)^{-1}V(\varepsilon - H + i\eta)^{-1}$$

or, using the definitions (9.51, 53)

$$G^+(\varepsilon) = G_0^+(\varepsilon) + G_0^+(\varepsilon) V G^+(\varepsilon) .$$ (9.54)

This equation, which relates the true and zeroth-order propagators, is a general form of the *Dyson equation* [*Dyson* 1949; *Fetter* and *Walečka* 1971, p. 105; *Roman* 1965, pp. 384, 481]. It is closely related to the wave-operator relation (9.29), and it can be used to generate the same kind of expansion

$$G^+(\varepsilon) = G_0^+(\varepsilon) + G_0^+(\varepsilon) V G_0^+(\varepsilon) + G_0^+(\varepsilon) V G_0^+(\varepsilon) V G_0^+(\varepsilon) + \cdots .$$ (9.55)

The spectral resolution of $G_0^+(\varepsilon)$ is in analogy with (9.17)

$$G_0^+(\varepsilon) = \sum_\beta \frac{|\phi^\beta\rangle\langle\phi^\beta|}{\varepsilon - E_0^\beta + i\eta} ,$$ (9.56)

where $\{\phi^\beta\}$ is a set of orthonormal eigenfunctions of H_0,

$$H_0 \phi^\beta = E_0^\beta \phi^\beta; \qquad \langle\phi^\beta|\phi^\gamma\rangle = \delta(\beta,\gamma) .$$

By comparing (9.56) and (9.17), we see that the resolvent T_E is identical to $G_0^+(\varepsilon)$ for $\varepsilon = E$ apart from the fact that T_E contains only functions in the orthogonal space. This gives the relation

$$T_E = Q G_0^+(E) .$$ (9.57)

Since the denominator of T_E can never vanish, no imaginary part is needed there. The resolvent (9.57), operating only in the orthogonal space, is sometimes called the *reduced resolvent* [*Löwdin* 1968, 1969].

9.3.2 The Green's Function in Coordinate Space

The resolvents $G^+(\varepsilon)$ and $G_0^+(\varepsilon)$ given above are *operators*. We can obtain the corresponding *functions* by considering the coordinate representation of these operators (Appendix B). The spectral resolution (9.56) of the operator $G_0^+(\varepsilon)$ then corresponds to the function

$$G_0^+(x, x', \varepsilon) = \langle x | G_0^+(\varepsilon) | x'\rangle = \sum_\beta \frac{\phi^\beta(x)\phi^{\beta*}(x')}{\varepsilon - E_0^\beta + i\eta} .$$ (9.58)

This function satisfies the equation

$$(\varepsilon - H_0 + i\eta) G_0^+(x, x', \varepsilon) = \sum_\beta \phi^\beta(x)\phi^{\beta*}(x') = \delta(x - x') ,$$ (9.59)

where H_0 operates on the unprimed coordinates. In the last step we have used the closure property (B.7) of the eigenfunctions (9.5). This equation is precisely the ordinary definition of the Green's function for the operator $(\varepsilon - H_0 + i\eta)$ [*Merzbacher* 1961, pp. 221, 490]. This implies that the solution of the equation

$$(\varepsilon - H_0 + i\eta)\Psi_0(x) = \phi(x)$$

can be written

$$\Psi_0(x) = \int G_0^+(x, x', \varepsilon)\phi(x')dx' ,$$

which can be verified by operating from the left with $(\varepsilon - H_0 + i\eta)$ using (9.59). In a similar way we find that the coordinate representation of the true Green's function operator $G^+(\varepsilon)$ corresponds to the function

$$G^+(x, x', \varepsilon) = \langle x | G^+(\varepsilon) | x' \rangle = \sum_n \frac{\Psi^n(x)\Psi^{n*}(x')}{\varepsilon - E^n + i\eta} , \qquad (9.60)$$

where $\{\Psi^n\}$ is a set of orthonormal eigenfunctions of H,

$$H\Psi^n = E^n\Psi^n; \qquad \langle \Psi^n | \Psi^m \rangle = \delta(m,n) .$$

This function satisfies the equation

$$(\varepsilon - H + i\eta)G^+(x, x', \varepsilon) = \delta(x - x') \qquad (9.61)$$

and it is the Green's function for the operator $(\varepsilon - H + i\eta)$. Thus, we have seen that *the Green's function is the coordinate representation of the corresponding resolvent or propagator*. Therefore, the two concepts are completely equivalent, and they differ only in their mathematical form. Using (9.61), the solution of (9.49) becomes

$$\Psi(x) = \lim_{\eta \to 0} \int G^+(x, x', E)\phi(x')dx' , \qquad (9.62)$$

which is the analogue of the operator expression (9.52).

9.4 General Rayleigh-Schrödinger Perturbation Theory

We turn now to the Rayleigh-Schrödinger perturbation theory, which will form the basis for our following many-body treatment.

We have seen that the Brillouin-Wigner perturbation theory treated above requires a self-consistency procedure, since each term of the perturbation expansion depends on the initially unknown exact energy. This makes it necessary to treat one energy level at a time. If, on the other hand, we are primarily interested in the energy-level *splitting* of an open-shell system, then the application of the Brillouin-Wigner perturbation theory would be rather tedious and impractical. In such cases it is more convenient to have a perturbation theory which is applicable to a group of levels simultaneously. The Rayleigh-Schrödinger perturbation theory provides such an energy-independent procedure. In addition, this procedure has the important property that in each order the energy has the correct linear N dependence, where N is the number of particles of the system. The Brillouin-Wigner theory does not have this property, which is of particular importance when quantities like the dissociation energy are calculated, where systems of different size are compared. This makes the Rayleigh-Schrödinger theory more suitable than the Brillouin-Wigner theory as a basis for a general many-body treatment.

The Rayleigh-Schrödinger theory can be derived from the energy-dependent Brillouin-Wigner theory by expanding the energy in the denominator, as shown by *Brandow* [1966, 1967]. Here, we shall instead derive the Rayleigh-Schrödinger theory more directly by eliminating the energy dependence at an early stage.

9.4.1 The Model Space

In the treatment of the Brillouin-Wigner perturbation theory, we assumed for simplicity that the eigenstates of the model Hamiltonian were nondegenerate. In our treatment of the Rayleigh-Schrödinger theory, on the other hand, we shall relax this restriction.

As before, we assume that the eigenfunctions of the model Hamiltonian are known,

$$H_0 \phi^\alpha = E_0^\alpha \phi^\alpha .$$

If there are several independent eigenfunctions corresponding to the same eigenvalue, then these functions are not automatically orthogonal (see Appendix B). Such functions, however, can be orthogonalized, for instance, by means of a Schmidt procedure [*Merzbacher* 1961, p. 145]. We assume that this is done, so that we have a complete orthonormal set of eigenfunctions of H_0,

$$\langle \phi^\alpha | \phi^\beta \rangle = \delta(\alpha, \beta) .$$

If there is no risk of ambiguity, we shall for convenience often use the symbols $|\alpha\rangle$, $|\beta\rangle$, ... to denote the basis functions $|\phi^\alpha\rangle$, $|\phi^\beta\rangle$ .... Then the relations above take the simpler form

$$H_0 |\alpha\rangle = E_0^\alpha |\alpha\rangle ; \qquad \langle \alpha | \beta \rangle = \delta(\alpha, \beta) . \tag{9.63}$$

For a degenerate system we cannot always determine the zeroth-order or model functions beforehand. How the degenerate eigenfunctions are mixed in zeroth order depends on the perturbation. Therefore, we can, in general, only confine the model function to a certain subspace of the entire functional space. We call this subspace the *model space* or more simply the *P space*. It is defined by means of eigenfunctions of H_0, corresponding to one or several eigenvalues. The remaining part of the functional space is called the *orthogonal space* or the *Q space*. We shall always assume that eigenfunctions of H_0 corresponding to the same eigenvalue belong to the same subspace.

The projection operators for the model and orthogonal spaces are

$$P = \sum_{\alpha \in P} |\alpha\rangle\langle\alpha|$$

$$Q = \sum_{\beta \notin P} |\beta\rangle\langle\beta| \ .$$

(9.64)

The P operator projects out of any function the component that lies in the model space and the Q operator projects out the component in the orthogonal space. As before,

$$P + Q = 1,$$

and the relations (9.10) are also still valid.

In the central-field approximation, the model space represents all functions associated with one or several *configurations*. The reason for including several configurations in the model space is that it is possible in this way to *take into account strongly interacting configurations to all orders* and treat the weakly interacting ones by means of a low-order expansion. A larger model space is therefore expected to lead to a better zeroth-order approximation and hopefully to better convergence, but at the expense of higher complexity. The proper choice of the model space may therefore be of vital importance for the success of the perturbation calculation.

As an illustration we may consider the beryllium atom, which in the central-field model has the ground configuration $1s^2 2s^2$. This configuration contains a single state (^1S), so if the model space is confined to the ground configuration, the model function would be given by the central-field wave function of that state. However, the configuration $1s^2 2p^2$, which also contains a ^1S state, is expected to interact strongly with the ground state. This is partly due to the strong overlap between the $2s$ and $2p$ radial orbitals. (These orbitals are shown for the carbon atom in Fig. 7.1, and the situation is quite similar for all the elements of the second row of the periodic table.) As a consequence, the matrix element of the noncentral part of the Coulomb interaction, which appears in the mixing amplitude, is quite large. Furthermore, the $2s$ and $2p$ orbitals are "quasi-degenerate" (very close in energy), which makes the energy denominator in the perturbation expansion quite small. The mixing of the two configurations may then be so strong that this leads to convergence problems. Instead of limiting the model space to the ground configuration, one can then extend this space to include also the $1s^2 2p^2$ configuration. It would then contain 16 states, but only the two ^1S states can be mixed by the electrostatic interaction. Then there will be two eigenstates of the full Hamiltonian which have a ^1S character and which have their major parts within the model space,

$$\Psi = a|1s^22s^2\ {}^1S\rangle + b|1s^22p^2\ {}^1S\rangle + \cdots$$
$$\Psi' = a'|1s^22s^2\ {}^1S\rangle + b'|1s^22p^2\ {}^1S\rangle + \cdots.$$

The model function corresponding to the first of these states is then

$$\Psi_0 = P\Psi = a|1s^22s^2\ {}^1S\rangle + b|1s^22p^2\ {}^1S\rangle.$$

If Ψ is the ground state, we expect $|a|$ to be larger than $|b|$. Due to the strong interaction, however, b has an appreciable magnitude in this case. Only mixtures outside these configurations are determined by means of the perturbation expansion. In this case the method with extended model space also has the advantage of yielding simultaneously the energy of the upper ^1S state. This will be discussed further in connection with the open-shell formalism in Chap. 13.

9.4.2 The Generalized Bloch Equation

We consider now a general model space, spanned by the eigenfunctions of the model Hamiltonian corresponding to one or several eigenvalues—for instance, the $1s^22s^2$ and $1s^22p^2$ configurations of the beryllium atom just mentioned. If the dimensionality of the model space is d, there are normally d well-defined eigenstates Ψ^a of the full Hamiltonian, which have their major part within the model space. In the beryllium example, these are the *exact* states with the conventional central-field notations $1s^22s^2\ {}^1S$ and $1s^22p^2\ {}^1S,\ {}^1D,\ {}^3P$. The projections of these exact states onto the model space

$$\boxed{\Psi_0^a = P\Psi^a \quad (a = 1, 2, ..., d)} \tag{9.65}$$

represent the *model functions*. We shall assume that the model functions are linearly independent, so that they span the entire model space. This means that there is a one-to-one correspondence between the d exact solutions of the Schrödinger equation and the model functions [*Kuo* et al. 1971; *Schucan* and *Weidenmüller* 1972, 1973]. We can then define a *single wave operator* Ω, which transforms *all* the model states back into the corresponding exact states

$$\boxed{\Psi^a = \Omega\Psi_0^a \quad (a = 1, 2, ... d)}. \tag{9.66}$$

We observe that in the Rayleigh-Schrödinger theory the wave operator is the same for all states considered.

The model functions (9.65) are generally *nonorthogonal*, since they are projections of orthogonal functions on a smaller space. If they are linearly independent, we can always find "dual" functions Ψ'^a_0 in the model space, satisfying the "biorthogonality condition"

$$\langle \Psi'^b_0 | \Psi^a_0 \rangle = \delta(a, b) \,. \tag{9.67}$$

We can then express the projection operator for the model space as

$$P = \sum_{b=1}^{d} | \Psi^b_0 \rangle \langle \Psi'^b_0 | \,. \tag{9.68}$$

This operator evidently destroys any function in the Q space, and it leaves all functions in the P space unaffected,

$$P | \Psi^a_0 \rangle = \sum_{b=1}^{d} | \Psi^b_0 \rangle \langle \Psi'^b_0 | \Psi^a_0 \rangle = | \Psi^a_0 \rangle \,.$$

Hence, it is identical to the P operator.

We shall now investigate the properties of the wave operator and derive a recursive formula for its generation.

Operating with P from the left on (9.66), using (9.65), we get in Dirac notation

$$| \Psi^a_0 \rangle = P\Omega | \Psi^a_0 \rangle \qquad (a = 1, 2, \dots d) \,.$$

This gives immediately

$$\sum_{a=1}^{d} | \Psi^a_0 \rangle \langle \Psi'^a_0 | = P\Omega \sum_{a=1}^{d} | \Psi^a_0 \rangle \langle \Psi'^a_0 |$$

or with (9.68)

$$\boxed{P\Omega P = P} \,. \tag{9.69}$$

In the following we shall find it convenient to use also another operator, χ, defined by

$$\boxed{\Omega = 1 + \chi} \,. \tag{9.70}$$

Operating on the model function, this operator generates the projection of the full wave function in the Q space

$$\chi \Psi^a_0 = (\Omega - 1)\Psi^a_0 = \Psi^a - P\Psi^a = Q\Psi^a \,.$$

This is identical to the correlation function, mentioned before [see (9.13)], and therefore χ will be referred to as the *correlation operator*. From (9.69) and the definition (9.70) it follows that this operator has the properties

$$P\chi P = 0; \qquad Q\Omega P = Q\chi P = \chi P, \tag{9.71}$$

which is another way of expressing the result above.

It should be observed that the definition (9.66) determines the wave operator only when it operates to the right on the model space, that is, only the ΩP part is defined. The remaining ΩQ part never appears in the theory, and therefore we shall leave it undefined here. In later chapters, when we express the wave operator in second quantization, we shall find it convenient to use a form which is defined also in the Q space.

We shall now derive an equation for the wave operator defined by (9.66). Since this operator is the same for all states in the model space, it cannot depend on the exact energy, as did the corresponding Brillouin-Wigner operator. Therefore, in order to obtain the desired result, we have to eliminate the energy from the basic equations. We start from the Schrödinger equation

$$H\Psi^a = E^a\Psi^a,$$

which we rewrite as

$$(E^a - H_0)\Psi^a = V\Psi^a,$$

using the partitioning (9.2). We operate on this equation from the left first with P

$$(E^a - H_0)\Psi_0^a = PV\Psi^a$$

and then with Ω,

$$E^a\Psi^a - \Omega H_0\Psi_0^a = \Omega PV\Omega\Psi_0^a,$$

where we have used the relations (9.65, 66) together with the fact that P commutes with H_0. We can now eliminate E^a by subtracting this equation from the Schrödinger equation above, which gives

$$(\Omega H_0 - H_0\Omega)\Psi_0^a = (V\Omega - \Omega PV\Omega)\Psi_0^a.$$

This equation holds for all model functions ($a = 1, 2, ..., d$), and we can then write it in the operator form [*Lindgren* 1974, 1978; *Kvasnička* 1974, 1977a, b]

$$\boxed{[\Omega, H_0] P = V\Omega P - \Omega PV\Omega P}. \tag{9.72}$$

This is a generalization of an equation derived by *Bloch* [1958a] for a completely degenerate model space. The equation above, however, is valid also for a model space containing several unperturbed energies, for instance, several atomic configurations. This equation will form the basis for our many-body treatment, and we shall refer to it as the *generalized Bloch equation*.

With the identity $P + Q = 1$ the first term in (9.72) can be rewritten as

$$V\Omega P = PV\Omega P + QV\Omega P \,,$$

and using the definition of the correlation operator (9.70) the second term on the right-hand side of (9.72) can be written

$$-\Omega PV\Omega P = -PV\Omega P - \chi PV\Omega P \,.$$

We then see that we can give the generalized Bloch equation the following alternative form:

$$\boxed{[\Omega, H_0] P = QV\Omega P - \chi PV\Omega P} \,. \tag{9.73}$$

Obviously, we have

$$[\chi, H_0] = [\Omega, H_0] \,,$$

so we can replace Ω by χ in the commutators above.

The generalized Bloch equation (9.72, 73) is exact and completely equivalent to the Schrödinger equation for the states considered. In the form given here it is very suitable for generating a perturbation expansion, called the *Rayleigh-Schrödinger* expansion, as we shall now demonstrate.

The correlation operator generates the component of the wave function in the orthogonal space and must therefore have at least one perturbation. It can then be expanded as

$$\chi = \Omega^{(1)} + \Omega^{(2)} + \cdots, \tag{9.74a}$$

where $\Omega^{(n)}$ contains n interactions with the perturbation V. Using the definition (9.70), we then get the following expansion of the wave operator

$$\Omega = 1 + \Omega^{(1)} + \Omega^{(2)} + \cdots. \tag{9.74b}$$

Substituting this expansion into (9.73) and identifying terms of the same order in V, we obtain the following sequence:

$$[\Omega^{(1)}, H_0] P = QVP \tag{9.75a}$$

$$[\Omega^{(2)}, H_0] P = QV\Omega^{(1)}P - \dot{\Omega}^{(1)}PVP \tag{9.75b}$$

$$[\Omega^{(3)}, H_0] P = QV\Omega^{(2)}P - \Omega^{(2)}PVP - \Omega^{(1)}PV\Omega^{(1)}P \tag{9.75c}$$

which can be generalized to

$$[\Omega^{(n)}, H_0] P = QV\Omega^{(n-1)} P - \sum_{m=1}^{n-1} \Omega^{(n-m)} PV\Omega^{(m-1)} P \ . \tag{9.76}$$

This equation will be used to generate the linked-diagram expansion for closed- as well as open-shell systems in later chapters.

As an alternative to the order-by-order expansion demonstrated above, we can solve the exact equation (9.73) in an iterative way. Suppose that we know the wave operator to a certain approximation

$$\Omega^n = 1 + \chi^n \ . \tag{9.77}$$

Then we can get the next higher approximation by substituting this into the right-hand side of (9.73)

$$[\Omega^{n+1}, H_0] P = QV\Omega^n P - \chi^n PV\Omega^n P \ . \tag{9.78}$$

This is a recursive formula which can be applied repeatedly, starting with $\Omega^0 = 1$ ($\chi^0 = 0$). This leads to a sequence of approximations which is different from the order-by-order expansion above and it may have different convergence properties. Together with the second-quantization formalism to be discussed in Chap. 11, this iterative procedure can be used to treat one-, two-, ... particle effects in a self-consistent manner, as we shall demonstrate in later chapters.

In order to convert the Rayleigh-Schrödinger expansion (9.75) to matrix form, we use the eigenfunctions of H_0 (9.63) as basis functions. The commutator in (9.73) is then

$$\langle \beta | \Omega H_0 - H_0 \Omega | \alpha \rangle = (E_0^\alpha - E_0^\beta)\langle \beta | \Omega | \alpha \rangle \ , \tag{9.79}$$

after operating with H_0 to the right in the first term and to the left in the second. The commutators in (9.75) can be treated similarly. Assuming $|\alpha\rangle$ to be in the model space (P) and $|\beta\rangle$ in the orthogonal space (Q), the matrix elements of QVP in (9.75a) become

$$\langle \beta | QVP | \alpha \rangle = \langle \beta | V | \alpha \rangle \ .$$

Together with (9.79) this gives

$$\langle \beta | \Omega^{(1)} | \alpha \rangle = \frac{\langle \beta | V | \alpha \rangle}{E_0^\alpha - E_0^\beta} \ . \tag{9.80a}$$

Similarly, we get from (9.75b), using the completeness relation (B.9),

$$\langle \beta | \Omega^{(2)} | \alpha \rangle = \sum_{\gamma} \frac{\langle \beta | V | \gamma \rangle \langle \gamma | \Omega^{(1)} | \alpha \rangle}{E_0^\gamma - E_0^\beta} - \sum_{\alpha'} \frac{\langle \beta | \Omega^{(1)} | \alpha' \rangle \langle \alpha' | V | \alpha \rangle}{E_0^\alpha - E_0^\beta} \ ;$$

where α, α' (β, γ) are in the $P(Q)$ space. By inserting the expression (9.80a) for $\Omega^{(1)}$, this leads to

$$\langle\beta|\Omega^{(2)}|\alpha\rangle = \sum_{\gamma}\frac{\langle\beta|V|\gamma\rangle\langle\gamma|V|\alpha\rangle}{(E_0^{\alpha} - E_0^{\beta})(E_0^{\alpha} - E_0^{\gamma})} - \sum_{\alpha'}\frac{\langle\beta|V|\alpha'\rangle\langle\alpha'|V|\alpha\rangle}{(E_0^{\alpha} - E_0^{\beta})(E_0^{\alpha'} - E_0^{\beta})}. \quad (9.80\text{b})$$

These equations represent the first two terms of the general Rayleigh-Schrödinger expansion. It should be observed that no assumption has been made concerning the unperturbed energies of the model space. In Sect. 9.5 we shall consider the more familiar case, where the model space is completely degenerate.

9.4.3 The Effective Hamiltonian

We now have a procedure for generating the Rayleigh-Schrödinger wave operator, which yields the exact wave function (9.66)—to the same accuracy as we have determined the wave operator—when operating on the model function. However, in the general case we do not know the model function initially, as discussed in Sect. 9.4.1. If there are several states within the model space which can be mixed by the perturbation, then we have to determine this mixture before the wave operator can be applied. This can be done by constructing an effective Hamiltonian, as we shall now demonstrate.

By operating with P from the left on the Schrödinger equation

$$H\Omega\Psi_0^a = E^a\Psi^a$$

we get

$$PH\Omega\Psi_0^a = E^a\Psi_0^a. \quad (9.81)$$

It then follows that the operator

$$H_{\text{eff}} = PH\Omega P, \quad (9.82)$$

which we call the *effective Hamiltonian* in Rayleigh-Schrödinger perturbation theory, operates entirely within the model space and satisfies the eigenvalue equation

$$\boxed{H_{\text{eff}}\Psi_0^a = E^a\Psi_0^a}. \quad (9.83)$$

This implies that

> the eigenvectors of the effective Hamiltonian represent the model functions and the eigenvalues are the exact energies of the corresponding true states.

Here we see one of the fundamental differences between the Brillouin-Wigner and the Rayleigh-Schrödinger formalisms. In the former case there is one effective Hamiltonian for each energy, while in the latter case a single operator yields all the model states and corresponding energies. This is a consequence of the fact that the effective Hamiltonian is energy dependent in the Brillouin-Wigner formalism but not in the Rayleigh-Schrödinger scheme.

The effective Hamiltonian (9.82) can also be written

$$H_{eff} = P(H_0 + V)\Omega P = PH_0 P + PV\Omega P , \qquad (9.84a)$$

using (9.69) and the fact that P commutes with H_0. Together with the definition $\Omega = 1 + \chi$, this operator can further be expressed as

$$H_{eff} = PHP + PV\chi P . \qquad (9.84b)$$

To construct the effective Hamiltonian and to solve the eigenvalue equation (9.83) are the basic problems in our approach to the many-body problem, and we shall be much concerned with that in the following. In principle, this problem can be treated in a brute-force manner by constructing large matrix equations as indicated above. Instead of doing so, however, we shall develop in the next few chapters a more effective formalism which is particularly suited to this kind of problem.

As mentioned, the treatment here is quite general. In order to establish its relation with the more familiar form of Rayleigh-Schrödinger perturbation theory, however, we shall consider also the special case where the model space is completely degenerate.

9.5 The Rayleigh-Schrödinger Expansion for a Degenerate Model Space

The general procedure described above can be simplified, if we assume that all the states of the model space correspond to the same eigenvalue (E_0) of the model Hamiltonian

$$H_0 \Psi_0^a = E_0 \Psi_0^a \quad (a = 1, 2, ..., d) . \qquad (9.85)$$

The left-hand side of the generalized Bloch equation (9.72) can then be replaced by

$$[\Omega, H_0]P = (\Omega H_0 - H_0\Omega)P = (E_0 - H_0)\Omega P , \qquad (9.86)$$

which gives

$$(E_0 - H_0)\Omega P = V\Omega P - \Omega P V\Omega P. \tag{9.87}$$

This is equivalent to the original Bloch equation [*Bloch* 1958a]. Similarly, (9.73) leads to

$$\boxed{(E_0 - H_0)\Omega P = QV\Omega P - \chi P V\Omega P}. \tag{9.88}$$

We can now treat these operator equations in the same way as we did in the Brillouin-Wigner theory. Equation (9.88) can be solved by means of a *resolvent* R, defined by

$$R(E_0 - H_0) = Q, \qquad RQ = R. \tag{9.89}$$

By operating from the left with R on (9.88), we get

$$Q\Omega P = R(V\Omega P - \chi P V\Omega P). \tag{9.90}$$

We then substitute into this equation the previous expansion (9.74)

$$\Omega = 1 + \chi = 1 + \Omega^{(1)} + \Omega^{(2)} + \cdots \tag{9.91}$$

and equate terms of the same order, which leads to

$$\begin{aligned}
\Omega^{(1)}P &= RVP \\
\Omega^{(2)}P &= R(V\Omega^{(1)}P - \Omega^{(1)}PVP) \\
\Omega^{(3)}P &= R(V\Omega^{(2)}P - \Omega^{(2)}PVP - \Omega^{(1)}PV\Omega^{(1)}P).
\end{aligned} \tag{9.92}$$

We may also obtain an explicit form of this expansion by successively substituting the expression for the terms of lower order into the higher order ones. In this way we obtain

$$\begin{aligned}
\Omega^{(1)}P &= RVP \\
\Omega^{(2)}P &= RVRVP - R^2VPVP \\
\Omega^{(3)}P &= RVRVRVP - RVR^2VPVP - R^2VRVPVP \\
&\quad + R^3VPVPVP - R^2VPVRVP
\end{aligned} \tag{9.93}$$

The resolvent (9.89) can be written in a form analogous to (9.16) as

$$R = \frac{Q}{E_0 - H_0}, \tag{9.94}$$

and it has the spectral resolution

$$R = \sum_{\beta \notin P} \frac{|\beta\rangle\langle\beta|}{E_0 - E_0^\beta}. \tag{9.95}$$

This gives a matrix form analogous to (9.80), the only difference being that all E_0^α are replaced by E_0, as a consequence of the fact that the model space is degenerate.

In order to obtain the wave functions from the wave operator, we have, as before, to know the model functions. When a small model space is used, however, it often happens that these functions are unambiguously determined by the symmetry of the state. This is the case, for instance, if there is only one term with a particular LS symmetry in the model space, since different LS states do not mix. Then the wave functions can be expanded directly, without first constructing the wave operator. The first few terms are then

$$\Psi^{(1)} = \Omega^{(1)}\Psi_0^a = \sum_\beta \frac{|\beta\rangle\langle\beta|V|\Psi_0^a\rangle}{E_0 - E_0^\beta}$$

$$\Psi^{(2)} = \Omega^{(2)}\Psi_0^a = \sum_{\beta\gamma} \frac{|\beta\rangle\langle\beta|V|\gamma\rangle\langle\gamma|V|\Psi_0^a\rangle}{(E_0 - E_0^\beta)(E_0 - E_0^\gamma)}$$

$$\qquad - \sum_{\alpha\beta} \frac{|\beta\rangle\langle\beta|V|\alpha\rangle\langle\alpha|V|\Psi_0^a\rangle}{(E_0 - E_0^\beta)^2}. \tag{9.96}$$

As before, α is in the model (P) and β, γ in the orthogonal space (Q). Note that the first-order wave function has the same form as the corresponding Brillouin-Wigner expression (9.23)—apart from the energy in the denominator. In the second-order expression, on the other hand, the Rayleigh-Schrödinger expansion contains a term which has no counterpart in the Brillouin-Wigner scheme. This additional term compensates partly for the difference in the first order.

If, on the other hand, the model functions are not known initially, it is not possible to expand the wave functions directly as above. Instead, the general procedure, described before has to be used, where the wave operator is evaluated and the effective Hamiltonian constructed.

The effective Hamiltonian (9.84) becomes for a degenerate model space, using (9.93),

$$H_{\text{eff}}^{(0)} = PH_0P$$

$$H_{\text{eff}}^{(1)} = PVP \tag{9.97}$$

$$H_{\text{eff}}^{(2)} = PV\Omega^{(1)}P = PVRVP$$

$$H_{\text{eff}}^{(3)} = PV\Omega^{(2)}P = PVRVRVP - PVR^2VPVP.$$

In general, the corresponding energies are obtained by diagonalizing this operator in the model space. If the model functions are known, then the energies can be obtained directly by means of (9.83),

$$E^a = \langle \Psi_0^a | H_{\text{eff}} | \Psi_0^a \rangle , \tag{9.98}$$

which leads to the well-known lowest-order results

$$E^{a(1)} = \langle \Psi_0^a | H_{\text{eff}}^{(1)} | \Psi_0^a \rangle = \langle \Psi_0^a | V | \Psi_0^a \rangle$$

$$E^{a(2)} = \langle \Psi_0^a | H_{\text{eff}}^{(2)} | \Psi_0^a \rangle = \sum_{\beta \notin P} \frac{\langle \Psi_0^a | V | \beta \rangle \langle \beta | V | \Psi_0^a \rangle}{E_0 - E_0^\beta} . \tag{9.99}$$

The first-order energy is the same as in the Brillouin-Wigner case (9.43a), which means that also in the present case the sum of the zeroth- and first-order energy contributions is equal to the expectation value of the full Hamiltonian in the unperturbed model state,

$$E_0 + E^{(1)} = \langle \Psi_0^a | H_0 + V | \Psi_0^a \rangle = \langle \Psi_0^a | H | \Psi_0^a \rangle . \tag{9.100}$$

Problem 9.1. Derive a matrix representation of the third-order energy contribution in Rayleigh-Schrödinger perturbation theory for a degenerate model space.

10. First-Order Perturbation for Closed-Shell Atoms

We shall now apply the perturbation theory, developed in the previous chapter, to determine the first-order wave function and the first-order energy of a closed-shell atom. This will provide a simple application of the formalism and serve as an introduction to the graphical procedure which will be extensively used in the following chapters.

We begin by dividing the atomic Hamiltonian (5.1)

$$H = -\frac{1}{2}\sum_{i=1}^{N} \nabla_i^2 - \sum_{i=1}^{N}\frac{Z}{r_i} + \sum_{i<j}^{N}\frac{1}{r_{ij}} \tag{10.1}$$

into a central-field part H_0 and a perturbation V

$$H = H_0 + V, \tag{10.2}$$

as discussed in Sect. 9.2. Here, we have neglected the magnetic interactions. H_0 describes N electrons moving in the field of the nucleus and an average central field due to the electrons,

$$H_0 = \sum_{i=1}^{N}\left[-\frac{1}{2}\nabla_i^2 - \frac{Z}{r_i} + u(r_i)\right] = \sum_{i=1}^{N} h_0(i). \tag{10.3}$$

The perturbation is then

$$V = H - H_0 = \sum_{i<j}^{N}\frac{1}{r_{ij}} - \sum_{i=1}^{N} u(r_i). \tag{10.4}$$

The wave function of a single electron moving in a central-field is of the form (6.5)

$$\varphi = \frac{P_{nl}(r)}{r} Y_m^l(\theta, \phi)\chi_{m_s}, \tag{10.5}$$

where the radial function $P_{nl}(r)$ satisfies the equation (6.7)

$$\left[-\frac{1}{2}\frac{d^2}{dr^2} + \frac{l(l+1)}{2r^2} - \frac{Z}{r} + u(r)\right]P(r) = \varepsilon P(r). \tag{10.6}$$

The eigenfunctions of H_0 may then be formed by constructing determinants of these single-electron functions,

$$|\alpha\rangle = | \{\varphi_i\varphi_j\varphi_k \ldots\} \rangle \tag{10.7}$$

in the way described in Sect. 5.2.

In the central-field approximation, a closed-shell state is represented by a single determinant

$$\Psi_0 = |\alpha\rangle . \tag{10.8}$$

This is an eigenfunction of the central-field Hamiltonian (10.3)

$$H_0|\alpha\rangle = E_0|\alpha\rangle$$

with the eigenvalue

$$E_0 = \sum_{i=1}^{N} \varepsilon_i . \tag{10.9}$$

The one function $|\alpha\rangle$ defines the model space (P) in this case, and all of the other possible determinants $|\beta\rangle$ span the orthogonal space (Q).

10.1 The First-Order Wave Function

According to (9.96), the first-order correction to the wave function $|\alpha\rangle$ is

$$\Psi^{(1)} = \sum_{\beta \neq \alpha} \frac{|\beta\rangle\langle\beta| V |\alpha\rangle}{E_0 - E_0^\beta} . \tag{10.10}$$

The perturbation V (10.4) has a one-electron part

$$-U = - \sum_{i=1}^{N} u(r_i) \tag{10.11a}$$

and a two-electron Coulomb part

$$C = \sum_{i<j}^{N} g_{ij} = \sum_{i<j}^{N} r_{ij}^{-1} . \tag{10.11b}$$

We consider now the effect of these two parts of the perturbation on the first-order wave function (10.10).

The one-body part of the perturbation gives the following contribution:

$$\Psi_U^{(1)} = \sum_{\beta \neq \alpha} \frac{|\beta\rangle\langle\beta| -U |\alpha\rangle}{E_0 - E_0^\beta} . \tag{10.12}$$

The unperturbed energy difference $E_0 - E_0^\beta$, which occurs in the denominator, is the difference between the sum of the single-electron energies of the orbitals occupied in $|\alpha\rangle$ and in $|\beta\rangle$, respectively. A one-body operator only joins determinant states which do not differ by more than a single orbital, as can be seen from (5.23, 25). So the states $|\beta\rangle$ which contribute to the summation must be single-substitution states of the kind $|\alpha_a^r\rangle$ (in the notation of Sect. 5.3.3). Equation (10.12) then becomes

$$\Psi_U^{(1)} = \sum_{ar} \frac{|\alpha_a^r\rangle\langle\alpha_a^r| - U|\alpha\rangle}{\varepsilon_a - \varepsilon_r}. \tag{10.13}$$

Here the orbital a is occupied in $|\alpha\rangle$ and r must be unoccupied (virtual).

We have here adopted the following notations, which will be used with only slight modifications throughout the book;

1) $a, b, c, \ldots$ for orbitals *occupied* in $|\alpha\rangle$, or *core* orbitals
2) $r, s, t, \ldots$ for orbitals *unoccupied* in $|\alpha\rangle$, or *virtual* orbitals
3) $i, j, k, \ldots$ for *unspecified* orbitals.

Using (5.23), the first-order wave function (10.13) can be written

$$\Psi_U^{(1)} = \sum_{ar} \frac{|\alpha_a^r\rangle\langle r| - u|a\rangle}{\varepsilon_a - \varepsilon_r}. \tag{10.14}$$

We shall represent each term in this summation by the same type of diagram as used in Chap. 5 to illustrate the interaction matrix element,

$$= \frac{|\alpha_a^r\rangle\langle r| - u|a\rangle}{\varepsilon_a - \varepsilon_r}. \tag{10.15}$$

This diagram describes a process where a single electron is excited from a core state by the interaction $(-u)$.

The following principles are used to draw this kind of diagram:

a) *the interaction is denoted by a dotted, horizontal line and the electron orbitals involved in that interaction by solid, vertical lines, connected with the interaction line to a vertex,*
b) *a core orbital is represented by a line directed* downwards *(hole line) and a virtual orbital by a line directed* upwards *(particle line),*
c) *the orbitals belonging to the* initial *state (to the right in the matrix element) have their arrows pointing toward the interaction vertex, those of the* final *state* away *from the vertex.*

This type of diagram, which is a modification of the original Feynman diagrams, was introduced by *Goldstone* [1957] in his treatment of the Brueckner perturbation theory [*Brueckner* 1955]. These diagrams are currently referred to as

Goldstone diagrams. We shall use them extensively in our treatment, and we shall later give general rules for their evaluation.

We now turn to the two-body part C of the perturbation (10.11b) and we shall find a similar representation of its contribution to the first-order wave function. According to (5.24, 26), a two-body operator can join determinantal states which differ by one or two single-electron orbitals. So in this case (10.10) becomes

$$\Psi_C^{(1)} = \sum_{ar} \frac{|\alpha_a^r\rangle\langle\alpha_a^r|C|\alpha\rangle}{\varepsilon_a - \varepsilon_r} + \frac{1}{4}\sum_{ab,rs} \frac{|\alpha_{ab}^{rs}\rangle\langle\alpha_{ab}^{rs}|C|\alpha\rangle}{\varepsilon_a + \varepsilon_b - \varepsilon_r - \varepsilon_s}.$$

In the second summation, each pair of core orbitals (a, b) and each pair of virtual orbitals (r, s) occur twice. The factor of $1/4$ ensures that each distinct excitation, which depends only on the *pairs* involved, is included only once. Using (5.24, 26), this equation may be written out explicitly in terms of two-particle matrix elements,

$$\Psi_C^{(1)} = \sum_{ar} \frac{|\alpha_a^r\rangle}{\varepsilon_a - \varepsilon_r}\sum_b (\langle rb|r_{12}^{-1}|ab\rangle - \langle br|r_{12}^{-1}|ab\rangle)$$

$$+ \frac{1}{4}\sum_{ab,rs}\frac{|\alpha_{ab}^{rs}\rangle}{\varepsilon_a + \varepsilon_b - \varepsilon_r - \varepsilon_s}(\langle rs|r_{12}^{-1}|ab\rangle - \langle sr|r_{12}^{-1}|ab\rangle).$$

The last term in the second summation may be shown to be equal to the first term by interchanging r and s in the sum and using the fact that determinantal states are antisymmetric with respect to an interchange of two orbitals. In this way we obtain the following expression for the two-body part of the first-order wave function;

$$\Psi_C^{(1)} = \sum_{ar} \frac{|\alpha_a^r\rangle}{\varepsilon_a - \varepsilon_r}\sum_b (\langle rb|r_{12}^{-1}|ab\rangle - \langle br|r_{12}^{-1}|ab\rangle)$$

$$+ \frac{1}{2}\sum_{abrs}\frac{|\alpha_{ab}^{rs}\rangle\langle rs|r_{12}^{-1}|ab\rangle}{\varepsilon_a + \varepsilon_b - \varepsilon_r - \varepsilon_s}. \tag{10.16}$$

In accordance with the Goldstone rules given above, each of the terms in these summations may be represented by the diagrams shown below,

$$= \frac{|\alpha_a^r\rangle\langle rb|r_{12}^{-1}|ab\rangle}{\varepsilon_a - \varepsilon_r} \tag{10.17a}$$

$$= -\frac{|\alpha_a^r\rangle\langle br|r_{12}^{-1}|ab\rangle}{\varepsilon_a - \varepsilon_r} \tag{10.17b}$$

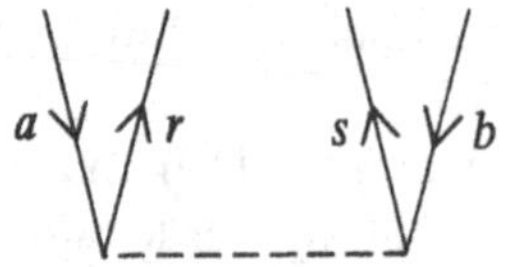

$$= \frac{1}{2} \frac{|\alpha_{ab}^{rs}\rangle \langle rs| r_{12}^{-1}|ab\rangle}{\varepsilon_a + \varepsilon_b - \varepsilon_r - \varepsilon_s} . \qquad (10.17c)$$

The dotted lines of these diagrams correspond to the Coulomb interaction between the electrons, and the orbitals involved in the interaction are as before represented by solid vertical lines. Since the Coulomb interaction is a two-body effect there are *two vertices*, joined together by the interaction line. Note that these diagrams distinguish between the direct and the exchange part of the Coulomb interaction. There exist other representations where this is not the case [*Hugenholtz* 1957], which can be used in order to reduce the number of diagrams.

The term (10.17a) of the perturbation expansion may be obtained from the term (10.17c) by setting $s = b$. This is represented diagrammatically by joining together the lines of that orbital. A similar diagram is obtained by setting instead $r = a$. These two diagrams are said to be *topologically equivalent*. By including only *distinct* diagrams (which are *not* topologically equivalent), we can discard the factor of 1/2, as done in the expression for the diagrams (10.17a). In the same way the exchange diagram (10.17b) can be obtained by setting either $a = s$ or $b = r$.

In the single-excitation part of the first-order wave function (10.16), there is a sum over the core orbital b. We shall adopt the convention that *there is always a summation over internal lines*. For instance,

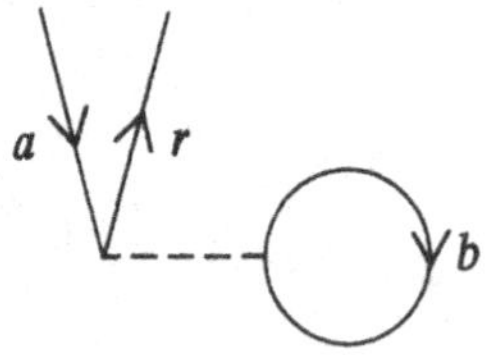

$$= \frac{|\alpha_a^r\rangle}{\varepsilon_a - \varepsilon_r} \sum_b^{core} \langle rb| r_{12}^{-1}|ab\rangle . \qquad (10.18)$$

Using this convention we can summarize the contributions to the first-order wave function, which are due to the one-body and the two-body parts of the perturbation, in the way shown in Fig. 10.1. Here, it is understood that there is a summation also over the external lines (with free ends).

$$\psi^{(1)} =$$

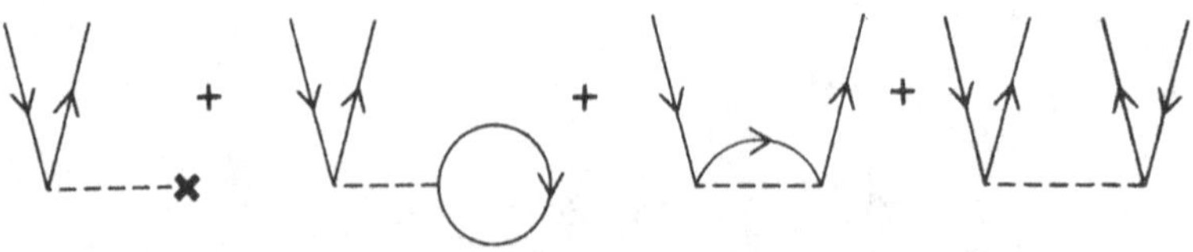

Fig. 10.1 Graphical representation of the first-order wave function for a closed-shell system

The first three diagrams in the figure represent *single* excitations, while the last diagram represents *double* excitations. It is often convenient to treat the former diagrams together, and we shall use a special symbol for them, defined in Fig. 10.2.

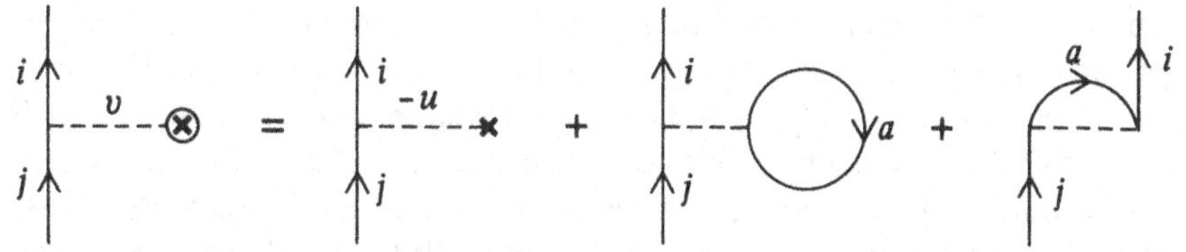

Fig. 10.2. Definition of the "effective" potential

The new interaction v represents an "effective" potential due to the single-particle potential $-u$ in the model Hamiltonian H_0 and the effective one-body part of the interaction with the electrons in the closed shells. The mathematical definition of this potential is

$$\langle i|v|j \rangle = \langle i|-u|j \rangle + \sum_a^{\text{core}} (\langle ia|r_{12}^{-1}|ja \rangle - \langle ai|r_{12}^{-1}|ja \rangle). \tag{10.19}$$

Using this notation, the complete first-order wave function can be expressed

$$\Psi^{(1)} = \sum_{ar} \frac{|\alpha_a^r\rangle\langle r|v|a \rangle}{\varepsilon_a - \varepsilon_r} + \frac{1}{2} \sum_{abrs} \frac{|\alpha_{ab}^{rs}\rangle\langle rs|r_{12}^{-1}|ab \rangle}{\varepsilon_a + \varepsilon_b - \varepsilon_r - \varepsilon_s}. \tag{10.20}$$

The corresponding graphical representation is shown in Fig. 10.3.

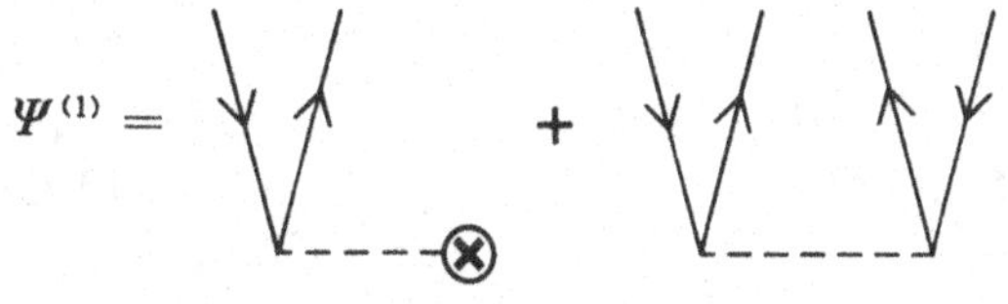

Fig. 10.3. Graphical representation of the first-order wave function (10.20). The first diagram represents the excitation due to the "effective" potential (10.19). This vanishes if Hartree-Fock orbitals are used (Brillouin's theorem). The second diagram represents "true" two-body effects. A summation over all orbital lines is assumed. Note also that there is a factor of 1/2 associated with the two-body diagram

The second term in (10.19) represents the matrix elements of the Hartree-Fock potential u_{HF} given in (5.33). Therefore, the elements of the effective potential v can also be written

$$\langle i|v|j\rangle = \langle i|-u|j\rangle + \langle i|u_{HF}|j\rangle . \tag{10.21}$$

We know from the treatment in Chap. 7 that the Hartree-Fock potential for a closed-shell system is spherically symmetric [see also (10.41)]. In particular, if we choose the central potential u in the model Hamiltonian H_0 (10.3) to be equal to the Hartree-Fock potential, all diagrams in Fig. 10.2 cancel. The first-order wave function (10.20) is then represented by a single diagram, which describes a two-particle excitation. This is a consequence of *Brillouin's theorem* [*Brillouin* 1933, 1934], mentioned in Sect. 5.4. If some other potential is used in H_0, then the single-particle diagrams serve as potential corrections, which yields approximate Hartree-Fock orbitals.

10.2 The First-Order Energy

In the previous section we have used perturbation theory to obtain the first-order wave function of a closed-shell system. We turn now to the problem of calculating the first-order energy contributions. According to (9.99) this is

$$E^{(1)} = \langle \alpha|V|\alpha\rangle .$$

The one-body part (10.11a) of the perturbation gives according to (5.20)

$$E_U^{(1)} = \langle \alpha|-U|\alpha\rangle = \sum_a^{\text{core}} \langle a|-u|a\rangle . \tag{10.22}$$

Using the above-mentioned sum convention, this energy can be represented by the diagram

$$= \sum_a^{\text{core}} \langle a|-u|a\rangle . \tag{10.23}$$

This can be obtained from the corresponding wave function diagram (10.15) by setting $r = a$ and joining the lines. Note that there is no excitation and hence no energy denominator associated with that type of diagram.

Similarly, the two-body part of $E^{(1)}$ is according to (5.22)

$$E_C^{(1)} = \langle \alpha|C|\alpha\rangle = \frac{1}{2}\sum_{ab}^{\text{core}} (\langle ab|r_{12}^{-1}|ab\rangle - \langle ba|r_{12}^{-1}|ab\rangle) , \tag{10.24}$$

where the terms can be represented in the following way:

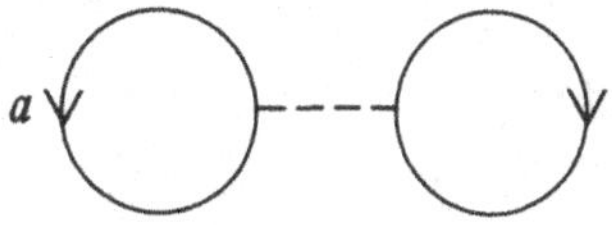

$$a \bigcirc ---\bigcirc b = \frac{1}{2} \sum_{ab}^{core} \langle ab | r_{12}^{-1} | ab \rangle \tag{10.25a}$$

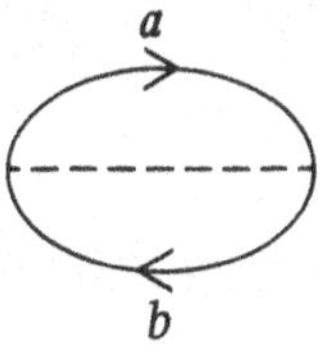

$$\overset{a}{\bigcirc} = -\frac{1}{2} \sum_{ab}^{core} \langle ba | r_{12}^{-1} | ab \rangle . \tag{10.25b}$$

The two factors of $1/2$ occur here because we allow the sums over a and b to run independently. Each distinct pair of orbitals should occur only once.

From (10.22, 24) the total first-order energy becomes

$$E^{(1)} = \langle \alpha | V | \alpha \rangle = \sum_{a}^{core} \langle a | -u | a \rangle + \frac{1}{2} \sum_{ab}^{core} (\langle ab | r_{12}^{-1} | ab \rangle - \langle ba | r_{12}^{-1} | ab \rangle). \tag{10.26}$$

This is represented graphically in Fig. 10.4 (top line).

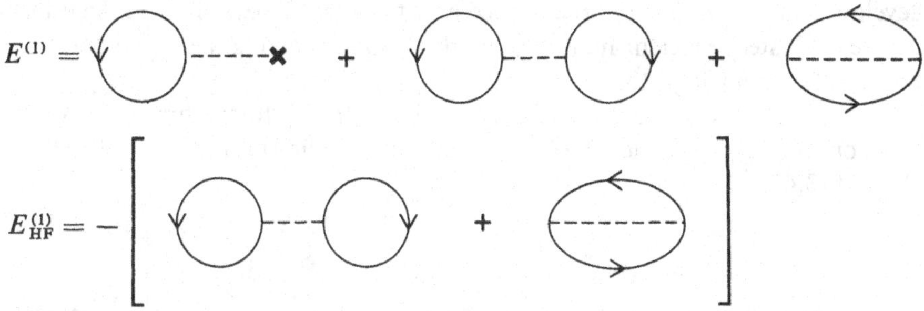

$$E^{(1)} = \bigcirc -----\times \ + \ \bigcirc ---\bigcirc \ + \ \bigcirc$$

$$E^{(1)}_{HF} = - \left[\bigcirc ---\bigcirc \ + \ \bigcirc \right]$$

Fig. 10.4. Graphical representation of the first-order energy in the general case (*top line*) and in the Hartree-Fock case (*bottom line*)

We have mentioned previously that the sum of the zeroth- and first-order energies is equal to the energy *expectation value* in the unperturbed state,

$$\langle E \rangle = \langle \alpha | H | \alpha \rangle = \langle \alpha | H_0 + V | \alpha \rangle = E_0 + E^{(1)} . \tag{10.27}$$

According to (10.3) we get

$$E_0 = \langle \alpha | H_0 | \alpha \rangle = \sum_{a}^{core} \langle a | -\frac{1}{2} \nabla^2 - \frac{Z}{r} + u | a \rangle$$

and $E^{(1)}$ is given by (10.26). If, in particular, u is the Hartree-Fock potential, then the potential part of E_0 is

$$\sum_a \langle a | u_{\mathrm{HF}} | a \rangle = \sum_{ab}^{\mathrm{core}} (\langle ab | r_{12}^{-1} | ab \rangle - \langle ba | r_{12}^{-1} | ab \rangle) .$$

Here, a and b run *independently* over all occupied orbitals, so this sum contains *the interaction between the electron pairs twice*. The first-order energy (10.26) is in this case

$$E_{\mathrm{HF}}^{(1)} = -\frac{1}{2} \sum_{ab}^{\mathrm{core}} (\langle ab | r_{12}^{-1} | ab \rangle - \langle ba | r_{12}^{-1} | ab \rangle) = -\frac{1}{2} \sum_a^{\mathrm{core}} \langle a | u_{\mathrm{HF}} | a \rangle, \quad (10.28)$$

which is illustrated graphically in Fig. 10.4 (bottom line). This term compensates for the double counting of the electron-electron interactions in E_0.

10.3 Evaluation of First-Order Diagrams

In the previous sections, we have seen that the first-order wave function and the first- and second-order expressions for the energy can be represented by simple diagrams. We shall now make contact with the angular-momentum theory developed in Chaps. 2–4 in order to see how the *angular* part of these diagrams can be evaluated. Efficient numerical methods for evaluating the radial part will be considered in Chap. 13.

We begin by considering the contribution (10.17c) to the first-order wave function. In Chap. 3 we derived a graphical representation of the Coulomb matrix element (3.42),

$$\langle ab | r_{12}^{-1} | cd \rangle = \delta(m_s^a, m_s^c) \delta(m_s^b, m_s^d) \sum_k X^k(ab, cd) \quad (10.29)$$

where

$$X^k(ab, cd) = X(k, l_a l_b l_c l_d) R^k(ab, cd) \quad (10.30)$$

and

$$X(k, l_a l_b l_c l_d) = (-1)^k \langle l_a \| C^k \| l_c \rangle \langle l_b \| C^k \| l_d \rangle$$

$$R^k(ab, cd) = \iint P_a(r_1) P_b(r_2) \frac{r_<^k}{r_>^{k+1}} P_c(r_1) P_d(r_2) \, dr_1 \, dr_2 . \quad (10.31)$$

[The reduced matrix elements of the C tensor are given by (2.127)]. So the contribution of the diagram (10.17c) is

$$
\begin{aligned}
&= \frac{1}{2} \frac{|\alpha_{ab}^{rs}\rangle \langle rs | r_{12}^{-1} | ab\rangle}{\varepsilon_a + \varepsilon_b - \varepsilon_r - \varepsilon_s} \\[2mm]
&= \frac{|\alpha_{ab}^{rs}\rangle}{\varepsilon_a + \varepsilon_b - \varepsilon_r - \varepsilon_s} \delta(m_s^r, m_s^a)\delta(m_s^s, m_s^b) \frac{1}{2} \sum_k X^k(rs, ab)
\end{aligned}
$$

(10.32)

Similarly, after summation over the m_l value of the internal line, diagrams (10.17a, b) give

$$
\begin{aligned}
\sum_{m_l^b} \quad &= \sum_{m_l^b} \frac{|\alpha_a^r\rangle \langle rb | r_{12}^{-1} | ab\rangle}{\varepsilon_a - \varepsilon_r} \\[2mm]
&= \frac{|\alpha_a^r\rangle}{\varepsilon_a - \varepsilon_r} \delta(m_s^r, m_s^a) \sum_k X^k(rb, ab)
\end{aligned}
$$

(10.33)

$$
\begin{aligned}
\sum_{m_l^b} \quad &= -\sum_{m_l^b} \frac{|\alpha_a^r\rangle \langle br | r_{12}^{-1} | ab\rangle}{\varepsilon_a - \varepsilon_r} \\[2mm]
&= -\frac{|\alpha_a^r\rangle}{\varepsilon_a - \varepsilon_r} \delta(m_s^r, m_s^a) \sum_k X^k(br, ab)
\end{aligned}
$$

(10.34)

Here, the corresponding lines of the angular-momentum diagrams are joined, in agreement with the sum convention introduced in Chap. 3.

In each of these cases *the angular-momentum diagram is topologically equivalent to the corresponding perturbation diagram*. We shall extend our graphical treatment of the perturbation expansion to higher orders in the following chapters, and it will then become clear that this topological equivalence is quite general [*Tolmachev* 1967; *Sandars* 1969].

The summation over m_s values of internal lines yields a simple multiplicative factor. For the direct diagram (10.33), m_s^b can be $\pm 1/2$, yielding a factor of 2. For the exchange diagram (10.34), on the other hand, m_s^b must be equal to $m_s^a = m_s^r$, which has a given value. Thus, the spin factor is equal to unity in this case. Alternatively, we could obtain the spin factor by means of a spin diagram as shown in Sect. 3.3.2.

Finally, we consider the potential diagram (10.15), which is quite trivial to evaluate. Since the potential is spherically symmetric, a general matrix element is given by

$$\langle i|-u|j\rangle = -\delta(m_s^i, m_s^j)\delta(l_i, l_j)\delta(m_l^i, m_l^j) \int P_i(r)uP_j(r)\, dr . \tag{10.35}$$

Here, the angular part is simply given by a single angular-momentum line (3.4).

In order to evaluate the angular-momentum diagrams appearing above, we use the graphical technique developed in Chaps. 3 and 4. According to (3.36) the diagram in (10.33) is

$$= \delta(l_a, l_r)\delta(m_l^a, m_l^r)\delta(k, 0)\,[l_b/l_a]^{1/2} . \tag{10.36}$$

This is nonzero only for $k = 0$ and $l_r = l_a$. The X factor in (10.33) is then

$$X(k, l_r l_b l_a l_b) = (-1)^k \langle l_r\|C^k\|l_a\rangle\langle l_b\|C^k\|l_b\rangle$$

$$= (-1)^{l_r+l_b}[l_r, l_a, l_b, l_b]^{1/2} \begin{pmatrix} l_r & 0 & l_a \\ 0 & 0 & 0 \end{pmatrix}\begin{pmatrix} l_b & 0 & l_b \\ 0 & 0 & 0 \end{pmatrix} \tag{10.37}$$

$$= [l_a, l_b]^{1/2} ,$$

where we have used (2.127) and (2.94) to evaluate the reduced matrix elements of the C tensor. Including the summation over the spin (which yields a factor of 2), we get

$$= \frac{|a_a'\rangle}{\varepsilon_a - \varepsilon_r}\, \delta(m_s^r, m_s^a)\delta(m_l^r, m_l^a)\delta(l_r, l_a)$$

$$\times (4l_b + 2)\, R^0(rb, ab) . \tag{10.38}$$

This diagram describes a process in which a single electron is excited by means of the *direct* interaction with all the electrons in a *filled* shell $n_b l_b$. The corresponding exchange interaction yields (see Problem 10.1)

$$= -\frac{|\alpha_a^r\rangle}{\varepsilon_a - \varepsilon_r} \delta(m_s^r, m_s^a)\delta(l_r, l_a)\delta(m_l^r, m_l^a)(2l_b + 1) \sum_k \begin{pmatrix} l_r & k & l_b \\ 0 & 0 & 0 \end{pmatrix}^2 R^k(br, ab) .$$

(10.39)

Note that the interaction with a filled shell cannot change any of the quantum numbers m_s, m_l or l. Only the principal quantum number n is allowed to change.

The sum of the direct and exchange diagrams above yields the excitation due to the Hartree-Fock potential (5.33)

(10.40)

where the sum over all occupied states b is understood. Identifying this last result with (10.15), we get

$$\langle r|u_{HF}|a\rangle = \delta(m_s^r, m_s^a)\delta(m_l^r, m_l^a)\delta(l_r, l_a)$$
$$\times \sum_{n_b l_b} (4l_b + 2)\left[R^0(rb, ab) - \frac{1}{2}\sum_k \begin{pmatrix} l_r & k & l_b \\ 0 & 0 & 0 \end{pmatrix}^2 R^k(br, ab)\right].$$

(10.41)

Thus, we have here recovered the Hartree-Fock potential for closed-shell systems obtained in Sect. 7.3. This shows that the potential is spherically symmetric, as we mentioned above.

Problem 10.1 Verify formula (10.39).

Problem 10.2 Separate the first-order energy (10.26) into radial and spin-angular parts and evaluate the latter graphically.

11. Second Quantization and the Particle-Hole Formalism

In the previous chapter we saw that the first-order wave function and the first-order energy can be represented in terms of diagrams. We would now like to extend our treatment to higher orders in the perturbation. Before treating higher-order effects, however, we shall introduce the formalism of second quantization. In order to see the need for this new formalism, we consider the first-order wave function (10.15)

$$
\cdots = \frac{|\alpha_a^r\rangle\langle r|-u|a\rangle}{\varepsilon_a - \varepsilon_r} . \tag{11.1}
$$

The state $|\alpha_a^r\rangle$ is one for which a single electron with the quantum numbers $n_a l_a m_r^a m_l^a$ of the determinant $|a\rangle$ is excited to a state having the quantum numbers $n_r l_r m_r^r m_l^r$. Intuitively, we know that the state $|\alpha_a^r\rangle$ is related to the state $|a\rangle$ in a simple way, but it is not possible to express this relation easily with the formalism we have been using. This is an indication that the formalism ought to be improved in some way.

In the matrix element which occurs in (11.1), the initial state is joined to the final state by means of the one-body part of the perturbation (10.4)

$$
V = - \sum_{i=1}^{N} u(r_i) + \sum_{i<j}^{N} \frac{1}{r_{ij}} . \tag{11.2}
$$

Here, the one- and two-body parts are written as sums over the individual electrons involved

$$
F = \sum_{i=1}^{N} f_i \tag{11.3a}
$$

$$
G = \sum_{i<j}^{N} g_{ij} \tag{11.3b}
$$

as in Sect. 5.3. From the point of view of perturbation theory, however, it would be more natural to decompose one- and two-particle operators in terms of basic excitations. As we shall see, this can be done with the formalism of *second quantization*. States of the kind $|\alpha_a^r\rangle$, $|\alpha_{ab}^{rs}\rangle$, ... for which one or several electrons are excited from the determinant $|\alpha\rangle$, are simply related to $|\alpha\rangle$ in this formalism. The requirement that the states of the system be antisymmetric is automatically satisfied, and the operators are given as a sum of terms, each representing an elementary excitation.

11.1 Second Quantization

A single-particle state φ_i having the quantum numbers $n_i l_i m_i^l m_i^s$ is represented in the formalism of second quantization as an operator $a_i^\dagger$ acting on the *vacuum state* $|0\rangle$ for which there are no electrons present

$$|\varphi_i\rangle \equiv a_i^\dagger|0\rangle . \tag{11.4}$$

Taking the hermitian adjoint of this equivalence, we have

$$\langle 0|a_i \equiv \langle\varphi_i| . \tag{11.5}$$

Since the operator $a_i^\dagger$ produces the state $|\varphi_i\rangle$ when it acts on the vacuum, it is called a *creation operator*. For reasons that we shall see presently, a_i is called an *absorption, destruction* or *annihilation operator*. Determinantal product states (10.7) may be built up by successively operating on the vacuum state with these operators

$$|\{\varphi_i\varphi_j\varphi_k \cdots\}\rangle = a_i^\dagger a_j^\dagger a_k^\dagger \cdots |0\rangle \tag{11.6}$$

and

$$\langle 0| \cdots a_k a_j a_i = \langle\{\varphi_i\varphi_j\varphi_k \cdots\}| . \tag{11.7}$$

The bra vector corresponding to $|\{\varphi_i\varphi_j\varphi_k \cdots\}\rangle$ is written $\langle\{\varphi_i\varphi_j\varphi_k \cdots\}|$, which is in agreement with the standard convention, although $\langle\{\cdots \varphi_k\varphi_j\varphi_i\}|$ would in a sense be more logical.

Since a determinant changes sign if two rows or columns are interchanged, the creation and absorption operators must satisfy the conditions

$$a_i^\dagger a_j^\dagger = -a_j^\dagger a_i^\dagger$$
$$a_i a_j = -a_j a_i$$

or

$$\{a_i^\dagger, a_j^\dagger\} = a_i^\dagger a_j^\dagger + a_j^\dagger a_i^\dagger = 0 \tag{11.8a}$$

$$\{a_i, a_j\} = a_i a_j + a_j a_i = 0 . \tag{11.8b}$$

These last equations are called anticommutation relations. Since it is not possible to create or absorb two electrons in the same state, these relations also hold for $i = j$.

The orthonormality of the determinantal product states imposes further conditions on these operators. For a two-particle system, we have

$$\langle \{\varphi_i \varphi_j\} \mid \{\varphi_k \varphi_l\} \rangle = \delta_{ik}\delta_{jl} - \delta_{il}\delta_{jk} . \tag{11.9}$$

The expression

$$\langle 0 \mid a_j a_i a_k^\dagger a_l^\dagger \mid 0 \rangle$$

will produce this result, if we suppose that

$$\{a_m, a_n^\dagger\} = a_m a_n^\dagger + a_n^\dagger a_m = \delta_{mn} \tag{11.10}$$

and also that[1]

$$a_m \mid 0 \rangle = 0 . \tag{11.11}$$

To see this we take the expression $a_j a_i a_k^\dagger a_l^\dagger$ above and use (11.8, 10) to transfer the operators a_j and a_i to the right

$$\begin{aligned}
a_j a_i a_k^\dagger a_l^\dagger &= \delta_{ik} a_j a_l^\dagger - a_j a_k^\dagger a_i a_l^\dagger \\
&= \delta_{ik}\delta_{jl} - \delta_{ik}a_l^\dagger a_j - \delta_{jk}a_i a_l^\dagger + a_k^\dagger a_j a_i a_l^\dagger \\
&= \delta_{ik}\delta_{jl} - \delta_{jk}\delta_{il} - \delta_{ik}a_l^\dagger a_j + \delta_{jk}a_l^\dagger a_i \\
&\quad + \delta_{il}a_k^\dagger a_j - \delta_{jl}a_k^\dagger a_i + a_k^\dagger a_l^\dagger a_j a_i .
\end{aligned} \tag{11.12}$$

Due to the condition (11.11) all terms except the first two terms give zero when operating on the vacuum. Thus (11.9) is reproduced.

From the equations above it follows that

$$a_i \mid \varphi_i \rangle = a_i a_i^\dagger \mid 0 \rangle = (1 - a_i^\dagger a_i) \mid 0 \rangle = \mid 0 \rangle .$$

This is the reason we call a_i an absorption operator. It absorbs an electron when operating to the right. Equation (11.11) may be interpreted to mean that an electron cannot be absorbed from the vacuum.

We can now summarize the anticommutation relations for the creation and absorption operators as follows:

$$\boxed{\begin{aligned}
\{a_i^\dagger, a_j^\dagger\} &= 0 \\
\{a_i, a_j\} &= 0 \\
\{a_i, a_j^\dagger\} &= \delta_{ij}
\end{aligned}} \tag{11.13}$$

---

[1] Note the difference between the vacuum state ($\mid 0 \rangle$) and the number zero (0).

In deriving these relations we have only used the fact that an electronic system must satisfy the exclusion principle. Therefore, the same relations hold for any fermion system. (For boson systems the anticommutation relations are replaced by the corresponding commutation relations).

In applying the formalism of second quantization to perturbation theory, we shall normally use a Slater determinant (10.7)

$$|\alpha\rangle = |\varphi_i \varphi_j \cdots\rangle = a_i^\dagger a_j^\dagger \cdots |0\rangle \tag{11.14}$$

as the starting point for absorbing and creating electrons. For closed-shell systems, this is normally chosen to be identical to the zeroth-order state, represented, for instance, by a Hartree-Fock determinant. Orbitals appearing in this determinant are referred to as *core* orbitals and the remaining, unoccupied ones as *virtual* orbitals. When operating on the reference state, it is not possible to create a particle in a core orbital or to absorb a particle from an virtual orbital. This means that

$$\boxed{a_{\text{core}}^\dagger |\alpha\rangle = a_{\text{virt}} |\alpha\rangle = 0 .} \tag{11.15}$$

Denoting, as before, core orbitals by a, b, ... and virtual orbitals by r, s, ..., the states $|\alpha_a^r\rangle$, $|\alpha_{ab}^{rs}\rangle$, for which one or two electrons are excited, can be expressed

$$|\alpha_a^r\rangle = a_r^\dagger a_a |\alpha\rangle \tag{11.16a}$$

$$|\alpha_{ab}^{rs}\rangle = a_r^\dagger a_s^\dagger a_b a_a |\alpha\rangle . \tag{11.16b}$$

Note that such excitations are always written with *the creation operators to the left of the absorption operators*[2]. Note also the order between the absorption operators in the second equation. In order to show that the order of the operators in this last equation is correct, we may use (11.14) to write the state $|\alpha\rangle$ explicitly in terms of creation operators, which gives

$$|\alpha_{ab}^{rs}\rangle = a_r^\dagger a_s^\dagger a_b a_a a_i^\dagger a_j^\dagger \cdots |0\rangle .$$

The anticommutation relations (11.13) may then be used to move first a_a and then a_b to the right. This will produce two vacancies in the line of creation operators following a_a. The creation operator $a_s^\dagger$ may then be moved into the vacancy formed by a_b and $a_r^\dagger$ into the vacancy produced by a_a without a change of sign[3].

---

[2] This ordering is referred to as the "normal ordering with respect to the vacuum" (see further section 11.2 below).

[3] Note that the symbol $|\alpha_{ab}^{rs}\rangle$ represents the determinant where the orbital a in $|\alpha\rangle$ has been replaced by r and the orbital b by s. If instead a is replaced by s and b by r, we will get a change in sign, corresponding to

$$|\alpha_{ab}^{sr}\rangle = -|\alpha_{ab}^{rs}\rangle = -|\alpha_{ba}^{sr}\rangle . \tag{10.17}$$

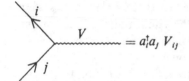

$$= a_i^\dagger a_j \, V_{ij}$$

Fig. 11.1. Graphical representation of a one-body interaction

The meaning of the formalism of second quantization can be illustrated by means of the simple scattering process in Fig. 11.1. Here, a particle in a state $|j\rangle$ is scattered by some potential V into another state $|i\rangle$. Mathematically, this can be represented by the expression $a_i^\dagger a_j V_{ij}$, indicating that a particle $|j\rangle$ is absorbed and a particle $|i\rangle$ is created. The probability for this process is given by the "amplitude" V_{ij}. These amplitudes are characteristic of the potential, and by considering all combinations of i and j, we have a complete description of the potential. The operator

$$V = \sum_{ij} a_i^\dagger a_j \, V_{ij} \tag{11.18}$$

can then be regarded as a mathematical representation of the scattering process. When V operates on an orbital $|j\rangle$, the result is

$$V|j\rangle = \sum_i a_i^\dagger \, V_{ij} |0\rangle = \sum_i V_{ij} |i\rangle \;.$$

Thus, the orbital $|i\rangle$ is produced with the desired amplitude V_{ij}. We also see that

$$\langle i|V|j\rangle = V_{ij} \;, \tag{11.19}$$

showing that the amplitude V_{ij} can be interpreted as a matrix element of the operator V.

We can easily show in a formal way that any single-particle operator (11.3a) can be written in analogy with (11.18) as

$$F = \sum_{ij} a_i^\dagger a_j \, \langle i|f|j\rangle \;. \tag{11.20}$$

In order to do this, we only have to show that the operator above has the same matrix elements between all combinations of Slater determinants as the operator (11.3a) treated previously.

Let $|\alpha\rangle$ and $|\beta\rangle$ be two Slater determinants, composed of the same set of orthonormal orbitals. Then using (11.20),

$$\langle \beta|F|\alpha\rangle = \sum_{ij} \langle \beta|a_i^\dagger a_j|\alpha\rangle \langle i|f|j\rangle \;. \tag{11.21}$$

We see that j must here represent a core orbital in order to contribute. Further, i must represent an orbital unoccupied in $a_j|\alpha\rangle$. This means that i can either be equal to j or represent a virtual orbital (not occupied in $|\alpha\rangle$). Using (11.16) and our convention that core orbitals are represented by the letters a, b, c, ..., (11.21) may be written

$$\langle\beta|F|\alpha\rangle = \sum_{ai} \langle\beta|\alpha_a^i\rangle\langle i|f|a\rangle . \tag{11.22}$$

When we evaluated the matrix elements of one- and two-particle operators in Chap. 5, we treated diagonal matrix elements and the cases for which α and β differed by the quantum numbers of one or two electrons separately. We shall follow this procedure again now and show that we obtain the same results.

In the diagonal case ($\beta = \alpha$) we must have $i = a$, and (11.22) becomes

$$\langle\alpha|F|\alpha\rangle = \sum_{ai} \langle\alpha|\alpha_a^i\rangle\langle i|f|a\rangle = \sum_a \langle a|f|a\rangle ,$$

which agrees with (5.20). If $|\beta\rangle$ differs from $|\alpha\rangle$ by a single orbital, then $|\beta\rangle$ is of the form $|\alpha_b^r\rangle$, where r is virtual, and (11.22) becomes

$$\langle\alpha_b^r|F|\alpha\rangle = \sum_{ai} \langle\alpha_b^r|\alpha_a^i\rangle\langle i|f|a\rangle = \langle\alpha_b^r|\alpha_b^r\rangle\langle r|f|b\rangle = \langle r|f|b\rangle .$$

This last equation agrees with (5.23). All matrix elements of F for states differing by more than one orbital are zero, since then $\langle\beta|\alpha_a^i\rangle = 0$ for all values of a and i.

We have then shown that all matrix elements of (11.20) agree with those of (11.3a) and hence the operators are identical.

The zeroth-order Hamiltonian H_0, defined in (10.3),

$$H_0 = \sum_{i=1}^{N} h_0(i) ,$$

is of the one-body type. Assuming that the orbitals are eigenfunctions of h_0,

$$h_0|i\rangle = \varepsilon_i|i\rangle ,$$

then h_0 will be diagonal in the corresponding representation, and the second-quantized form of H_0 becomes

$$H_0 = \sum_i a_i^\dagger a_i \varepsilon_i . \tag{11.23}$$

Note that in the last expression i runs over *all* orbitals.

The ideas applied above can easily be extended to two-body operators. In analogy to the potential scattering process in Fig. 11.1, we can illustrate the two-body interaction by means of the scattering process shown in Fig. 11.2. This process can be represented by the operator

$$a_i^\dagger a_j^\dagger a_l a_k \, W_{ijkl} \,,$$

and the general two-body operator, defined in (11.3b), can be expressed in analogy to (11.20) as

$$G = \frac{1}{2} \sum_{ijkl} a_i^\dagger a_j^\dagger a_l a_k \, \langle ij|g|kl \rangle \,. \tag{11.24}$$

The problem of showing that (11.24) has the same matrix elements as (11.3b) is left as an exercise (Problem 11.1).

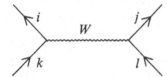

Fig. 11.2.
Graphical representation of a two-body interaction

Since all physical two-body interactions are symmetrical with respect to an interchange of the two particles, we have

$$\langle ij|g|kl \rangle = \langle ji|g|lk \rangle \,,$$

and since the indices in (11.24) run independently, each physical amplitude appears twice in that sum. This is the reason for the factor of 1/2. Each term in (11.24) describes a process in which one particle makes a transition from the state k to the state i and the other particle from l to j. Since the particles are indistinguishable, one can only talk about transitions from one two-particle state to another. The amplitude for such a transition contains a direct and an exchange part. Formally, however, we can distinguish between these parts of the amplitude. As we have seen, the exclusion principle is automatically taken care of by the anticommutation rules (11.13).

Summarizing, we can express one- and two-body operators in two equivalent ways as follows:

$$F = \sum_{i=1}^{N} f_i = \sum_{ij} a_i^\dagger a_j \langle i | f | j \rangle$$
$$G = \sum_{i<j}^{N} g_{ij} = \frac{1}{2} \sum_{ijkl} a_i^\dagger a_j^\dagger a_l a_k \langle ij | g | kl \rangle$$

(11.25)

In particular, this gives the following forms of the model Hamiltonian (10.3) and of the perturbation (11.2):

$$H_0 = \sum_{i=1}^{N} h_0(i) = \sum_{i} a_i^\dagger a_i \, \varepsilon_i$$
$$V = -\sum_{i=1}^{N} u_i + \sum_{i<j}^{N} r_{ij}^{-1}$$
$$= -\sum_{ij} a_i^\dagger a_j \langle i | u | j \rangle + \frac{1}{2} \sum_{ijkl} a_i^\dagger a_j^\dagger a_l a_k \langle ij | r_{12}^{-1} | kl \rangle$$

(11.26)

In the second-quantization formulas the summations run over all orbitals.

The idea of second quantization was introduced by *Dirac* [1927] in treating quantized fields of radiation. Shortly afterwards it was realized that a similar formalism could also be applied to fermion systems [*Jordan* and *Klein* 1927; *Jordan* and *Wigner* 1928]. This application represents just a reformulation of the ordinary quantum mechanics, and no new feature, not present in the Schrödinger equation, has been added. Therefore, the term "second quantization" is somewhat inappropriate in this case but it is now generally accepted.

Although the formalism of second quantization was developed in the early days of quantum mechanics, it is only during the last two decades that it has been commonly used as a tool in many-body perturbation theory. In the present section we have given only the fundamentals of the formalism needed for such applications. More general treatments of second quantization can be found in several books, such as [*Berezin* 1966; *Judd* 1967; *Avery* 1976]. In the following sections of the present chapter, we shall develop the particle-hole formalism, which is a form of second quantization particularly suitable for many-body applications (see, for instance, [*March* et al. 1967; *Fetter* and *Walecka* 1971; *Linderberg* and *Öhrn* 1973]). Wick's theorem will be derived in a time-independent form, following *Bogoliubov* and *Shirkov* [1959] (see also [*Paldus* and *Čížek* 1975]), and this will form the basis for the graphical analysis which we shall apply in the following chapters.

Problem 11.1. Show that the matrix elements of the operator (11.24) between Slater determinants are the same as those of the two-electron operator in Chap. 5, (5.22, 24, 26).

11.2 Operators in Normal Form

As we have mentioned, the idea of second quantization is to express the operators in terms of elementary (single, double, ...) excitations. We can easily see, however, that the operators (11.25) do not always have such a simple interpretation.

As an illustration, we consider the one-body operator

$$F = \sum_{ij} a_i^\dagger a_j \langle i|f|j \rangle . \qquad (11.27)$$

As before, we choose a particular Slater determinant $|\alpha\rangle$ as the reference and define core and virtual orbitals with respect to this state. If $i(=a)$ in (11.27) represents a core orbital, then we can rewrite the operators by means of the anticommutation rules (11.13) as

$$a_a^\dagger a_j = -a_j a_a^\dagger + \delta_{aj} .$$

When operating on $|\alpha\rangle$, the first term on the right-hand side does not contribute due to (11.15), while the last term reproduces $|\alpha\rangle$ for $j = a$. So in this case $a_i^\dagger a_j$ does not represent any excitation at all.

In order to see in a more systematic way which terms do correspond to real excitations when operating on $|\alpha\rangle$, we shall rearrange the creation and absorption operators so that

$$a_{\text{core}}^\dagger \quad \text{and} \quad a_{\text{virt}} \quad \textit{appear to the right of} \quad a_{\text{core}} \quad \text{and} \quad a_{\text{virt}}^\dagger .$$

The terms which contain the former operators will then give zero when operating on $|\alpha\rangle$, while the remaining ones will produce real excitations. A second-quantized operator written in this way is said to be in *normal order* or in *normal form* with respect to the state $|\alpha\rangle$. This state will form our new *"vacuum"* state. If this state is chosen to be the real vacuum state (11.4), then there will be no core orbitals and the normal order will be identical to the ordering used in the previous section. In the following, however, we shall always assume that the normal form is defined with respect to some conveniently chosen Slater determinant. For closed-shell systems, to be discussed in detail in the next chapter, we shall assume that this determinant is identical to the zeroth-order wave function. In the open-shell formalism, on the other hand, different choices can be made, which gives some additional flexibility, as will be further discussed in Chap. 13.

We shall now use the definition above and the anticommutation rules to transform the perturbation V in (11.26) into normal form with respect to a vacuum state $|\alpha\rangle$. As we shall see, this leads to a decomposition of V into elementary excitations with respect to this state, each of which can be conveniently represented by means of a Goldstone diagram of the kind introduced in the previous chapter.

We consider first the one-body part of the perturbation (11.26)

$$-U = \sum_{ij} a_i^\dagger a_j \langle i|-u|j\rangle \qquad (11.28)$$

and separate this into four sums depending upon whether the orbitals are core orbitals $(a, b, \ldots)$ or virtual orbitals $(r, s, \ldots)$

$$-U = \sum_{rs} a_r^\dagger a_s \langle r|-u|s\rangle + \sum_{ar} a_r^\dagger a_a \langle r|-u|a\rangle$$
$$+ \sum_{ar} a_a^\dagger a_r \langle a|-u|r\rangle + \sum_{ab} a_a^\dagger a_b \langle a|-u|b\rangle . \qquad (11.29)$$

Here, it is only the last term that is not in normal form according to the definitions given above. Moving $a_a^\dagger$ to the right in this term gives

$$a_a^\dagger a_b = \delta_{ab} - a_b a_a^\dagger \qquad (11.30)$$

and

$$\sum_{ab} a_a^\dagger a_b \langle a|-u|b\rangle = \sum_{a} \langle a|-u|a\rangle - \sum_{ab} a_b a_a^\dagger \langle a|-u|b\rangle .$$

We now introduce a special notation for operators in normal form

$$\{a_i^\dagger a_j^\dagger \cdots a_l a_k\} . \qquad (11.31)$$

This notation means that the operators within the curly brackets are rearranged so that the $a_{\text{core}}^\dagger$ and the a_{virt} operators appear to the right. There is a phase associated with this rearrangement depending on the number of transpositions necessary to transform the operator into normal form (or the parity of the required permutation). The phase factor is $+1$ for an even permutation (even number of transpositions) and -1 for an odd permutation.

The normal form of a collection of creation and absorption operators is defined only within a permutation of the $a_{\text{core}}^\dagger$, a_{virt} operators and a permutation of the a_{core}, $a_{\text{virt}}^\dagger$. The quantity in (11.31) is always well defined, however, when the above-mentioned phase convention is applied.

As a simple illustration we consider the terms appearing in (11.29). The first term is already in normal form, and there is no alternative way of expressing this form. Thus,

$$\{a_r^\dagger a_s\} = a_r^\dagger a_s . \tag{11.32a}$$

The second and third terms, on the other hand, are in normal form regardless of the order between the operators, which gives

$$\{a_r^\dagger a_a\} = a_r^\dagger a_a = -a_a a_r^\dagger \tag{11.32b}$$

$$\{a_a^\dagger a_r\} = a_a^\dagger a_r = -a_r a_a^\dagger . \tag{11.32c}$$

Finally, in the last term in (11.29) the operators have to be interchanged in order to transform them into normal form,

$$\{a_a^\dagger a_b\} = -a_b a_a^\dagger . \tag{11.32d}$$

It is obvious that a permutation within the normal product (11.31) will only affect the phase. For instance, from (11.32a) it follows that

$$\{a_s a_r^\dagger\} = -a_r^\dagger a_s = - \{a_r^\dagger a_s\}$$

and from (11.32b) that

$$\{a_a a_r^\dagger\} = a_a a_r^\dagger = - \{a_r^\dagger a_a\}$$

and so on.

Using (11.32d), the relation (11.30) can now be written

$$a_a^\dagger a_b = \delta_{ab} + \{a_a^\dagger a_b\} \tag{11.33}$$

and we can express the entire operator (11.28, 29) in normal form as follows:

$$-U = \sum_{ij} a_i^\dagger a_j \langle i | -u | j \rangle \tag{11.34}$$

$$= \sum_a^{core} \langle a | -u | a \rangle + \sum_{ij} \{a_i^\dagger a_j\} \langle i | -u | j \rangle .$$

As usual, a runs over core orbitals only, while i and j run over all orbitals. The first sum is a number—or a zero-body operator—while the second sum represents a normal-ordered, one-body operator. When operating on the vacuum state $|\alpha\rangle$, each operator in the second summation either gives zero or produces a determinantal state of the form $|\alpha_a^r\rangle$ for which one of the orbitals is virtual. We have thus separated $-U$ into a zero-body term, which reproduces the original state $|\alpha\rangle$, and a normal-ordered, one-body operator, which gives rise to single excitations.

Applying the same procedure to the zeroth-order Hamiltonian, we get

$$H_0 = \sum_{ij} a_i^\dagger a_j \langle i|h_0|j\rangle = \sum_a^{occ} \varepsilon_a + \sum_i \{a_i^\dagger a_i\}\,\varepsilon_i\,. \tag{11.35}$$

The first term represents the zeroth-order energy (10.9). This is the only part of H_0 which contributes when operating on the vacuum state.

The second term of the perturbation (11.26)

$$C = \frac{1}{2}\sum_{ijkl} a_i^\dagger a_j^\dagger a_l a_k \,\langle ij|r_{12}^{-1}|kl\rangle \tag{11.36}$$

can be normal-ordered in a similar way as the one-body operator (11.28). It should be clear from the previous discussion, however, that it is sufficient to make the separation according to the character of the *creation* operators only. If those creation operators which correspond to core orbitals are moved to the right, then the terms will be in normal form regardless of the character of the absorption operators (possibly apart from a trivial transposition of the latter). We then write (11.36) as

$$C = \frac{1}{2}\sum_{rskl} a_r^\dagger a_s^\dagger a_l a_k \,\langle rs|r_{12}^{-1}|kl\rangle + \frac{1}{2}\sum_{askl} a_a^\dagger a_s^\dagger a_l a_k \,\langle as|r_{12}^{-1}|kl\rangle$$
$$+ \frac{1}{2}\sum_{arkl} a_a^\dagger a_r^\dagger a_l a_k \,\langle ra|r_{12}^{-1}|kl\rangle + \frac{1}{2}\sum_{abkl} a_a^\dagger a_b^\dagger a_l a_k \,\langle ab|r_{12}^{-1}|kl\rangle\,.$$

Here, the first term is normal ordered, while the remaining ones are not. It is left as an exercise (Problem 11.2) to show that normal-ordering this expression leads to

$$C = \frac{1}{2}\sum_{ijkl} \{a_i^\dagger a_j^\dagger a_l a_k\} \langle ij|r_{12}^{-1}|kl\rangle$$
$$+ \sum_{ij} \{a_i^\dagger a_j\} \sum_a (\langle ia|r_{12}^{-1}|ja\rangle - \langle ai|r_{12}^{-1}|ja\rangle) \tag{11.37}$$
$$+ \frac{1}{2}\sum_{ab} (\langle ab|r_{12}^{-1}|ab\rangle - \langle ba|r_{12}^{-1}|ab\rangle)\,.$$

Combining (11.34) and (11.37), we see that the one-body terms can be conveniently expressed by means of the effective potential, introduced in the previous chapter (10.19),

$$\langle i|v|j\rangle = \langle i|-u|j\rangle + \sum_a^{core} (\langle ia|r_{12}^{-1}|ja\rangle - \langle ai|r_{12}^{-1}|ja\rangle)\,. \tag{11.38}$$

The perturbation may thus be separated into the following normal-ordered zero-, one- and two-body parts:

$$V = V_0 + V_1 + V_2$$

$$V_0 = \sum_a^{core} \langle a | -u | a \rangle + \frac{1}{2} \sum_{ab}^{core} (\langle ab | r_{12}^{-1} | ab \rangle - \langle ba | r_{12}^{-1} | ab \rangle)$$

$$V_1 = \sum_{ij} \{a_i^\dagger a_j\} \langle i | v | j \rangle$$

$$V_2 = \frac{1}{2} \sum_{ijkl} \{a_i^\dagger a_j^\dagger a_l a_k\} \langle ij | r_{12}^{-1} | kl \rangle .$$

$$(11.39)$$

Here a and b run over core orbitals, while i, j, k and l run over all orbitals—in agreement with our conventions. This form of the perturbation agrees with that given by Čížek [1966, 1969], and it is the form we shall use in the following. The separate terms V_0, V_1 and V_2 represent true zero-, one- and two-body operators in the sense that V_0 operating on the vacuum state $|\alpha\rangle$ gives back the same state, while V_1 and V_2 produce single and double excitations, respectively— or a null result. It should be noted that the effective potential, appearing in the one body operator, vanishes if Hartree-Fock orbitals are used [see (10.21)]. In this case then, the perturbation contains only zero- and two-body parts

$$V_{HF} = V_0 + V_2. \tag{11.40}$$

Problem 11.2. Verify the normal form (11.37) of the Coulomb interaction (11.36).

11.3 The Particle-Hole Formalism

The normal product we have introduced in the previous section can be expressed in a clear way by means of "*particle-hole*" creation and absorption operators. With our definition, an operator is transformed into normal form by moving the operators $a_{core}^\dagger$ and a_{virt} to the right. The underlying reason for doing this is that these operators satisfy the relations (11.15)

$$a_{core}^\dagger |\alpha\rangle = a_{virt} |\alpha\rangle = 0 . \tag{11.41}$$

When operating on the "vacuum" state $|\alpha\rangle$, it is not possible to absorb a particle from a virtual orbital which is unoccupied in this state, and due to the exclusion principle it is not possible to create a particle in a core orbital which is already occupied.

The relations (11.41) and expressions like (11.32) for operators in normal form can be written in a symmetric fashion by means of particle-hole operators. These operators are defined as follows:

$$\begin{cases} b^\dagger = a^\dagger, & b = a & \textit{for virtual orbitals ("particles")} \\ b^\dagger = a, & b = a^\dagger & \textit{for core orbitals ("holes").} \end{cases} \tag{11.42}$$

The new particle-hole operators $(b^\dagger, b)$ are identical to the previous creation and absorption operators $(a^\dagger, a)$ for virtual orbitals ("particle states"), but they have the opposite interpretation for core orbitals ("hole states"). It is easy to show that the b operators satisfy the same anticommutation relations as do the a operators.

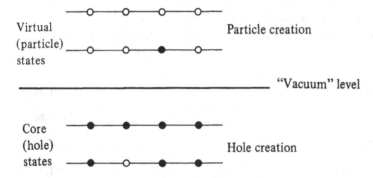

Fig. 11.3. Simple illustration of the particle-hole creation operators. For virtual orbitals (particle states) $b^\dagger = a^\dagger$ can create a particle and for core orbitals (hole states) $b^\dagger = a$ can absorb a particle or create a hole. The b operators have the opposite effect. It can be imagined that the particle and hole states are separated by some "vacuum" or "Fermi" level

The meaning of the particle-hole creation operators is illustrated in Fig. 11.3. The $b^\dagger$ operator can create a particle in a virtual orbital, which is not occupied in the vacuum state $|\alpha\rangle$, and it can create a hole in a core orbital which is occupied in this state. The b operator can absorb a particle in a virtual orbital and it can fill or absorb a hole in a core orbital. Normally, the core orbitals have lower orbital energies than the virtual ones, and the two categories can be separated by means of some "vacuum" or "Fermi" level, as indicated in the figure. This is by no means required by the formalism, however, so it should not be taken too literally.

In terms of the particle-hole operators, the relation (11.41) becomes

$$b|\alpha\rangle = 0 , \tag{11.43}$$

which means simply that *there are no particles or holes in the "vacuum" state* $|\alpha\rangle$. It is just the particle-hole *absorption* operators b which are moved to the right in transforming an operator into the normal form defined above. We can say then that an operator is in normal form if *the particle-hole absorption operators appear to the right of the particle-hole creation operators*. For instance, the equations (11.32) can be written

$$\{b_r^\dagger b_s\} = b_r^\dagger b_s$$

$$\{b_r^\dagger b_a^\dagger\} = b_r^\dagger b_a^\dagger = -b_a^\dagger b_r^\dagger$$

$$\{b_a b_r\} = b_a b_r = -b_r b_a$$

$$\{b_a b_b^\dagger\} = -b_b^\dagger b_a \ .$$

The main advantage of the particle-hole formalism is that we do not need to distinguish between core orbitals and virtual orbitals in the formal treatment. For instance, the following relations hold for arbitrary combinations of orbitals i and j:

$$\{b_i^\dagger b_j^\dagger\} = b_i^\dagger b_j^\dagger \tag{11.44a}$$

$$\{b_i b_j\} = b_i b_j \tag{11.44b}$$

$$\{b_i^\dagger b_j\} = b_i^\dagger b_j \tag{11.44c}$$

$$\{b_i b_j^\dagger\} = -\, b_j^\dagger b_i \ . \tag{11.44d}$$

11.4 Graphical Representation of Normal-Ordered Operators

In the previous chapter we introduced Goldstone diagrams to represent the first-order terms in the perturbation expansion of the wave function and the energy. We shall now use the same kind of diagrams to represent general *operators in normal form*. Virtual orbitals (particle states) are represented by lines directed *upwards* and core orbitals (hole states) by lines directed *downwards*. As shown in Fig. 11.4, a *creation* operator ($a^\dagger$) is furthermore associated with lines directed *away* from the vertex and an *absorption* operator (a) with lines directed *toward* the vertex.

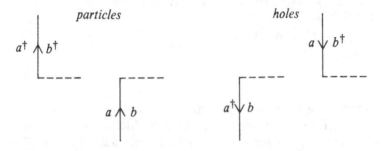

Fig. 11.4. Graphical representation of the ordinary creation and absorption operators ($a^\dagger$, a) and the particle-hole creation and absorption operators ($b^\dagger$, b). The two figures on the left correspond to particle states and the figures on the right to hole states.

In the figure the corresponding particle-hole operators (11.42) are also shown, and we observe that

> *the particle-hole creation operators are associated with lines above the vertex and particle-hole absorption operators with lines below the vertex.*

Thus, operators which appear to the left in the normal form are associated with lines *above* the vertex and operators which appear to the *right* are associated with lines *below* the vertex.

We shall represent graphically now each of the parts of the normal-ordered perturbation (11.39). The zero-body part V_0 is just a number. According to (10.26), it corresponds to the first-order energy for the vacuum state $|\alpha\rangle$

$$V_0 = \langle \alpha | V | \alpha \rangle .$$

This is represented graphically by the diagrams in Fig. 10.4.

Since the Goldstone representation we use distinguishes between core orbitals and virtual orbitals, we separate the one-body part of the perturbation (11.39) as we did in (11.29). Two examples are given below,

$$= \{a_r^\dagger a_a\} \langle r|v|a \rangle = a_r^\dagger a_a \langle r|v|a \rangle \tag{11.45a}$$

$$= \{a_a^\dagger a_b\} \langle a|v|b \rangle = -a_b a_a^\dagger \langle a|v|b \rangle . \tag{11.45b}$$

The two-body part of (11.39) can be treated in a similar way. For instance,

$$\begin{aligned} &= \frac{1}{2} \{a_r^\dagger a_s^\dagger a_u a_t\} \langle rs|r_{12}^{-1}|tu \rangle \\ &= \frac{1}{2} a_r^\dagger a_s^\dagger a_u a_t \langle rs|r_{12}^{-1}|tu \rangle \end{aligned} \tag{11.46a}$$

$$\begin{aligned} &= \frac{1}{2} \{a_a^\dagger a_r^\dagger a_c a_b\} \langle ar|r_{12}^{-1}|bc \rangle \\ &= -\frac{1}{2} a_r^\dagger a_c a_b a_a^\dagger \langle ar|r_{12}^{-1}|bc \rangle \end{aligned} \tag{11.46b}$$

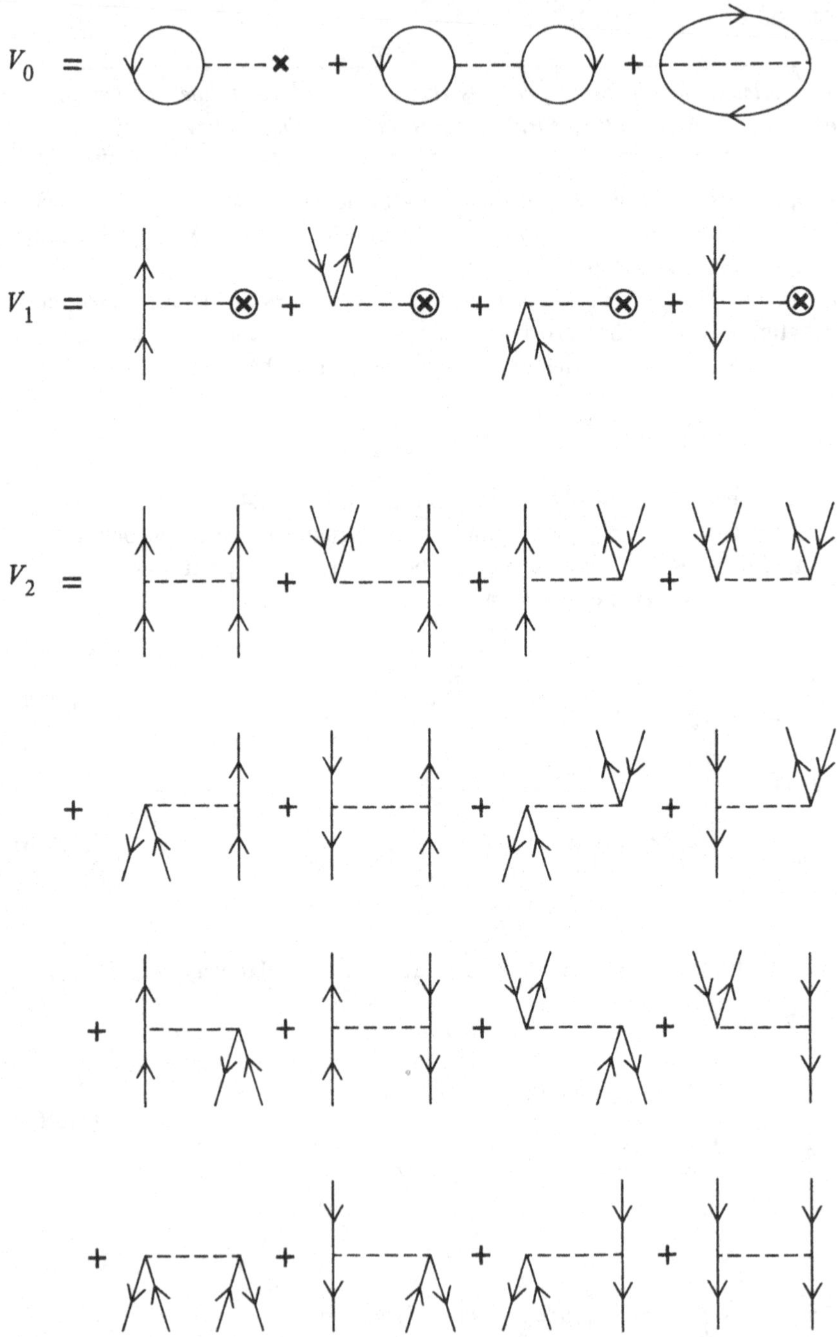

Fig. 11.5. Graphical representation of the normal-ordered zero-, one- and two-body parts of the perturbation (11.39)

$$c \qquad \qquad d \qquad = \frac{1}{2} \{a_a^\dagger a_b^\dagger a_d a_c\} \langle ab | r_{12}^{-1} | cd \rangle$$

(11.46c)

$$a \qquad \qquad b \qquad = \frac{1}{2} a_d a_c a_a^\dagger a_b^\dagger \langle ab | r_{12}^{-1} | cd \rangle .$$

These examples should be sufficient to show how normal-ordered operators are represented graphically. We see that each diagram corresponds to a product of creation and absorption operators in normal form and a matrix element. The final phase factor of each diagram depends upon the number of transpositions necessary to transform the operators into normal order. All diagrams of the normal-ordered perturbation (11.39) are shown in Fig. 11.5.

In this section we have assumed that all terms originating from the two-body interaction in (11.39) are represented by separate diagrams. Hence, all diagrams have the numerical factor 1/2. As mentioned before, however, some diagrams may be equivalent in the sense that they yield the same result, when the summation is performed over the free lines (of course, restricted to the categories indicated by the different lines). It is sufficient to consider only distinct (nonequivalent) diagrams, provided that these are supplied with the appropriate weights, as will be discussed at some length in the next chapter.

11.5 Wick's Theorem

Thus far, we have used the anticommutation rules (11.13) to transform operators into normal form. As we shall see, however, this can be done more easily by means of a theorem due to *Wick* [1950]. Originally, Wick's theorem was used in time-dependent perturbation theory to express "time-ordered" operators in normal form (see, for instance [*Fetter* and *Walečka* 1971]). The theorem can easily be reformulated, however, so that it applies to the kind of time-independent analysis we use here (see, for instance, [*Bogoliubov* and *Shirkov* 1959; *Paldus* and *Čižek* 1975]).

10.5.1 Statement of the Theorem

In order to formulate Wick's theorem, we shall first define the *contraction* of two creation/absorption operators as follows:

If x and y are arbitrary creation or absorption operators, then the contraction of x and y is defined as the difference between the ordinary and the normal product of the operators, i.e.,

$$\overline{x\,y} = x\,y - \{x\,y\} .$$

(11.47)

From the relations (11.44 a-c) it follows immediately that

$$\overset{\frown}{b_i^\dagger b_j^\dagger} = \overset{\frown}{b_i b_j} = \overset{\frown}{b_i^\dagger b_j} = 0 \tag{11.48a}$$

and by means of the anticommutation relations we get from (11.44d)

$$\overset{\frown}{b_i b_j^\dagger} = b_i b_j^\dagger + b_j^\dagger b_i = \delta_{ij} . \tag{11.48b}$$

These relations hold for any combination of i and j. We observe that the contraction is nonvanishing only for the combination of absorption/creation operators, for which the absorption operator appears to the left of the creation operator and the two operators are associated with the same orbital. The corresponding relations for the original operators $(a, a^\dagger)$ are obtained by means of the definitions (11.42).

Wick's theorem can now be formulated as follows:

If A is a product of creation and absorption operators, then

$$\boxed{A = \{A\} + \{\overset{\frown}{A}\}} , \tag{11.49}$$

where $\{A\}$ represents the normal form of A and $\{\overset{\frown}{A}\}$ represents the sum of the normal-ordered terms obtained by making all possible single, double, ... contractions within A.

The phase of the first term $\{A\}$ depends on the parity of the permutation required to bring the operator into normal form. Similarly, the phase of the contracted terms depends on the permutation required to bring the operators to be contracted into contact with each other (the b operator immediately to the left of the $b^\dagger$ operator), and to transform the remaining operators into normal form.

As an illustration of Wick's theorem we consider the following example:

$$b_i b_j^\dagger b_k b_l^\dagger = \{b_i b_j^\dagger b_k b_l^\dagger\} + \{\overset{\frown}{b_i b_j^\dagger} b_k b_l^\dagger\}$$

$$+ \{\overset{\frown}{b_i b_j^\dagger b_k} b_l^\dagger\} + \{b_i b_j^\dagger \overset{\frown}{b_k b_l^\dagger}\} + \{\overset{\frown}{b_i b_j^\dagger} \overset{\frown}{b_k b_l^\dagger}\}$$

$$= -b_j^\dagger b_l^\dagger b_i b_k - \delta_{ij} b_l^\dagger b_k + \delta_{il} b_j^\dagger b_k - \delta_{kl} b_j^\dagger b_i + \delta_{ij}\delta_{kl} .$$

Here, only the nonvanishing contractions (11.48b) have been indicated, that is, contractions between b and $b^\dagger$ operators with the b operator to the left. It is left as an exercise to verify that the result is in agreement with the anticommutation rules.

Problem 11.3 a) Use the anticommutation rules (11.13) to verify the result above. b) Verify (11.12) by applying Wick's theorem with normal ordering with respect to the *real vacuum*.

11.5.2 Proof of the Theorem

We shall now prove the time-independent form (11.49) of Wick's theorem essentially following *Bogoliubov* and *Shirkov* [1959]. We assume that the theorem is valid for an operator A which is a product of N particle-hole creation/absorption operators $(x_1, x_2, \ldots x_N)$

$$A = \{A\} + \{\overset{\frown}{A}\} \tag{11.50}$$

$$A = x_1 x_2 \ldots x_N,$$

and we shall show that this implies that the theorem is valid also for the operator B

$$B = \{B\} + \{\overset{\frown}{B}\}, \tag{11.51}$$

where B is

$$B = A x_{N+1} = x_1 x_2 \ldots x_N x_{N+1}$$

and x_{N+1} is a particle-hole creation or absorption operator. Then the general theorem follows from a simple inductive argument.

The term $\{\overset{\frown}{A}\}$ in (11.50) represents the sum of all normal-ordered terms with single, double, ... contractions within A, or more explicitly

$$
\{\overset{\frown}{A}\} = \{\overset{\frown}{x_1 x_2} x_3 \ldots x_N\} + \{\overset{\frown}{x_1 x_2 x_3} \ldots x_N\} + \{x_1 \overset{\frown}{x_2 x_3} \ldots x_N\} + \ldots
$$

$$
+ \{\overset{\frown}{x_1 x_2} \overset{\frown}{x_3 x_4} \ldots x_N\} + \{\overset{\frown}{x_1 \overset{\frown}{x_2 x_3} x_4} \ldots x_N\}
$$

$$
+ \{\overset{\frown}{x_1 \overset{\frown}{x_2 x_3} x_4} \ldots x_N\} + \ldots + \{\overset{\frown}{x_1 x_2} \overset{\frown}{x_3 x_4} \overset{\frown}{x_5 x_6} \ldots x_N\} + \ldots .
$$

From the assumption (11.50) and the definition of B we have

$$B = A x_{N+1} = \{A\} x_{N+1} + \{\overset{\frown}{A}\} x_{N+1} \tag{11.52}$$

and we shall now consider separately the two cases for which x_{N+1} is a particle-hole absorption and a particle-hole creation operator.

If x_{N+1} is a particle-hole *absorption* operator (b), it should be placed to the right in normal form, and it can be directly drawn inside the curly brackets of (11.52)

$$\{A\} b = \{A b\}$$

$$\{\overset{\frown}{A}\} b = \{\overset{\frown}{A} b\}.$$

According to (11.48) all contractions vanish when b appears to the right, and thus $\{\overline{Ab}\}$ represents the sum of *all* contracted terms within $B = Ab$, and the statement (11.51) is proved in this case.

If x_{N+1} is a *creation* operator ($b^\dagger$) it should be moved to the left in transforming $Ab^\dagger$ into normal form. It follows from (11.47) and the anticommutation relations that

$$\overline{x b^\dagger} = x b^\dagger + b^\dagger x ,$$

where x is a creation or an absorption operator. Hence

$$x_1 x_2 \ldots x_N b^\dagger = - x_1 x_2 \ldots b^\dagger x_N + x_1 x_2 \ldots \overline{x_N b^\dagger}$$

and similarly

$$x_1 x_2 \ldots x_{N-1} b^\dagger x_N = - x_1 x_2 \ldots b^\dagger x_{N-1} x_N + x_1 x_2 \ldots \overline{x_{N-1} b^\dagger} x_N .$$

The last term can also be written

$$-x_1 x_2 \ldots \overline{x_{N-1} x_N b^\dagger} .$$

By continuing this process, we find that

$$x_1 x_2 \ldots x_N b^\dagger = (-1)^N b^\dagger x_1 x_2 \ldots x_N + \overline{x_1 x_2 \ldots x_N b^\dagger}$$
$$+ x_1 \overline{x_2 \ldots x_N b^\dagger} + \ldots + x_1 x_2 \ldots \overline{x_N b^\dagger} .$$

Without loss of generality we can assume that $x_1 x_2 \ldots x_N$ is the normal form of A. We then get

$$\{A\} b^\dagger = (-1)^N b^\dagger \{A\} + \{\overline{Ab^\dagger}\} , \tag{11.53}$$

where the last term represents the sum of all normal-ordered terms with contractions between $b^\dagger$ and the operators of A. Since the creation operators appear to the left in normal form, the $b^\dagger$ operator in the first term on the right-hand side can be drawn inside the curly brackets. Furthermore, moving the $b^\dagger$ operator to the right inside this bracket can only introduce a phase factor depending on the parity of the permutation required. Thus,

$$b^\dagger \{A\} = \{b^\dagger A\} = (-1)^N \{Ab^\dagger\} ,$$

and (11.53) can be written

$$\{A\} b^\dagger = \{Ab^\dagger\} + \{\overline{Ab^\dagger}\} . \tag{11.54}$$

In the same way it can be shown that

$$\{\overline{A}\}b^\dagger = \{\overline{Ab^\dagger}\} + \{\overline{Ab^\dagger}\} \,, \tag{11.55}$$

where the last term represents the sum of all normal-ordered terms with contractions between $b^\dagger$ and the uncontracted operators of $\overline{A}$. Combining (11.52)
with (11.54, 55), we find that

$$Ab^\dagger = \{Ab^\dagger\} + \{\overline{Ab^\dagger}\} + \{\overline{Ab^\dagger}\} + \{\overline{Ab^\dagger}\} \,. \tag{11.56}$$

The last three terms here represent the sum of all normal-ordered terms with
single, double, ... contractions within $B = Ab^\dagger$, which proves the statement
(11.51) also in the case when x_{N+1} is a creation operator.

We have now shown that if Wick's theorem (11.49) holds for an arbitrary
product of N creation/absorption operators, then it holds also for an arbitrary
product of $N + 1$ such operators. Since the theorem obviously holds for $N = 1$,
it follows that it holds in general.

Problem 11.4. Derive the result (11.37) by means of Wick's theorem.

11.5.3 Wick's Theorem for Operator Products

Our analysis in the following chapters will be based on Wick's theorem. This
theorem assumes a very simple form for products of normal-ordered operators,
and this is the underlying reason why we shall express our operators in normal
form.

If A and B are each arbitrary products of creation and absoption operators
in normal form, then it follows directly from Wick's theorem (11.49) that

$$\boxed{AB = \{AB\} + \{\overline{AB}\}} \,, \tag{11.57}$$

where the last term represents the sum of the normal-ordered terms with all
possible contractions between the operators in A and those in B. Since A and
B are assumed to be in normal form, there are no nonvanishing contractions
within A or within B. This theorem is sometimes called the "generalized Wick's
theorem" [*Paldus* and *Čižek* 1975].

In this form, Wick's theorem can easily be interpreted in terms of diagrams.
It follows from (11.48) that only contractions between b operators in A and $b^\dagger$
operators in B can be nonvanishing. Furthermore, a contraction can be nonvanishing only if the operators involved represent the same orbital. If we assume
that there is a summation over the orbitals represented by the creation/absorp-
tion operators, then there will be a single summation over the orbital involved in
the contraction. This gives us immediately the following graphical interpretation
of Wick's theorem in the form (11.57).

> *If A and B are two operators in normal form represented by Goldstone dia-grams, then the graphical representation of the operator product AB is the sum of diagrams obtained by joining zero, one, two, . . . lines at the bottom of the diagram of A with lines at the top of the diagram of B in all possible ways so that the direction of the arrows is continuous.*

This rule is illustrated in Fig. 11.6. Diagram (a) represents the first term $\{AB\}$ without contractions, and the remaining diagrams represent the contracted term $\{\overline{AB}\}$. Diagrams (b-d) correspond to single contractions and (e, f) to double contractions.

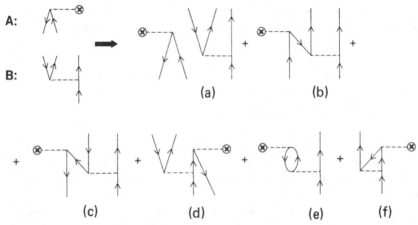

Fig. 11.6a-f. Illustration of Wick's theorem for operator products (11.57). Diagram (a) represents $\{AB\}$ without contractions and (b-f) represent $\{\overline{AB}\}$. Diagrams (b-d) correspond to single contractions and (e, f) to double contractions

11.6 The Wave Operator in Graphical Form

Our intention is to use Wick's theorem to express the Rayleigh-Schrödinger perturbation theory in second-quantized form and to represent the terms in the expansion by diagrams. For this purpose we need also a second-quantized form of the wave operator (or the correlation operator).

The correlation operator (9.70) generates the projection of the full wave function on the Q space. For a closed-shell system the components of the wave function in this space can be uniquely separated into single, double, . . . excitations, and we may then make a similar separation of the correlation operator, in analogy with the perturbation (11.39),

$$\chi = \sum_{ij} \{a_i^\dagger a_j\} x_j^i + \frac{1}{2} \sum_{ijkl} \{a_i^\dagger a_j^\dagger a_l a_k\} x_{kl}^{ij} + \frac{1}{3!} \sum \cdots . \tag{11.58}$$

The wave operator then becomes

$$\boxed{\Omega = 1 + \sum_{ij} \{a_i^\dagger a_j\} x_j^i + \frac{1}{2} \sum_{ijkl} \{a_i^\dagger a_j^\dagger a_l a_k\} x_{kl}^{ij} + \frac{1}{3!} \sum \cdots }. \tag{11.59}$$

The second term is a one-body operator, which generates single excitations, the third term a two-body operator, which generates double excitations, and so on. This kind of separation can be made also in the open-shell case, as will become clearer when we treat such systems in more detail later. Letting Ω_n represent the n-body part, we then get

$$\Omega = 1 + \chi = 1 + \Omega_1 + \Omega_2 + \cdots , \tag{11.60}$$

where

$$\begin{cases} \Omega_1 = \sum_{ij} \{a_i^\dagger a_j\} x_j^i \\[2mm] \Omega_2 = \frac{1}{2} \sum_{ijkl} \{a_i^\dagger a_j^\dagger a_l a_k\} x_{kl}^{ij} \\[2mm] \Omega_3 = \frac{1}{3!} \sum_{ijklmn} \{a_i^\dagger a_j^\dagger a_k^\dagger a_n a_m a_l\} x_{lmn}^{ijk} . \end{cases} \tag{11.61}$$

In connection with the introduction of the wave operator in Chap. 9, we mentioned that this operator was defined only when operating to the right on the model space, which means that only the ΩP part was defined. The second-quantized form (11.59), however, is defined in the entire space, once the coefficients are determined. These coefficients are obtained from the Bloch equation (9.72, 73), which we shall write in the following as

$$[\Omega, H_0] = [\chi, H_0] = V\Omega - \Omega P V\Omega = QV\Omega - \chi P V\Omega , \tag{11.62}$$

leaving out the projection operator to the right. Only terms which give non-zero results when operating on the model space will be considered, but these do not necessarily give zero result when operating on the Q space.

For *closed-shell systems* the general expression (11.59) can be considerably simplified. The wave operator then operates to the right on a closed-shell state $|a\rangle$, and it follows from (11.43) that the normal-ordered form cannot contain any b operator. In other words, the absorption operators must represent hole (core) states and the creation operators particle (virtual) states. The wave operator (11.59) then reduces to

$$\Omega = 1 + \sum_{ar} a_r^\dagger a_a x_a^r + \frac{1}{2} \sum_{abrs} a_r^\dagger a_s^\dagger a_b a_a x_{ab}^{rs} + \cdots \qquad (11.63)$$

and the separation (11.61) to

$$\begin{cases} \Omega_0 = 1 \\[4pt] \Omega_1 = \sum_{ar} a_r^\dagger a_a x_a^r \\[4pt] \Omega_2 = \frac{1}{2} \sum_{abrs} a_r^\dagger a_s^\dagger a_b a_a x_{ab}^{rs} \\[4pt] \Omega_3 = \frac{1}{3!} \sum_{abcrst} a_r^\dagger a_s^\dagger a_t^\dagger a_c a_b a_a x_{abc}^{rst} \\[4pt] \text{etc.} \end{cases} \qquad (11.64)$$

Here, the curly brackets are omitted, since the operators are automatically in normal form.

The one-, two-, ... body terms can be represented by means of diagrams analogous to those of the perturbation in Fig. 11.5. The main difference is that the dotted line, which represents the Coulomb interaction, has to be replaced by some other symbol representing the coefficients in the operator. For closed-shell systems we introduce the following notations:

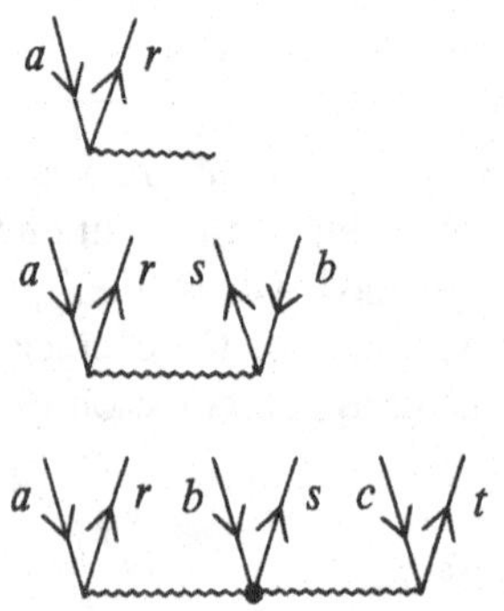

$$= a_r^\dagger a_a x_a^r$$

$$= \frac{1}{2} a_r^\dagger a_s^\dagger a_b a_a x_{ab}^{rs} \qquad (11.65)$$

$$= \frac{1}{3!} a_r^\dagger a_s^\dagger a_t^\dagger a_c a_b a_a x_{abc}^{rst}.$$

Since there are no b operators in these terms, there are no orbital lines below the vertices (see Fig. 11.4). These terms form the correlation operator (11.60), which thus has the graphical representation shown in Fig. 11.7.

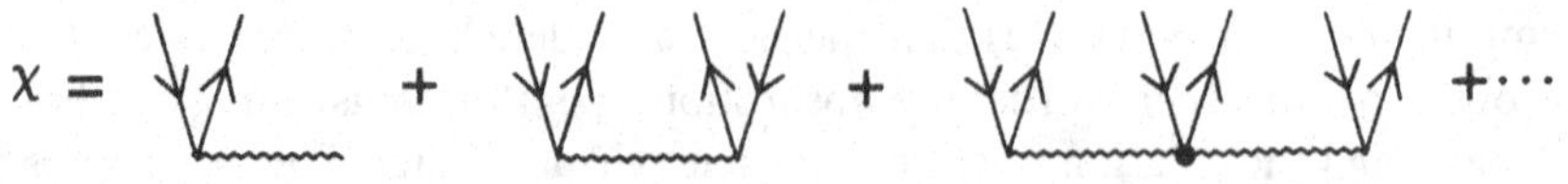

Fig. 11.7. Graphical representation of the correlation operator in the closed-shell case. A summation over the free lines is implemented

In order to solve the Bloch equation (11.62) using second quantization, we first express the right-hand side in normal form, using Wick's theorem, and then identify the terms with those obtained from the commutator on the left-hand side.

With the definition (9.70) we get

$$V\Omega = V + V\chi \,. \tag{11.66}$$

V is given in normal form by (11.39) and the operator product $V\chi$ can be normal-ordered by means of Wick's theorem, starting from the normal forms of V and χ.

The procedure indicated above can also be performed directly in a graphical way by means of diagrams and the graphical form of Wick's theorem, illustrated in Fig. 11.6. This graphical procedure forms the basis of the many-body perturbation formalism we shall develop in the following chapters. Before applying this procedure we shall derive a normal-ordered representation also of the commutator of (11.62).

In order to find the normal form of the commutator $[\Omega, H_0]$, we consider an arbitrary matrix element, using the eigenfunctions of H_0 as the basis functions

$$\langle \beta | [\Omega, H_0] | \alpha \rangle = \langle \beta | \Omega H_0 - H_0 \Omega | \alpha \rangle \,.$$

By operating with H_0 to the right in the first term and to the left in the second term, this can be transformed as in (9.79) to

$$\langle \beta | [\Omega, H_0] | \alpha \rangle = (E_0^\alpha - E_0^\beta) \langle \beta | \Omega | \alpha \rangle \,, \tag{11.67}$$

where E_0^α and E_0^β are the eigenvalues of H_0 (9.63).

If Ω is represented by the one-body term in (11.61)

$$\Omega_1 = \sum_{ij} \{a_i^\dagger a_j\} x_j^i \,,$$

we get with (11.67)

$$\langle \beta | [\Omega_1, H_0] | \alpha \rangle = (E_0^\alpha - E_0^\beta) \sum_{ij} \langle \beta | \{a_i^\dagger a_j\} | \alpha \rangle x_j^i \,.$$

Obviously, the matrix element on the right-hand side is nonzero only if

$$| \beta \rangle = \{a_i^\dagger a_j\} | \alpha \rangle \,,$$

which represents a single excitation $j \to i$. Then the energy difference is simply the difference between the orbital energies

$$E_0^\alpha - E_0^\beta = \varepsilon_j - \varepsilon_i \,,$$

and we get

$$\langle \beta | [\Omega_1, H_0] | \alpha \rangle = \sum_{ij} (\varepsilon_j - \varepsilon_i) \langle \beta | \{a_i^\dagger a_j\} | \alpha \rangle x_j^i .$$

This relation holds for any $|\alpha\rangle$ and $|\beta\rangle$, and therefore it can be written as an operator identity

$$[\Omega_1, H_0] = \sum_{ij} (\varepsilon_j - \varepsilon_i) \{a_i^\dagger a_j\} x_j^i . \tag{11.68}$$

Similar arguments can be applied to the two-, three-, ... body parts of the correlation operator, which yields the following second-quantized form of the commutator:

$$
\begin{aligned}
[\Omega, H_0] = &\sum_{ij} \{a_i^\dagger a_j\} (\varepsilon_j - \varepsilon_i) x_j^i \\
&+ \frac{1}{2} \sum_{ijkl} \{a_i^\dagger a_j^\dagger a_l a_k\} (\varepsilon_k + \varepsilon_l - \varepsilon_i - \varepsilon_j) x_{kl}^{ij} + \cdots
\end{aligned}
\tag{11.69}
$$

Like (11.59), this expression is completely general, and for closed-shell systems it reduces to

$$[\Omega, H_0] = \sum_{ar} a_r^\dagger a_a (\varepsilon_a - \varepsilon_r) x_a^r + \frac{1}{2} \sum_{abrs} a_r^\dagger a_s^\dagger a_b a_a (\varepsilon_a + \varepsilon_b - \varepsilon_r - \varepsilon_s) x_{ab}^{rs} .$$

The relation (11.69) can also be derived in a more formal way using the anti-commutation rules (or Wick's theorem) and the second-quantized form of H_0 given in (11.35) (see problem 11.5).

The commutator can be represented graphically by means of the same diagrams as the correlation operator itself (Fig. 11.7), if we associate with each diagram the energy

$$D = \sum (\varepsilon_{in} - \varepsilon_{out}), \tag{11.70}$$

where ε_{in} (ε_{out}) represent the eigenvalues associated with incoming (outgoing) orbital lines. This energy corresponds to the *excitation energy* associated with the particular term.

In principle, we have now developed all the tools we need in order to express the Rayleigh-Schrödinger perturbation theory in second-quantized form and hence also in graphical form. In the next chapter we shall apply this procedure to the order-by-order expansion for closed-shell systems, leading to the linked-diagram expansion. In Chap. 13 the same procedure will be applied to open-shell systems.

Problem 11.5 a) Use Wick's theorem to verify the commutation relation

$$[\{a_i^\dagger a_j\}, \{a_k^\dagger a_k\}] = (\delta_{jk} - \delta_{ik}) \{a_i^\dagger a_j\} .$$

b) Use this result to verify the first term in (11.69).

12. Application of Perturbation Theory to Closed-Shell Systems

In the remaining chapters of the book we shall apply perturbation theory to different kinds of atomic systems. In the present chapter, we shall generate the diagrammatic expansion for closed-shell systems. This will enable us to represent the wave operator and the energy for such systems in a simple way by means of diagrams. The evaluation of diagrams will be considered in detail, and we shall formulate general rules for their evaluation. We shall also discuss the separation of the diagrams into radial and spin-angular parts, taking advantage of the angular-momentum graphs developed in the earlier chapters.

As an example of a perturbation calculation based on a diagrammatic analysis we shall consider the calculation of *Kelly* of the ground-state energy of the beryllium atom, which was the first application of many-body perturbation theory to an atomic system.

12.1 The First-Order Contributions to the Wave Function and the Energy

We begin our treatment of closed-shell systems by rederiving the first-order wave function obtained in Chap. 10, using the more general formalism developed in the previous chapter. As before, we assume that the zeroth-order or model function is a single determinant (10.8)

$$\Psi_0 = |\alpha\rangle \,, \tag{12.1}$$

containing only orbitals from filled shells. This is the only function in the model space (P space), and the projection operators corresponding to the P and Q spaces are

$$
\begin{aligned}
P &= |\alpha\rangle\langle\alpha| \\
Q &= \sum_{\beta\neq\alpha} |\beta\rangle\langle\beta| = 1 - P \,.
\end{aligned}
\tag{12.2}
$$

The first-order wave function (9.96) is

$$\Psi^{(1)} = \Omega^{(1)} |\alpha\rangle \,,$$

where the first-order wave operator satisfies the equation (9.75a)

$$[\Omega^{(1)}, H_0] \, P = QVP \, . \tag{12.3}$$

The procedure we now use is to express the right-hand side of (12.3) in normal form and equate this with the normal form of the commutator (11.69).

The perturbation V is given in normal form by (11.39). The zero-body term V_0 is a number and it cannot contribute to the right-hand side of (12.3) due to the projection operators. The one-body part of V gives the following contribution:

$$QV_1 P = Q \sum_{ij} \{a_i^{\dagger} a_j\} \langle i|v|j \rangle \, P \, , \tag{12.4}$$

where i and j, as usual, run over all orbitals. Since P is here of the form (12.2), only terms for which j is a core orbital and i is virtual contribute [see (11.63)]. We can then write (12.4) as

$$QV_1 P = Q \sum_{ar} a_r^{\dagger} a_a \langle r|v|a \rangle \, P \, , \tag{12.5}$$

where a runs over core orbitals and r over virtual orbitals.

The first-order wave operator can be separated into one- and two-body parts in analogy with (11.64)

$$\Omega^{(1)} = \Omega_1^{(1)} + \Omega_2^{(1)} \, ,$$

where

$$\Omega_1^{(1)} = \sum_{ar} a_r^{\dagger} a_a x_a^{r(1)}$$

$$\Omega_2^{(1)} = \frac{1}{2} \sum_{abrs} a_r^{\dagger} a_s^{\dagger} a_b a_a \, x_{ab}^{rs(1)} \, . \tag{12.6}$$

The commutator with H_0 (11.69) gives for the one-body part of $\Omega^{(1)}$

$$[\Omega_1^{(1)}, H_0] = \sum_{ar} a_r^{\dagger} a_a (\varepsilon_a - \varepsilon_r) x_a^{r(1)} \, . \tag{12.7}$$

Equating this with (12.5) gives immediately

$$x_a^{r(1)} = \frac{\langle r|v|a \rangle}{\varepsilon_a - \varepsilon_r} \tag{12.8}$$

and the full one-body part of the first-order wave operator becomes

$$\Omega_1^{(1)} = \sum_{ar} a_r^{\dagger} a_a \frac{\langle r|v|a \rangle}{\varepsilon_a - \varepsilon_r} \, . \tag{12.9}$$

According to (10.21) this term vanishes when Hartree-Fock orbitals are used.

Similarly, the two-body part of the perturbation gives

$$QV_2P = \frac{1}{2} Q \sum_{abrs} a_r^\dagger a_s^\dagger a_b a_a \langle rs | r_{12}^{-1} | ab \rangle P \tag{12.10}$$

and the corresponding part of the commutator is

$$[\Omega_2^{(1)}, H_0] = \frac{1}{2} \sum_{abrs} a_r^\dagger a_s^\dagger a_b a_a (\varepsilon_a + \varepsilon_b - \varepsilon_r - \varepsilon_s) x_{ab}^{rs(1)} . \tag{12.11}$$

Equating these two expressions leads to

$$x_{ab}^{rs(1)} = \frac{\langle rs | r_{12}^{-1} | ab \rangle}{\varepsilon_a + \varepsilon_b - \varepsilon_r - \varepsilon_s} \tag{12.12}$$

and

$$\Omega_2^{(1)} = \frac{1}{2} \sum_{abrs} a_r^\dagger a_s^\dagger a_b a_a \frac{\langle rs | r_{12}^{-1} | ab \rangle}{\varepsilon_a + \varepsilon_b - \varepsilon_r - \varepsilon_s} . \tag{12.13}$$

We observe that the energy *numerator* in the commutators (12.7, 11) gives rise to an energy *denominator* in the corresponding contributions to the wave operator or the wave function.

It should be observed that the identification procedure used above to obtain (12.12) is not unique. Each physical excitation occurs four times in the second-quantized form of the two-body operator, and obviously only the sum of these four terms has any significance. The anticommutation relations (11.13) may be used to show that only the antisymmetric part of the coefficients $x_{ab}^{rs(1)}$ contributes to the final result. The identification (12.12) above, however, is the most natural one in the graphical procedure we employ, and we shall evaluate higher-order contributions in a similar manner.

The first-order contributions (12.9, 13) may be summarized as follows:

$$\begin{aligned}
\Omega^{(1)} &= \sum_{ar} a_r^\dagger a_a x_a^{r(1)} + \frac{1}{2} \sum_{abrs} a_r^\dagger a_s^\dagger a_b a_a x_{ab}^{rs(1)} \\
&= \sum_{ar} a_r^\dagger a_a \frac{\langle r | v | a \rangle}{\varepsilon_a - \varepsilon_r} + \frac{1}{2} \sum_{abrs} a_r^\dagger a_s^\dagger a_b a_a \frac{\langle rs | r_{12}^{-1} | ab \rangle}{\varepsilon_a + \varepsilon_b - \varepsilon_r - \varepsilon_s} .
\end{aligned} \tag{12.14}$$

Operating with $\Omega^{(1)}$ on the unperturbed state $|\alpha\rangle$, we recover the first-order result obtained in (10.20).

The first-order results above can also be derived graphically. The right-hand side of (12.3) is represented by those diagrams of V_1 and V_2 which have no free lines at the bottom, namely

$$QVP = \quad a \bigvee \bigwedge r \quad + \quad a \bigvee \bigwedge r \quad s \bigvee \bigvee b \tag{12.15}$$

Here it is understood that a summation is performed over the orbitals represented by the free lines and furthermore that there is a P operator to the right. The diagrams correspond to the following expressions:

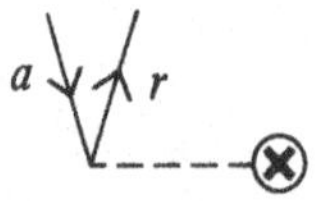

$$= \{a_r^\dagger a_a\} \langle r|v|a \rangle = a_r^\dagger a_a \langle r|v|a \rangle \tag{12.16a}$$

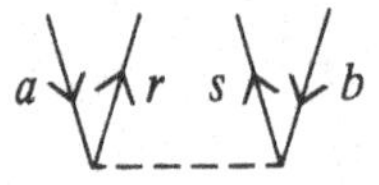

$$= \frac{1}{2} \, {}_a a_s^\dagger a_b a_a \langle rs|r_{12}^{-1}|ab \rangle \ . \tag{12.16b}$$

The first-order wave operator $\Omega^{(1)}$ is obtained by supplying to the diagrams above the energy denominator associated with the commutator. Representing the resulting expressions with the same diagrams, we obtain the graphical representation shown in Fig. 12.1.

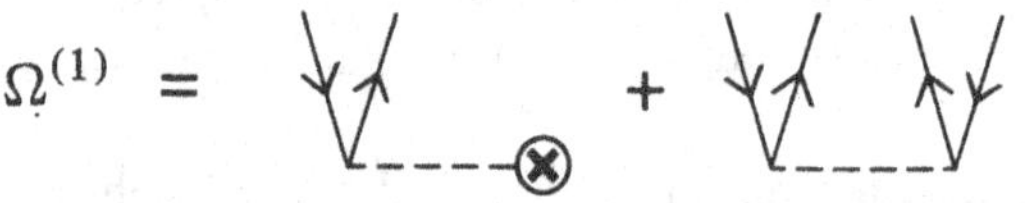

$$\Omega^{(1)} \ = \qquad + $$

Fig. 12.1. Graphical representation of the first-order wave operator (12.14). This is the same as the corresponding representation of the first-order wave function in Fig. 10.3. It is understood that an energy denominator is associated with each diagram and that a summation is performed over all orbital lines

Finally, we consider the first-order energy. According to (9.99) this is

$$E^{(1)} = \langle \alpha| H_{\text{eff}}^{(1)}|\alpha \rangle = \langle \alpha|V|\alpha \rangle \ . \tag{12.17}$$

Only the zero-body part of the perturbation (11.39) contributes here, giving

$$E^{(1)} = \langle \alpha|V|\alpha \rangle = \langle \alpha|V_0|\alpha \rangle = V_0 \ . \tag{12.18}$$

The graphical form of V_0 is given in the first line of Fig. 11.5, which is identical to the representation of $E^{(1)}$ given in Fig. 10.4.

12.2 The Second-Order Wave Operator

12.2.1 Construction and Evaluation of Diagrams

The second-order contribution to the wave operator satisfies the equation (9.75b)

$$[\Omega^{(2)}, H_0] P = QV\Omega^{(1)}P - \Omega^{(1)}PVP \ . \tag{12.19}$$

The last term in this equation can be written

$$-\Omega^{(1)}PVP = -\Omega^{(1)}|a\rangle\langle a|V|a\rangle\langle a| = -V_0\Omega^{(1)}P,\tag{12.20}$$

where we have used the form (12.2) of the projection operator together with (12.18). This term cancels the V_0 part of $QV\Omega^{(1)}P$, and (12.19) reduces to

$$[\Omega^{(2)}, H_0] = Q(V_1 + V_2)\Omega^{(1)},\tag{12.21}$$

leaving out the P operator to the right. We shall now express V_1, V_2 and $\Omega^{(1)}$ in normal form and then use Wick's theorem to evaluate the operator products $V_1\Omega^{(1)}$ and $V_2\Omega^{(1)}$.

Using (11.39) and (12.14), the right-hand side of (12.21) becomes

$$Q\left(\sum_{ij}\{a_i^\dagger a_j\}\langle i|v|j\rangle + \frac{1}{2}\sum_{ijkl}\{a_i^\dagger a_j^\dagger a_l a_k\}\langle ij|r_{12}^{-1}|kl\rangle\right)$$
$$\times\left(\sum_{ar}a_r^\dagger a_a\frac{\langle r|v|a\rangle}{\varepsilon_a - \varepsilon_r} + \frac{1}{2}\sum_{abrs}a_r^\dagger a_s^\dagger a_b a_a\frac{\langle rs|r_{12}^{-1}|ab\rangle}{\varepsilon_a + \varepsilon_b - \varepsilon_r - \varepsilon_s}\right).\tag{12.22}$$

This expression can easily be transformed into normal form by means of Wick's theorem (11.57). The coefficients of $\Omega^{(2)}$ can then be evaluated by equating the various terms which arise in this way with the corresponding terms in the commutator (11.69). This provides an additional energy denominator, associated with the excitation process in the second interaction. We shall now consider the construction of the diagrams of $\Omega^{(2)}$ in some detail.

The diagrams corresponding to (12.22) are obtained by joining the lines at the bottom of the diagrams of V_1 and V_2 (Fig. 11.5) with the orbital lines of $\Omega^{(1)}$ (Fig. 12.1) in all possible ways. For each diagram of V this gives rise to one disconnected diagram for which the lines V are not joined to the lines of $\Omega^{(1)}$, and a number of diagrams for which one or more lines of V and $\Omega^{(1)}$ are joined. Since Ω always operates to the right on the P space, any resulting diagram with a free line at the bottom will give a null result. So *all orbital lines at the bottom of V_1 and V_2 must be connected to orbital lines of $\Omega^{(1)}$*. Furthermore, since the final state is in the Q space, all diagrams must have at least one pair of free lines at the top. It should also be observed that the intermediate state must be in the Q space, but this condition is automatically satisfied with the graphical procedure we use here.

The construction of a few second-order wave-operator diagrams is illustrated in Fig. 12.2. The first two diagrams in the figure originate from the term

$$Q\sum_{ij}\{a_i^\dagger a_j\}\langle i|v|j\rangle\sum_{ar}a_r^\dagger a_a\frac{\langle r|v|a\rangle}{\varepsilon_a - \varepsilon_r}\tag{12.23}$$

in (12.22). In diagram (a), i and j represent virtual states. In the diagram to the right j and r are connected, which corresponds to the contraction

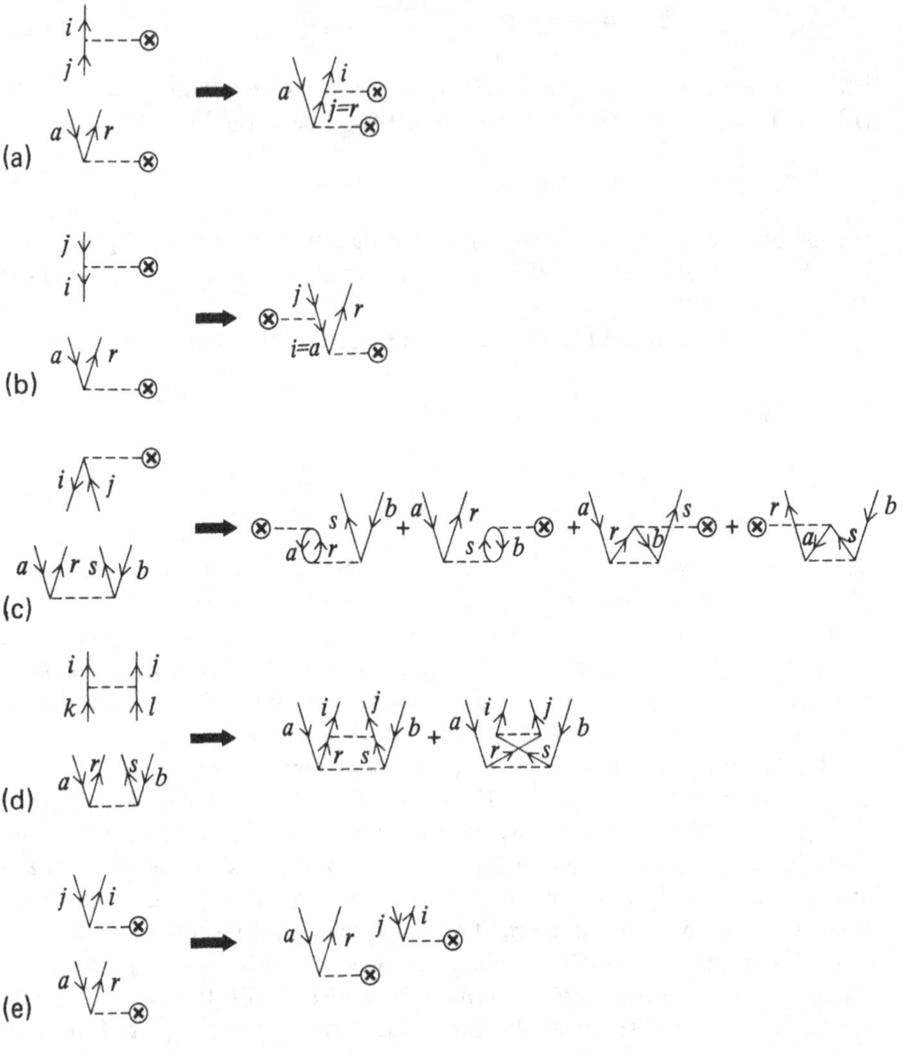

Fig. 12.2 a-e. Illustration of the construction of some second-order wave-operator diagrams. The algebraic expressions for these diagrams are obtained by identifying them with the corresponding terms in (12.22)

$$Q \sum_{ij} \{a_i^\dagger a_j\} \langle i|v|j\rangle \sum_{ar} a_r^\dagger a_a \frac{\langle r|v|a\rangle}{\varepsilon_a - \varepsilon_r}$$

$$= Q \sum_{ai} a_i^\dagger a_a \sum_r \frac{\langle i|v|r\rangle\langle r|v|a\rangle}{\varepsilon_a - \varepsilon_r}. \tag{12.24}$$

According to (11.69), the one-body part of the commutator in (12.21) is

$$\sum_{as} a_s^\dagger a_a (\varepsilon_a - \varepsilon_s)\, x_a^{s(2)}. \tag{12.25}$$

The algebraic expression for diagram (a) is then obtained by identifying (12.24), which originates from the right-hand side of (12.21), with the corresponding term in the commutator (12.25). As can be seen, this gives an additional energy denominator, which is associated with the final excitation. After some change in the notation, we obtain the following expression for diagram (a):

$$= a_s^\dagger a_a \sum_r \frac{\langle s|v|r\rangle\,\langle r|v|a\rangle}{(\varepsilon_a - \varepsilon_s)\,(\varepsilon_a - \varepsilon_r)}. \tag{12.26}$$

Similarly, in the diagram to the right in Fig. 12.2b, i and a are joined, which corresponds to the contraction

$$Q \sum_{ij} \{a_i^\dagger a_j\} \langle i|v|j\rangle \sum_{ar} a_r^\dagger a_a \frac{\langle r|v|a\rangle}{\varepsilon_a - \varepsilon_r}$$

$$= -Q \sum_{jr} a_r^\dagger a_j \sum_a \frac{\langle a|v|j\rangle\,\langle r|v|a\rangle}{\varepsilon_a - \varepsilon_r}. \tag{12.27}$$

An even permutation is required to bring $a_i^\dagger$ and a_a in contact with each other, which does not change the phase. Since j represents an occupied orbital, we reverse the order of a_j and $a_r^\dagger$ so that they appear in the same order as in the first-order expression (12.9). This leads to the negative sign in the last line of (12.27). The expression for the corresponding wave-operator diagram is obtained by adding an energy denominator for the final excitation, as before. Again, after changing the notations to our usual conventions, we get

$$= - a_r^\dagger a_b \sum_a \frac{\langle a|v|b\rangle\,\langle r|v|a\rangle}{(\varepsilon_b - \varepsilon_r)\,(\varepsilon_a - \varepsilon_r)}. \tag{12.28}$$

The problem of evaluating the remaining diagrams in Fig. 12.2 is left as an exercise (see Problems 12.1, 12.2). These results will later be used to motivate general rules for evaluating diagrams of this kind.

Problem 12.1. Apply the procedure illustrated above to verify the following results:

$$= \frac{1}{2} a_s^\dagger a_b \sum_{ar} \frac{\langle a|v|r\rangle \langle rs|r_{12}^{-1}|ab\rangle}{(\varepsilon_b - \varepsilon_s)(\varepsilon_a + \varepsilon_b - \varepsilon_r - \varepsilon_s)} \qquad (12.29a)$$

$$= \frac{1}{2} a_r^\dagger a_a \sum_{bs} \frac{\langle b|v|s\rangle \langle rs|r_{12}^{-1}|ab\rangle}{(\varepsilon_a - \varepsilon_r)(\varepsilon_a + \varepsilon_b - \varepsilon_r - \varepsilon_s)}. \qquad (12.29b)$$

Problem 12.2. Evaluate in a similar way the remaining diagrams in Fig. 12.2.

12.2.2 Equivalent Diagrams and Weight Factors

In considering the diagrams which appear above, we see that some diagrams are mirror images of each other. This is the case, for instance, with the diagrams (12.29). Such diagrams are said to be (topologically) *equivalent*. More generally, we shall say that

two diagrams are (topologically) equivalent, if they can be transformed into one another by a reflection in a vertical plane or a distortion which keeps all vertices on the same horizontal level.

Diagrams which are not equivalent in this sense are said to be nonequivalent or *distinct*.

It is obvious that equivalent diagrams give the same contributions when a summation is performed also over the orbitals corresponding to the free lines. Therefore, it is sufficient to include distinct diagrams in the perturbation expansion, provided that a numerical factor is assigned to each diagram corresponding to the number of equivalent diagrams it represents. Together with the factor of 1/2 associated with each Coulomb interaction, this defines the "*weight factor*" or simply the "*weight*" of the diagram.

This means, for instance, that if we include only one of the diagrams (12.29) in the expansion, then we should give it the weight of 1 and associate with it the expression

$$= a_s^\dagger a_b \sum_{ar} \frac{\langle a|v|r\rangle \langle rs|r_{12}^{-1}|ab\rangle}{(\varepsilon_b - \varepsilon_s)(\varepsilon_a + \varepsilon_b - \varepsilon_r - \varepsilon_s)}. \quad (12.30a)$$

In a similar way, the corresponding exchange diagram becomes (see Problem 12.2)

$$= -a_s^\dagger a_a \sum_{br} \frac{\langle b|v|r\rangle \langle rs|r_{12}^{-1}|ab\rangle}{(\varepsilon_a - \varepsilon_s)(\varepsilon_a + \varepsilon_b - \varepsilon_r - \varepsilon_s)} \quad (12.30b)$$

The two diagrams in Fig. 12.2d are also equivalent, since they can be distorted into one another. (The labels on the free lines have no significance here, since a summation will be carried out finally over them). These diagrams originate from the combination of the two two-body terms in (12.22), each of which has a factor of 1/2 associated with it. Including only one of the diagrams, we assign to it the weight of 1/2 and the algebraic expression

$$= \frac{1}{2} a_t^\dagger a_u^\dagger a_b a_a \sum_{rs} \frac{\langle tu|r_{12}^{-1}|rs\rangle \langle rs|r_{12}^{-1}|ab\rangle}{(\varepsilon_a + \varepsilon_b - \varepsilon_t - \varepsilon_u)(\varepsilon_a + \varepsilon_b - \varepsilon_r - \varepsilon_s)} \quad (12.31)$$

This diagram is referred to as a particle-particle interaction diagram, since the two electrons interact in the virtual (particle) states.

As an additional example we consider some particle-hole diagrams, for which one electron in a virtual state and one in a core state interact

$$= -\frac{1}{2} a_t^\dagger a_s^\dagger a_b a_c \sum_{ar} \frac{\langle at|r_{12}^{-1}|cr\rangle \langle rs|r_{12}^{-1}|ab\rangle}{(\varepsilon_c + \varepsilon_b - \varepsilon_t - \varepsilon_s)(\varepsilon_a + \varepsilon_b - \varepsilon_r - \varepsilon_s)} \quad (12.32a)$$

$$= -\frac{1}{2} a_r^\dagger a_t^\dagger a_c a_a \sum_{bs} \frac{\langle tb|r_{12}^{-1}|sc\rangle \langle rs|r_{12}^{-1}|ab\rangle}{(\varepsilon_a + \varepsilon_c - \varepsilon_r - \varepsilon_t)(\varepsilon_a + \varepsilon_b - \varepsilon_r - \varepsilon_s)} \quad (12.32b)$$

These two diagrams are topologically equivalent in the same sense as the diagrams (12.29), and considering only distinct diagrams we include only one of them with the weight of 1.

In Sect. 12.4 we shall formulate general rules which make it possible to write down the algebraic expression associated with a particular diagram directly without going through the fairly tedious identification procedure employed here.

12.2.3 Choices of Single-Particle States

If *Hartree-Fock* orbitals are used, there is no one-body part of the perturbation (11.40) and no single excitations in the first-order wave function. In that case, the number of second-order diagrams is considerably reduced. The remaining diagrams are given in Fig. 12.3.

We see that single excitations, which are eliminated in first order by using Hartree-Fock orbitals, reappear in second order (diagrams a-d). It is possible, however, to choose another set of orbitals so that single excitations are eliminated to *all* orders of perturbation theory.

Using the form (11.64) of the correlation operator for closed-shell systems, we can write the exact wave function as

$$\Psi = \Omega|\alpha\rangle = (1 + \chi)|\alpha\rangle$$
$$= |\alpha\rangle + \sum_{ar} x_a^r|\alpha_a^r\rangle + \frac{1}{2}\sum_{abrs} x_{ab}^{rs}|\alpha_{ab}^{rs}\rangle + \cdots .$$

Single excitations disappear, if $x_a^r = 0$ for all a and r. These coefficients can be expressed

$$x_a^r = \langle \alpha_a^r|\Psi\rangle ,$$

so that the condition for vanishing single excitations is

$$\langle \alpha_a^r|\Psi\rangle = 0 .$$

Using the Schrödinger equation, $H\Psi = E\Psi$, this can also be written

$$\langle \alpha_a^r|H|\Psi\rangle = \langle \alpha|a_a^\dagger a_r H|\Psi\rangle = 0 , \tag{12.33}$$

which is usually referred to as the *Brillouin-Brueckner condition*. This is analogous to the ordinary *Brillouin condition* (5.30) [*Brillouin* 1933, 1934]

$$\langle \alpha_a^r|H|\alpha\rangle = \langle \alpha|a_a^\dagger a_r H|\alpha\rangle = 0 , \tag{12.34}$$

which leads to the Hartree-Fock equations. As mentioned in Sect. 10.1, the Brillouin condition eliminates single excitations in first order, while the Brillouin-Brueckner condition eliminates them to all orders. Orbitals satisfying this condition are known as *Brueckner orbitals*. They are sometimes also referred to as

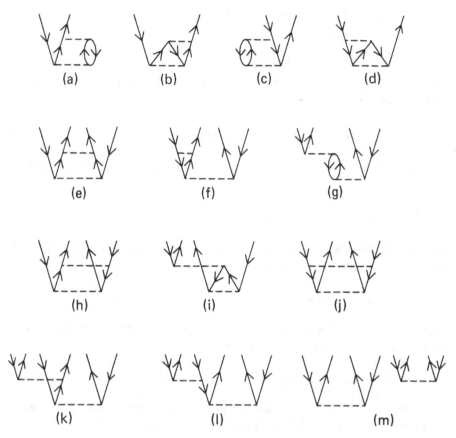

(a)   (b)   (c)   (d)

(e)   (f)   (g)

(h)   (i)   (j)

(k)   (l)   (m)

Fig. 12.3. The distinct second-order wave-operator diagrams when Hartree-Fock orbitals are used (no effective-potential interactions). Diagrams (a-d) represent single, (e-j) double, (k-l) triple, and diagram (m) quadruple excitations. Single excitations disappear if Brueckner orbitals are used

"*maximum-overlap orbitals*", since it can be shown that they lead to the maximum overlap between the exact wave function Ψ and the (normalized) zeroth-order wave function Ψ_0,

$$\langle \Psi_0 | \Psi \rangle = \text{maximum}$$

[*Brenig* 1957; *Nesbet* 1958; *Löwdin* 1962; *Kobe* 1971; *Schäfer* and *Weidenmüller* 1971].

For closed-shell systems, it has been found that single excitations are relatively unimportant, which means that the Brueckner orbitals are quite close to the Hartree-Fock orbitals. For open shells, on the other hand, single excitations can be quite significant, as we shall demonstrate in the following chapters.

12.3 Perturbation Expansion of the Energy

12.3.1 The Correlation Energy

Using the form (9.84b) of the effective Hamiltonian, the exact energy of a closed-shell system can be written

$$E = \langle \alpha | H_{\text{eff}} | \alpha \rangle = \langle \alpha | H | \alpha \rangle + \langle \alpha | V \chi | \alpha \rangle . \tag{12.35}$$

The first term is the expectation value of the energy of the zeroth-order state $|\alpha\rangle$,

$$\langle E \rangle = \langle \alpha | H | \alpha \rangle = \langle \alpha | H_0 + V | \alpha \rangle = E_0 + E^{(1)} ,$$

and the last term in (12.35) represents the *energy shift* caused by second- and higher-order perturbations

$$E - \langle E \rangle = \langle \alpha | V \chi | \alpha \rangle = E^{(2)} + E^{(3)} + \cdots .$$

Since χ operates to the left on the Q space, the zero-body part of V does not contribute, and the energy shift becomes

$$E - \langle E \rangle = \langle \alpha | (V_1 + V_2) \chi | \alpha \rangle . \tag{12.36}$$

Expanding the correlation operator in powers of the perturbation, we obtain the following form of the kth-order energy contribution:

$$E^{(k)} = \langle \alpha | (V_1 + V_2) \Omega^{(k-1)} | \alpha \rangle \qquad k \geq 2 . \tag{12.37}$$

The Hartree-Fock orbitals are defined to minimize the expectation value of the energy with a determinantal wave function. When Hartree-Fock orbitals are used, the energy shift is referred to as the *correlation energy*

$$\boxed{\begin{aligned} E_{\text{corr}} &= E - \langle E \rangle_{\text{HF}} \\ &= \langle \alpha | V_2 \chi | \alpha \rangle_{\text{HF}} = E_{\text{HF}}^{(2)} + E_{\text{HF}}^{(3)} + \cdots \end{aligned}} \tag{12.38}$$

Since the energy shift is always negative, the correlation energy is numerically the smallest of all energy shifts obtained with independent-particle functions.

12.3.2 The Second-Order Energy

According to (12.37), the second-order energy of a closed-shell system can be written

$$E^{(2)} = \langle \alpha | (V_1 + V_2) \Omega^{(1)} | \alpha \rangle .$$

This can be expressed as (12.22) using the normal forms of the perturbation (11.39) and the first-order wave operator (12.14)

$$E^{(2)} = \langle \alpha | \left(\sum_{ij} \{a_i^\dagger a_j\} \langle i|v|j \rangle + \frac{1}{2} \sum_{ijkl} \{a_i^\dagger a_j^\dagger a_l a_k\} \langle ij|r_{12}^{-1}|kl \rangle \right)$$

$$\times \left(\sum_{ar} a_r^\dagger a_a \frac{\langle r|v|a \rangle}{\varepsilon_a - \varepsilon_r} + \frac{1}{2} \sum_{abrs} a_r^\dagger a_s^\dagger a_b a_a \frac{\langle rs|r_{12}^{-1}|ab \rangle}{\varepsilon_a + \varepsilon_b - \varepsilon_r - \varepsilon_s} \right) |\alpha \rangle . \tag{12.39}$$

The graphical representation of the second-order energy can be obtained in the same way as for the second-order wave operator. The only difference here is that the final state should be in the P space. This means that no free lines whatsoever are allowed—or, in other words, that the diagrams of $\Omega^{(1)}$ should be *closed* by the diagrams of V.

The one-body part of $\Omega^{(1)}$ can be closed only by a one-body diagram of V_1 with both free lines *below* the vertex

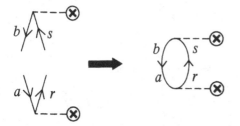

Identifying this diagram with the corresponding term in (12.39) leads to the following expression:

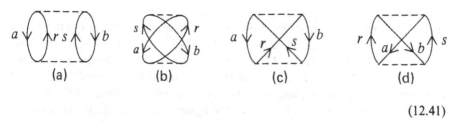

$$= \sum_{ar} \frac{\langle a|v|r \rangle \langle r|v|a \rangle}{\varepsilon_a - \varepsilon_r} . \tag{12.40}$$

The two-body part of $\Omega^{(1)}$ can be closed by V_2 in four different ways, which we can illustrate in the following way:

(a)   (b)   (c)   (d)

$$\tag{12.41}$$

Here, diagrams (a) and (b) are equivalent, as are diagrams (c) and (d). Since there is a factor of 1/2 associated with each of the Coulomb interactions, each diagram gets the weight of 1/4. Considering only distinct diagrams, the weights become 1/2, which gives the algebraic expressions

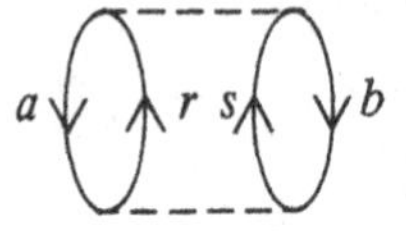

$$
a \bigcirc\!\!\!\!\bigcirc\, r\ s\, \bigcirc\!\!\!\!\bigcirc\, b \;=\; \frac{1}{2} \sum_{abrs} \frac{\langle ab|r_{12}^{-1}|rs\rangle\langle rs|r_{12}^{-1}|ab\rangle}{\varepsilon_a + \varepsilon_b - \varepsilon_r - \varepsilon_s} \tag{12.42a}
$$

$$
a \bigcirc\!\!\!\times\!\!\!\bigcirc\, b \;=\; -\frac{1}{2} \sum_{abrs} \frac{\langle ab|r_{12}^{-1}|sr\rangle\langle rs|r_{12}^{-1}|ab\rangle}{\varepsilon_a + \varepsilon_b - \varepsilon_r - \varepsilon_s}. \tag{12.42b}
$$

The graphical representation of the complete second-order energy is given in Fig. 12.4.

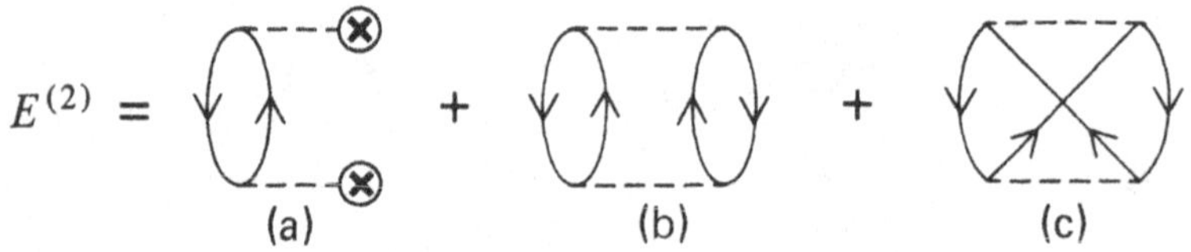

$$
E^{(2)} = \; \text{(a)} \quad + \quad \text{(b)} \quad + \quad \text{(c)}
$$

Fig. 12.4 a-c. Graphical representation of the second-order energy for closed-shell systems. Diagram (a) vanishes if Hartree-Fock orbitals are used

---

Problem 12.3. Verify the evaluations of the second-order energy diagrams (12.40) and (12.42).

12.4 The Goldstone Evaluation Rules

In the previous two sections we have evaluated the second-order contributions to the wave operator and to the energy. The basic equations we have used for the wave operator are (12.21, 22). For the energy the corresponding equations are (12.37, 39). In each case the quantity of interest is expressed in terms of operators in normal form, and then the graphical representation of Wick's theorem is used to expand it in terms of diagrams. The resulting perturbation diagrams have been evaluated by identifying them with the corresponding terms in the algebraic expression. This procedure of evaluating the terms in the perturbation expansion leads to a set of rules originally due to *Goldstone* [1957]. These rules enable us to write down immediately the algebraic expression for a particular diagram, and they make the perturbation procedure used in the previous sections simple and easy to apply.

The Goldstone rules can be formulated as follows:

For each wave-operator diagram there is

a) *a creation (absorption) operator for each free outgoing (incoming) orbital line. These are written in normal form*

$$\{a_i^\dagger a_j^\dagger a_k^\dagger \ldots a_{k'} a_{j'} a_{i'}\} \ ;$$

where $(a_i^\dagger a_{i'})$, *etc., originate from the same vertex or from vertices connected by orbital lines;*

b) *a matrix element for each interaction line;*

c) *an energy denominator for each interaction line*

$$D = \sum(\varepsilon_{\text{down}} - \varepsilon_{\text{up}}),$$

where $\varepsilon_{\text{down}}$ (ε_{up}) *is the single-particle eigenvalue associated with the down- (up-) going orbital lines, cut by a line immediately above the interaction line;*

d) *a summation over all internal orbital lines;*

e) *a phase factor equal to*

$$(-1)^{h+l},$$

where h *is the number of internal core (hole) lines and* l *the number of closed loops of orbital lines;*

f) *a factor of* $1/2$ *for each two-particle interaction and an equivalence factor equal to the number of equivalent diagrams represented by the diagram considered.*

The rules above hold also for the energy diagrams, except that rule a) does not apply, and there is

g) *no energy denominator associated with the last interaction.*

All of these rules should be fairly obvious from the discussion above, except the phase rule e), which is derived formally in Sect. 12.8. It is easy to verify, however, that the results above are consistent with this rule. Rule f), concerning the weight factor, requires some special care, and we shall consider this rule in more detail.

We observe that distinct diagrams which have the weight of $1/2$, like (12.16b, 31, 42), are symmetrical in the sense that a reflection in a vertical plane transforms each diagram into itself. Diagrams with the weight of 1, like (12.30), do not have this property. For the disconnected diagram in Fig. 12.3m, there is a factor of 1/2 associated with each interaction, and the weight is 1/4. This diagram is transformed into itself when the disconnected pieces are reflected separately in a vertical plane. From these considerations we can draw the conclusion that there is a factor of 1/2 for each symmetry operation—like reflection in a vertical plane or interchange of the vertices of an interaction line—which transforms the diagram into itself (disregarding the labels of the orbital lines).

We have assumed here that only distinct diagrams are included in the expansion. We shall see later that it is sometimes desirable to include some equivalent diagrams for symmetry reasons. Suppose, for instance, that we want to include the diagram (12.29a) as well as (12.29b) in the expansion. Then each diagram will obviously get the weight of 1/2. Similarly, if we include both diagrams in Fig. 12.2d, each of them will have the weight of 1/4. This is also the case if we represent the two-body part of the second-order energy by means of all four diagrams (12.41) instead of only the two distinct diagrams (12.42).

This leads us to the following rule for the weight factor:

> f') *There is a factor of 1/2 for each symmetry operation—like reflection in a vertical plane or interchange of the vertices of an interaction line—which transforms the diagram into itself or to any other diagram appearing in the expansion.*

One symmetry operation, namely reflection, can be performed on the distinct diagrams (12.31, 42), which is consistent with the weight of 1/2. If both diagrams in Fig. 12.2d or all four diagrams (12.41) are included, on the other hand, then two symmetry operations can be performed, namely interchange of the vertices associated with both interactions, or, alternatively, reflection and interchange of one pair of vertices. This gives the weight of 1/4.

Rule f') should now be fairly obvious. If we include *all* diagrams obtained in the application of Wick's theorem, then there is a factor of 1/2 for each two-body interaction. But all such interactions are symmetrical, so an interchange of the two vertices of an interaction line will then always lead to a diagram in the expansion. If, on the other hand, a more selective group of diagrams is chosen, such as distinct diagrams, then certain vertex interchanges will yield diagrams *outside* the expansion, and the corresponding factor of 1/2 is removed.

As a further example of the perturbation procedure described above and the Goldstone evaluation rules, we consider the third-order energy $E^{(3)}$. According to (12.37), this can be written

$$E^{(3)} = \langle \alpha | (V_1 + V_2) \Omega^{(2)} | \alpha \rangle . \tag{12.43}$$

The graphical representation of this equation is obtained by closing the diagrams of $\Omega^{(2)}$ with the perturbation. Obviously, only the one- and two-body parts of $\Omega^{(2)}$ can be closed in this way, so the third-order energy can be written

$$E^{(3)} = \langle \alpha | V_1 \Omega_1^{(2)} + V_2 \Omega_2^{(2)} | \alpha \rangle \; .$$

Again, if Hartree-Fock orbitals are used, V_1 vanishes and $E^{(3)}$ reduces to

$$E_{\text{HF}}^{(3)} = \langle \alpha | V_2 \Omega_2^{(2)} | \alpha \rangle \; . \tag{12.44}$$

We can now use the graphical form of Wick's theorem to express $E_{\text{HF}}^{(3)}$ in terms of diagrams and use the Goldstone rules to evaluate them. The diagrams of $E_{\text{HF}}^{(3)}$ are obtained by closing the diagrams of $\Omega_2^{(2)}$ (Fig. 12.3e-j) with those of V_2. The distinct diagrams obtained in this way are shown in Fig. 12.5.

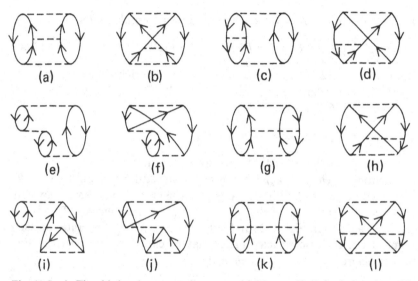

Fig. 12.5 a-l. The third-order energy diagrams with Hartree-Fock basis functions. These diagrams are obtained by "closing" the second-order wave-operator diagrams in Fig. 12.3. Since there is no effective one-body part of the perturbation, only wave-operator diagrams representing double excitations can contribute

As an example of diagram evaluation using the Goldstone rules we consider diagram (a) in Fig. 12.5,

$$\text{(diagram)} = \frac{1}{2} \sum_{abrstu} \frac{\langle ab | r_{12}^{-1} | tu \rangle \langle tu | r_{12}^{-1} | rs \rangle \langle rs | r_{12}^{-1} | ab \rangle}{(\varepsilon_a + \varepsilon_b - \varepsilon_t - \varepsilon_u)(\varepsilon_a + \varepsilon_b - \varepsilon_r - \varepsilon_s)} \; . \tag{12.45}$$

This diagram is symmetric with respect to a reflection in a vertical plane, and hence the weight is 1/2. It has two internal core lines and two closed orbital loops, which makes the phase factor positive.

Problem 12.4. Use the Goldstone rules to evaluate diagrams (b) and (a) in Fig. 12.5.

12.5 The Linked-Diagram Expansion

The procedure we have developed here can easily be extended to higher orders. In this section we shall consider the wave operator in third order, where so-called *unlinked diagrams* appear for the first time.[1] It will be shown that all such diagrams disappear in that order, and this result will be used to formulate the general linked-diagram theorem for closed-shell systems.

12.5.1 Cancellation of Unlinked Diagrams in Third Order

We start from the third-order Rayleigh-Schrödinger formula (9.75c)

$$[\Omega^{(3)}, H_0] P = Q V \Omega^{(2)} P - \Omega^{(2)} P V P - \Omega^{(1)} P V \Omega^{(1)} P . \tag{12.46}$$

As in second order (12.19–21), the second term on the right-hand side cancels the V_0 part of the first term. Furthermore, using (12.37), the last term becomes $-E^{(2)} \Omega^{(1)}$ and (12.46) reduces to

$$[\Omega^{(3)}, H_0] = Q(V_1 + V_2) \Omega^{(2)} - E^{(2)} \Omega^{(1)} , \tag{12.47}$$

as before, leaving out the projection operator to the right.

The diagrams of $Q(V_1 + V_2)\Omega^{(2)}$ can be constructed by operating with V_1 and V_2 on the diagrams of $\Omega^{(2)}$ in the same way as before. A new feature appears here, however, because some of the $\Omega^{(2)}$ diagrams are *disconnected*. This is the case with the diagrams in Fig. 12.2e and in Fig. 12.3m. When V operates on such a diagram, one of the disconnected pieces may be closed, as illustrated in Fig. 12.6. Such a diagram is said to be *unlinked*.

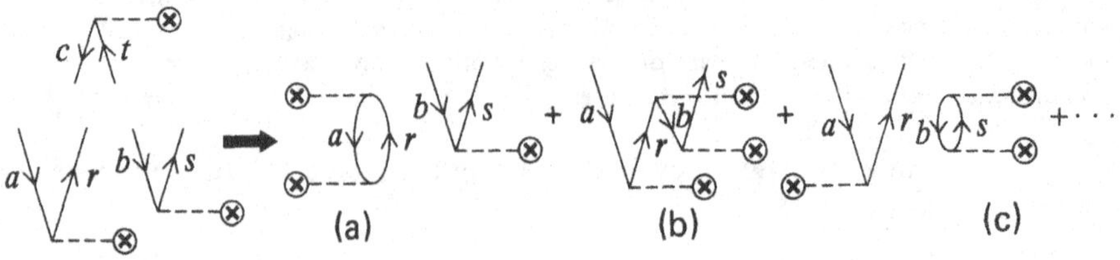

Fig. 12.6. Unlinked diagrams (a, c) can be formed when V operates on a disconnected wave-operator diagram

---

[1] It should be observed that unlinked diagrams appear already in the second order in the more traditional approach (see, for instance, [*Kelly* 1963]). In our approach, however, unlinked second-order diagrams are eliminated by expressing the operators in normal form with respect to some suitable "vacuum" state (see p. 232; also [*Čížek* 1969] and [*Paldus* and *Čížek* 1975]).

Generally, we say that

a diagram is closed *if it has no free orbital lines*

and that

a diagram is unlinked *if it has a disconnected, closed part.*

A diagram which is not unlinked is said to be *linked*. It should be observed that the words "linked" and "unlinked" are not synonymous with "connected" and "disconnected" in this nomenclature. A diagram like that in Fig. 12.2e is disconnected, since there is no connection between the two parts of the diagram. But since both parts are *open* (with free lines), the diagram is linked according to the definition given here.

The algebraic expressions for the unlinked diagrams in Fig. 12.6a,c are

$$\sum_{ar} a_s^\dagger a_b \frac{\langle a|v|r\rangle\langle s|v|b\rangle\langle r|v|a\rangle}{(\varepsilon_a + \varepsilon_b - \varepsilon_r - \varepsilon_s)(\varepsilon_a - \varepsilon_r)}$$

$$\sum_{bs} a_r^\dagger a_a \frac{\langle b|v|s\rangle\langle s|v|b\rangle\langle r|v|a\rangle}{(\varepsilon_a + \varepsilon_b - \varepsilon_r - \varepsilon_s)(\varepsilon_a - \varepsilon_r)}.$$

If we change the summation indices $(a \leftrightarrow b, r \leftrightarrow s)$ in the last expression, we get

$$\sum_{ar} a_s^\dagger a_b \frac{\langle a|v|r\rangle\langle r|v|a\rangle\langle s|v|b\rangle}{(\varepsilon_a + \varepsilon_b - \varepsilon_r - \varepsilon_s)(\varepsilon_b - \varepsilon_s)}$$

and we see that this differs from the first expression only in the energy denominator. From the algebraic identity

$$\frac{1}{(A+B)A} + \frac{1}{(A+B)B} = \frac{1}{AB} \tag{12.48}$$

it then follows that the sum of the two diagrams can be factorized, as illustrated in Fig. 12.7. This is a consequence of the general *factorization theorem* [*Hugenholtz* 1957; *Frantz* and *Mills* 1960; *Brandow* 1967, 1977; *Baker* 1971; *Sandars* 1969; *Lindgren* 1974] which says that

the denominators of disconnected diagram parts can be evaluated separately when the sum of diagrams with all possible orderings of the interactions of the disconnected parts is considered.

When this rule is applied to diagrams with a closed part, it is assumed that the "closing" interaction is applied last. This means that it always appears at the top, as in Fig. 12.7.

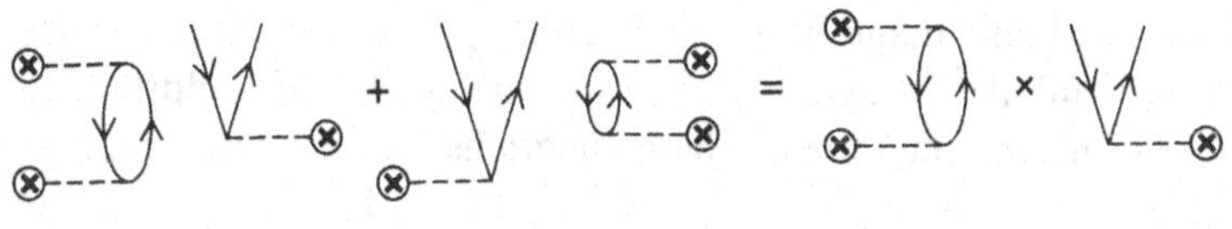

Fig. 12.7. Illustration of the factorization of disconnected wave-operator diagrams

The factorized wave-operator diagram in Fig. 12.7 is a product of a first-order wave-operator diagram (Fig. 12.1) and a second-order energy diagram (Fig. 12.4). Therefore, this will be cancelled by the corresponding part of the term $-E^{(2)}\Omega^{(1)}$ in (12.47).

One important observation should be made here. The disconnected $\Omega^{(2)}$ diagram on the left in Fig. 12.6 is nonzero only if $a \neq b$ and $r \neq s$, due to the exclusion principle. In the product $E_1^{(2)}\Omega_1^{(1)}$, on the other hand, there is no such restriction

$$
E_1^{(2)}\Omega_1^{(1)} \;=\; \sum_{ar} \quad \times \quad \sum_{bs} \qquad\qquad (12.49)
$$

since the two sums run independently. The situation is similar for unlinked diagrams involving two-body interactions. Therefore, in order to get a complete cancellation between $E^{(2)}\Omega^{(1)}$ and the corresponding unlinked diagrams of $Q(V_1 + V_2)\Omega^{(2)}$ in (12.47), we must *disregard the exclusion principle in the disconnected diagrams of* $\Omega^{(2)}$. Formally, we can include terms violating the exclusion principle in the second-quantized expression for the correlation operator (11.58). The anticommutation relations ensure that such terms do not contribute. Consequently, *exclusion-principle-violating* (EPV) *diagrams* can be included in the graphical representation of $\Omega^{(2)}$, provided that this is done consistently. This means that we should include also the corresponding *linked* diagrams which violate the exclusion principle. If $a = b$, for instance, in Fig. 12.6, then the $\Omega^{(2)}$ diagram vanishes, and all diagrams formed by operating with V on it add to zero. The unlinked diagrams (a) and (c) are used to cancel (12.49), and so the linked EPV diagrams must be retained in the graphical representation (see Fig. 12.8).

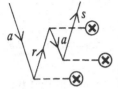

Fig. 12.8. Example of an EPV diagram which has to be retained in the linked-diagram expansion

As a further illustration of the formation of unlinked diagrams, we consider the disconnected four-body diagram of $\Omega^{(2)}$ (Fig. 12.3m). Operating with V_2 gives rise to four unlinked diagrams, as shown in Fig. 12.9 (in addition to a large number of linked diagrams). These diagrams can be factorized as

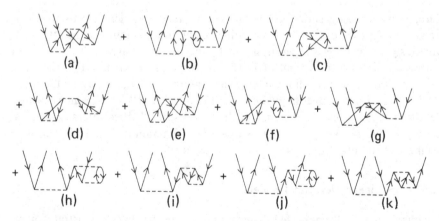

Fig. 12.9. Unlinked diagrams formed by operating with V on the disconnected four-body diagram of $\Omega^{(2)}$. The corresponding linked diagrams are shown in Fig. 12.10

before, provided that EPV diagrams are included. The result can be expressed

$$\left[\text{diagram} + \text{diagram}\right] \times \text{diagram} = E_2^{(2)} \text{diagram}$$

(12.50)

where $E_2^{(2)}$ is the two-body part of the second-order energy (Fig. 12.4). This will cancel the corresponding part of $-E^{(2)}\Omega^{(1)}$ in (12.47). As before, the *linked* diagrams formed in this process should be maintained in the graphical representation also when the underlying $\Omega^{(2)}$ diagram violates the exclusion principle. These diagrams are shown in Fig. 12.10.

(a)   (b)   (c)

(d)   (e)   (f)   (g)

(h)   (i)   (j)   (k)

Fig. 12.10 a–k. Linked two-body diagrams formed by operating with V on the disconnected four-body diagrams of $\Omega^{(2)}$. These diagrams should be included in the linked-diagram expansion, also when the underlying diagram of $\Omega^{(2)}$ violates the exclusion principle (EPV diagrams)

The diagrams in Fig. 12.10 can be evaluted by means of the Goldstone rules given in the previous section. As illustrations we consider diagrams (a, b, e, h)

$$= \frac{1}{2} a_r^\dagger a_s^\dagger a_b a_a \sum \frac{1}{D} \langle cd | r_{12}^{-1} | tu \rangle$$
$$\times \langle rs | r_{12}^{-1} | cd \rangle \langle tu | r_{12}^{-1} | ab \rangle \qquad (12.51a)$$

$$= a_r^\dagger a_s^\dagger a_b a_a \sum \frac{1}{D} \langle cd | r_{12}^{-1} | tu \rangle$$
$$\times \langle us | r_{12}^{-1} | db \rangle \langle rt | r_{12}^{-1} | ac \rangle \qquad (12.51b)$$

$$= a_r^\dagger a_s^\dagger a_b a_a \sum \frac{1}{D} \langle cd | r_{12}^{-1} | tu \rangle$$
$$\times \langle ru | r_{12}^{-1} | cb \rangle \langle ts | r_{12}^{-1} | ad \rangle \qquad (12.51e)$$

$$= - a_r^\dagger a_s^\dagger a_b a_a \sum \frac{1}{D} \langle cd | r_{12}^{-1} | tu \rangle$$
$$\times \langle tu | r_{12}^{-1} | bd \rangle \langle rs | r_{12}^{-1} | ac \rangle . \qquad (12.51h)$$

Here, D is used as a shorthand notation for the energy denominators, which are given by means of (11.70). Again, the weight factors require some attention. Obviously no symmetry operation can be applied on diagrams (b, e, h), and hence they all have the weight of 1. This corresponds to the fact that there are eight possible ways of forming each of the diagrams. In diagram (a), on the other hand, we can interchange the two open paths "atcr" and "buds", which is equivalent to a *simultaneous* interchange of the vertices of all three interaction lines. Hence, the weight factor is 1/2. This is the only diagram in Fig. 12.10 with this weight. All the other diagrams have the weight of 1.

12.5.2 The Linked-Diagram Theorem

By considering all disconnected diagrams of $\Omega^{(2)}$ in the way demonstrated here, one can show that the unlinked diagrams of $Q(V_1 + V_2)\Omega^{(2)}$ exactly cancel the entire term $-E^{(2)}\Omega^{(1)}$ in (12.47), provided that the exclusion principle is discarded between the disconnected parts of the $\Omega^{(2)}$ diagrams. This equation can then be written

$$[\Omega^{(3)}, H_0] = (Q(V_1 + V_2)\Omega^{(2)})_{\text{linked}} \, .$$

Since V_0 can never form any linked diagrams when operating on $\Omega^{(2)}$, we can write this equation also as

$$[\Omega^{(3)}, H_0] = (QV\Omega^{(2)})_{\text{linked}} \, .$$

The subscript "linked" indicates that only linked diagrams—including EPV diagrams—are included.

It can be shown in a similar way that the unlinked diagrams cancel in higher orders. Recalling that $\Omega^{(0)} \equiv 1$, this leads to

$$\boxed{[\Omega^{(n)}, H_0] = (QV\Omega^{(n-1)})_{\text{linked}}} \tag{12.52}$$

for all values of $n \geq 1$. This is the *linked-diagram theorem* for closed-shell systems.

When the wave operator is known in a certain order, the energy can be obtained in the next higher order by means of (12.37)

$$\boxed{E^{(n+1)} = \langle \alpha | (V_1 + V_2)\Omega^{(n)} | \alpha \rangle} \, . \tag{12.53}$$

The relations (12.52,53) lead to the following rules for constructing the diagrams of the wave operator and of the energy order by order:

a) *take a wave-operator diagram of order $(n-1)$;*
b) *operate on it with the diagrams of the perturbation V, using the graphical form of Wick's theorem;*
c) *discard unlinked diagrams, i.e., diagrams which have a disconnected, closed part;*
e) *the open diagrams represent the wave operator and the closed diagrams the energy in order n;*
f) *repeat the procedure for all wave-operator diagrams of order $(n-1)$, including all disconnected wave-operator diagrams which violate the exclusion principle.*

It follows from this procedure that the energy—in contrast to the wave operator— is represented only by *connected* diagrams.

Using the expansions (9.74)

$$\Omega = \Omega^{(0)} + \Omega^{(1)} + \Omega^{(2)} + \cdots \qquad (\Omega^{(0)} \equiv 1)$$

we can sum the linked-diagram formula (12.52) to all orders, and express the result in the form

$$\boxed{[\Omega,\, H_0] = (QV\Omega)_{\text{linked}}}.\tag{12.54}$$

Expanding this equation order by order yields the linked-diagram expansion in the same way as the Rayleigh-Schrödinger expansion is obtained from the Bloch equation (9.72).

12.6 Separation of Goldstone Diagrams into Radial and Spin-Angular Parts

We have now developed a procedure for generating diagrams of the wave operator and the energy for closed-shell systems to arbitrary order. The algebraic expressions associated with these diagrams are obtained by the "Goldstone rules" given in Sect. 12.4. This procedure is applicable to molecules as well as atoms—and with small modifications also to nuclei. For atomic systems which are approximately spherically symmetric, we can go one step further and separate the diagrams into radial, angular, and spin parts by using the graphical angular-momentum theory developed in Chaps. 3 and 4. We shall demonstrate this procedure by a number of examples. (For molecular systems a similar technique can be used to separate out the *spin* part, as discussed, for instance, by *Paldus* [1977]).

12.6.1 Examples of Diagram Evaluations

As a first illustration we consider the second-order energy diagram (12.42a)

$$a\underset{}{\bigcirc}r\; s\underset{}{\bigcirc}b \;\; = \tfrac{1}{2}\sum\tfrac{1}{D}\langle ab|r_{12}^{-1}|rs\rangle\langle rs|r_{12}^{-1}|ab\rangle\,,\tag{12.55}$$

where D is the excitation energy of the intermediate state, given by (11.70). Using the graphical representation (3.42) for each of the Coulomb interactions, we can represent the *angular* part of the expression by the following *orbital* angular-momentum diagram:

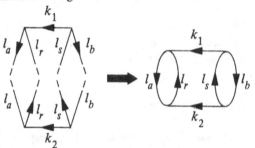

after summing over the m_l values. This diagram is topologically identical to the Goldstone diagram above.

Since the Coulomb interaction is independent of spin, we must have

$$m_s^a = m_s^r \quad \text{and} \quad m_s^b = m_s^s .$$

Therefore, the summation over the spin quantum numbers yields a factor of 4 in this case. Alternatively, we can use the graphical form (3.45), in which case the spin diagram becomes just two closed loops, each of which has the numerical value of 2.

The diagram (12.55) can now be expressed

$$a \underbrace{}_{r\ s} b = \frac{1}{2} \sum \frac{1}{D} X^{k_1}(ab, rs) X^{k_2}(rs, ab) \times 4 \times \underbrace{}_{k_2}^{k_1} \tag{12.56}$$

where $X^k(ab, cd)$ is the reduced matrix element of the Coulomb interaction (3.43, 44)

$$
\begin{aligned}
X^k(ab, cd) &= X(k, l_a l_b l_c l_d)\, R^k(ab, cd) \\
&= (-1)^k \langle l_a \| C^k \| l_c \rangle \langle l_b \| C^k \| l_d \rangle R^k(ab, cd) .
\end{aligned}
\tag{12.57}
$$

The sum in (12.56) runs over all n, l, and k values.

The exchange diagram (12.42b) can be evaluated in the same way, and the result is

$$a \underbrace{}_{r\quad s} b = -\frac{1}{2} \sum \frac{1}{D} \langle ab | r_{12}^{-1} | sr \rangle \langle rs | r_{12}^{-1} | ab \rangle \tag{12.58}$$

$$= -\frac{1}{2} \sum \frac{1}{D} X^{k_1}(ab, sr) X^{k_2}(rs, ab) \times 2 \times \underbrace{}_{k_2}^{k_1}$$

Again, the orbital diagram is toplogically identical to the Goldstone diagram. The spin sum here gives a factor of 2, since all four m_s values must be equal. Alternatively, we get a spin diagram consisting of a single (but twisted) loop,

$$s \underbrace{\bowtie}{} \quad = \quad s \bigcirc \quad = \quad 2$$

The orbital diagrams above can be evaluated by means of the standard technique developed in Chaps. 3 and 4 (see, for instance, Problems 3.9, 4.2 and 12.5).

It should be noted that in order to permit a factorization of the type considered here, it is necessary that the sums over the internal orbital lines run independently, *disregarding the exclusion principle*. As mentioned before, this is possible to do since this principle is automatically fulfilled in the formalism of second quantization on which the graphical representation is based. The exclusion-principle-violating part of a diagram is cancelled by the corresponding part of an exchange diagram. This is illustrated in Problem 12.5.

The corresponding treatment of the effective potential interaction is trivial, since the potential is spherically symmetric. According to (10.21) we have

$$\langle i|v|j\rangle = \langle i|-u|j\rangle + \langle i|u_{\mathrm{HF}}|j\rangle ,$$

where the matrix elements of $-u$ and u_{HF} are given by (10.35) and (10.41), respectively. We may then represent the v interaction in the following way:

$$\langle i|v|j\rangle = \langle i|v|j\rangle_{\mathrm{rad}} \quad \begin{vmatrix} l_i & s_i \\ & \\ l_j & s_j \end{vmatrix} \tag{12.59}$$

where $\langle i|v|j\rangle_{\mathrm{rad}}$ is the radial part of the matrix element,

$$\langle i|v|j\rangle_{\mathrm{rad}} = -\int P_i(r)uP_j(r)\, dr \\ + \sum_{n_b l_b}(4l_b + 2)\left[R^0(ib, jb) - \frac{1}{2}\sum_k \begin{pmatrix} l_i & k & l_j \\ 0 & 0 & 0 \end{pmatrix}^2 G^k(bi, jb)\right], \tag{12.60}$$

and l_i and s_i in the diagram represent $l_i m_i^l$ and $s m_s^l$, respectively.

Applying this result to the second-order energy diagram (12.40)

$$= \sum \frac{1}{D}\langle a|v|r\rangle\langle r|v|a\rangle \tag{12.61}$$

we get simply a spin and an orbital loop from the $m_s m_l$ summations. Writing the result in conformity with the previous results, we get

$$= \sum \frac{1}{D}(\langle a|v|r\rangle_{\mathrm{rad}})^2 \times 2 \times \quad l_a \bigcirc l_r \quad . \tag{12.62}$$

The orbital loop has the value $\delta(l_a, l_r)(2l_a + 1)$.

Problem 12.5. Evaluate the diagrams (12.56) and (12.58) for a $2s^2 \to 2p^2$ excitation—that is, with $l_a = l_b = 0$ and $l_r = l_s = 1$—and show that the exclusion-principle-violating parts of the diagrams cancel.

Next, we consider some wave-operator diagrams, and we leave it as an exercise for the reader to verify the following results:

$$= a_r^\dagger a_a \sum \frac{1}{D} \langle rb | r_{12}^{-1} | as \rangle \langle s | v | b \rangle$$

$$= a_r^\dagger a_a \sum \frac{1}{D} X^k(rb, as) \langle s | v | b \rangle_{\text{rad}} G_1 , \tag{12.63}$$

where G_1 is the angular-momentum diagram

$$G_1 = \quad = \delta(k, 0)\, \delta(l_b, l_s)\, [l_b/l_a]^{1/2} \quad \begin{vmatrix} l_r \\ \\ l_a \end{vmatrix} .$$

$$= a_r^\dagger a_a \sum \frac{1}{D} \langle bc | r_{12}^{-1} | as \rangle \langle sr | r_{12}^{-1} | bc \rangle$$

$$= a_r^\dagger a_a \sum \frac{1}{D} X^{k_1}(bc, as)\, X^{k_2}(sr, bc)\, G_2 \tag{12.64}$$

$$G_2 = \quad = [l_a]^{-1} \begin{Bmatrix} l_b & k_1 & l_r \\ l_c & k_2 & l_s \end{Bmatrix} \quad \begin{vmatrix} l_r \\ \\ l_a \end{vmatrix}$$

$$= \frac{1}{2} a_r^\dagger a_s^\dagger a_b a_a \sum \frac{1}{D} \langle rs | r_{12}^{-1} | tu \rangle \langle tu | r_{12}^{-1} | ab \rangle$$

$$= \frac{1}{2} a_r^\dagger a_s^\dagger a_b a_a \sum \frac{1}{D} X^{k_1}(rs, tu)\, X^{k_2}(tu, ab)\, G_3 \tag{12.65}$$

$$G_3 = \quad = \sum_k [k] \begin{Bmatrix} l_r & k & l_a \\ k_2 & l_t & k_1 \end{Bmatrix} \begin{Bmatrix} l_b & k & l_s \\ k_1 & l_u & k_2 \end{Bmatrix} \quad + \quad -$$

Note that k may here take *any* integral value consistent with the triangular conditions—that is, not only those in which the sum of the angular momenta at each vertex is even, For that reason the vertex signs are shown explicitly in the last diagram.

....

Problem 12.6. Verify the evaluations above.

12.6.2 Evaluation Rules for the Radial, Spin and Angular Factors

From the examples considered here we can formulate a set of rules for evaluating the separate radial, spin, and angular parts of an energy or a wave-operator diagram. The *angular factor* is given by an orbital angular-momentum diagram which is obtained by means of the following rules [*Sandars* 1969; *Tolmachev* 1969]:

    a) *remove all potential interaction lines;*
    b) *replace each orbital line by the corresponding angular-momentum line and add an arrow on outgoing lines from each interaction;*
    c) *replace each Coulomb interaction line by a directed angular-momentum line k.*

The *spin factor* contains

    d) *a factor of 2 for each closed orbital loop.*

The *radial* part of the diagram is obtained by means of the Goldstone rules given in Sect. 12.4 with the modification that the matrix element of the Coulomb interaction is replaced by the reduced element

$$\langle ij | r_{12}^{-1} | mn \rangle \Rightarrow X^k(ij, mn) = X(k, l_i l_j l_m l_n) \, R^k(ij, mn)$$

and the potential interaction by the radial factor (12.60)

$$\langle i | v | j \rangle \Rightarrow \langle i | v | j \rangle_{\text{rad}} \, .$$

The summations over the internal lines run over the nl values and, in addition, there is a summation over all k values.

12.7 The Correlation Energy of the Beryllium Atom

The first atomic application of the many-body perturbation technique based on the linked-diagram theorem was made by *Kelly* [1963, 1964b] on the beryllium atom, and we shall discuss this calculation in some detail as an illustration of a closed-shell calculation. Later, a large number of calculations were performed on this atom in order to test various approaches to the correlation problem, and we shall return to a comparison between such approaches in Chap. 15.

    The primary problem in the procedure employed by *Kelly* is to generate a "complete" set of single-particle states in order to evalute the matrix elements and perform the summations/integrations involved in the perturbation expansion.

In his first paper *Kelly* used Hartree-Fock orbitals generated in the potential of the neutral atom as basis orbitals for core states as well as virtual states. For the core the exchange interaction cancels exactly one of the direct terms in the potential

$$\langle r|u_{\mathrm{HF}}|a\rangle = \sum_{b}^{\mathrm{core}} [\langle rb|r_{12}^{-1}|ab\rangle - \langle br|r_{12}^{-1}|ab\rangle]$$

$$= \sum_{b\neq a}^{\mathrm{core}} [rb|r_{12}^{-1}|ab\rangle - \langle br|r_{12}^{-1}|ab\rangle] \,, \tag{12.66}$$

which means that an electron in a core orbital "feels" only the potential from the nucleus and the remaining $(N-1)$ electrons. In a virtual orbital, on the other hand, there is no similar cancellation. Since the exchange becomes negligible at large distances, this has the consequence that an electron in a virtual orbital at large distances moves in the potential of *all* N electrons. This is an "unphysical" situation and leads to "unphysical" virtual orbitals.

Kelly found that orbitals of this kind led to convergence problems in the perturbation expansion, and for that reason he modified the potential in his second paper to a so-called V^{N-1} *potential*, which is the Hartree-Fock potential of the system with one electron removed. In order to get a completely orthogonal set, also the occupied orbitals were generated in this potential. This had the consequence that the zeroth-order state was not the normal Hartree-Fock state for the neutral atom, but that only led to small corrections which could easily be taken care of.

For virtual, *bound* states the Hartree-Fock orbitals can be determined by solving an eigenvalue equation with the same boundary conditions as for core orbitals (7.13). For unbound states in the energy continuum, on the other hand, the solutions approach at large distances that of a free particle, which in spherical coordinates can be expressed by means of spherical Bessel and Neumann functions [*Arfken* 1970, Chap. 11]. The actual solutions are obtained by integrating outwards numerically in the atomic potential and then at a certain radius matching the solution to that of a free particle. Since every value of the energy in the continuum is allowed, such solutions are determined in a certain energy grid. These solutions together with those of the bound states form a "complete" set of single-particle states. Summations over excited states appearing in the perturbation formulas are then replaced by summations over bound states and integration over unbound states in the standard way.

The first-order contribution to the wave operator is in our notations given by (12.14, 16)

$$a \bigvee \bigwedge_r s \bigwedge \bigvee_b \quad = \frac{1}{2} a_r^\dagger a_s^\dagger a_b a_a \frac{1}{D} \langle rs|r_{12}^{-1}|ab\rangle \,, \tag{12.67}$$

where

$$D = \varepsilon_a + \varepsilon_b - \varepsilon_r - \varepsilon_s \,,$$

when Hartree-Fock orbitals are used. (When the V^{N-1} potential is used, there are also small single excitations in first order.)

In the next order there is a considerable number of diagrams (Fig. 12.3). *Kelly* found, however, that the most important was the hole-hole interaction diagram (j), particularly the diagonal type

$$= \frac{1}{2} a_r^\dagger a_s^\dagger a_b a_a \frac{1}{D^2} \langle ab | r_{12}^{-1} | ab \rangle \langle rs | r_{12}^{-1} | ab \rangle \quad (12.68)$$

and the corresponding exchange variant. We see that this diagram can be written as the factor

$$\frac{1}{D} \langle ab | r_{12}^{-1} | ab \rangle \quad (12.69)$$

times the first-order diagram (12.67). The same factor appears in successively higher orders, so diagrams of this kind form a geometric series which can be summed to all orders. From the relation

$$1 + \frac{1}{D} \langle ab | r_{12}^{-1} | ab \rangle + \left(\frac{1}{D} \langle ab | r_{12}^{-1} | ab \rangle \right)^2 + \cdots \quad (12.70)$$

$$= \frac{D}{D - \langle ab | r_{12}^{-1} | ab \rangle}$$

we see that this summation simply has the effect of modifying the energy denominator of the first-order diagram (12.67)

$$\frac{1}{2} a_r^\dagger a_s^\dagger a_b a_a \frac{1}{D'} \langle rs | r_{12}^{-1} | ab \rangle \,,$$

where

$$D' = D - \langle ab | r_{12}^{-1} | ab \rangle \,.$$

The inclusion of the exchange variants leads similarly to

$$D' = D - \langle ab | r_{12}^{-1} | ab \rangle + \langle ba | r_{12}^{-1} | ab \rangle \,. \quad (12.71)$$

The unperturbed denominator (12.67) represents the excitation energy in zeroth order, while (12.71) is the corresponding first-order value, corrected for the mutual Coulomb interaction between the electrons being excited.

Kelly referred to diagrams of the form (12.68) as EPV diagrams (of the first kind). In our procedure, based on the normal ordering, however, these diagrams do not violate the exclusion principle and we shall not refer to them as EPV diagrams.

Kelly also found in his calculation that other third-order contributions to the wave function, besides those included in the summations above, are quite significant. These are of the type represented by the diagrams in Fig. 12.10. Again, the "diagonal" contributions are the most important, such as

$$\tag{12.72}$$

These diagrams, referred to by *Kelly* as EPV diagrams of the second kind, are "true" EPV diagrams, since the exclusion principle is violated in the intermediate state.

It can be shown that also this kind of diagram can be summed to all orders, and this gives rise to a modification of the energy denominator in the first-order diagram (12.67) by the part of the second-order energy which depends on a and b. By continuing this process one would obtain denominators of the kind appearing in various pair-correlation procedures [*Kelly* 1964b; *Sinanuğlu* 1964, 1969; *Nesbet* 1965, 1969; *Meyer* 1973, 1976], which we shall return to in Chap. 15.

The Be atom has the ground state $1s^2 2s^2\, {}^1S$, and the correlation energy can be separated into three parts, depending on the occupied orbitals being excited in the final excitation

$$E_{corr} = \varepsilon(1s^2) + \varepsilon(1s, 2s) + \varepsilon(2s^2) \,. \tag{12.73}$$

Furthermore, each contribution can be separated according to the angular momenta involved.

For the $1s^2$ pair, *Kelly* found the following second-order contributions (in Hartrees, see Appendix A):

$$\varepsilon^{(2)}(1s^2 \rightarrow s^2) = -0.0141$$
$$\varepsilon^{(2)}(1s^2 \rightarrow p^2) = -0.0252$$
$$\varepsilon^{(2)}(1s^2 \rightarrow d^2) = -0.0041$$

$$\overline{\varepsilon^{(2)}(1s^2 \rightarrow s^2, p^2, d^2) = -0.0434}$$

The denominator modification (12.71) has only a minor effect here due to the large excitation energies. By estimating remaining third-order effects, the value

$$\varepsilon^{(2)}(1s^2) = -0.0421$$

was obtained.

In second order the $1s$, $2s$ correlation energy was found to be

$$\varepsilon^{(2)}(1s, 2s) = -0.0050.$$

The higher-order contributions to this pair energy were neglected.

The correlation between the $2s$ electrons is more difficult to estimate accurately, and here it is necessary to consider higher-order effects in more detail. In addition to the hole-hole interactions, which led to the denominator modification (12.71), the particle-particle "ladder" diagrams (Fig. 12.3e) were also found to be important. The accurate calculation of such effects is quite tedious, since four virtual states are involved. *Kelly* found, however, that these diagrams formed an approximate geometric series and therefore could be included in an approximate way by a further modification of the energy denominator. In this way the following results were obtained:

$$\varepsilon(2s^2 \rightarrow s^2) = -0.0037$$
$$\varepsilon(2s^2 \rightarrow p^2) = -0.0499$$
$$\varepsilon(2s^2 \rightarrow d^2) = -0.0053$$

The contribution of third- and fourth-order nondiagonal ladder diagrams was estimated to be $+0.0141$, which gave the total correlation energy of the $2s^2$ pair

$$\varepsilon(2s^2) = -0.0449.$$

By adding the three contributions the following total correlation energy was obtained

$$E_{corr} = -0.0920,$$

which should be compared with the experimental value

$$(E_{corr})_{exp} = -0.0943.$$

Thus, *Kelly's* calculation gave between 97 and 98% of the true correlation energy. Even if this result is partly fortuitous in view of the approximations involved, it clearly shows that this form of many-body perturbation theory is capable of describing the main features of the intriguing electron-correlation problem.

In Chap. 15 we shall return to the correlation problem of Be and other simple systems and give some results of later, more accurate calculations.

12.8 Appendix. The Goldstone Phase Rule

In this appendix we shall derive the phase rule e) of the Goldstone evaluation rules given in Sect. 12.4.

According to rule a), the phase of a diagram is given by the normal form of the operators associated with the free lines

$$\{a_1^\dagger a_2^\dagger a_3^\dagger \cdots a_3 a_2 a_1\} \ ,$$

where $a_i^\dagger$ and a_i are associated with the same vertex or vertices connected by orbital lines. The sign is positive or negative depending on the parity of the permutation required to transform the operators to normal form. The expression above can also be written

$$\{a_1^\dagger a_1 a_2^\dagger a_2 a_3^\dagger a_3 \cdots\}$$

after an even permutation of the operators, which does not change the phase.

We consider now the product of two operators in normal form

$$A = \{a_1^\dagger a_1 a_2^\dagger a_2 \cdots\}$$
$$B = \{a_1'^\dagger a_1' a_2'^\dagger a_2' \cdots\} \ .$$

According to Wick's theorem (11.57), this product is equal to the sum of all normal products of A and B with zero, one, two, ... contractions, and we shall now analyze the phase of these contracted terms.

We consider first a single contraction of an absorption operator in A and a creation operator in B, for instance,

$$\{a_1^\dagger \overbrace{a_1} a_2^\dagger a_2 \cdots \overbrace{a_1'^\dagger} a_1' a_2'^\dagger a_2' \cdots\}$$

which corresponds to the connection of two lines representing virtual orbitals (particle lines). Then an even permutation will bring the operators to be contracted in contact with each other. To get the phase the remaining unpaired operators (in this example $a_1^\dagger$ and a_2') should form a pair, since they now belong to vertices connected by an orbital line. In this pair the creation operator precedes the absorption operator. Also this permutation is even. Therefore, the connection of two particle lines does not introduce any additional phase factor.

Next, we consider a contraction of a creation operator in A and an absorption operator in B, for instance,

$$\{\overbrace{a_1^\dagger} a_1 a_2^\dagger a_2 \cdots a_1'^\dagger \overbrace{a_1'} a_2'^\dagger a_2' \cdots\}$$

which corresponds to the connection of two core (hole) lines. Again, an even permutation will bring the operators to be contracted in contact with each other, but an odd permutation is now needed to form a pair of the unpaired operators (a_1 should appear to the right of $a_2'^\dagger$). Therefore, a *negative* sign is associated with the connection of two *hole* lines.

For repeated contractions the phase is multiplicative with one exception. If there is a double contraction between two pairs of operators, associated with the same vertices, for instance,

$$\{a_1^\dagger a_1 a_2^\dagger a_2 \cdots a_1'^\dagger a_1' a_2'^\dagger a_2' \cdots\} \ ,$$

then we see that the total phase is positive. In this case a closed orbital loop is formed.

We have then shown that a negative phase is introduced each time two hole lines are connected, provided that an orbital loop is not formed. Another way of expressing this is that the phase is given by the sum of the number of internal hole lines and the number of orbital loops, as in the Goldstone rule e).

13. Application of Perturbation Theory to Open-Shell Systems

Open-shell systems are characterized by the fact that the unperturbed energy levels are degenerate. This degeneracy shows itself physically in the general appearance of the optical spectra and in the sensitivity of open-shell systems to the effect of an external field. The degeneracy also affects the way in which perturbation theory is applied to such systems. Since the model space contains several Slater determinants, it is no longer possible to classify the single-particle states into occupied and unoccupied states as in the closed-shell case. We need in addition a third category of states, representing open-shell states or valence states, which are partly occupied in the model space. In order to be able to use the particle-hole formalism developed in the previous chapters, we shall still separate the single-particle states into *two* categories, which can be referred to as "particle" and "hole" states, respectively. In this formalism we have the option of treating the valence shell(s) either as particles or as holes, which gives some additional flexibility.

Another problem for open-shell systems is associated with the determination of the zeroth-order or model functions. A closed-shell system is in zeroth order described by a single Slater determinant, which is then identical to the model function. For a general open-shell system, on the other hand, the function space associated with the energy levels is multidimensional, and the model functions are not always known initially. As discussed in Chap. 9, these functions can be determined by diagonalizing the effective Hamiltonian. Therefore, the idea of an effective interaction plays a fundamental role in the treatment of open-shell systems.

As we have seen in Chaps. 5 and 6, the LS term splitting of a configuration is due principally to the noncentral part of the Coulomb interaction among the electrons. In the present chapter we shall consider this splitting in some detail. In the formalism we shall also allow for an additional perturbation, such as the hyperfine interaction or the interaction with an external field. The formalism of perturbation theory developed so far is obviously valid in this case, if we replace the perturbation V by

$$V = H - H_0 = V_{es} + h, \tag{13.1}$$

where V_{es} is the Coulomb perturbation (11.39) and h is the additional perturbation.

We begin this chapter by redefining the particle-hole concept and the corresponding graphical representation in a way that is suitable for the open-shell case. This will lead to the same form of Wick's theorem as before, and we can derive the diagrammatic expansion of the wave operator from the generalized Bloch equation using the same procedure as in the closed-shell case. The unlinked diagrams, which have a different interpretation in the open-shell case, are cancelled in the wave-operator expansion, if a new kind of diagram, called "folded" or "backwards," is introduced. A diagrammatic expansion of the effective Hamiltonian is presented and used to describe the effective interaction of the valence electrons which is responsible for the energy-level structure of open-shell systems. Reference is made to early and more recent calculations.

13.1 The Particle-Hole Representation

13.1.1 Classification of Single-Particle States

As we have mentioned, we need a new classification of the single-particle states (orbitals) for open-shell systems. As an illustration we consider the ground configuration $1s^2 2s^2 2p^2$ of the neutral carbon atom. This configuration contains $\binom{6}{2} = 15$ determinants, some of which are indicated in Table 13.1.

Table 13.1. Determinants in the ground configuration of the carbon atom

$1s0^+$	$1s0^-$	$2s0^+$	$2s0^-$	$2p1^+$	$2p0^+$	$2p-1^+$	$2p1^-$	$2p0^-$	$2p-1^-$	$3s0^+$	$3s0^-$	$4s0^+$ ...
×	×	×	×	×	×							
×	×	×	×	×		×						
×	×	×	×	×			×					
×	×	×	×	×				×				
×	×	×	×	×					×			
×	×	×	×		×	×						
×	×	×	×		×		×					
—	—	—	—	—	—							

| Core states, occupied in all determinants of the model space | | | | Valence states, occupied in some determinants of the model space | | | | | | Virtual states, unoccupied in all determinants of the model space | | |

If we let these determinants define the model space, then the $1s$ and $2s$ orbitals are occupied in all, the $2p$ orbitals occupied in some, and the $3s$, $3p$, $4s$, ... orbitals unoccupied in all determinants of the model space. These orbitals will then represent core, valence and virtual single-particle states, respectively. As discussed before, we may have several configurations in the model space, which is particularly useful if the configurations are so strongly interacting that a perturbation expansion may lead to convergence problems. For the carbon atom we may in addition to the ground configuration include the low-lying configuration $1s^2 2p^4$. Then the $2s$ shell will be only partially filled in the model space and we would consider it as an *open* shell. The only core states—filled in all determinants in the model space—would then be the $1s$ states.

The arguments above lead us to the following convenient classification of the orbitals in the open-shell case [*Sandars* 1969]:

a) core orbitals, *occupied in all determinants of the model space;*
b) valence orbitals, *which may be occupied or unoccupied in the model space;*
c) virtual orbitals, *unoccupied in all determinants of the model space.*

We shall further assume that the model space is *"complete"* in the sense that

the model space contains all configurations which can be formed by distributing a certain number of (valence) electrons among the open shells.

Considering the $2s$ and $2p$ shells as valence shells in the carbon example above, this assumption implies that the configurations $1s^2 2s^2 2p^2$, $1s^2 2s 2p^3$, and $1s^2 2p^4$ should be included in the model space. The second configuration, however, will not mix with the other two for parity reasons, and hence it can be treated separately from the other configurations. On the other hand, if we study a system like the Cr atom with the ground configuration [Ar] $3d^5 4s$ and consider the $3d$ and $4s$ shell as open, then all configurations formed by distributing six electrons in these shells ($3d^6$, $3d^5 4s$, $3d^4 4s^2$) have the same parity. According to the rule given here, all these configurations should then be included in the model space and considered simultaneously. Sometimes this rule may lead to a model space which is unnecessarily large. It is then possible to work with a smaller model space, for instance, in the Cr example limited to the single $3d^5 4s$ configuration. This requires certain modifications of the formal development, however, which we shall not consider in this book. (The perturbation expansion for *incomplete* model spaces has been treated in the literature by *Hose* and *Kaldor* [1979, 1980, 1982], *Jeziorski* and *Monkhorst* [1981], *Haque* and *Mukherjee* [1984] and *Lindgren* [1985 b].)

13.1.2 The Particle-Hole Representation

According to (9.64), the projection operator for the model space (P space) is

$$P = \sum_{\alpha \in P} |\alpha\rangle\langle\alpha| \, , \tag{13.2a}$$

where the sum runs over all determinants of the model space. The projection operator for the orthogonal Q space is then

$$Q = \sum_{\beta \notin P} |\beta\rangle\langle\beta| = 1 - P \, . \tag{13.2b}$$

When operating on the model space, it is not possible to create a particle in a core orbital or to absorb a particle in a virtual orbital, as follows directly from the definitions given above. Formally, we can express this

$$a^\dagger_{core} P = a_{virt} P = 0 .$$ (13.3)

This corresponds to the relation (11.15) in the closed-shell case. It should be observed, however, that there is no analogous relation involving valence orbitals. It is possible to create as well as absorb valence particles in the model space. This is the fundamental difference between closed- and open-shell systems in our formalism, and it leads to some freedom as to how we define the normal form.

In the closed-shell case we defined normal form as the ordering for which the operators $a^\dagger_{core}$ and a_{virt} appear to the right, the reason being that they give zero when operating on the unperturbed state. In view of (13.3) it is then obvious that these operators should appear to the right in the normal form also for open-shell systems. On the other hand, it is less obvious how the valence operators should be treated, and, as a matter of fact, we have here several options. If there are several valence shells, it is even possible to treat some of the valence states as particle states and some of them as hole states. In this book, however, we shall for simplicity treat all valence states as particle states. We then define

the normal form as the ordering for which the $a^\dagger_{core}$, a_{val} and a_{virt} operators appear to the right.

Alternatively, we can define normal ordering by means of *particle-hole* operators, in analogy with those introduced in the closed-shell case (11.42). Our convention here would then correspond to the definition

$$b^\dagger = a^\dagger, b = a \quad \text{for \textit{virtual} and \textit{valence} orbitals (``particle states'')}$$

$$b^\dagger = a, b = a^\dagger \quad \text{for core orbitals. (``hole states'')}$$ (13.4)

As before, normal form is the ordering for which all the particle-hole absorption operators (b) appear to the right of the creation operators ($b^\dagger$).

In the closed-shell case, we defined the particle and hole states in relation to a determinantal state $|\alpha\rangle$, which then had the character of a "*vacuum state*". In the open-shell case we choose as our new "vacuum state" a determinant with all the *core orbitals occupied* and all the remaining ones unoccupied. Then the core orbitals correspond to the hole states and the valence and virtual orbitals to the particle states, as indicated in (13.4).

The wave operator can be written in the normal form (11.59) also in the open-shell case

$$\Omega = 1 + \sum_{ij} \{a^\dagger_i a_j\} x^i_j + \frac{1}{2} \sum_{ijkl} \{a^\dagger_i a^\dagger_j a_l a_k\} x^{ij}_{kl} + \cdots ,$$ (13.5)

and the electrostatic part of the perturbation, V_{es}, is given by (11.39)

$$V_{es} = V_0 + V_1 + V_2$$

$$V_0 = \sum_a^{\text{holes}} \langle a | -u | a \rangle + \frac{1}{2} \sum_{ab}^{\text{holes}} (\langle ab | r_{12}^{-1} | ab \rangle - \langle ba | r_{12}^{-1} | ab \rangle) \qquad (13.6)$$

$$V_1 = \sum_{ij} \{a_i^\dagger a_j\} \langle i | v | j \rangle$$

$$V_2 = \frac{1}{2} \sum_{ijkl} \{a_i^\dagger a_j^\dagger a_l a_k\} \langle ij | r_{12}^{-1} | kl \rangle .$$

Here, the sums over a and b run over the hole states, that is, the orbitals occupied in the determinant we have chosen as the "vacuum" state. It should be observed that also the effective potential (10.19) appearing in V_1 contains a summation over such orbitals only, that is,

$$\langle i | v | j \rangle = \langle i | -u | j \rangle + \sum_a^{\text{holes}} (\langle ia | r_{12}^{-1} | ja \rangle - \langle ai | r_{12}^{-1} | ja \rangle) . \qquad (13.7)$$

If the additional perturbation h is a one-body operator, then it can be written in analogy with the one-body parts of V_{es} as

$$h = h^0 + h_1 \qquad (13.8a)$$

$$h^0 = \sum_a^{\text{holes}} \langle a | h | a \rangle \qquad (13.8b)$$

$$h_1 = \sum_{ij} \{a_i^\dagger a_j\} \langle i | h | j \rangle . \qquad (13.8c)$$

13.1.3 Graphical Representation

It follows directly from the arguments of the preceding subsection that the graphical representation developed for closed-shell systems can be applied also in the open-shell case, provided that we let

upgoing lines represent particle states and downgoing lines represent hole states.

Treating the valence states as particle states, this rule implies that a line directed down represents a core orbital ("hole state"), while *a line directed up can represent either a virtual or a valence orbital,* i.e. a general "particle state".

In the graphical representation we shall usually not distinguish between virtual and valence orbitals. This implies that a single arrow directed upwards represents general particle states, that is, virtual as well as valence states. Sometimes it is necessary to indicate that an upgoing line represents valence states only, and for that purpose we shall use the notation of *Sandars* [1969] of representing valence orbitals by upgoing lines with *double arrows*. This is illustrated in Fig. 13.1.

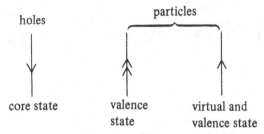

Fig. 13.1. Graphical representation of the different kinds of orbitals in the open-shell formalism. Note that a line with a single arrow directed upwards can represent virtual as well as valence orbitals, while a line with a double arrow directed upwards represents valence orbitals only

With the conventions introduced here, the Coulomb perturbation $V_{es} = V_0 + V_1 + V_2$ will be represented as before by the diagrams in Fig. 11.5. It should then be remembered that the internal lines in V represent hole states or *core orbitals only*. Also the graphical form of the effective potential (13.7) is the same as before (Fig. 10.2), if we let internal lines represent core states. This means that in order to eliminate the effective potential, the orbitals must be generated in the *Hartree-Fock potential of the closed-shell core*. This may be a realistic potential if there are only one or two particles in the open shells. The formalism developed here, however, is not limited to this particular choice of potential. From a physical point of view it might be quite natural to include the monopole ($k = 0$) term of the interaction with the valence electrons in the potential. If a potential different from the Hartree-Fock potential is used, then the effective potential (13.7) is nonvanishing and we have to keep the one-body part (V_1) of the Coulomb perturbation in the expansion. In many cases this may be necessary in order to achieve a reasonable convergence.

$$h^0 = \text{[diagram]}$$

$$h_1 = \text{[diagram]} + \text{[diagram]} + \text{[diagram]} + \text{[diagram]}$$

Fig. 13.2. Graphical representation of the normal-ordered zero- and one-body parts of the additional perturbation (13.8). The representation of the Coulomb perturbation (13.6) is given in Fig. 11.5

The additional perturbation h, defined in (13.8), is represented by the diagrams in Fig. 13.2. The algebraic form of three of these diagrams is given below as an illustration.

$$a \bigcirc \!\!-\!\! \overset{h}{-}\!-\!\!\triangleleft \; = \langle a|h|a\rangle \tag{13.9a}$$

$$\overset{a}{\diagdown}\!\!\overset{r}{\diagup}\!\!\overset{h}{-}\!-\!\!\triangleleft \; = \{a_r^\dagger a_a\}\langle r|h|a\rangle = a_r^\dagger a_a\langle r|h|a\rangle \tag{13.9b}$$

$$\overset{b}{\big\downarrow}\,\overset{h}{-}\!-\!\!\triangleleft \; = \{a_a^\dagger a_b\}\langle a|h|b\rangle = -a_b a_a^\dagger\langle a|h|b\rangle . \tag{13.9c}$$

As before, $a, b, \ldots$ are used to represent core orbitals and $r, s, \ldots$ will be used to represent general particle states (virtual and valence orbitals).

The normal form of the wave operator (13.5) is the same as in the closed-shell case. The graphical representation in Fig. 11.7, on the other hand, is not generally valid. According to (11.43), no b operators can operate to the right on the model space in the closed-shell case, and as a consequence the wave-operator diagrams have no free lines at the bottom. The situation is different in the open-shell case, since (11.43) does not apply for valence orbitals. In the particle-hole formulation we can express the corresponding relation (13.3) as

$$b_{\text{core}}P = b_{\text{virt}}P = 0 , \tag{13.10}$$

while $b_{\text{val}}P$ is not necessarily zero. It then follows that

the wave-operator diagrams may have free valence lines at the bottom.

This is illustrated in Fig. 13.3.

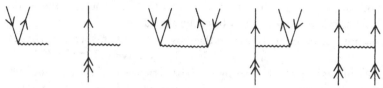

Fig. 13.3. Examples of wave-operator diagrams for open-shell systems. Free valence lines are allowed at the bottom. Outgoing lines represent valence or virtual orbitals (see Fig. 13.1). The diagrams must be open, which means that all in- and outgoing lines cannot be valence lines

In the following we shall also need new definitions of *closed* and *open* diagrams.

A diagram is said to be closed if it has no other free lines than valence lines.

A diagram which is not closed is said to be *open*.

As mentioned before, we assume that the model space is complete in the sense that it contains all configurations which can be formed by distributing the valence electrons among the valence shells. It then follows that all closed diagrams will operate within the P space (as in the closed-shell case) and the open diagrams will connect the P and Q spaces. Formally, we can express this as follows:

For any operator O the diagrams of POP are closed and the diagrams of QOP and POQ are open.

This means in particular that

the correlation operator is represented by open *diagrams and the effective Hamiltonian by* closed *diagrams.*

(It is obvious that this simple classification holds only if the model space is complete in the sense defined above.)

With the particle-hole formalism introduced here, Wick's theorem as stated in Chap.11 [(11.49, 57)] will be valid in the open-shell case. Also the graphical interpretation is the same as before (Fig. 11.6), if we let the lines directed upwards represent general particle states. Of course, only lines of the same kind can be joined; virtual to virtual, valence to valence, and core to core.

We can now apply the general procedure for generating the perturbation expansion in graphical form, essentially as in Chap. 12. The Rayleigh-Schrödinger expansion (9.75, 76) is generally valid, and the basic idea is to express the right-hand side in graphical form by means of Wick's theorem and to identify the result with the commutator on the left-hand side. This will lead to evaluation rules which are almost identical to those developed for closed-shell systems in Sect. 12.4.

The main difference between the closed- and open-shell procedures is that in the latter case valence lines are allowed at the bottom when operating on the model space. This means that in forming the diagrams of $V\Omega^{(n-1)}$, say, not all lines at the bottom of V need to be joined to lines of $\Omega^{(n-1)}$. Generally, this leads to more diagrams but the procedure is basically the same.

Another important difference between the closed- and open-shell procedures originates from the remaining terms of (9.75, 76) of the type $\Omega^{(n-m)} PV\Omega^{(m-1)}$. In the closed-shell case, normal ordering of these terms is trivial, since the last part, $PV\Omega^{(m-1)}$, represents just a number. In the open-shell case, on the other hand, this is an *operator*, and the terms have to be treated by means of Wick's theorem. We shall see that this leads to a new type of diagram, called "folded" or "backwards", which first appears in the second-order wave operator.

In the next section we shall consider the effective Hamiltonian to second order, which requires the wave function only to first order. In the following section we shall indicate how higher-order terms can be generated and formulate the linked-diagram theorem in the general open-shell case.

13.2 The Effective Hamiltonian

As shown in Chap. 9, the eigenvectors of the effective Hamiltonian represent the model functions, and the eigenvalues are the exact energies of the corresponding states,

$$H_{\text{eff}} \Psi_0^a = E^a \Psi_0^a . \tag{13.11}$$

Thus, if a model function is known, we can get the exact energy of the corresponding state by taking the expectation value of the effective Hamiltonian for that model state,

$$E^a = \langle \Psi_0^a | H_{\text{eff}} | \Psi_0^a \rangle . \tag{13.12}$$

According to (9.82, 84), the effective Hamiltonian is

$$H_{\text{eff}} = PH\Omega P = PH_0 P + PV\Omega P , \tag{13.13a}$$

where

$$PV\Omega P = PVP + PV\chi P . \tag{13.13b}$$

The term PH_0P represents the zeroth-order Hamiltonian H_0 operating in the model space, which reproduces the zeroth-order energy (10.9). We shall not need a graphical representation of this term. The first term of (13.13b), PVP, is the first-order contribution to the effective Hamiltonian, and it is represented by the closed diagrams of V according to the definitions given in the previous section. The last term in (13.13b), $PV\chi P$, represents the higher-order contributions, and the corresponding diagrams are obtained by "closing" the diagrams of χ by means of the perturbation so that the final state is in the model space. Such a diagram has no other free lines than valence lines.

In the following subsections we shall first consider the first-order term PVP and then generate the first-order wave operator $\Omega^{(1)}$ in order to construct the second-order contributions to the effective Hamiltonian, $PV\Omega^{(1)}P$.

13.2.1 The First-Order Effective Hamiltonian

With the substitution (13.1) the first-order effective Hamiltonian becomes

$$H_{\text{eff}}^{(1)} = PVP = PV_{\text{es}}P + PhP. \tag{13.14}$$

This is represented graphically by the closed diagrams of V_{es} and h, for which the initial as well as the final state is in the model space. Such diagrams can have no free lines other than valence lines. Diagrams without any free lines at all, which appear also in the closed-shell case (see, for instance, Fig. 11.5), represent the zero-body part of the perturbation. They are shown in Fig. 13.4. The first three

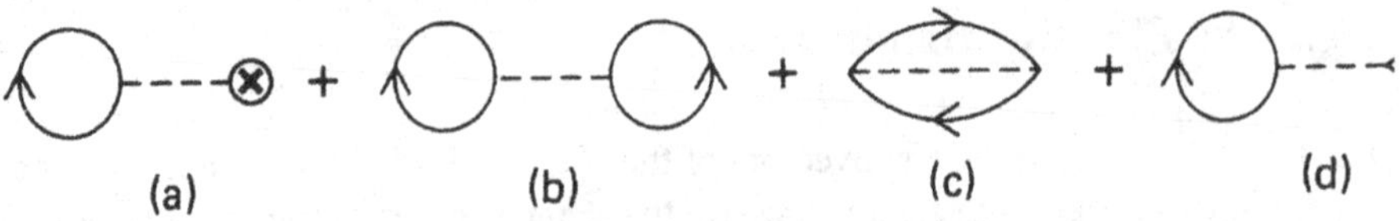

Fig. 13.4 a-d. Graphical representation of the zero-body parts of the effective Hamiltonian (13.14). The orbital lines represent core orbitals only

diagrams represent the zero-body part, V_0, of the Coulomb interaction and diagram (d) represents the zero-body part, h^0, of the additional perturbation (13.8). In addition, for open-shell systems also V_1, V_2, and h_1 have closed parts, that is, diagrams which operate within the model space. These diagrams, which have only free valence lines, are given below.

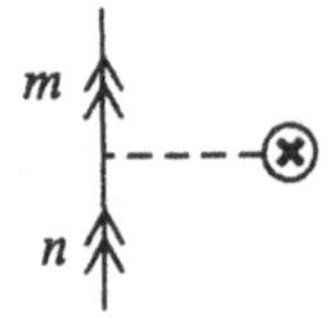

$$= a_m^\dagger a_n \langle m|v|n \rangle \qquad (13.15a)$$

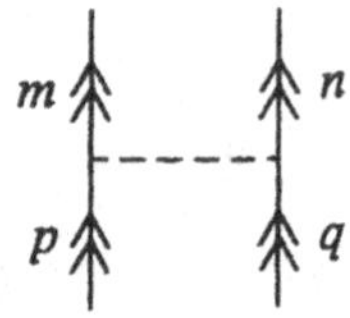

$$= \frac{1}{2} a_m^\dagger a_n^\dagger a_q a_p \langle mn|r_{12}^{-1}|pq \rangle \qquad (13.15b)$$

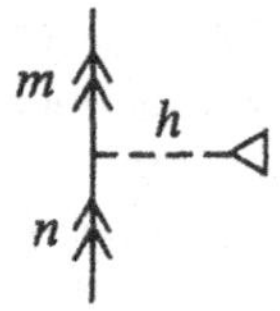

$$= a_m^\dagger a_n \langle m|h|n \rangle . \qquad (13.15c)$$

Here we have adopted the convention of letting m, n, p, q *represent valence orbitals*. As before, a, b, c, ... represent core orbitals (hole states) and r, s, t, ... general particle states (virtual as well as valence orbitals). The graphical representation of the first-order effective Hamiltonian (13.14) is summarized in Fig. 13.5.

$$H_{\text{eff}}^{(1)} = V_0 + h^0 + \quad \text{(a)} \quad + \quad \text{(b)} \quad + \quad \text{(c)}$$

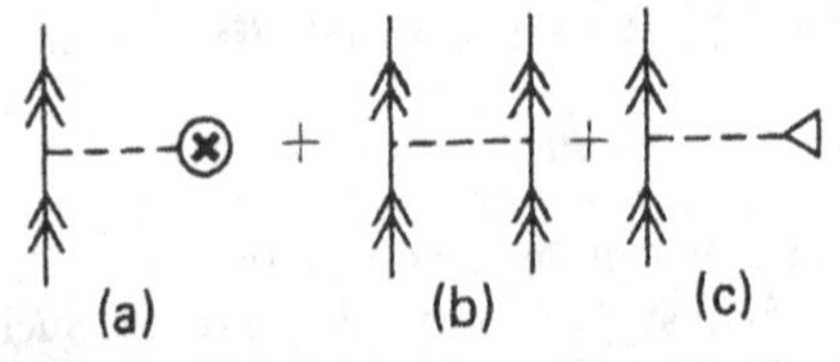

Fig. 13.5 a-c. The first-order effective Hamiltonian is represented by the *closed* diagrams of V, with no other free lines than valence lines. The zero-body part, $V_0 + h^0$, is represented by the diagrams in Fig. 13.4

According to (13.12), the first-order energy contribution in the state ψ^a is given by

$$E^{a(1)} = \langle \Psi_0^a | H_{\text{eff}}^{(1)} | \Psi_0^a \rangle .$$

The constant term $V_0 + h^0$ of $H_{\text{eff}}^{(1)}$ is the first-order energy contribution for the closed-shell system we have chosen as the "vacuum" state. This term is the same for all configurations which have the same closed-shell structure. The open-shell part of the first-order effective Hamiltonian is represented by diagrams (a-c) in Fig. 13.5. As mentioned, the potential diagram (a) vanishes if the orbitals are generated in the Hartree-Fock potential of the closed-shell core. If not, this diagram will shift the energy of the valence orbital towards the corresponding Hartree-Fock value. Diagram (b), which describes the Coulomb interaction between valence electrons, is responsible for the first-order LS term splitting of the configuration.

The part of the effective Hamiltonian which depends on the additional perturbation h is called the *effective operator*, h_{eff}, associated with the perturbation h. It follows directly from (13.12) that the energy contribution due to this perturbation is given by

$$\Delta E^a = \langle \Psi_0^a | h_{\text{eff}} | \Psi_0^a \rangle . \tag{13.16}$$

In first order, h_{eff} is given by

$$h_{\text{eff}}^{(1)} = PhP = P\left(h^0 + \sum_{mn}^{\text{val}} a_m^\dagger a_n \langle m | h | n \rangle\right) P . \tag{13.17}$$

These two terms are represented by the diagrams in Figs. 13.4d and 13.5c, respectively. Of course, only the second, one-body term can lead to any energy-level splitting.

13.2.2 The First-Order Wave Operator

It follows from (13.13) that the second-order contribution to the effective Hamiltonian is obtained by closing the diagrams of the first-order wave operator by the perturbation, so that the final state falls in the model space. Therefore, before we can construct these diagrams, we shall need the diagrams of the first-order wave operator.

The open-shell wave-operator diagrams can be constructed in essentially the same way as in the closed-shell case. We start again with the first-order equation (9.75a)

$$[\Omega^{(1)}, H_0] P = QVP = Q(V_{\text{es}} + h)P , \tag{13.18}$$

using the perturbation in the form (13.1). We can then separate $\Omega^{(1)}$ into a Coulomb part, $\Omega_{\text{es}}^{(1)}$, independent of h, and an h-dependent part, $\Omega_h^{(1)}$,

$$\Omega^{(1)} = \Omega_{es}^{(1)} + \Omega_h^{(1)} , \tag{13.19}$$

which satisfy the equations

$$[\Omega_{es}^{(1)}, H_0]P = QV_{es}P \tag{13.20a}$$

$$[\Omega_h^{(1)}, H_0]P = QhP . \tag{13.20b}$$

The diagrams of $QV_{es}P$ and QhP are the open diagrams of V_{es} and h, respectively, with no other free lines at the bottom than valence lines. As before, the commutator yields the excitation energy (11.70)

$$D = \sum(\varepsilon_{in} - \varepsilon_{out}) , \tag{13.21}$$

where ε_{in} (ε_{out}) represent the eigenvalues associated with the incoming (outgoing) orbital lines. We then get the graphical representation of $\Omega^{(1)}$ given in Fig. 13.6.

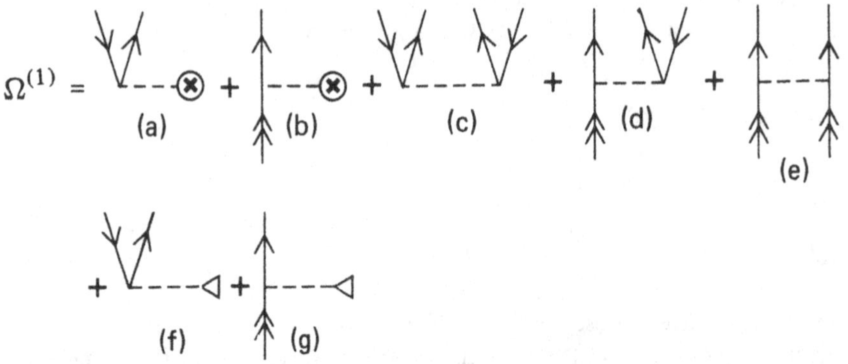

Fig. 13.6 a-g. The first-order wave-operator diagrams for open-shell systems. The notation is the same as in Fig. 13.3. Diagrams (a-e) represent the Coulomb part $\Omega_{es}^{(1)}$, and diagrams (f-g) the h-dependent part Ω_h

It should be observed now that the incoming lines which appear in the denominator (13.21) may be valence lines which approach the vertex from below. Note also that the outgoing lines represent virtual as well as valence orbitals.

As before, we get the corresponding algebraic expressions by means of the Goldstone rules given in Sect. 12.4. Here, only the rule for the denominators (c) needs some extra consideration. For closed-shell systems we gave the rule

$$D = \sum (\varepsilon_{down} - \varepsilon_{up}) , \tag{13.22}$$

where ε_{down} (ε_{up}) represent the energies associated with orbital lines directed down (up), cut by a line immediately above the interaction line. Since free orbital lines can approach the vertex from below in the open-shell case, this rule has to be replaced by (13.21). Alternatively, we can "bend" the incoming valence lines upwards, for instance,

$$(13.23)$$

and then the previous rule (13.22) can be applied. It should be observed, however, that this is allowed only in order to get the denominator. It is necessary to draw the valence lines upwards in order to be able to use Wick's theorem in the same simple way as before.

We shall now illustrate the application of the evaluation rules for open-shell systems by giving the expressions for diagrams (d, e, f) in Fig. 13.6.

$$= a_r^\dagger a_s^\dagger a_a a_m \frac{\langle rs | r_{12}^{-1} | ma \rangle}{\varepsilon_m + \varepsilon_a - \varepsilon_r - \varepsilon_s} \tag{13.24d}$$

$$= \frac{1}{2} a_r^\dagger a_s^\dagger a_n a_m \frac{\langle rs | r_{12}^{-1} | mn \rangle}{\varepsilon_m + \varepsilon_n - \varepsilon_r - \varepsilon_s} \tag{13.24e}$$

$$= a_r^\dagger a_a \frac{\langle r | h | a \rangle}{\varepsilon_a - \varepsilon_r} \tag{13.24f}$$

The factor of 1/2 for diagram (e) is in agreement with rule f') in Sect 12.4. The two vertices are here equivalent in the sense that interchanging them gives back the same diagram, apart from labelling.

We now have rules for constructing the first-order wave operator, and we shall proceed by constructing the second-order effective Hamiltonian.

13.2.3 The Second-Order Effective Hamiltonian

According to (13.13) the second-order effective Hamiltonian is

$$H_{\text{eff}}^{(2)} = PV\Omega^{(1)} P . \tag{13.25}$$

Separating the wave operator according to (13.19) and the perturbation according to (13.1), this leads to

$$H_{\text{eff}}^{(2)} = PV_{\text{es}}\,\Omega_{\text{es}}^{(1)}P + Ph\Omega_{\text{es}}^{(1)}P + PV_{\text{es}}\Omega_{\text{h}}^{(1)}P + Ph\Omega_{\text{h}}^{(1)}P \tag{13.26}$$

We consider first the coulombic term $PV_{\text{es}}\Omega_{\text{es}}^{(1)}\,P$. The Coulomb interaction V_{es} is represented by the same diagrams as before (Fig. 11.5), provided that lines directed upwards represent virtual as well as valence orbitals. The graphical representation of $PV_{\text{es}}\Omega_{\text{es}}^{(1)}P$ is then obtained by operating with V_{es} on the diagrams of $\Omega_{\text{es}}^{(1)}$, using Wick's theorem (Fig. 11.6) in the same way as in the closed-shell case. Since H_{eff} operates only in the model space, no free lines other than valence lines are allowed. This procedure gives rise to zero-, one-, two- and three-body diagrams with zero, one, two and three pairs of free valence lines, respectively. The zero-body diagrams are the same as in the closed-shell case (Fig. 12.4). The one- and two-body diagrams obtained in this way are given in Fig. 13.7. The one-body diagrams (a-j) with a single pair of free lines shift the single-particle energies, and the two-body diagrams (k-t) with two pairs of such lines lead to a modification of the first-order Coulomb interaction between the valence electrons, represented by diagram (b) in Fig. 13.5.

In closing a diagram, all free lines which are not valence lines must be connected. Since a wave-operator diagram contains at least one such line, it follows that the diagrams of the second-order effective Hamiltonian must all be *connected*. We shall see later that this, in fact, holds to all orders.

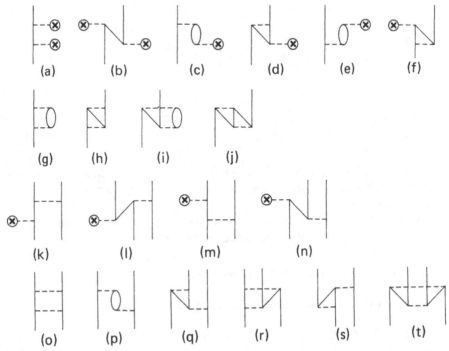

Fig. 13.7 a-t. The one- and two-body diagrams of the Coulomb part of the second-order effective Hamiltonian. All free lines are valence lines. For simplicity, all arrows have been left out, but the directions of the lines are determined uniquely by the fact that all valence lines are directed upwards and that the arrows on connected lines are continuous

In order to illustrate the evaluation of diagrams of the effective Hamiltonian, we give the expressions for diagrams (g, o, p) Fig. 13.7,

$$= a_m^\dagger a_n \sum_{ars} \frac{\langle ma|r_{12}^{-1}|rs\rangle\langle rs|r_{12}^{-1}|na\rangle}{\varepsilon_n + \varepsilon_a - \varepsilon_r - \varepsilon_s} \qquad (13.27\text{g})$$

$$= \frac{1}{2} a_m^\dagger a_n^\dagger a_q a_p \sum_{rs} \frac{\langle mn|r_{12}^{-1}|rs\rangle\langle rs|r_{12}^{-1}|pq\rangle}{\varepsilon_p + \varepsilon_q - \varepsilon_r - \varepsilon_s}. \qquad (13.27\text{o})$$

$$= a_m^\dagger a_n^\dagger a_q a_p \sum_{ar} \frac{\langle ma|r_{12}^{-1}|pr\rangle\langle rn|r_{12}^{-1}|aq\rangle}{\varepsilon_a + \varepsilon_q - \varepsilon_r - \varepsilon_n} \qquad (13.27\text{p})$$

The denominators are the same as in the corresponding first-order wave-operator diagrams (13.24), that is, given by the rule (13.21), or by the closed-shell rule (13.22), if we imagine that the incoming valence lines are bent upwards. This explains, for instance, that in diagram (p) the orbital q appears in the denominator while the orbital p does not.

The last three terms in the second-order effective Hamiltonian (13.26), which depend on the additional perturbation h, represent the corresponding second-order effective operator

$$h_{\text{eff}}^{(2)} = Ph\Omega_{\text{es}}^{(1)}P + PV_{\text{es}}\Omega_{\text{h}}^{(1)}P + Ph\Omega_{\text{h}}^{(1)}P. \qquad (13.28)$$

The graphical representation of the terms linear in h, $Ph\Omega_{\text{es}}^{(1)}P$ and $PV_{\text{es}}\Omega_{\text{h}}^{(1)}P$, is obtained by closing the diagrams of $\Omega_{\text{es}}^{(1)}$ (Fig. 13.6a-e) with h and the diagrams of $\Omega_{\text{h}}^{(1)}$ (Fig. 13.6f, g) with V_{es}, so that no free lines other than valence lines appear. The diagrams obtained in this way are shown in Fig. 13.8. They lead to a modification of the first-order interaction of the valence electrons with the external perturbation, described by diagram (c) in Fig. 13.5.

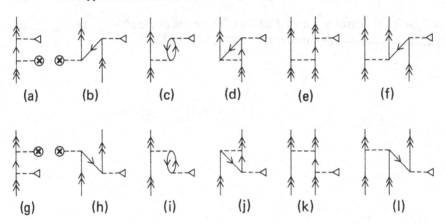

Fig. 13.8 a-l. Graphical representation of the second-order effective operator, linear in h. Diagrams (a-f) represent $Ph\Omega_{es}^{(1)} P$ and (g-l) $PV_{es}\Omega_{h}^{(1)}P$. Diagrams (a-d, g-j) are one-body and (e,f,k,l) two-body diagrams

The algebraic expressions for these diagrams are obtained in analogy with the Coulomb diagrams in (13.27). As examples we consider diagrams (c) and (i),

$$= a_m^\dagger a_n \sum_{ar} \frac{\langle a|h|r\rangle\langle mr|r_{12}^{-1}|na\rangle}{\varepsilon_n + \varepsilon_a - \varepsilon_m - \varepsilon_r} \qquad (13.29c)$$

$$= a_n^\dagger a_m \sum_{ar} \frac{\langle na|r_{12}^{-1}|mr\rangle\langle r|h|a\rangle}{\varepsilon_a - \varepsilon_r}. \qquad (13.29i)$$

If there is only one open shell, then ε_m is always equal to ε_n and the denominators in the two diagrams become equal. In this case, the two diagrams are the hermitian adjoints of each other. Since the operators h and r_{12}^{-1} are hermitian, it then follows that the two diagrams yield identical contributions to the effective operator after summing over the valence states. The same is true for the remaining pairs of diagrams in Fig. 13.8.

So far, we have not considered the last term in (13.28), which is quadratic in h. Often, this is negligible compared to the linear term. Apart from cases where extreme accuracy is required, this is the case in dealing with the hyperfine interaction. The quadratic term will be significant, of course, if the linear terms vanish for some reason. This is the case if the additional perturbation represents the dipole interaction with an external electric field, which is of odd parity, and if all states of the model space have the same parity (odd or even). The quadratic term $Ph\Omega_h^{(1)}P$ in (13.28) then represents the static polarizability (quadratic Stark effect). The diagrams corresponding to this term are shown in Fig. 13.9.

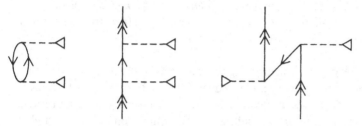

Fig. 13.9. Graphical representation of the terms in the second-order effective Hamiltonian quadratic in the additional perturbation h. If h is the electric dipole operator, these diagrams represent the static polarizability (quadratic Stark effect) in the lowest nonvanishing order

We have now studied the first- and second-order contributions to the effective Hamiltonian, and we can start to see a general pattern for the diagrammatic representation of this operator. It is convenient to separate the diagrams according to the number of pairs of free valence lines. Diagrams with no free lines at all represent numbers or zero-body operators, and diagrams with one, two, ... pairs of free lines represent one-, two-, ... body operators. Furthermore, the diagrams can be classified according to the number of interactions of the additional perturbation h. This leads to the classes of diagrams indicated in Fig. 13.10.

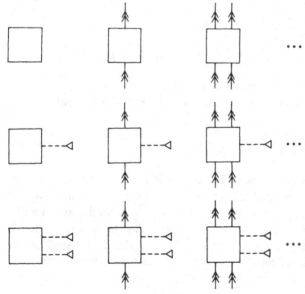

Fig. 13.10. Classification of the diagrams of the effective Hamiltonian H_{eff}. The first row represents the Coulomb part of the effective Hamiltonian and the following rows the parts which are linear, quadratic...in the additional perturbation (h). The latter rows represent the effective operator, h_{eff}, associated with the perturbation h. The columns represent effective zero-, one-, two-,...body operators

13.3 Higher-Order Perturbations. The Linked-Diagram Theorem

In the previous section we have considered the first-order wave operator and the first- and second-order effective Hamiltonian. The procedure we have used is quite similar to the procedure used in the closed-shell case. In the present section we shall generate the higher-order contributions to the wave operator. This will lead to a new type of diagram, called "*folded*" or "*backwards*". When these diagrams are included, all unlinked diagrams are eliminated from the expansion of the wave operator and of the effective Hamiltonian. This is the open-shell form of the linked-diagram theorem.

We begin this section by constructing the diagrams of the second-order wave operator. The discussion is then generalized to higher orders, and the third-order diagrams of the effective Hamiltonian are considered in some detail. Our intention is to provide a general understanding of the perturbation expansion and to show how the difficulties which arise may be overcome.

13.3.1 The Second-Order Wave Operator

In order to construct the second-order wave operator, we start as in the closed-shell case from the second-order Rayleigh-Schrödinger formula (9,75b, 12.19)

$$[\Omega^{(2)}, H_0]P = QV\Omega^{(1)}P - \Omega^{(1)}PVP . \tag{13.30}$$

In analogy with the closed-shell treatment (12.19–21), we can eliminate the zero-body part of V from this equation by observing that

$$Q(V_0 + h^0)\Omega^{(1)}P = \Omega^{(1)}P(V_0 + h^0)P . \tag{13.31}$$

Equation (13.30) then reduces to

$$[\Omega^{(2)}, H_0] = Q(V_1 + V_2 + h_1)\Omega^{(1)} - \Omega^{(1)}P(V_1 + V_2 + h_1)P , \tag{13.32}$$

leaving out the P operator to the right of Ω.

We consider first the Coulomb part of the wave operator, which satisfies the equation

$$[\Omega_{es}^{(2)}, H_0] = Q(V_1 + V_2)\Omega_{es}^{(1)} - \Omega_{es}^{(1)}P(V_1 + V_2)P . \tag{13.33}$$

The first term on the right-hand side is the same as in the corresponding closed-shell equation (12.21). Therefore, the second-order diagrams obtained in the closed-shell case (Figs. 12.2,3) will appear also in the open-shell case. In addition, there will be diagrams with one or several open-shell lines at the bottom, which we can get by replacing the incoming lines of the closed-shell wave-operator diagrams by valence lines. Obviously, this leads to a large number of diagrams, a few of which are shown in Fig. 13.11.

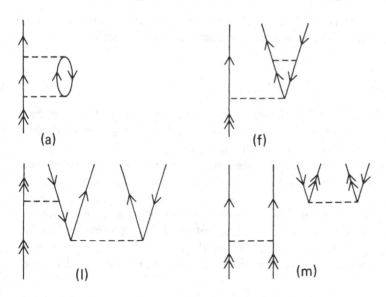

Fig. 13.11. Examples of second-order wave-operator diagrams. These diagrams are obtained by replacing one or several incoming lines of the closed-shell diagrams in Fig. 12.3 by valence lines. The labels refer to the corresponding diagrams in Fig. 12.3

The wave-operator diagrams are evaluated in essentially the same way as the effective-operator diagrams in Fig. 13.7. As examples, we consider the following diagrams:

$$= a_r^\dagger a_m \sum \frac{\langle ra | r_{12}^{-1} | st \rangle \langle st | r_{12}^{-1} | ma \rangle}{(\varepsilon_m - \varepsilon_r)(\varepsilon_m + \varepsilon_a - \varepsilon_s - \varepsilon_t)} \qquad (13.34a)$$

$$= -a_r^\dagger a_s^\dagger a_a a_m \sum \frac{\langle sb | r_{12}^{-1} | ta \rangle \langle rt | r_{12}^{-1} | mb \rangle}{(\varepsilon_m + \varepsilon_a - \varepsilon_r - \varepsilon_s)(\varepsilon_m + \varepsilon_b - \varepsilon_r - \varepsilon_t)} .$$

$$(13.34f)$$

The last factor in the denominator of these two expressions is the same as for $\Omega_{es}^{(1)}$ and it represents the *intermediate* excitation energy. The first factor is due to the commutator in (13.33) and represents the *final* excitation energy.

The last term of (13.33) has no counterpart in the corresponding closed-shell equation (12.21). The reason for this is that in the closed-shell case we have the identity

$$PVP = PV_0P,\tag{13.35}$$

and the last term of (12.19) is eliminated by means of (12.20). In the open-shell case, on the other hand, also V_1 and V_2 have closed parts, represented by diagrams (a, b) in Fig. 13.5. Therefore, the last term of (13.33) is nonzero in the open-shell case, and this fact constitutes one of the major differences between the closed-shell and open-shell formalisms. We shall now consider this term in some detail.

In order to construct the diagrams of the last term in (13.33), we use Wick's theorem as before (Fig. 11.6). This means that we should connect the lines at the bottom of $\Omega_{es}^{(1)}$ with the lines at the top of $P(V_1 + V_2) P$ in all possible ways, including no connection of all. This procedure is illustrated in Fig. 13.12 (top line).

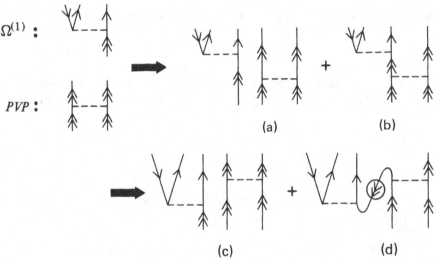

Fig. 13.12 a-d. Construction of some diagrams of $\Omega^{(1)} PVP$ by means of Wick's theorem. In order to be able to apply the ordinary Goldstone rules for the energy denominators, the diagrams are redrawn as shown in the bottom line of the figure. Diagram (c) is unlinked, since it has a disconnected closed part, and diagram (d) is "folded" or "backwards"

In our treatment of closed-shell systems in the previous chapter, a diagram was said to be *unlinked* if it had a disconnected, closed part. We shall use the same definition here, but a "closed" part now has the wider meaning introduced above, namely as a part with no free lines other than valence lines. With this definition, diagram (a) in Fig. 13.12 is unlinked. The same diagram, however, appears in the expansion of $QV\Omega_{es}^{(1)}$, when V is closed, as we shall presently demonstrate.

The algebraic expressions corresponding to diagrams (a, b) in Fig. 13.12 are obtained in the standard way, apart from the denominators. We observe that there is a denominator associated with the wave-operator diagram but not with the diagram of PVP. In order to be able to use the same denominator rule as before (13.21, 23), we interchange the ordering of the two interactions, as illustrated in the bottom line of the figure. The connected diagram (d) then appears with a valence line running downwards, and therefore this kind of diagram is referred to as *folded* [*Brandow* 1967] or *backwards* [*Sandars* 1969].

We can see that the unlinked diagram (c) in Fig. 13.12 also appears in the expansion of the first term of (13.33) as illustrated in Fig. 13.13. Due to the minus sign in the equation, these diagrams cancel. Diagram (d), on the other hand, remains in the expansion and appears with a minus sign. We shall now show in a formal way that all unlinked diagrams in the second-order wave operator cancel in the same way.

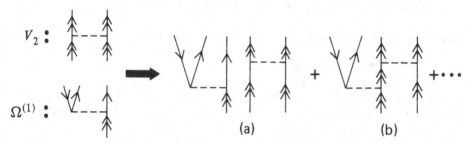

Fig. 13.13 a-b. Construction of some diagrams of $Q V \Omega^{(1)}$. Diagram (a) is unlinked and cancels diagram (c) in Fig. 13.12. Note that the outgoing lines of $\Omega^{(1)}$ can represent virtual as well as valence orbitals. In the latter case such a line can be joined to a valence line at the bottom of V

Using Wick's theorem (11.57), we can express the second-order equation (13.30) as

$$[\Omega^{(2)}, H_0]P = Q\{V\Omega^{(1)}\}P + Q\{\overline{V\Omega^{(1)}}\}P - \{\Omega^{(1)}PVP\} - \{\overline{\Omega^{(1)}P\overline{V}P}\}. \quad (13.36)$$

The first and third terms on the right-hand side are represented by *disconnected* diagrams (no connection between $\Omega^{(1)}$ and V), while the second and fourth terms are represented by connected diagrams. The third term, furthermore, represents *unlinked* diagrams, since the disconnected part, PVP, is closed. Identical unlinked diagrams appear in the first term, when V is closed, however, which we can see formally as follows.

We separate the perturbation V into closed (V_{cl}) and open (V_{op}) parts,

$$V = V_{cl} + V_{op}, \quad (13.37)$$

where V_{cl} is represented by the closed and V_{op} by the open diagrams. The open diagrams connect the model space with the complementary space and they give no contribution if the initial as well as the final states lie in the model space. Therefore, we have the relation

$$PVP = PV_{cl}P . \tag{13.38}$$

Furthermore, since a closed diagram cannot connect the P and Q spaces, it follows that

$$QV_{cl}P = 0 , \tag{13.39}$$

which means that

$$PVP = PV_{cl}P = V_{cl}P . \tag{13.40}$$

Substituting (13.37) into the first term of (13.36), we get

$$Q\{V\Omega^{(1)}\} P = Q\{V_{cl}\Omega^{(1)}\} P + Q\{V_{op}\Omega^{(1)}\} P \tag{13.41}$$

and using (13.40) the third term of (13.36) can be written

$$-\{\Omega^{(1)} PVP\} = -\{\Omega^{(1)} V_{cl}\} P . \tag{13.42}$$

Apart from sign, the latter term is identical to the first term of (13.41), since the operators can be permuted within a normal product. Therefore these terms cancel and the expression (13.36) reduces to

$$[\Omega^{(2)}, H_0] = Q\{V_{op}\Omega^{(1)}\} + Q\{\overline{V\Omega^{(1)}}\} - \{\overline{\Omega^{(1)} V_{cl}}\} , \tag{13.43}$$

leaving out the P operators. The first term is disconnected, but since both parts are open, the diagrams are linked according to our definition. The remaining terms are obviously represented by connected diagrams, and hence no terms on the right-hand side contain any disconnected, closed parts. This means that the corresponding diagrams are all *linked*.

The first two terms in (13.43) represent the linked diagrams of $QV\Omega^{(1)}$—disconnected and connected, respectively—and the last term represents the connected (linked) part of $\Omega^{(1)} V_{cl}$ or $\Omega^{(1)} PVP$, that is, the folded diagrams. Equation (13.43) can then be written

$$[\Omega^{(2)}, H_0] = (QV\Omega^{(1)} - \Omega^{(1)} PVP)_{linked} . \tag{13.44}$$

This is identical to the second-order Rayleigh-Schrödinger formula (13.30), apart from the fact that only the linked part of the right-hand side appears. We have then proved the linked-diagram theorem for open-shell systems to second order.

The perturbation V appearing here may contain an additional perturbation h. By separating V and $\Omega^{(1)}$ into Coulomb and h-dependent parts according to (13.1) and (13.19), we can then separate (13.44) as follows into h-independent and h-dependent parts

$$[\Omega_{es}^{(2)}, H_0] = (QV_{es}\Omega_{es}^{(1)} - \Omega_{es}^{(1)}PV_{es}P)_{\text{linked}} \tag{13.45a}$$

$$[\Omega_{h}^{(2)}, H_0] = (Qh\Omega_{es}^{(1)} + QV_{es}\Omega_{h}^{(1)} - \Omega_{h}^{(1)}PV_{es}P - \Omega_{es}^{(1)}PhP)_{\text{linked}} , \tag{13.45b}$$

leaving out terms quadratic in h. The diagrams of the first two terms of (13.45b) are constructed in the standard way. The last two terms lead to a new kind of folded diagram, illustrated in Fig. 13.14.

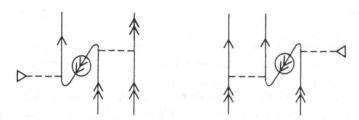

Fig. 13.14. Folded diagrams appearing in the h-dependent part of the second-order wave operator

For the remainder of this chapter we shall concentrate on the Coulomb interaction. In the next chapter we shall treat the effect of an additional perturbation—particularly the hyperfine interaction—in some detail.

13.3.2 The Linked-Diagram Expansion

The procedure used above to demonstrate the elimination of the unlinked diagrams in the second-order wave operator can be extended to show that the unlinked part of the Rayleigh-Schrödinger expansion cancels in each order. This can be expressed

$$[\Omega^{(n)}, H_0] = (QV\Omega^{(n-1)} - \sum_{m=1}^{n-1} \Omega^{(n-m)}PV\Omega^{(m-1)})_{\text{linked}} \tag{13.46}$$

which is the *linked-diagram theorem for open-shell systems*.

Equation (13.46) is identical to the corresponding closed-shell result (12.52), apart from the last term, which represents the folded diagrams. In the closed-shell case, the diagrams of $PV\Omega^{(m-1)}$ have no free lines, so it is not possible to join them to the diagrams of $\Omega^{(n-m)}$, which is necessary in order to form linked diagrams. Therefore, this term does not appear in the closed-shell case.

In analogy with (12.54), we can sum (13.46) to all orders, which yields

$$[\Omega, H_0] = (QV\Omega - \chi PV\Omega)_{\text{linked}} . \tag{13.47}$$

This is the linked part of the generalized Bloch equation (9.73), and expanding this equation order by order obviously yields the linked-diagram expansion (13.46) in exactly the same way as the Rayleigh-Schrödinger expansion is generated from the Bloch equation.

When the nth-order contribution to the wave operator is known, we can form the contribution to the effective Hamiltonian of order $(n + 1)$ according to (13.13) by closing the diagrams with the perturbation V,

$$H_{\text{eff}}^{(n+1)} = PV\Omega^{(n)}P . \tag{13.48}$$

The corresponding energy contributions are obtained by means of (13.12), if the model functions are known. If this is not the case, the matrix of H_{eff} in the model space has to be formed in order to solve the eigenvalue equation (13.11).

We have mentioned before that the first proof of the linked-diagram theorem for nondegenerate systems was given by *Goldstone* [1957], using a time-dependent field-theoretical approach. An early attempt to generalize this procedure to degenerate systems was made by *Morita* [1963]. The first complete proof of the theorem in this case, however, was given by *Brandow* [1966, 1967], using a time-independent procedure. Later, equivalent results using time-dependent formalisms have been obtained by several groups [*Oberlechner* et al. 1970; *Kuo* et al. 1971; *Johnson* and *Baranger* 1971]. The proof indicated here follows mainly the procedures of *Sandars* [1969] and *Lindgren* [1974, 1978].

13.3.3 The Third-Order Effective Hamiltonian

In the next section we shall consider the problem of calculating the LS term structure of a configuration of equivalent electrons. As we shall see, the term structure depends upon the effective interaction between the valence electrons. Before taking up this interesting and complex problem, we consider here the third-order diagrams of the effective Hamiltonian. This will enable us to see how important higher-order contributions to the effective electron-electron interaction can be included in a systematic way. Our discussion here follows closely a recent article by *Morrison* and *Salomonson* [1980], in which a calculation of the effective interaction of the $2p$ electrons of carbon is given. This problem will be further discussed in the final section of Chap. 15, where some numerical results are given.

According to (13.48) the third-order effective Hamiltonian is equal to

$$H_{\text{eff}}^{(3)} = PV\Omega^{(2)}P . \tag{13.49}$$

This means that we can get its graphical representation by closing the diagrams of the second-order wave operator with the perturbation V, in analogy with the treatment of the second-order effective Hamiltonian in Sect. 13.2.3. A few examples of the large number of possible diagrams of this kind are shown in

Fig. 13.15. These diagrams were obtained by closing three of the wave-operator diagrams in Fig. 13.11.

In order to get a more complete idea of what is involved, we consider the diagrams of the second-order wave operator in more detail. As we have said, the diagrams of $\Omega^{(2)}$ in the closed-shell case (Fig. 12.3) will also appear in the open-shell case. In addition, there will be diagrams for which one or several of the free lines have been replaced by valence orbital lines, and folded or backwards diagrams, which occur only for the open-shell problem. The folded diagrams of $\Omega^{(2)}$ are shown in Fig. 13.16.

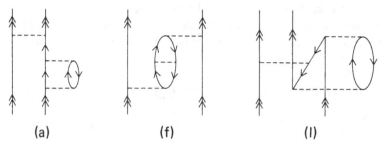

(a)   (f)   (l)

Fig. 13.15. Examples of third-order diagrams of the effective Hamiltonian, obtained by closing three of the second-order wave-operator diagrams in Fig. 13.11

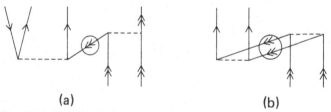

(a)   (b)

Fig. 13.16a,b. Folded diagrams of the Coulomb part of the second-order wave operator, assuming no effective-potential interactions

Apparently, there is a large number of diagrams of $\Omega^{(2)}$, and it is natural to look for some way of limiting the number of diagrams, so that the problem is more manageable. We shall generally limit our discussion to wave-operator diagrams for which no more than two particles are excited simultaneously. This restriction is reasonable from a physical point of view, since it means that we include the important two-particle scattering processes but neglect processes for which more than two particles are involved simultaneously. As we shall see, diagrams of two-particle type can be evaluated by solving a two-particle equation iteratively. The simplest physical systems having an electrostatic term structure are those for which there are two electrons outside of closed shells. So we shall also limit our discussion here to the zero-, one-, and two-body parts of the effective Hamiltonian. The matrix elements of an effective three-body operator are zero for a configuration of two valence electrons.

A summary of the second-order diagrams of the wave operator which contribute to the two-body part of $H_{\text{eff}}^{(3)}$ is given in Fig. 13.17. According to (13.49) the diagrams of $H_{\text{eff}}^{(3)}$ are obtained by closing off the diagrams of $\Omega^{(2)}$ with the perturbation. There are a total of 157 distinct two-body diagrams of $H_{\text{eff}}^{(3)}$ which can be formed in this way, omitting potential interactions. As shown in Fig. 13.17, however, these diagrams can be separated into general classes according to the number of free lines of the wave-operator diagram from which they are constructed. We expect the first family of diagrams with a single pair of free lines to have a small effect, as is the case for closed-shell systems discussed in Chap. 12. So we consider first the second general class of diagrams for which a pair of core or valence states is excited and there is an additional interaction involving the holes and the particles in the excited state. There are 52 distinct two-body diagrams of $H_{\text{eff}}^{(3)}$ which can be obtained from the wave-operator

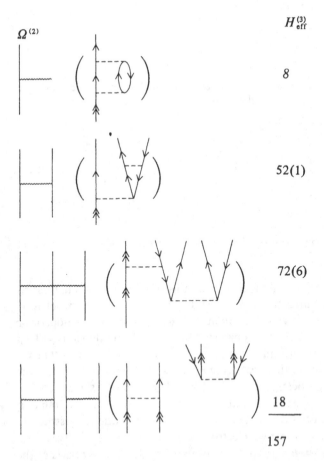

Fig. 13.17. Classification of the second-order wave-operator diagrams which contribute to the two-body part of the third-order effective Hamiltonian

diagrams in Fig. 12.3 in this way and one diagram which can be formed from the folded diagram (b) in Fig. 13.16. As we shall show in Chap 14, this particular family of diagrams can be included to infinite orders by solving a two-particle equation iteratively until self-consistency is achieved. The third family of diagrams shown in Fig. 13.17, for which an additional valence orbital polarizes a core, valence or virtual orbital, was found to give rise to convergence problems in the Pr^{3+} calculation to be discussed in the next section. The iteration procedure mentioned previously may be extended to include these diagrams. Alternatively, the model space can be extended so that it includes several strongly interacting configurations. The diagram which is given in Fig. 13.17 as an example of this family of diagrams will not occur if the core orbitals that appear are treated as valence orbitals. An example of a calculation using a multiconfigurational model space will be given at the end of this chapter.

The last family of diagrams shown in Fig. 13.17 represents two independent pair excitations. This kind of excitation can be treated by means of the coupled-cluster approach to be discussed in Chap. 15.

13.4 The *LS* Term Splitting of an $(nl)^N$ Configuration

13.4.1 The Slater and Trees Parameters

As a first illustration of the open-shell formalism developed here we consider the problem of evaluating to second order the *LS* term splitting of an $(nl)^N$ configuration, that is, a configuration for which there is a single valence shell (nl) of N electrons moving outside a closed-shell core.

It is well known that the Hartree-Fock model—or any other central-field model—usually serves to reproduce the term structure of an atom only in a qualitative way. As an example, we consider the energy levels of the $4f^2$ configuration of the Pr^{3+} ion, which has been the subject of a previous study [*Morrison* et al. 1970; *Morrison* and *Rajnak* 1971]. The Hartree-Fock (HF) and experimental energy levels of this configuration are shown in Fig. 13.18. The HF values were obtained using the configuration-average procedure described in Sect. 6.7 and evaluating the two-electron Slater integrals and the spin-orbit interaction constants. The energy matrix of the Coulomb and the spin-orbit perturbation was then formed and diagonalized. From the comparison in Fig. 13.18 we see that there is an obvious correlation between the HF and the experimental levels. However, the HF energy level scheme is considerably expanded with respect to what is found experimentally, and there are a number of crossovers.

Instead of using HF values of the Slater integrals, we may regard them as adjustable parameters ("Slater parameters") and perform a least-squares fit to the experimental levels. Table 13.2 shows the Hartree-Fock values of the Slater integrals and the fitted values of the corresponding parameters.

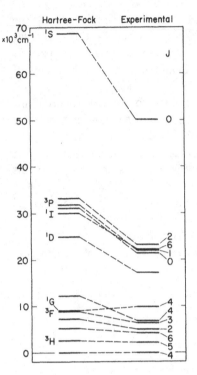

Fig. 13.18. Comparison of experimental and Hartree-Fock energy levels of the $4f^2$ configuration of Pr^{3+}

Table 13.2. Comparison of Hartree-Fock and experimental values of the Slater and Trees parameters for Pr^{3+} [cm^{-1}]

	F^2	F^4	F^6	α	β	γ
Hartree-Fock	104089	65507	47192	0	0	0
Experimental	75549	53874	35973	24	—586	728

The fact that the HF integrals are much larger than the empirical values of the corresponding parameters can easily be understood qualitatively. The Hartree-Fock model is an independent-particle model in which each electron moves in an average field of the other electrons. The electrons, however, correlate with each other in their mutual Coulomb field. They try to avoid each other, and this has the effect of reducing their effective interaction.

For an $(nl)^N$ configuration only Slater integrals of *even* rank appear. Hence, also the Slater parameters must be of even rank. It was found empirically by *Trees* [1951a, b; 1952], however, that the fit of experimental energy levels can be improved considerably by adding effective operators of *odd* rank to the Hamiltonian. The empirical values of parameters of this kind, usually called *Trees parameters* and denoted α, β, γ, are also shown in Table 13.2. The parameter α, which gives an energy contribution $\alpha L(L + 1)$, is mainly responsible for the fact that the high-angular-momentum states tend to cross over the low-angular-momentum states in going from the Hartree-Fock scheme to experimental levels. The properties of these additional parameters have been analyzed by *Racah*

[1952] and by *Racah* and *Stein* [1967]. As we shall see in the next subsection, these properties can be easily understood from the point of view of effective operators (see, for instance, [*Wybourne* 1965b] and [*Rajnak* and *Wybourne* 1963]).

13.4.2 Evaluation of the First- and Second-Order Contributions to the Term Splitting

We shall now evaluate the first- and second-order contributions to the term splitting of an $(nl)^N$ configuration. We shall neglect the spin-orbit interaction and any other additional perturbation and consider only the effect of the electrostatic perturbation (10.4)

$$V_{es} = -\sum_{i=1}^{N} u(r_i) + \sum_{i<j} \frac{1}{r_{ij}}, \tag{13.50a}$$

where according to (2.102)

$$\frac{1}{r_{ij}} = \sum_k \frac{r_<^k}{r_>^{k+1}} C^k(i) \cdot C^k(j). \tag{13.50b}$$

This leads to the first-order diagrams of the effective Hamiltonian given in Figs. 13.4 a-c and 13.5 a,b and the second-order diagrams in Fig. 13.7.

Denoting the model function of an LS state by $|\gamma SL\rangle$, it follows from (13.12) that the exact energy of that state is

$$E_{SL} = \langle \gamma SL | H_{eff} | \gamma SL \rangle. \tag{13.51}$$

It is obvious that the zero-body part of H_{eff} does not contribute to the splitting within a configuration. This part is just a number, and hence it shifts all energies of the configuration by the same amount. We can also easily verify that the one-body diagrams of H_{eff}, with a single incoming and outgoing valence line, do not contribute to the splitting either. This follows from the fact that the angular part of a Goldstone diagram is given by an angular-momentum diagram topologically identical to the Goldstone diagram. A one-body diagram of the Coulomb part of H_{eff} (no additional perturbation) will then lead to an angular-momentum diagram with two free lines. According to the theorem JLV2 in (4.5,6), such a diagram can be reduced to

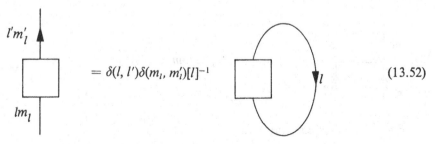

$$= \delta(l, l')\delta(m_l, m_l')[l]^{-1} \tag{13.52}$$

This implies that diagrams of this kind are diagonal in m_l and also independent of the m_l value. Hence, for an $(nl)^N$ configuration they contribute only to the diagonal elements of the effective Hamiltonian, and all these contributions are identical. Therefore, single-particle diagrams shift all levels of such a configuration by the same amount and do not contribute to the splitting.

From the arguments above it follows that the first-order term splitting is determined solely by diagram (b) in Fig. 13.5. After summing over the valence orbitals, this diagram represents the expression

$$\sum_{mnpq} \quad \begin{array}{c} m \uparrow \qquad \uparrow n \\ \vert\text{----}\vert \\ p \uparrow \qquad \uparrow q \end{array} \quad = \frac{1}{2} \sum_{mnpq} a_m^\dagger a_n^\dagger a_q a_p \langle mn \vert r_{12}^{-1} \vert pq \rangle \tag{13.53}$$

$$= \frac{1}{2} \sum_{mnpq} a_m^\dagger a_n^\dagger a_q a_p \sum_k R^k(mn, pq) \langle mn \vert C^k(1) \cdot C^k(2) \vert pq \rangle .$$

This is just the second-quantized form of the part of the Coulomb interaction (13.6) which operates within the valence shells. For a configuration of the kind $(nl)^N$, the orbitals m, n, p, and q all have the same nl values, and the first-order expression (13.53) can be written

$$\frac{1}{2} \sum_{mnpq} a_m^\dagger a_n^\dagger a_q a_p \sum_k^{\text{even}} F^k(nl, nl) \langle mn \vert C^k(1) \cdot C^k(2) \vert pq \rangle \tag{13.54}$$

$$= \sum_k^{\text{even}} F^k(nl, nl) \sum_{i<j}^{\text{val}} C^k(i) \cdot C^k(j) ,$$

where $F^k (nl, nl)$ is the Slater integral (3.47). The Coulomb interaction between the valence electrons is given here in second quantization (11.24) as well as in the form used in earlier chapters [see, for instance, (11.26)]. According to (2.127), the reduced matrix elements of the C^k tensor are nonzero in this case only if k is even. So to first order in perturbation theory the electrostatic term splitting of the $(nl)^N$ configuration depends only upon the Slater integrals $F^k(nl, nl)$ for which k is even.

Using the graphical representation (3.42) of the Coulomb interaction, the diagram in (13.53) can be expressed

$$\begin{array}{c} m \uparrow \qquad \uparrow n \\ \vert\text{----}\vert \\ p \uparrow \qquad \uparrow q \end{array} \quad = \frac{1}{2} a_m^\dagger a_n^\dagger a_q a_p \sum_k X^k(mn, pq) G_1(k) , \tag{13.55}$$

where, for an $(nl)^N$ configuration,

$$X^k(mn, pq) = (-1)^k \langle l_m \| C^k \| l_p \rangle \langle l_n \| C^k \| l_q \rangle R^k(mn, pq)$$
$$= \langle l \| C^k \| l \rangle^2 F^k(nl, nl) \tag{13.56}$$

and $G_1(k)$ is the angular-momentum graph

$$G_1(k) = \qquad\qquad\qquad\qquad\qquad\qquad\qquad\qquad \tag{13.57}$$

The diagram (13.55) can thus be expressed

$$= \frac{1}{2} a_m^\dagger a_n^\dagger a_q a_p \sum_k^{\text{even}} F^k(nl, nl) \langle l \| C^k \| l \rangle^2 G_1(k) . \tag{13.58}$$

We shall now see that much of the second-order contribution to the term splitting can be expressed in the same form and hence included in the first-order expression (13.58) by modifying the radial integrals $F^k(nl, nl)$. This means that we replace the first-order expression (13.54) by an *effective operator*, which has the same tensor structure

$$\frac{1}{2} \sum_{mnpq} a_m^\dagger a_n^\dagger a_q a_p \sum_k^{\text{even}} F^k \langle mn | C^k(1) \cdot C^k(2) | pq \rangle$$
$$= \sum_k^{\text{even}} F^k \sum_{i<j}^{\text{val}} C^k(i) \cdot C^k(j) . \tag{13.59}$$

The F^k coefficients, which appear here, are the Slater parameters introduced empirically, as mentioned above. So by evaluating the effective operator (13.59) using perturbation theory, we can make direct contact with the way in which the experimental data are conventionally analyzed.

As examples of the second-order contributions to the term splitting of an $(nl)^N$ configuration we consider the two-body diagrams (o, p) in Fig. 13.7 with the algebraic expressions (13.27o, p). Separating the expression into radial, spin and angular parts using the procedure developed in Sect. 12.6, we then get for diagram (p)

$$= a_m^\dagger a_n^\dagger a_q a_p \, 2 \sum_{ar} \sum_{k_1 k_2} \frac{X^{k_2}(ma, pr) X^{k_1}(rn, aq)}{\varepsilon_a - \varepsilon_r} G_2 \tag{13.60}$$

observing that $\varepsilon_q = \varepsilon_n$. The factor of 2 is the spin factor and G_2 is the angular-momentum diagram

$$= \delta(k_1, k_2)\,[k_1]^{-1}\,G_1(k_1). \qquad (13.61)$$

Since $k_1 = k_2$ in (13.61) must be even, the second-order contribution (13.60) can be written in the same form as the first-order expression (13.55)

$$= a_m^\dagger a_n^\dagger a_q a_p\, 2\sum_k^{\text{even}} [k]^{-1} \sum_{ar} \frac{X^k(nl\,a,\,nl\,r)X^k(r\,nl,\,a\,nl)}{\varepsilon_a - \varepsilon_r}\, G_1(k).$$
$$(13.62)$$

It then follows that this second-order contribution can be included in the first-order expression (13.58) by modifying the Slater parameters F^k by

$$\Delta F^k = 4[k]^{-1} \sum_{ar} \frac{X^k(nl\,a,\,nl\,r)\,X^k(r\,nl,\,a\,nl)}{(\varepsilon_a - \varepsilon_r)\,\langle l\|C^k\|l\rangle^2}. \qquad (13.63)$$

Here, as well as in the previous formulas, the summations run only over the nl values of the orbitals. It should also be observed that a represents core states only, while r represents general particle states, including valence states.

Next, we consider the diagram (13.27o). Separating the expression using the same procedure as before, we get

$$= \frac{1}{2} a_m^\dagger a_n^\dagger a_q a_p \sum_{rs} \frac{X^{k_2}(nl\,nl,\,rs)\,X^{k_1}(rs,\,nl\,nl)}{2\varepsilon_{nl} - \varepsilon_r - \varepsilon_s}\, G_3, \qquad (13.64)$$

where G_3 is the angular-momentum diagram (see Problem 4.3)

$$\text{(diagram)} = \sum_k (-1)^k [k] \begin{Bmatrix} l & l & k \\ k_1 & k_2 & l_r \end{Bmatrix} \begin{Bmatrix} l & l & k \\ k_1 & k_2 & l_s \end{Bmatrix} G_1(k) \qquad (13.65)$$

and $G_1(k)$ is given by (13.57). When k is even, this is also of the same form as the first-order expression (13.58) and it can be included in the Slater parameters. In the diagram (13.64), however, k *is not restricted to even integers*. k can take all values satisfying the triangular conditions. The contributions for which k is odd cannot be treated in the same way, since the reduced matrix element $\langle l \| C^k \| l \rangle$ vanishes in that case, as follows from (2.127).

In order to include the two-body effects of odd rank, we use the unit tensor operator u^k, introduced in Chap. 8 [(8.27)]. The reduced matrix elements of this operator are

$$\langle l \| u^k \| l' \rangle = \delta(l, l') \qquad (13.66)$$

for all values of k. We then construct an effective two-body operator in analogy with (13.59)

$$T = \frac{1}{2} \sum_{mnpq} a_m^\dagger a_n^\dagger a_q a_p \sum_k^{\text{odd}} c_k \langle mn | u^k(1) \cdot u^k(2) | pq \rangle$$
$$= \sum_k^{\text{odd}} c_k \sum_{i<j}^{\text{val}} u^k(i) \cdot u^k(j) . \qquad (13.67)$$

The graphical representation of this operator is the same as for the Coulomb interaction (13.55) after replacing the Slater integrals by the c_k parameters and setting the reduced matrix elements equal to unity,

$$T = \frac{1}{2} \sum_{mnpq} a_m^\dagger a_n^\dagger a_q a_p \sum_k^{\text{odd}} (-1)^k c_k G_1(k) . \qquad (13.68)$$

We now see that this operator can be used to represent the diagram (13.64) for odd values of k, and by identification we find that the contribution to the c_k parameters due to this diagram is

$$\Delta c_k = [k] \sum_{rs} \frac{X^{k_2}(nl \, nl, rs) \, X^{k_1}(rs, nl \, nl)}{2\varepsilon_{nl} - \varepsilon_r - \varepsilon_s} \begin{Bmatrix} l & l & k \\ k_1 & k_2 & l_r \end{Bmatrix} \begin{Bmatrix} l & l & k \\ k_1 & k_2 & l_s \end{Bmatrix} . \qquad (13.69)$$

Of course, this procedure can be used for *all* values of k, but traditionally the contributions for even k are included in the Slater parameters and the additional operator (13.67) is restricted to *odd* k values.

The remaining two-body diagrams of the second-order effective Hamiltonian in Fig. 13.7 can be treated in a similar way. It then follows that to second order the LS term splitting can be represented exactly by the Slater parameters for even k and the c_k parameters for odd k. The conventional Trees parameters α, β, γ, discussed previously, are linear combinations of the c_k coefficients, corresponding to the Casimir operators of the continuous groups introduced by *Racah* [1949] to classify atomic states. For f electrons, the two-body operator (13.67) can be written in two equivalent ways

$$T = \sum_k^{\text{odd}} c_k \sum_{i<j} u^k(i) \cdot u^k(j)$$
$$= \alpha G(R_3) + \beta G(G_2) + \gamma G(R_7) + \delta,$$

where $G(R_3)$, $G(G_2)$, and $G(R_7)$ are the Casimir operators of the continuous groups R_3, G_2, and R_7, respectively [*Judd* 1963a]. Apart from a constant term, which is the same for all the states of the configuration, the Trees parameters for f electrons are related to the c_k coefficients by the equations

$$
\begin{cases}
\alpha = \dfrac{1}{56}\left(\dfrac{c_1}{3} - \dfrac{c_5}{11}\right) \\[2mm]
\beta = 2\left(\dfrac{c_5}{11} - \dfrac{c_3}{7}\right). \\[2mm]
\gamma = \dfrac{5}{2}\dfrac{c_3}{7}
\end{cases}
\tag{13.70}
$$

13.4.3 Application to the Pr^{3+} Ion

One way to carry out the summations/integrations involved in the second-order diagrams above is to generate a complete set of single-particle states in the potential used to define the zeroth-order Hamiltonian (10.3). In the present section we shall illustrate this procedure with some results from the Pr^{3+} calculation of *Morrison* and *Rajnak*, mentioned earlier. Another possibility, which we shall consider in the next section, is to set up inhomogeneous one- and two-particle equations for the particular linear combinations of the virtual states which contribute to the Goldstone diagrams.

The contributions to the Slater and Trees parameters from the bound-state configurations which interact most strongly with $4f^2$ are given in Table 13.3. The most important excitation is the one in which two electrons are excited from the $4d$ core into the $4f$ shell, represented by the diagram (t) in Fig. 13.7, but also other single and double excitations to bound states are of importance.

Next, we shall consider the excitations for which a single electron is excited to the continuum (see Table 13.4). The most striking feature of these contributions is the importance of continuum states for which $l > 3$. The largest single contribution comes from the excitation of the $4d$ to a continuum g state ($4d \rightarrow$

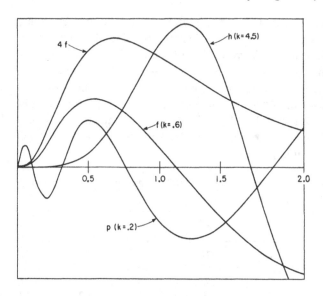

Fig. 13.19. p, f and h continuum functions, which give the largest contribution to the Slater parameters. These functions have large overlaps with the $4f$ function

kg). The excitations $4p \rightarrow kh$ and $4f \rightarrow kf$ are also important. The continuum p, f, and h functions which give the largest contribution are shown together with the $4f$ function in Fig. 13.19. The figure shows that part of the functions which contributes the most to the matrix elements involved. At large distances from the nucleus the $4f$ function decreases monotonically, and the contributions from the oscillating continuum functions tend to cancel. The f continuum function which contributes the most has a wave number $k = 0.6$. In the free region this corresponds to a wavelength $\lambda = 10.5$ atomic units. Near the origin the continuum f function is drawn in by the attractive field of the nucleus and oscillates more rapidly, but as it moves out, its wavelength increases to its value in the free region. The h function has a much larger centrifugal potential than the f function. For $k = 0.6$ its first maximum occurs far outside the maximum of the $4f$. As the wave number and energy of the function increase, though, the h wave moves in, until it overlaps the $4f$ function very favorably. So the excited states with large angular momentum produce resonances which can be deep in the continuum.

Table 13.3. Second-order contributions to the Slater and the Trees parameters from bound-state configurations [cm^{-1}]

	F^2	F^4	F^6	α	β	γ
$4d^2 \rightarrow 4f^2$	-3550	-5563	-5111	52	-761	1508
$5p^2 \rightarrow 4f^2$	$-\ 821$	-1024	-1921	4	-161	540
$4p^2 \rightarrow 4f^2$	$-\ 659$	$-\ 778$	-1404	4	-146	427
$4f^2 \rightarrow 5d^2$	$-\ 385$	$-\ 787$	$-\ 773$	5	$-\ 94$	211
$4p \rightarrow 4f$	$-\ 458$	4196	1206	13	24	-686
$5p \rightarrow 4f$	$-\ 441$	1648	402	5	68	-311

In order to understand the structure of the bound and continuum excitations, it is best to consider the excitations from a single core orbital to all states having a particular angular momentum. From Tables 13.3 and 13.4 we can see, for instance, that the excitation $4p \rightarrow 4f$ contributes more to the effective interaction between the two $4f$ electrons than do the continuum excitations $4p \rightarrow kf$. The reason for this is that the $4p$ and $4f$ orbitals are largely located in the same region of space. By contrast, an excitation like $4d \rightarrow 5g$ has a very small effect since the $5g$ function is very diffuse. Most of the contribution from the $4d \rightarrow g$ series comes from energetic continuum waves.

Table 13.4. Second-order contributions to the Slater and the Trees parameters from single and double excitation to the continuum [cm^{-1}]

	F^2	F^4	F^6.	α	β	γ
$4p \rightarrow kf$	−254	564	172	2	−10	−79
h	538	973	1013	−8	150	−255
k	66	19	193	−1	19	−24
$4d \rightarrow kd$	−72	−12	39	0	−2	10
g	−5695	−2450	−4540	−77	1088	−1883
i	70	−878	−510	−2	22	−62
$4f \rightarrow kp$	22	−34	−36	0	−1	−18
f	−1508	−1646	−1625	−1	− 77	475
h	−86	−245	−431	−4	93	−304
$4f^2 \rightarrow kdk'd$	−280	−755	− 811	3	−68	201
g	−348	−148	−199	−4	92	−210
$4f^2 \rightarrow kfk'f$	−3285	−3009	−2666	−1	−32	311
h	−65	−34	−174	0	20	−100
$4f^2 \rightarrow kgk'g$	−3095	−3165	−3142	38	−560	1163
i	−29	−14	−222	0	19	−114

Table 13.4 also shows the contribution to the Slater and Trees parameters from some two-particle excitations into the continuum. Again, excitations to f and g states are the most important. The excitation of two $4f$ electrons to continuum g states gives a contribution which is about one-half of the single-particle excitation $4d \rightarrow kg$, while the double excitation $4f^2 \rightarrow kfk'$ is more important than the corresponding single excitation $4f \rightarrow kf$.

The Pr^{3+} calculation of *Morrison* and *Rajnak* was carried out to third order in perturbation theory. As we have mentioned, the diagrams in the third family of Fig. 13.17, for which the valence $4f$ orbital polarizes the $4d$ core, were quite large and raised serious doubts about the convergence properties of the Slater parameters in this early calculation. The diagrams mentioned do not contribute to the Trees parameters α, β, and γ, however, and it was possible to obtain reliable values of these parameters. The values for the Trees parameters obtained in this way are compared with the experimental values in Table 13.5.

Table 13.5. Trees parameters for Pr^{3+} [cm^{-1}]

	α	β	γ
Second order	41	-655	1533
Third order	-8	7	128
Magnetic interaction	-4	32	-50
Total calculated	28	-616	1612
Experimental	24	-586	728

The experimental value of γ is determined empirically by the position of a single level ^1S which lies far above all of the other levels of the $4f^2$ configuration. This level is inconsistent with this calculation and with the crystal data [*Carnall* et al. 1969], and it seems unlikely that the ^1S level has been properly identified (for a recent tabulation of the rare-earth free-ion data, see [*Martin* et al. 1978]).

The most difficult part of this calculation was to carry out the summations and integrations over the excited states. In the next section we shall derive inhomogeneous one- and two-particle equations for the particular linear combination of excited states which contribute to the Goldstone diagrams. This greatly simplifies the numerical calculation.

13.5 The Use of One- and Two-Particle Equations

In the previous section, we evaluated the contributions to the effective Hamiltonian by explicitly constructing a complete set of virtual states. We note, however, that a summation is performed over the bound states and the continuum is integrated over, so that it is actually a wave packet of the virtual states which contributes to the final result. As an example we consider diagram (13.60). The radial part of this diagram is

$$\sum_{n_a l_a n_r l_r} \frac{X^{k_2}(ma, pr) X^{k_1}(rn, aq)}{\varepsilon_a - \varepsilon_r} \tag{13.71}$$

and the contribution of the diagram to the Slater parameter is according to (13.63)

$$\Delta F^k = 4[k]^{-1} \sum_{n_a l_a n_r l_r} \frac{X^k(nl\, a,\, nl\, r) X^k(r\, nl,\, a\, nl)}{(\varepsilon_a - \varepsilon_r)\langle l \| C^k \| l \rangle^2} . \tag{13.72}$$

Here, the X^k factor is defined as before by (13.56)

$$X^k(ab,\, cd) = X(k,\, l_a l_b l_c l_d) R^k(ab,\, cd) \tag{13.73}$$

where

$$X(k, l_a l_b l_c l_d) = (-1)^k \langle l_a \| C^k \| l_c \rangle \langle l_b \| C^k \| l_d \rangle \text{ and} \tag{13.74}$$

$$R^k(ab, cd) = \iint P_a(r_1) P_b(r_2) \frac{r_<^k}{r_>^{k+1}} P_c(r_1) P_d(r_2) \, dr_1 dr_2 , \tag{13.75}$$

Equations (13. 71, 72) may now be expressed more simply in terms of a single-particle function

$$\rho_d^k(a \rightarrow l_r, r_1) = \sum_{n_r}^{\text{part}} \frac{P_r(r_1) X^k(rn, aq)}{\varepsilon_a - \varepsilon_r} , \tag{13.76}$$

where n_r runs over the n values of the particle states (valence and virtual states) with the orbital angular-momentum quantum number l_r. The radial part (13.71) can then be written

$$\sum_{n_a l_a l_r} X^{k_2}(ma, p\rho_d^{k_1}(a \rightarrow l_r)) \tag{13.77}$$
$$= \sum_{n_a l_a l_r} X(k_2, l_m l_a l_p l_r) R^{k_1}(ma, p\rho_d^{k_1}(a \rightarrow l_r)) .$$

Here, the orbital r in the integrals $X^{k_1}(ma, pr)$ and $R^{k_1}(ma, pr)$ is replaced by the single-particle function (13.76). Similarly, the contribution (13.72) to the Slater parameter can be expressed

$$\Delta F^k = 4[k]^{-1} \langle l \| C^k \| l \rangle^{-2} \sum_{n_a l_a l_r} X^k(nl \, a, nl \, \rho_d^k(a \rightarrow l_r)) \tag{13.78}$$
$$= 4[k]^{-1} \sum_{n_a l_a l_r} \frac{\langle l_a \| C^k \| l_r \rangle}{\langle l \| C^k \| l \rangle} R^k(nl \, a, nl \, \rho_d^k(a \rightarrow l_r)) .$$

13.5.1 The Single-Particle Equation

We shall now derive an equation for the single-particle function introduced above. In order to be able to deal with the general case with several valence shells, we shall no longer assume that the orbital energies of the valence shells are the same. The denominator in diagram (13.60) will then be

$$\varepsilon_a + \varepsilon_q - \varepsilon_r - \varepsilon_n \tag{13.79}$$

as in (13.27p). We then generalize the single-particle function (13.76) to

$$\rho_d^k(a \rightarrow l_r, r_1) = \sum_{n_r}^{\text{part}} \frac{P_r(r_1) X^k(rn, aq)}{\varepsilon_a + \varepsilon_q - \varepsilon_r - \varepsilon_n} . \tag{13.80}$$

The radial function $P_r(r_1)$ satisfies the equation (6.7)

$$h_0^{l_r} P_r(r_1) = \varepsilon_r P_r(r_1) , \tag{13.81}$$

where

$$h_0^l = -\frac{1}{2}\frac{d^2}{dr_1^2} + \frac{l(l+1)}{2r_1^2} - \frac{Z}{r_1} + u(r_1).$$ (13.82)

It then follows that the function (13.80) satisfies the equation

$$(\varepsilon_a + \varepsilon_q - \varepsilon_n - h_0^l)\rho_a^k(a \to l_r; r_1) = \sum_{n_r}^{\text{part}} P_r(r_1)X^k(rn, aq)$$

$$= \sum_{n_r}^{\text{part}} P_r(r_1)X(k, l_r l_n l_a l_q) \, R^k(rn, aq).$$ (13.83)

The radial functions for a given l form a complete set in one dimension and hence they satisfy the closure property

$$\sum_{n_r}^{\text{all}} P_r(r_1)P_r(r_1') = \delta(r_1 - r_1'),$$ (13.84)

in analogy with the general, three-dimensional relation in Appendix B. Here n_r runs over the principal quantum numbers of all states with a given l, particle as well as hole (core) states.

We can now use the closure property to eliminate the infinite sum in (13.83) in the following way. The radial part of the right-hand side is

$$\sum_{n_r}^{\text{part}} P_r(r_1)R^k(rn, aq)$$

$$= \sum_{n_r}^{\text{part}} P_r(r_1) \iint P_r(r_1')P_n(r_2') \frac{r_<^k}{r_>^{k+1}} P_a(r_1')P_q(r_2') \, dr_1'dr_2'.$$ (13.85)

From the closure property (13.84) we get

$$\sum_{n_r}^{\text{part}} P_r(r_1)P_r(r_1') = \sum_{n_r}^{\text{all}} P_r(r_1)P_r(r_1') - \sum_{n_r}^{\text{holes}} P_r(r_1)P_r(r_1')$$

$$= \delta(r_1 - r_1') - \sum_{n_r}^{\text{holes}} P_r(r_1)P_r(r_1'),$$ (13.86)

where the last sum runs over hole states. Equation (13.85) can now be transformed to

$$\sum_{n_r}^{\text{part}} P_r(r_1)R^k(rn, aq) = \iint \delta(r_1 - r_1')P_n(r_2') \frac{r_<^k}{r_>^{k+1}} P_a(r_1')P_q(r_2') \, dr_1'dr_2'$$

$$- \sum_{n_r}^{\text{holes}} P_r(r_1)R^k(rn, aq).$$ (13.87)

Integration over the delta function simply implies that the integration variable r_1' is replaced by r_1 and the first term becomes

$$\int P_n(r_2') \frac{r_<^k}{r_>^{k+1}} P_q(r_2') \, dr_2' P_a(r_1) \, .$$

Using the Hartree function (7.5) introduced previously

$$Y_k(n, q; r_1) = r_1 \int P_n(r_2') \frac{r_<^k}{r_>^{k+1}} P_q(r_2') \, dr_2' \, , \tag{13.88}$$

we can write (13.87) in the form

$$\sum_{n_r}^{part} P_r(r_1) R^k(rn, aq) = \frac{1}{r_1} Y_k(n, q; r_1) P_a(r_1) - \sum_{n_b}^{holes} P_b(r_1) R^k(bn, aq) \tag{13.89}$$

and the single-particle equation (13.83) becomes

$$
\boxed{
\begin{aligned}
&(\varepsilon_a + \varepsilon_q - \varepsilon_n - h_0^l r) \rho_d^k(a \rightarrow l_r; r_1) \\
&= X(k, l_r l_n l_a l_q) \left[\frac{1}{r_1} Y_k(n, q; r_1) P_a(r_1) - \sum_{n_b}^{holes} R^k(bn, aq) P_b(r_1) \right] .
\end{aligned}
}
\tag{13.90}
$$

The last term in the equation has the effect of making the right-hand side orthogonal to all hole states with the given orbital angular momentum $l_r = l_b$. This can easily be verified, recalling that the Slater integral can be written

$$R^k(ab, cd) = \int P_a(r_1) \frac{1}{r_1} Y_k(b, d; r_1) P_c(r_1) \, dr_1 \, . \tag{13.91}$$

Therefore, these terms are usually referred to as *"orthogonality terms."* In practice, they are most easily generated by removing the overlap of the first term with the hole states in a numerical way without using the explicit form of these terms.

Next, we consider an exchange variant of the diagram (13.60), given in Fig. 13.7s,

$$= - a_m^\dagger a_n^\dagger a_q a_p \sum_{ar} \frac{\langle ma | r_{12}^{-1} | pr \rangle \langle rn | r_{12}^{-1} | qa \rangle}{\varepsilon_a + \varepsilon_q - \varepsilon_r - \varepsilon_n} \, , \tag{13.92}$$

which has the radial part

$$\sum_{n_a l_a n_r l_r} \frac{X^{k_2}(ma, pr) X^{k_1}(rn, qa)}{\varepsilon_a + \varepsilon_q - \varepsilon_r - \varepsilon_n} \, . \tag{13.93}$$

In order to evaluate this sum we can use the "exchange" single-particle function

$$\rho_e^k(a \rightarrow l_r, r_1) = \sum_{n_r}^{\text{part}} \frac{P_r(r_1)X^k(rn, qa)}{\varepsilon_a + \varepsilon_q - \varepsilon_r - \varepsilon_n}, \tag{13.94}$$

which satisfies the differential equation

$$(\varepsilon_a + \varepsilon_q - \varepsilon_n - h_0^l r)\rho_e^k(a \rightarrow l_r; r_1)$$
$$= X(k, l_r l_n l_q l_a) \frac{1}{r_1} Y_k(n, a; r_1) P_q(r_1) - \sum_{n_b}^{\text{hole}} X^k(bn, qa) P_b(r_1). \tag{13.95}$$

Equations (13.90, 95) are inhomogeneous differential equations, and solving them is equivalent to evaluating the infinite sums in (13.80, 94). This offers a very convenient and efficient way of performing perturbative calculations, since it eliminates the need for generating a "complete" set of single-particle states and performing the corresponding summations/integrations. Computer programs are now available which solve inhomogeneous equations of this kind in a small fraction of a second.

Single-particle equations of the kind considered here have been used for a long time by *Sternheimer* [1950, 1951, 1972] and *Sternheimer* and *Peierls* [1971a, b] to evaluate the effect of core polarization on the quadrupole hyperfine coupling. Early applications of this technique were also made by *Dalgarno* and *Lewis* [1955] for studying the dipole polarizability of atomic systems (see also [*Dalgarno* 1959, 1962]). In recent years there has been an increasing interest in using single-particle equations in many-body calculations of different kinds [*Morrison* 1972; *Chang* and *Poe* 1975; *Garpman* et al. 1975, 1976; *Carter* and *Kelly* 1977].

As an illustration of the use of single-particle functions we return to the Pr^{3+} example considered previously [*Morrison* 1972]. Two of the functions

$$\frac{\rho_d^k(a \rightarrow l_r, r_1)}{X(k, l, 3l_a 3)} = \sum_{n_r}^{\text{part}} \frac{P_r(r_1)R^k(r\,4f, a\,4f)}{\varepsilon_a - \varepsilon_r} \tag{13.96}$$

which describe the excitations $4d \rightarrow g$ and $5s \rightarrow d$ are shown in Fig. 13.20. These functions have the form of the first unfilled shell with the corresponding value of angular momentum. The $4d \rightarrow g$ function has the form of a $5g$ orbital, and $5s \rightarrow d$ has the form of a $5d$. However, these functions are drawn in toward the nucleus with respect to ordinary $5g$ and $5d$ functions in Pr^{3+}. For instance, the position of the maximum of the g function corresponds to a hydrogenic $5g$ function with a screening constant of 26. By contrast, the screening constant of the $4f$ is 40. This feature of the excited functions may be understood by considering the form of the continuum function which contribute to the summation over excited states (13.80). The $R^2(kg4f, 4d4f)$ integral has its maximum value for g functions with a wave number $k = 4.0$. Energetic g waves of this kind increase much more rapidly as r moves away from the origin than does the $5g$ function. When the summation is performed over the virtual states, the oscillating portions of the continuum functions destructively interfere, and one obtains a function which

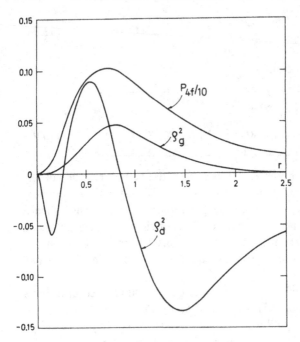

Fig. 13.20. Single-particle functions representing the core excitations $4d \to g$ and $5s \to d$ for Pr^{3+}. The Hartree-Fock radial $4f$ function is shown for comparison

has the form of the $5g$ but which is drawn in toward the origin. The contribution of the $5s \to 5d$ excitation is large compared to the contributions of the other members of the d series [*Morrison* and *Rajnak* 1971; *Morrison* 1972]. So the ρ_d^2 function has a large $5d$ component. The inclusion of the more highly excited d states has the effect of diminishing the magnitude of this function and drawing it in toward the nucleus. The single-particle d and g functions are localized in the same region of space as the $4f$.

13.5.2 The Pair Equation

The procedure developed above for evaluating diagrams with single excitations can easily be extended to diagrams with double excitations. As an example we consider the diagram (13.64). Again, allowing for the orbital energies to be different, we write the radial part of this diagram as

$$\sum_{n_r l_r n_s l_s} \frac{X^{k_2}(mn, rs)X^{k_1}(rs, pq)}{\varepsilon_p + \varepsilon_q - \varepsilon_r - \varepsilon_s} . \tag{13.97}$$

Here, the sums run over the nl values of the particle states r and s. The intermediate state must lie in the Q space, which means that r or s—but not both— can be valence states.

If we introduce the *pair function*

$$\rho^{k_1}(pq \rightarrow l_r l_s;\ r_1 r_2) = \sum_{n_r n_s}^{\text{part}} \frac{P_r(r_1) P_s(r_2)\ X^{k_1}(rs,\ pq)}{\varepsilon_p + \varepsilon_q - \varepsilon_r - \varepsilon_s}, \tag{13.98}$$

the radial part (13.97) can be written

$$\sum_{l_r l_s} X^{k_2}(mn,\ \rho^{k_1}(pq \rightarrow l_r l_s))$$
$$= \sum_{l_r l_s} X(k_2,\ l_m l_n l_r l_s)\ R^{k_2}(mn,\ \rho^{k_1}(pq \rightarrow l_r l_s)) . \tag{13.99}$$

Here, the pair function replaces two orbitals r and s in the X^k factor and in the Slater integral.

We shall now derive an equation for the pair function (13.98). In analogy with (13.83), we find that this function satisfies the equation

$$(\varepsilon_p + \varepsilon_q - h_0^{l_r}(1) - h_0^{l_s}(2))\ \rho^k(pq \rightarrow l_r l_s;\ r_1 r_2)$$
$$= \sum_{n_r n_s}^{\text{part}} P_r(r_1) P_s(r_2)\ X^k(rs,\ pq) . \tag{13.100}$$

As in the single-particle case we can remove the infinite summations by means of the closure property (13.84). Extending the summations over *all* principal quantum numbers gives

$$\sum_{n_r n_s}^{\text{all}} P_r(r_1) P_s(r_2) R^k(rs,\ pq)$$
$$= \sum_{n_r n_s}^{\text{all}} \iint \delta(r_1 - r_1')\delta(r_2 - r_2') \frac{r_<^k}{r_>^{k+1}} P_p(r_1') P_q(r_2')\ dr_1'\ dr_2' \tag{13.101}$$
$$= \frac{r_<^k}{r_>^{k+1}} P_p(r_1) P_q(r_2) .$$

In order to get the restricted sum in (13.100) we have to remove the terms where r and/or s are core states and also the terms where r and s are valence states. The terms where one of these orbitals is a virtual orbital and the other is a valence orbital, on the other hand, should be included in the sum according to the argument above. These exclusions can be made by making (13.101) orthogonal to the core orbitals individually and to the *pair* (p, q). We then get the pair equation

$$\boxed{\begin{aligned} &[\varepsilon_p + \varepsilon_q - h_0^{l_r}(1) - h_0^{l_s}(2)]\ \rho^k(pq \rightarrow l_r l_s;\ r_1 r_2) \\ &= \frac{r_<^k}{r_>^{k+1}} P_p(r_1) P_q(r_2) - \text{orthogonality terms.} \end{aligned}} \tag{13.102}$$

It is only fairly recently that accurate methods have become available for solving two-particle excitations of this kind [*Musher* and *Schulman* 1968; *McKoy* and *Winter* 1968]. Such equations have been applied in perturbation calculations

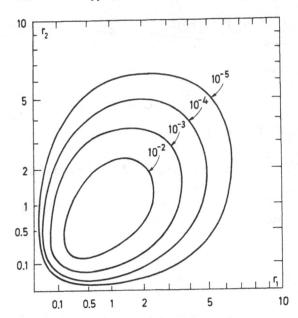

Fig. 13.21. Radial pair function for the excitation $4f^2 \rightarrow g^2$ in Pr^{3+}

by *Schulman* and *Lee* [1972], *Morrison* [1973] and *Garpman* et al. [1975]. In Fig. 13.21 we show a plot of the pair function for the $4f^2 \rightarrow g^2$ excitation in Pr^{3+}. Although the real g states are very diffuse, we see that the pair function is well localized in space. As is the case for the corresponding single-particle function, shown in Fig. 13.20, this is due to the fact that the contributions from the outer parts of the excited wave functions tend to cancel.

13.6 The Beryllium Atom Treated as an Open-Shell System

In Sect. 13.4 we saw that certain diagrams lead to serious convergence problems in the calculation of the *LS* term structure. As we have said, one way to remedy the situation is to extend the iteration procedure mentioned previously to include the particular family of diagrams which cause the convergence difficulties. Another possibility is to extend the model space so that the most important excitations are included in the zeroth-order function. An extended model-space calculation in which several shells are treated as valence shells can easily be included with the formalism developed here. So far, this procedure has been applied only to a few light atomic systems, and we shall illustrate it by means of a second-order calculation performed on the Be atom by *Salomonson* et al. [1980].

In this calculation we regard the $1s$ shell as the core and the $2s$ and $2p$ shells as valence shells. The model space then contains all configurations which can be formed by distributing the two valence electrons among the valence shells, that is, $1s^2 2s^2$, $1s^2 2s 2p$, and $1s^2 2p^2$. The second configuration does not interact with the remaining configurations and can be treated separately. The $1s^2 2s^2$ and $1s^2 2p^2$ configurations, on the other hand, are very strongly interacting and are treated simultaneously. This means that their mixing can in principle be included to all orders in the zeroth-order wave function. In order to find this mixing, we have to evaluate the effective Hamiltonian and solve the eigenvalue problem (13.11). This is a good illustration of the fact that it is necessary to work with *operators* rather than with wave functions and energies in this formalism.

By including the $1s^2 2p^2$ configuration in the model space, we may expect the perturbation expansion to converge much faster. In fact, we shall see in this section that such a calculation gives a quite accurate value of the ground-state energy already in second order. In addition, this procedure gives the energy of the excited state, designated $1s^2 2p^2\,{}^1S$.

13.6.1 The Energy Matrix

We consider now a model space containing the configurations $1s^2 2s^2$ and $1s^2 2p^2$. The configuration $1s^2 2s2p$ does not mix with these two and can be omitted. The configuration $1s^2 2s^2$ can only give rise to a ${}^1S$ term, while the $1s^2 2p^2$ configuration can couple to the terms ${}^3P$, ${}^1D$, and ${}^1S$. Since the Coulomb interaction commutes with S and L, it follows that the ${}^3P$ and ${}^1D$ terms are not affected by the $1s^2 2s^2$ configuration. The two ${}^1S$ terms, on the other hand, can mix, and these are the states we shall investigate.

We assume that the orbitals are generated in the Hartree-Fock potential of the $1s^2$ core, which means that the effective-potential interaction (13.7) vanishes. The diagrams of the effective Hamiltonian relevant in this case are shown schematically in Fig. 13.22. According to (13.52), in the one-body diagrams a,b the incoming and outgoing lines must respresent the same orbital. The two-body diagrams (c-f), on the other hand, can connect the two configurations, yielding diagonal as well as nondiagonal contributions.

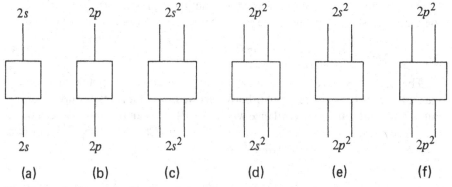

Fig. 13.22 a-f. Schematic effective-operator diagrams for the beryllium atom using an extended model space, $1s^2 2s^2 + 1s^2\,2p^2$. The one-body diagrams are diagonal, while the two-body diagrams can connect the two configurations. All lines shown are open-shell lines

In principle, we can now evaluate the diagrams of the effective Hamiltonian to the desired order and form the corresponding matrix within the model space, using Slater determinants, for instance, as basis functions. A model space with the two configurations $1s^2 2s^2$ and $1s^2 2p^2$ contains 16 states. If we choose coupled LS functions as basis functions, however, only the two ^1S states mix. As a consequence, the 16×16 matrix reduces to 14 one-dimensional and one two-dimensional matrix blocks along the main diagonal. In order to get the energies of the two ^1S states, we need only consider the two-dimensional submatrix

$$\begin{pmatrix} \langle 2s^2 \, {}^1S | H_{\text{eff}} | 2s^2 \, {}^1S \rangle & \langle 2s^2 \, {}^1S | H_{\text{eff}} | 2p^2 \, {}^1S \rangle \\ \langle 2p^2 \, {}^1S | H_{\text{eff}} | 2s^2 \, {}^1S \rangle & \langle 2p^2 \, {}^1S | H_{\text{eff}} | 2p^2 \, {}^1S \rangle \end{pmatrix}. \tag{13.103}$$

The eigenvalues of this submatrix gives the energies of the two ^1S states to the accuracy with which we have evaluated the effective Hamiltonian. The eigenvectors give the corresponding model or zeroth-order functions, that is, the mixture between the $2s^2 \, {}^1S$ and the $2p^2 \, {}^1S$ states. We shall now consider a second-order calculation of this submatrix.

13.6.2 The First-Order Results

The zeroth-order Hamiltonian H_0 yields the zeroth-order energies (6.11), which in the present case become[1]

$$\begin{cases} E_0(2s^2) = 2\varepsilon_{1s} + 2\varepsilon_{2s} = -12.6665 \\ E_0(2p^2) = 2\varepsilon_{1s} + 2\varepsilon_{2p} = -12.3732. \end{cases} \tag{13.104}$$

It should be observed that this gives different contributions to the two diagonal elements.

The first-order effective Hamiltonian is given in Fig. 13.5. The zero-body part V_0 is the same as for closed-shell systems, and in the present case it contains only the closed $1s$ shell. This gives the same contribution to both states, namely

$$V_0 = -R^0(1s1s, 1s1s) = -2.2770. \tag{13.105}$$

The one-body part of $H_{\text{eff}}^{(1)}$ vanishes with the choice of potential we have made. In order to evaluate the two-body part we have to form the coupling of the initial and final states in accordance with (13.103). We can then use the technique introduced in Sect. 4.4 and further developed in Chap. 8. The diagram can then be evaluated in the following way:

---

[1] All values in this section are in Hartrees (see Appendix A).

$$= \sum_k X^k(mn, pq) \, G, \qquad (13.106)$$

where G is the angular-momentum graph

$$G = \; = (-1)^{l_n + l_p + L + k} \begin{Bmatrix} l_m & l_n & L \\ l_q & l_p & k \end{Bmatrix} \qquad (13.107)$$

evaluated in Problem 4.2b. With the Hartree-Fock orbitals we are using, this gives the following result:

$$2s^2 \rightarrow 2s^2 : \; R^0(2s2s, 2s2s) = 0.3964$$

$$\left.\begin{matrix} 2s^2 \rightarrow 2p^2 \\ 2p^2 \rightarrow 2s^2 \end{matrix}\right\} : \; -\frac{1}{\sqrt{3}} R^1(2s2s, 2p2p) = -0.1438 \qquad (13.108)$$

$$2p^2 \rightarrow 2p^2 : \; R^0(2p2p, 2p2p) + \frac{2}{5} R^2(2p2p, 2p2p) = 0.4565$$

Adding (13.104, 105, 108), we get the following first-order energy matrix:

$$H_0 + H_{\text{eff}}^{(1)} = \begin{pmatrix} -14.5471 & -0.1438 \\ -0.1438 & -14.1938 \end{pmatrix}, \qquad (13.109)$$

where the first row/column represents the $2s^2 \; {}^1S$ state. According to (9.100), the diagonal elements are the expectation values of the ordinary Hamiltonian, which is the same as the Hartree-Fock energies of the two states (with the particular choice of Hartree-Fock potential). It should be observed that the first-order effective Hamiltonian has also *nondiagonal* matrix elements.

As mentioned, the eigenvalues of the effective Hamiltonian yield the energies of the true states corresponding to the unperturbed states in the model space and the eigenvectors yield the corresponding model functions. Applying this to the first-order matrix above, we get the energies

$$\begin{cases} E'(2s^2) = -14.5983 \\ E'(2p^2) = -14.1427 \end{cases} \tag{13.110}$$

and the corresponding model functions

$$\begin{cases} |2s^2\ {}^1S\rangle' = 0.942|2s^2\ {}^1S\rangle + 0.335|2p^2\ {}^1S\rangle \\ |2p^2\ {}^1S\rangle' = -0.335|2s^2\ {}^1S\rangle + 0.942|2p^2\ {}^1S\rangle . \end{cases} \tag{13.111}$$

This corresponds to a perturbative expansion to all orders, where only these two states are considered (see Problem 13.1).

Problem 13.1. Compare the results above with a perturbative expansion to first (second) order in the wave function and second (third) order in the energy, considering no states other than those indicated.

13.6.3 The Second-Order Results

We turn now to the second-order effective Hamiltonian. The zero-body part is the same as in the closed-shell case (Fig. 12.4) and it gives the same contribution for the two states. These diagrams have been evaluated in Sect. 12.6. The summations over the excited states were performed by solving the inhomogeneous one- and two-particle equations described above, which gave the following results:

$$E_0^{(2)}(2s^2\ {}^1S) = E_0^{(2)}(2p^2\ {}^1S) = -0.0417 . \tag{13.112}$$

The one-body diagrams (Fig. 13.7g-j) must have the same single-particle state in and out, as mentioned above. Considering the diagram (13.27 g), for instance, this means that $m = n$. The operator $a_m^+ a_n$ then represents the "number operator", which simply counts the number of particles in the open shell. So in this case the contribution from this diagram is twice the sum in (13.27g) with $m = n$ set equal to $2s$ and $2p$, respectively. This gives

$$\begin{cases} E_1^{(2)}(2s^2\ {}^1S) = -0.0058 \\ E_1^{(2)}(2p^2\ {}^1S) = -0.0079 . \end{cases} \tag{13.113}$$

The two-body diagrams of the effective Hamiltonian (Fig. 13.7o-t) must be evaluated for the coupled ${}^1S$ state. This can be accomplished in the same way as in first order. For instance, the expression (13.60) for diagram (p) becomes

$$= 2\sum_{ar} \frac{X^{k_2}(ma,\,pr)X^{k_1}(rn,\,aq)}{\varepsilon_a + \varepsilon_n - \varepsilon_r - \varepsilon_q}\, G\,, \qquad (13.114)$$

where G is the angular-momentum graph

$$= \delta(k_1,\,k_2)[k_1]^{-1} \qquad (13.115)$$

The graph to the right is identical to the first-order diagram (13.107). In a similar way the remaining two-body diagrams can be evaluated. The results are given in the following matrix, where as before the first row/column represents $2s^2\,{}^1S$:

$$H_2^{(2)} = \begin{pmatrix} -0.0434 & 0.0757 \\ 0.0437 & -0.1231 \end{pmatrix}. \qquad (13.116)$$

By combining the second-order results given here with the first-order results (13.109), we obtain the following energy matrix:

$$H_0 + H_{\text{eff}}^{(1)} + H_{\text{eff}}^{(2)} = \begin{pmatrix} -14.6380 & -0.0681 \\ -0.1001 & -14.3665 \end{pmatrix}. \qquad (13.117)$$

This has the following eigenvalues:

$$\begin{cases} E''(2s^2\,{}^1S) = -14.661 \\ E''(2p^2\,{}^1S) = -14.343 \end{cases} \qquad (13.118)$$

and the corresponding eigenvectors

$$\begin{cases} |2s^2\ {}^1S\rangle'' = 0.947\,|2s^2\ {}^1S\rangle + 0.322\,|2p^2\ {}^1S\rangle \\ |2p^2\ {}^1S\rangle'' = -0.225\,|2s^2\ {}^1S\rangle + 0.974\,|2p^2\ {}^1S\rangle\,. \end{cases} \qquad (13.119)$$

It should be observed that the matrix (13.117) is *nonhermitian*. This is a general property of the effective Hamiltonian. As a consequence, the eigenvectors are nonorthogonal. The eigenvalues, however, are real, as they should be, since they represent true energies.

The ground state of the beryllium atom has the total nonrelativistic energy -14.6674 [*Bunge* 1976a]. Thus, we see that the second-order result above, -14.661, is quite close to the correct value. As a comparison, we could mention that a similar calculation, where the $2p$ shell is considered as virtual and only $2s$ as valence shell, yields the energy -14.710, which is quite far off the correct value. Even if the former result may be partly fortuitous, one would expect it to be one order of magnitude better than the latter, since it includes the important mixtures $2s^2 \to 2p^2$ to higher orders. We see from the eigenvectors (13.119) that this mixture is of the order of 30 %, which makes it difficult to treat it by ordinary perturbation-expansion techniques.

Another advantage with the extended-model-space approach demonstrated here is that it also yields automatically the energy of the excited ^1S state with a dominating $2p^2$ component. This state has recently been observed experimentally [*Fassett* et al. 1985] and found to lie about 1000 cm^{-1} above the $2s$ ionization limit.

In Chap. 15 we shall consider still another approach to the correlation problem of the Be atom, namely by treating the atom as a closed-shell system but evaluating the pair-correlation effects to all orders.

14. The Hyperfine Interaction

In this chapter we shall apply the formalism of effective operators, introduced in the previous chapter, to the hyperfine interaction. The idea of using effective operators to analyze experimental hyperfine data has gradually developed over the years. It has long been known that a significantly improved fit to the experimental results can be obtained if the radial integrals appearing in the hyperfine Hamiltonian are treated as free parameters in a way similar to our treatment of the Slater parameters in the previous chapter.

It was first pointed out by *Abragam* and *Pryce* [1951] (see also [*Abragam et al.* 1955]) that the polarization of the electron core due to the Coulomb interaction with the open shell(s) can have an appreciable effect on the hyperfine interaction. In particular, this can cause a net contact interaction with the nucleus also for systems with no open s shell. This type of effect—now known as the spin-polarization—can be described by including a contact term in the hyperfine Hamiltonian also for non-s electrons. Later, it was found by *Harvey* [1965] in her analysis of oxygen and fluorine hyperfine data that a considerably improved fit to the experimental data could be obtained if a third parameter was added to the Hamiltonian. The result of her analysis for oxygen is reproduced in Table 14.1. The experimental data are corrected for a slight breakdown of LS coupling. The table also contains some "off-diagonal" hyperfine constants, which affect the splitting at strong external magnetic fields, where J is no longer a good quantum number. This analysis shows the significant improvement obtained by a single additional parameter.

This kind of empirical procedure can now be understood using the theory of effective operators [*Judd* 1963a, b, 1967; *Sandars* 1969], and the corresponding parameters can be evaluated by means of many-body perturbation theory.

Table 14.1. Experimental and theoretical, fitted values for the magnetic dipole interaction constants in oxygen [MHz] (from [*Harvey* 1965])

	Exptl.	Two-parameter fit	Three-parameter fit
$A(^3P_2)$	$-219.61(5)$	-219.61	-219.61
$A(^3P_{2,1})$	$-127.5(20)$	-127.5	-128.5
$A(^3P_1)$	$4.738(36)$	-18.4	4.74
$A(^3P_{1,0})$	$-91.7(72)$	-106.5	-88.7

In this chapter we shall use the one- and two-particle functions introduced in the previous chapter to evaluate the diagrams of the effective hyperfine operator. Single-particle functions are used to describe the important polarization effect. Correlation effects are evaluated using pair functions. Furthermore, one-particle effects and certain families of two-particle effects are included to all orders by solving the corresponding equations iteratively. This will be demonstrated by means of several numerical examples.

14.1 The Hyperfine Interaction

The hyperfine interaction is caused by the interaction between the electrons and the electromagnetic multipole moments of the nucleus, and, therefore, from an electronic point of view it is a one-body interaction.

14.1.1 The Hyperfine Operator

The lowest multipole orders of the hyperfine interaction—magnetic dipole and electric quadrupole—can be represented by means of the following operators [*Armstrong* 1971; *Lindgren* and *Rosén* 1974]:

$$\begin{cases} h_D = \alpha^2 \sum_{i=1}^{N} \left[l r^{-3} - \sqrt{10}(sC^2)^1 r^{-3} + \frac{8\pi}{3} s\delta(r) \right]_i \cdot \boldsymbol{\mu}_I & (14.1a) \\ h_Q = -\frac{1}{2} \sum_{i=1}^{N} C_i^2 \cdot Q^2 \, r^{-3} \,, & (14.1b) \end{cases}$$

using atomic units (see Appendix A). Here l and s are the orbital and spin angular-momentum operators, respectively, of the electron, $(sC^2)^1$ is a tensor product of s and C^2 of rank 1 (2.95), and $\delta(r)$ is a three-dimensional delta function. $\boldsymbol{\mu}_I$ is the nuclear magnetic dipole moment and Q^2 the electric quadrupole tensor. The conventional nuclear moments are defined as the expectation values of these operators in the state with the maximum component of the nuclear spin, $M_I = I$, that is,

$$\begin{cases} \mu_I = \langle \gamma_I II | \mu_{Iz} | \gamma_I II \rangle & (14.2a) \\ Q = \langle \gamma_I II | Q_0^2 | \gamma_I II \rangle \,, & (14.2b) \end{cases}$$

where γ_I and γ_J are additional quantum numbers necessary to specify the nuclear and electronic states, respectively.

The electronic part of the dipole operator (14.1a) represents the magnetic field due to the electrons at the site of the nucleus. The first term is caused by the orbital motion of the electrons and is called the *orbital term*. The second term represents the dipole field due to the spin motion of the electrons and is called the *spin-dipole* term. These two terms, which have an r^{-3} radial dependence, contribute only for non-s electrons ($l \neq 0$). The last term in (14.1a) represents the *contact interaction* between the nucleus and the electron spin. This contributes only for s electrons.

14.1.2 The Hyperfine Splitting

The hyperfine interaction leads to a coupling of the electronic (J) and the nuclear (I) angular momenta. These angular momenta are then no longer constants of the motion separately, only their resultant (see Fig. 14.1),

$$F = I + J.$$

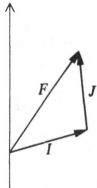

Fig. 14.1 Vector-coupling of the nuclear and electronic angular momenta

The coupled state can be represented by

$$\Psi = |\gamma_I \gamma_J (IJ) F M_F \rangle , \tag{14.3}$$

which is the correct zeroth-order state when the hyperfine interaction is considered as the only perturbation. According to first-order perturbation theory (9.99), the leading energy contribution due to the hyperfine interaction is given by the expectation value of the perturbation for that state

$$W_F = N \langle \gamma_I \gamma_J (IJ) F M_F | h | \gamma_I \gamma_J (IJ) F M_F \rangle . \tag{14.4}$$

The normalization constant

$$N = \langle \gamma_I \gamma_J (IJ) F M_F | \gamma_I \gamma_J (IJ) F M_F \rangle^{-1} \tag{14.5}$$

is included here, since we do not necessarily assume the function to be normalized.

As mentioned, we can in most applications neglect the contributions due to multiple hyperfine interactions. In that case, the expression (14.4) yields the exact hyperfine splitting. Although this is a first-order result, it should not be confused with the first-order effect discussed in Sect. 13.2.1. The latter is equivalent to

$$W_F^0 = {}_0\langle \gamma_I \gamma_J (IJ) FM_F | h | \gamma_I \gamma_J (IJ) FM_F \rangle_0 , \qquad (14.6)$$

where

$$\Psi_0 = | \gamma_I \gamma_J (IJ) FM_F \rangle_0 \qquad (14.7)$$

represents the zeroth-order state when h *as well as* V are considered as perturbations. This corresponds to the hyperfine splitting evaluated by means of a central-field function, like Hartree-Fock, and is referred to as the *zeroth-order hyperfine splitting*. In (14.4), on the other hand, the *exact* wave function is used, so this expression is equivalent to a perturbation expansion, where h is treated to first order and V *is treated to all orders*. This result can also be expressed by means of the effective operator (13.16)

$$W_F = {}_0\langle \gamma_I \gamma_J (IJ) FM_F | h_{\text{eff}} | \gamma_I \gamma_J (IJ) FM_F \rangle_0 . \qquad (14.8)$$

Here, we employ the intermediate normalization (9.11), so that the *zeroth-order* function Ψ_0 is normalized.

14.1.3 The Hyperfine Interaction Constants

Generally, we can express the hyperfine operator in terms of scalar products of electronic (T^K) and nuclear (M^K) tensor operators [*Schwartz* 1955]

$$h = \sum_{K=1}^{\infty} T^K \cdot M^K . \qquad (14.9)$$

M^K represents the 2^K-pole moment of the nucleus and T^K the corresponding electronic field at the nucleus. For parity reasons, magnetic interactions give rise only to odd K (dipole, octupole, ...) and electric interactions only to even K (quadrupole, hexadecapole, ...).

T^K is a one-body operator, and we can write it in analogy with (5.14) as

$$T^K = \sum_{i=1}^{N} t_i^K . \qquad (14.10)$$

For $K = 1$ and 2 we get the T and M operators by identification with (14.1),

$$t^1 = \alpha^2 \left[l r^{-3} - \sqrt{10} (sC^2)^1 r^{-3} + \frac{8\pi}{3} s\delta(r) \right] \qquad (14.11a)$$

$$t^2 = -\frac{1}{2}\,C^2 r^{-3} \tag{14.11b}$$

$$M^1 \equiv \mu_I; \qquad M^2 = Q^2\ . \tag{14.11c}$$

In evaluating the energy shift (14.4) with the operator in the form (14.9), we can apply formula (4.22) for a scalar product in order to determine the F dependence

$$
\begin{aligned}
W_F &= \sum_K N \langle \gamma_I \gamma_J (IJ) F M_F | \, T^K \cdot M^K \, | \gamma_I \gamma_J (IJ) F M_F \rangle \\
&= N(-1)^{I+J+F} \sum_K \begin{Bmatrix} I & J & F \\ J & I & K \end{Bmatrix} \langle \gamma_J J \| T^K \| \gamma_J J \rangle \langle \gamma_I I \| M^K \| \gamma_I I \rangle\ .
\end{aligned}
\tag{14.12}
$$

For the dipole interaction ($K = 1$), the matrix element (14.12) has the same F dependence as the scalar product $I \cdot J$,

$$
\begin{aligned}
&\langle (IJ) F M_F | \, I \cdot J \, | (IJ) F M_F \rangle \\
&= (-1)^{I+J+F} \begin{Bmatrix} I & J & F \\ J & I & 1 \end{Bmatrix} \langle J \| J \| J \rangle \langle I \| I \| I \rangle\ .
\end{aligned}
\tag{14.13}
$$

Hence, we will produce the result (14.12), if we replace the dipole operator (14.1a) by

$$h_D = A I \cdot J\ , \tag{14.14}$$

where

$$A = \frac{\langle \gamma_J J \| T^1 \| \gamma_J J \rangle \langle \gamma_I I \| \mu_I \| \gamma_I I \rangle}{\langle J \| J \| J \rangle \langle I \| I \| I \rangle}\ . \tag{14.15}$$

The operator (14.14) is an *"equivalent" operator* for the dipole interaction, and it produces correct results for matrix elements *diagonal with respect to I and J*. The A factor is called the *dipole interaction constant*.

The F dependence of the dipole energies can now easily be obtained from the 6-j symbol or by means of the equivalent operator (14.14) and the identity

$$I \cdot J = \frac{1}{2}\,(F^2 - I^2 - J^2)\ . \tag{14.16}$$

This gives

$$
\begin{aligned}
W_F^{\mathrm{dip}} &= A \langle (IJ) F M_F | \, I \cdot J \, | (IJ) F M_F \rangle \\
&= \frac{A}{2}\,[F(F+1) - I(I+1) - J(J+1)]\ .
\end{aligned}
\tag{14.17}
$$

The same result can be obtained, of course, by using the explicit expression for the 6-j symbol in (14.12) (see Appendix C).

By means of the Wigner-Eckart theorem the dipole interaction constant (14.15) can be written

$$A = \frac{\langle \gamma_J JJ | T_0^1 | \gamma_J JJ \rangle \langle \gamma_I II | \mu_{Iz} | \gamma_I II \rangle}{\langle JJ | J_z | JJ \rangle \langle II | I_z | II \rangle}. \tag{14.18}$$

The nuclear matrix element in the numerator is by definition the nuclear dipole moment μ_I (14.2a). We assume that the nuclear wave function is normalized but the electronic one is not, which gives

$$\langle II | I_z | II \rangle = I \tag{14.19}$$
$$N \langle JJ | J_z | JJ \rangle = J,$$

where

$$N = \langle \gamma_J JJ | \gamma_J JJ \rangle^{-1}. \tag{14.20}$$

Then the dipole constant becomes

$$A = \frac{\langle \gamma_J JJ | T_0^1 | \gamma_J JJ \rangle}{\langle \gamma_J JJ | \gamma_J JJ \rangle} \frac{\mu_I}{IJ}. \tag{14.21}$$

In a similar way it can be shown that the quadrupole energy becomes [*Kopfermann* 1958; *Armstrong* 1971]

$$W_F^{\text{quadr}} = B \frac{3/2\, C(C+1) - 2I(I+1)\, J(J+1)}{2I(2I-1)2J(2J-1)}, \tag{14.22}$$

where C is the expression in the square brackets in (14.17) and

$$B = \frac{1}{2} \frac{\langle \gamma_J JJ | T_0^2 | \gamma_J JJ \rangle}{\langle \gamma_J JJ | \gamma_J JJ \rangle} Q \tag{14.23}$$

is the *electric quadrupole constant*.

By analyzing the hyperfine splitting, the constants A and B can be determined, and—if the experimental accuracy is high enough—also some higher-order constants. In order to obtain information about the nuclear moments, it is necessary to evaluate the corresponding electronic part of the constants. It is now known that the Hartree-Fock model is in many cases quite insufficient for this purpose, and it is necessary to go to quite high orders in perturbation theory—or to use any other equivalent procedure, such as multiconfigurational Hartree-Fock—in order to obtain reasonable accuracy. It is the purpose of this chapter to demonstrate how the formalism built up in the previous chapters can be used for such an evaluation. This will hopefully also give some deeper insight into the different types of "distortions" which appear in the electronic shells due to an "external" perturbation of this kind.

The nuclear moments—particularly the dipole moment—can in many cases be measured "directly", without considering the hyperfine interaction, for instance, by means of nuclear magnetic resonance (nmr) or optical pumping. If the corresponding hyperfine constant is also known, it is possible in such a case to obtain experimental information about the electronic part of the interaction constant. This value can then be used to check the theoretical calculation. This procedure will be used in the examples discussed in this chapter.

14.2 The Zeroth-Order Hyperfine Constants

14.2.1 The Effective Hyperfine Operator

Suppose for a moment that instead of the hyperfine interaction we consider a pure electronic perturbation h. If this is sufficiently weak, then the corresponding energy shift would be in analogy with (14.4)

$$\Delta E_J = N\langle \gamma_J JJ | h | \gamma_J JJ \rangle , \tag{14.24}$$

where N is again the normalization constant (14.20). But the same shift can also be expressed by means of the corresponding *effective* operator h_{eff}, and the *unperturbed* (normalized) function

$$\Delta E_J = {}_0\langle \gamma_J JJ | h_{\text{eff}} | \gamma JJ \rangle_0 \tag{14.25}$$

in analogy with (14.8).

Returning to the hyperfine problem, we see that the dipole interaction constant (14.21) is proportional to the expectation value of T_0^1 in the exact electronic state

$$A \propto N\langle \gamma_J JJ | T_0^1 | \gamma_J JJ \rangle , \tag{14.26}$$

which is of the same form as (14.24). In analogy with (14.25) we can then construct a corresponding effective operator T_{eff}^1 so that

$$N\langle \gamma_J JJ | T_0^1 | \gamma_J JJ \rangle = {}_0\langle \gamma_J JJ | T_{\text{eff}}^1 | \gamma_J JJ \rangle_0 \tag{14.27}$$

and

$$A = {}_0\langle \gamma_J JJ | T_{\text{eff}}^1 | \gamma_J JJ \rangle_0 \frac{\mu_I}{IJ} . \tag{14.28a}$$

Similarly, we can express the quadrupole interaction constant (14.23)

$$B = \frac{1}{2}\,{}_0\langle\gamma_J JJ \,|\, T^2_{\text{eff}} \,|\, \gamma_J JJ \rangle_0 \, Q \,.\tag{14.28b}$$

The operators T^K_{eff} evidently contain all of the electronic many-body effects on the hyperfine interaction. By substituting T^K_0 for h in the procedure developed in the previous chapter, we get the order-by-order expansion of these operators. Before considering higher orders, however, we shall analyze the hyperfine interaction in the zeroth-order (central-field) approximation using the formalism of second quantization.

14.2.2 The Zeroth-Order Hyperfine Constants in Second Quantization

The operator T^K_0 is defined in (14.10,11), and it has the second-quantized form, in analogy with (11.20),

$$T^K_0 = \sum_{ij} a^\dagger_i a_j \langle i | t^K_0 | j \rangle \,.\tag{14.29}$$

The zero-body term (13.8b), which contains a summation over closed shells, vanishes in the hyperfine case, and only the normal-ordered one-body term (13.8c) remains,

$$T^K_0 = \sum_{ij} \{a^\dagger_i a_j\} \langle i | t^K_0 | j \rangle \,.\tag{14.30}$$

The first-order effective operator is according to (13.17)

$$T^{K(1)}_{\text{eff}} = PT^K_0 P = P \sum_{mn}^{\text{val}} a^\dagger_m a_n \langle m | t^K_0 | n \rangle P \,,\tag{14.31}$$

which is represented by the diagram (13.15c)

$$= a^\dagger_m a_n \langle m | t^K_0 | n \rangle \,.\tag{14.32}$$

Substituted into (14.28), this operator yields the *zeroth-order hyperfine constant*, for instance,

$$A^{(0)} = {}_0\langle\gamma_J JJ \,|\, T^{1(1)}_{\text{eff}} \,|\, \gamma_J JJ \rangle_0 \frac{\mu_I}{IJ} = {}_0\langle\gamma_J JJ \,|\, T^1_0 \,|\, \gamma_J JJ \rangle_0 \frac{\mu_I}{IJ}\tag{14.33}$$

or

$$A^{(0)} = \frac{\mu_I}{IJ} \sum_{mn}^{\text{val}} {}_0\langle \gamma_J JJ \,|\, a_m^\dagger a_n \,|\, \gamma_J JJ \rangle_0 \langle m \,|\, t_0^1 \,|\, n \rangle \,. \tag{14.34}$$

This is equivalent to the zeroth-order formula (14.6), and represents the hyperfine constant in the central-field approximation.

Evaluation of the zeroth-order constants is not always trivial, even if the wave function is of central-field type. In general, this requires the whole machinery of angular-momentum theory, fractional-parentage coefficients, and possibly intermediate-coupling coefficients (see, for instance, [*Lindgren* and *Rosén* 1974]). A graphical procedure for this kind of evaluation has been developed in the first part of the book, and for that reason we shall not be further concerned with this problem here. Our main interest now is to evaluate the higher-order perturbations, and we shall show later that many of these perturbations contain the same many-electron matrix elements as do the zeroth-order constants. This makes it possible to incorporate important higher-order effects formally into the zeroth-order result by modifying the single-particle matrix elements $\langle m \,|\, t_0^K \,|\, n \rangle$. When these modifications have been made, the hyperfine constant can be evaluated exactly as in the zeroth-order approximation. This is the basic idea behind the effective-operator formalism. Before we turn to the higher-order modifications, however, we shall analyze the spin-angular structure of the single-particle matrix elements.

14.2.3 The Spin-Angular Structure of the Hyperfine Operator

A component t_Q^K of the hyperfine operator (14.10, 11) is generally a sum of terms of the form

$$t_Q^{(\kappa\lambda)K} = (s^\kappa v^\lambda)_Q^K f_{(\kappa\lambda)K}(r), \tag{14.35}$$

where

$$\begin{cases} s^0 \equiv 1 \\ s^1 \equiv s \end{cases} \tag{14.36}$$

is the spin part and v^λ is the angular part of the interaction. We consider now one particular term of this kind.

Using the definition of a tensor product (2.95), we can write the operator above as

$$t_Q^{(\kappa\lambda)K} = \sum_{\mu\nu} \langle \kappa\mu, \lambda\nu \,|\, KQ \rangle s_\mu^\kappa v_\nu^\lambda f_{(\kappa\lambda)K}(r) \,. \tag{14.37}$$

Applying the Wigner-Eckart theorem in graphical form to the spin and the orbital parts separately as in (8.24), we can express an arbitrary matrix element as

$$\langle i|t_Q^{(\kappa\lambda)K}|j\rangle = \langle i\|t^{(\kappa\lambda)K}\|j\rangle \, F, \tag{14.38}$$

where F is a spin-angular factor

$$F = \sum_{\mu\nu} \langle \kappa\mu, \lambda\nu|KQ\rangle \quad \tag{14.39a}$$

which contains all the dependence on the magnetic quantum numbers, and $\langle i\|t^{(\kappa\lambda)K}\|j\rangle$ is a reduced matrix element

$$\langle i\|t^{(\kappa\lambda)K}\|j\rangle = \langle s\|s^\kappa\|s\rangle \, \langle l_i\|v^\lambda\|l_j\rangle \, \langle i|f_{(\kappa\lambda)K}|j\rangle \tag{14.39b}$$
$$\langle i|f_{(\kappa\lambda)K}|j\rangle = \int P_i(r) f_{(\kappa\lambda)K}(r) P_j(r) \, dr \, .$$

As before, l_i and s_i represent $l_i m_i^l$ and $sm_{s_i}^i$, respectively. Using the graphical form (3.12) of the vector-coupling coefficient, we may also represent F as a single diagram

$$F = \quad \tag{14.39c}$$

According to (2.108) and (2.109) we have for $\kappa = 0$ and 1, respectively,

$$\langle s\|s^0\|s\rangle = \langle \tfrac{1}{2}\|1\|\tfrac{1}{2}\rangle = \sqrt{2} \tag{14.40}$$
$$\langle s\|s^1\|s\rangle = \langle s\|s\|s\rangle = \sqrt{3/2} \, .$$

It should be observed that the reduced matrix element of $s^0 = 1$ is *not* equal to unity. This is compensated, however, by the spin graph in that case, since the complete matrix element $\langle sm_s|1|sm_s\rangle$ obviously is unity (see the discussion in connection with the Wigner-Eckart theorem for double tensors in Sect. 8.2.3).

14.2.4 Single Open Shell

We consider now the important special case of an atom with a single open shell (nl). For non-s electrons, the orbital [(01)1], spin-dipole [(12)1], and quadrupole [(02)2] parts of the hyperfine operator contribute, and they give rise to the same radial integral (14.39b)

$$\langle nl|f_{(\kappa\lambda)K}|nl\rangle = \int P_{nl}(r) r^{-3} P_{nl}(r) \, dr = \langle r^{-3}\rangle_{nl} \, , \tag{14.41a}$$

that is, the expectation value of r^{-3} for the open shell. For s electrons only the contact interaction contributes, and the radial integral is then

$$\langle ns | f_{(10)1} | ns \rangle = \frac{2}{3} \int P_{ns}(r) \frac{\delta(r)}{r^2} P_{ns}(r) \, dr = \frac{2}{3} \left\langle \frac{\delta(r)}{r^2} \right\rangle_{ns}. \tag{14.41b}$$

In the latter case we have used the fact that an integration of the three-dimensional delta function over the angular coordinates yields

$$\int \delta(\mathbf{r}) \, d\Omega = 4\pi r^2 \delta(r), \tag{14.42}$$

where $\delta(r)$ is the *radial* delta function. This means that the term $(8\pi/3) \delta(\mathbf{r})$ in the dipole operator (14.1a) can be replaced by $(2/3) [\delta(r)/r^2]$.

In the case of a single open shell we can obviously perform the radial integrations in advance and write the hyperfine operators (14.11) in the form

$$\begin{cases} \mathbf{t}^1 = \alpha^2 [\mathbf{l} - \sqrt{10}(s\mathbf{C}^2)^1] \langle r^{-3} \rangle_{nl} & (l \neq 0) & \tag{14.43a} \\ \mathbf{t}^1 = \frac{2}{3} \alpha^2 \mathbf{s} \left\langle \frac{\delta(r)}{r^2} \right\rangle_{ns} & (l = 0) & \tag{14.43b} \end{cases}$$

$$\mathbf{t}^2 = -\frac{1}{2} \mathbf{C}^2 \langle r^{-3} \rangle_{nl}. \qquad (l \neq 0) \tag{14.43c}$$

As we have mentioned, important higher-order perturbations can be included in this form by modifying the radial parameters. This leads to the following *effective hyperfine operator*:

$$\begin{cases} \mathbf{t}^1 = \alpha^2 [\mathbf{l} \langle r^{-3} \rangle_l - \sqrt{10}(s\mathbf{C}^2)^1 \langle r^{-3} \rangle_{sd} + \mathbf{s} \langle r^{-3} \rangle_c] \\ \mathbf{t}^2 = -\frac{1}{2} \mathbf{C}^2 \langle r^{-3} \rangle_q. \end{cases} \tag{14.44}$$

This is the kind of operator which was introduced by *Harvey* [1965] in an empirical way and is nowadays used in essentially all analyses of experimental hyperfine data [*Lindgren* and *Rosén* 1974]. In order to find empirical values of all parameters of the dipole operator, it is necessary to know the hyperfine interaction for at least three states of the configuration.

As an illustration of the use of an effective operator of the kind (14.44), we consider again the oxygen atom. The experimental values of the radial parameters determined by *Harvey* are shown in Table 14.2, together with the Hartree-Fock values and theoretical values obtained by *Kelly* [1968, 1969a], using a many-body procedure similar to that described in Sect. 12.7. Relativistic effects, which are here less than 1 %, are neglected. These results show that the parameters of the effective Hamiltonian can be well reproduced by means of many-body perturbation theory, but that it is necessary to go beyond second order to achieve accurate results.

Table 14.2. Radial hyperfine parameters for the oxygen atom [atomic units]. Experimental results from [*Harvey* 1965] and theoretical results from [*Kelly* 1968, 1969a]

	$\langle r^{-3}\rangle_l$	$\langle r^{-3}\rangle_{sd}$	$\langle r^{-3}\rangle_c$	$\langle r^{-3}\rangle_q$
Experimental	4.56(4)	5.18(1)	0.48(4)	—
Hartree-Fock	4.975	4.975	—	4.975
Second order	4.746	5.212		4.375
Higher order	4.547	5.126		4.205

It should be mentioned that calculations on oxygen, comparable in accuracy to those by *Kelly*, have been performed by several groups, using either a more direct configurational-mixing approach [*Bessis* et al. 1962; *Schaefer* and *Klemm* 1969], unrestricted Hartree-Fock [*Larsson* 1970, 1971] or multiconfigurational Hartree-Fock [*Bagus* and *Bauche* 1973]. Our point here, however, is to demonstrate the order-by-order effect, which does not appear in these nonperturbative approaches.

14.3 First-Order Core Polarization

We shall now apply the perturbation formalism developed here in order to investigate how the hyperfine interaction is affected by higher-order perturbations due to the nonspherical symmetry of the atom.

We consider first the second-order effective operator in Fig. 13.8. This contains one hyperfine interaction and one Coulomb interaction and hence represents the next approximation beyond the zeroth-order (central-field) approximation discussed in the previous section. These diagrams can be of one-body or two-body type. If there is only one valence electron, then only the one-body diagrams contribute. Also for systems with several valence electrons, it has been found that the effective one-body contributions often dominate. Therefore, we shall concentrate on these effects here.

Generally, it can be shown that all one-body effects, represented by diagrams with a single pair of free valence lines, can be included in the effective operator (14.44) with the same parameters for the entire configuration. In order to represent effective two-body diagrams in a similar manner, a much more complex operator is needed [*Bauche-Arnoult* 1971, 1973]. Alternatively, two-body effects can be included in the operators of the type (14.44), if the parameters are allowed to be term dependent, that is, constant only for a particular *LS* term. In order to find experimental evidence for effective two-particle contributions, it is thus necessary to analyse the hyperfine interaction for at least two terms of the same configuration. We shall return briefly to this problem, which has not been much studied, at the end of this chapter.

14.3.1 Evaluation of the Polarization Diagrams

Assuming that the orbitals are generated in the Hartree-Fock potential of the core, so that the effective potential (10.19) vanishes, then there are only four one-body, second-order effective operator diagrams (see Fig. 13.8). Their analytical expressions are

$$= a_m^\dagger a_n \sum \frac{\langle a|t_Q^K|r\rangle\langle mr|r_{12}^{-1}|na\rangle}{\varepsilon_n + \varepsilon_a - \varepsilon_m - \varepsilon_r} \tag{14.45a}$$

$$= -a_m^\dagger a_n \sum \frac{\langle a|t_Q^K|r\rangle\langle mr|r_{12}^{-1}|an\rangle}{\varepsilon_a + \varepsilon_n - \varepsilon_m - \varepsilon_r} \tag{14.45b}$$

$$= a_n^\dagger a_m \sum \frac{\langle na|r_{12}^{-1}|mr\rangle\langle r|t_Q^K|a\rangle}{\varepsilon_a - \varepsilon_r} \tag{14.45c}$$

$$= -a_n^\dagger a_m \sum \frac{\langle an|r_{12}^{-1}|mr\rangle\langle r|t_Q^K|a\rangle}{\varepsilon_a - \varepsilon_r} \tag{14.45d}$$

These diagrams can be evaluated in the same way as ordinary Goldstone diagrams, using the technique described in Sect. 12.6. The hyperfine operator is generally spin dependent [see (14.35, 39)], and for that reason we shall use the representation (3.45) of the Coulomb interaction, where also the spin dependence is given in graphical form.

The evaluation of the first "direct" polarization diagram (14.45a) then leads to

$$= a_m^\dagger a_n \sum \frac{1}{D} \langle a \| t^{(\kappa\lambda)K} \| r \rangle X^k(mr, na)$$
$$\times \sum \langle \kappa\mu, \lambda\nu | KQ \rangle G_o^d G_s^d ,$$

(14.46)

where D is the excitation energy in the intermediate state, and G_o^d and G_s^d represent the orbital- and spin-angular-momentum graphs, respectively,

$$G_o^d = \qquad = \delta(k, \lambda)[\lambda]^{-1}$$

$$G_s^d = \qquad = \delta(\kappa, 0) \sqrt{2}$$

The reduced hyperfine element is defined in (14.39b). The spin diagram can also be expressed in analogy with the orbital diagram as

$$G_s^d = 2\delta(\kappa, 0)$$

using (3.20).

In a similar way one finds that the "exchange" diagram (14.45b) becomes

$$= -a_m^\dagger a_n \sum \frac{1}{D} \langle a \| t^{(\kappa\lambda)K} \| r \rangle X^k(mr, an)$$
$$\times \sum_{\mu\nu} \langle \kappa\mu, \lambda\nu | KQ \rangle G_o^e G_s^e ,$$

(14.47)

where

$$G_0^e = (-1)^{l_m+\lambda+l_n}\begin{Bmatrix} l_m & \lambda & l_n \\ l_r & k & l_a \end{Bmatrix}$$

$$G_s^e = -$$

As mentioned before, the inverted diagrams with the hyperfine interaction at the bottom are, apart from the energy denominators, the hermitian adjoints of those considered here. They can be evaluated in a similar way and yield the same contributions in the case of a single open shell.

The spin and angular diagrams obtained here can be combined with the vector-coupling coefficient to form a single diagram (14.39c) in the same way as in first order. In any case, the important point is that the spin-angular structure—that is, the dependence on the magnetic quantum numbers—is the same in second order as in first order. This is the underlying reason for the use of effective operators in perturbation theory, as we shall discuss in some more detail in the following sections.

14.3.2 Contributions to the Effective Hyperfine Operators

The second-order contributions to the hyperfine constants are obtained by substituting the expressions above into the general formulas (14.28) and evaluating the expectation values in the unperturbed states considered. Then we see that the results are quite similar to the first-order contributions, leading to the zeroth-order constants (14.34). The same many-electron matrix element, $_0\langle \gamma_J JJ | a_m^\dagger a_n | \gamma_J JJ \rangle_0$, reappears in higher orders, and hence the perturbed hyperfine constants can be expressed by means of the same formula after modifying the single-particle matrix elements.

Furthermore, since the single-particle elements have the same spin-angular structure as the corresponding elements in the zeroth-order hyperfine constants, it follows that these perturbations can be included by modifying the *radial* parameters only.

Another observation we can make from the second-order results above is that the direct polarization (14.46) does not contribute to spin-dependent perturbations—that is, to the spin-dipole and contact interactions (14.1). The reason for this is that the contributions from the two spin orientations cancel, when the summation is performed over the loop. The direct polarization does not contribute to the orbital interaction either, as we can see in the following way.

In diagram (14.46) the tensor rank λ, which in the orbital term is 1, must be equal to k. But the operator l is diagonal with respect to l, so l_r must be equal to l_a. This implies that the vertex sum $l_a + k + l_r$ is odd, and the diagram vanishes. In other words, *the direct polarization diagram does not contribute to the magnetic dipole interaction*. It does, however, contribute to the quadrupole interaction which is spin independent and of even rank. The exchange diagram contributes to all operators. Since only the exchange interaction contributes to the magnetic dipole interaction, this effect is often referred to as "*exchange polarization*."

Summarizing the results above, we find that the direct diagram (14.46) yields

$$a_m^\dagger a_n \sum \frac{1}{D} \langle a \| t^{(\kappa\lambda)K} \| r \rangle X^k(mr, na) \frac{2}{[\lambda]} \delta(\kappa, 0)\delta(k, \lambda)F \tag{14.48}$$

and the exchange diagram (14.47)

$$-a_m^\dagger a_n \sum \frac{1}{D} \langle a \| t^{(\kappa\lambda)K} \| r \rangle X^k(mr, an)(-1)^{l_m+\lambda+l_n} \begin{Bmatrix} l_m & \lambda & l_n \\ l_r & k & l_a \end{Bmatrix} F, \tag{14.49}$$

where F is the spin-angular factor (14.39a). The remaining, inverted diagrams (14.45c, d) differ only in their denominators.

These results should now be compared with the diagram (14.32) appearing in the zeroth-order constants. Using the formulas (14.38, 39), this becomes

$$a_m^\dagger a_n \langle m | t_Q^{(\kappa\lambda)K} | n \rangle = a_m^\dagger a_n \langle m \| t^{(\kappa\lambda)K} \| n \rangle F \tag{14.50}$$
$$= a_m^\dagger a_n \langle s \| s^\kappa \| s \rangle \langle l_m \| v^\lambda \| l_n \rangle \langle m | f_{(\kappa\lambda)K} | n \rangle F.$$

In the case of a single open shell we can use the simplified hyperfine operator (14.43) and include the effect of first-order core polarization in the form (14.44). The inverted diagrams yield identical contributions and we can express the result in the following way:

$$\begin{cases} \langle r^{-3} \rangle_l = \langle r^{-3} \rangle_{nl} + \Delta_{nl}^{(01)1} \\ \langle r^{-3} \rangle_{sd} = \langle r^{-3} \rangle_{nl} + \Delta_{nl}^{(12)1} \\ \langle r^{-3} \rangle_c = \frac{2}{3} \left\langle \frac{\delta(r)}{r^2} \right\rangle_{ns} + \Delta_{ns}^{(10)1} \\ \langle r^{-3} \rangle_q = \langle r^{-3} \rangle_{nl} + \Delta_{nl}^{(02)2}, \end{cases} \tag{14.51}$$

where

$$\Delta_{nl}^{(\kappa\lambda)K} = \delta(\kappa, 0)\frac{4}{2K+1}\sum_{n_a l_a n_r l_r}\frac{\langle l_a \|v^K\| l_r\rangle X^K(nl\,r,\,nl\,a)\langle a|f_{(0K)K}|r\rangle}{\langle l\|v^K\|l\rangle(\varepsilon_a-\varepsilon_r)}$$

$$-2(-1)^\lambda\sum_{n_a l_a n_r l_r k}\frac{\langle l_a\|v^\lambda\|l_r\rangle X^k(nl\,r,\,a\,nl)\langle a|f_{(\kappa\lambda)K}|r\rangle}{\langle l\|v^\lambda\|l\rangle(\varepsilon_a-\varepsilon_r)}\begin{Bmatrix} l & \lambda & l \\ l_r & k & l_a \end{Bmatrix}.$$

Substituted into (14.44) this yields the effective hyperfine operator corrected for first-order core polarization. The corresponding hyperfine constants can then be evaluated as in zeroth order.

14.3.3 Evaluation of the Radial Parts

The summations over the particle states appearing in the core polarization expressions (14.46, 47) can be evaluated by generating a complete set of single-particle states and performing the summations and integrations involved, as described in Sect. 12.7. It is more convenient, however, to use the idea of in-homogeneous differential equations introduced in the previous chapter. This is the procedure used by *Sternheimer* in his first-order calculations, but it can easily be extended to higher orders, as we shall demonstrate later.

The direct polarization diagram (14.46) contains the "radial" part

$$\sum_{n_a l_a n_r l_r k}\frac{\langle a\|t^{(\kappa\lambda)K}\|r\rangle X^k(mr,\,na)}{\varepsilon_n+\varepsilon_a-\varepsilon_m-\varepsilon_r}. \tag{14.52}$$

The summation over n_r can be performed by means of the single-particle function (13.80)

$$\rho_a^k(a\to l_r;\,r_1) = \sum_{n_r}^{\text{part}}\frac{P_r(r_1)X^k(mr,\,na)}{\varepsilon_n+\varepsilon_a-\varepsilon_m-\varepsilon_r}. \tag{14.53}$$

Then the sum (14.52) can be expressed as a *finite* sum of integrals (14.39b)

$$\sum\langle a\|t^{(\kappa\lambda)K}\|\rho_a^k(a\to l_r)\rangle$$
$$= \sum_{n_a l_a l_r k}\langle s\|s^\kappa\|s\rangle\langle l_a\|v^\lambda\|l_r\rangle\langle a|f_{(\kappa\lambda)K}|\rho_a^k(a\to l_r)\rangle. \tag{14.54}$$

In a similar way we can express the radial part of the exchange diagram (14.47)

$$\sum_{n_a l_a l_r k}\langle a\|t^{(\kappa\lambda)K}\|\rho_e^k(a\to l_r)\rangle \tag{14.55}$$

using an exchange Coulomb function similar to (13.94).

The inverted diagrams with the hyperfine interaction at the bottom differ from those considered here only in their denominators. Therefore, we get the corresponding single-particle functions simply by modifying the left-hand sides of the equations.

By means of the four single-particle functions considered here, we can evaluate all first-order core-polarization diagrams. In the case of a single open shell (nl), only two functions are required. Then the first-order corrections (14.51) to the radial hyperfine parameters can be expressed as

$$\Delta_{nl}^{(\kappa\lambda)K} = \delta(\kappa, 0) \frac{4}{2K+1} \sum_{n_a l_a l_r} \frac{\langle l_a \| v^K \| l_r \rangle}{\langle l \| v^K \| l \rangle} \langle a | f_{(0K)K} | \rho_d^K(a \rightarrow l_r) \rangle$$

$$-2(-1)^\lambda \sum_{n_a l_a l_r k} \frac{\langle l_a \| v^\lambda \| l_r \rangle}{\langle l \| v^\lambda \| l \rangle} \begin{Bmatrix} l & \lambda & l \\ l_r & k & l_a \end{Bmatrix} \langle a | f_{(\kappa\lambda)K} | \rho_e^k(a \rightarrow l_r) \rangle . \tag{14.56}$$

Problem 14.1. Consider the radial part of the exchange diagram (14.47) and show that it can be written in the form (14.55), where the exchange Coulomb function $p_e^k(a \rightarrow l_r)$ satisfies an equation analogous to (13.95).

As an alternative to the single-particle Coulomb functions discussed above, we can use corresponding *hyperfine* functions. For simplicity, we assume that there is a single open shell, that is, $\varepsilon_m = \varepsilon_n$. By means of the single-particle function

$$\rho_h(a \rightarrow l_r; r_1) = \sum_{n_r}^{\text{part}} P_r(r_1) \frac{\langle r \| t^{(\kappa\lambda)K} \| a \rangle}{\varepsilon_a - \varepsilon_r} \tag{14.57}$$

we can then express the radial part (14.52) of the direct diagram (14.46) as well as of its hermitian adjoint as

$$X^k(na, m\rho_h(a \rightarrow l_r)) \tag{14.58a}$$

and similarly for the exchange diagram

$$X^k(an, m\rho_h(a \rightarrow l_r)) . \tag{14.58b}$$

The function (14.57) satisfies the equation

$$(\varepsilon_a - h_0^{l_r})\rho_h(a \rightarrow l_r; r_1) = \sum_{n_r}^{\text{part}} P_r(r_1)\langle r \| t^{(\kappa\lambda)K} \| a \rangle . \tag{14.59}$$

By means of the closure property (13.84) and the definition (14.39b), we get

$$\sum_{n_r}^{\text{all}} P_r(r_1)\langle r | f_{(\kappa\lambda)K} | a \rangle = \sum_{n_r}^{\text{all}} P_r(r_1) \int P_r(r_1') f_{(\kappa\lambda)K}(r_1') P_a(r_1') \, dr_1'$$

$$= f_{(\kappa\lambda)K}(r_1)P_a(r_1) \tag{14.60}$$

and the equation above reduces to

$$(\varepsilon_a - h_0^{lr})\rho_h(a \rightarrow l_r; r_1)$$

$$= \langle s\|s^\kappa\|s\rangle\langle l_a\|v^\lambda\|l_r\rangle f_{(\kappa\lambda)K}(r_1)P_a(r_1) - \text{orthogonality terms} \qquad (14.61)$$

in analogy with the Coulomb equations (13.90) and (13.95). The orthogonality terms make the right-hand side orthogonal to all core shells with $l = l_r$.

The single-particle Coulomb functions represent first-order core polarization of the wave function caused by the Coulomb interaction. These are represented by diagrams (a, b) in Fig. 14.2. The single-particle hyperfine function (14.57) represents first-order core polarization due to the hyperfine interaction (Fig. 14.2c). This corresponds to the core-excitation part of the operator $\Omega_h^{(1)}$ introduced in the previous chapter (Fig. 13.6f). The two ways discussed here of evaluating the first-order polarization effect correspond to evaluating the first two terms in (13.28). In the case of a single open shell these terms are identical.

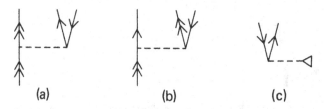

(a)   (b)   (c)

Fig. 14.2 a-c. The single-particle Coulomb functions (13.76, 94) represent single excitations caused by the Coulomb interaction (diagrams a,b) and the single-particle hyperfine function (14.57) represents single excitations due to the hyperfine interaction (diagram c)

14.3.4 Interpretation of the Core Polarization

The first-order polarization considered here involves only single excitations from the closed shells (core). As we have mentioned, this can be regarded as a first-order modification of the core orbitals, and is therefore referred to as the first-order *core polarization*. Generally, we define the core polarization as the perturbations which in each order can be described by means of single excitations from the core. The remaining effects—which involve at least one multiple excitation—will be referred to as (pure) *correlation* effects.

It should be observed that the correlation effect is sometimes defined to include all perturbations beyond restricted Hartree-Fock, that is, effects due to single as well as multiple excitations. For physical and computational reasons, however, it is convenient to separate the terms in the perturbation expansion into the two categories mentioned here, and therefore we shall restrict the interpretation of correlation effects to effects involving *multiple* excitations only.

Polarization affects different orbitals within the same shell differently, which means that the shell structure is partly broken up. The exchange interaction (14.47), for instance, can excite only core electrons having the same spin as the open-shell electron involved in the interaction. If the open shell is completely spin polarized—as in the Hund's-rule ground state for an atom with a less-than-half-filled valence shell—then the exchange polarization affects only one half the closed shells, leaving the other half unaffected.

The exchange interaction is attractive, since it reduces the Coulomb repulsion. This means that the core electrons with the same spin orientation as the valence electrons would be pulled towards the valence shell, compared to the core electrons with the opposite spin orientation (see Fig. 14.3). This is the important *spin polarization*, mentioned in the introduction to this chapter. It has the effect that the two s electrons in the same shell do not have the same density at the nucleus, leading to a net contact interaction. Since the inner s electrons have very high density at the nucleus, a very small unbalance is sufficient to cause a net interaction which is comparable to that of the open shell.

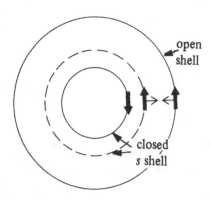

Fig. 14.3. Simple illustration of the spin polarization. Electrons with the same spin orientation are attracted to each other due to the exchange effect. If the valence shell is (partly) spin polarized, this leads to a spin polarization of the core

The spin polarization can be taken into account by a modification of the ordinary Hartree-Fock procedure, where the radial functions associated with the two spin orientations are allowed to be different. This represents a simplified form of the general "unrestricted Hartree-Fock" (see Chap. 5), known as "*spin-polarized Hartree-Fock*". Here, the ordinary self-consistent procedure leads to a spin polarization if the open shell is spin polarized. This technique yields reasonably good hyperfine results when the spin polarization is the dominating perturbation.

As an illustration of the spin polarization we consider the hyperfine structure in the ground state of the Mn^{++} ion, $[Ar] 4s^2 3d^5 \, ^6S$. In the central-field approximation this state is spherically symmetric, and hence there will be no magnetic field at the nucleus due to the orbital or spin motions of the electrons. Furthermore, since the valence electrons have $l \neq 0$, there is no net contact interaction either in zeroth order. Experimentally, however, it is found that this ion exhibits an appreciable hyperfine interaction, corresponding to a magnetic field at the nucleus of about 60 Tesla (600 kGauss). The effect of spin polarization has been calculated by *Freeman* and *Watson* [1965], using spin-polarized Hartree-Fock functions, and their results are shown in Table 14.3.

Table 14.3. Spin polarization of the Mn^{++} ion (from [*Freeman* and *Watson* 1965])

Shell	Field at the nucleus [Tesla]	Net field [Tesla]
1s up	250 284	
down	−250 287	−3
2s up	22 667	
down	−22 808	−141
3s up	3 121	
down	−3 047	+74
Total		−70

From this table it can be seen that the magnetic field produced at the nucleus by the inner s electrons is enormous, and therefore a relatively small unbalance can give rise to an appreciable net effect. The total field obtained in this calculation is in reasonable agreement with the experimental value, which indicates that spin polarization is the dominating perturbation in this case.

Besides the spin polarization, there is an *orbital polarization,* which is caused by the nonspherical charge distribution of the valence shell. This affects the core electrons differently, depending on their m_l quantum numbers, and as a first approximation it can be interpreted as a quadrupole distortion of the closed shells (see Fig. 14.4). This distortion gives an additional quadrupole interaction with the nucleus, which is known as the *Sternheimer effect.*

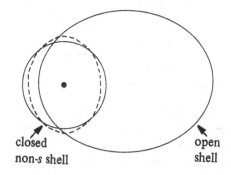

closed non-s shell   open shell

Fig. 14.4. Simple illustration of the orbital polarization of the electron core. The nonspherical charge distribution of the open shell causes a quadrupole distortion of the closed shells. This leads to an additional quadrupole interaction with the nucleus, which is the Sternheimer effect. This polarization affects also the magnetic interaction

This has been studied by *Sternheimer* for a long time, using second-order perturbation theory, in order to obtain correction factors to the nuclear quadrupole moments determined from the hyperfine interaction using Hartree-Fock values of $\langle r^{-3} \rangle$ [*Sternheimer* 1950, 1951, 1972; *Sternheimer* and *Peierls* 1971a, b]. *Sternheimer* expressed his correction in the form

$$Q_{corr} = \frac{1}{1-R} Q_{HF} \tag{14.62}$$

and his correction factor, R, is identical to $-\Delta_{2l}^{(02)}$ given in (14.51). The orbital polarization affects also the magnetic dipole interaction, although it is there usually less important than the spin polarization.

The orbital and the spin polarizations are included to all orders in the completely unrestricted Hartree-Fock procedure, and several successful hyperfine calculations have been performed in this way [*Larsson* 1970, 1971]. Equivalent results are obtained by solving coupled single-particle equations as we shall demonstrate later.

As an illustration of the relative importance of the different kinds of perturbations, we consider the lithium atom in the $2p\,^2P$ state, where a number of accurate calculations have been performed [*Nesbet* 1970; *Ahlenius* and *Larsson* 1973; *Garpman* et al. 1975, 1976; *Glass* and *Hibbert* 1976]. Table 14.4, taken from the work of *Garpman* et al., gives the contributions due to spin and orbital polarization (to all orders) as well as the lowest-order correlational contribution. (A more complete calculation, including also pair-correlation effects to all orders – see Sects. 15.2.3 and 15.4.2 – has recently been performed by *Lindgren* [1984, 1985 a].)

Table 14.4. Contributions to the hyperfine interaction in the $2p\,^2P$ states of Li (from [*Garpman* et al. 1976])

	$A(^2P_{1/2})$ [MHz]	$A(^2P_{3/2})$ [MHz]	$B(^2P_{3/2})/Q$ [MHz/barn]
Hartree-Fock	32.30	6.47	5.50
Spin polarization	10.46	−10.46	—
Orbital polarization	1.15	−0.11	−0.65
First-order correlation	1.30	1.39	0.45
Total (theoretical)	45.22	−2.71	5.30
Experimental	45.914(25)	−3.055(14)	—

We see from this table that the spin polarization is by far the most dominating perturbation and it is entirely responsible for the inversion of the hyperfine splitting in the $^2P_{3/2}$ state. The orbital polarization has a smaller—but still quite significant—effect. In the quadrupole interaction there is, of course, no spin-polarization effect. The orbital polarization changes the Hartree-Fock value of the quadrupole interaction by approximately 10%, which is a typical value for p electrons. For higher angular momenta, the relative effect can be considerably larger, as we shall demonstrate later,

We can also see from Table 14.4 that the effect of correlation is comparable to that of the orbital polarization, which is not an unusual situation. In the case of the quadrupole interaction the two effects almost cancel each other, which means that the Hartree-Fock value is actually more accurate than the "Sternheimer-corrected" value, where only the polarization is included. This demonstrates the difficulty in obtaining accurate quadrupole moments from the hyperfine interaction and the inadequacy of low-order perturbation theory for this kind of calculation.

14.4 Second-Order Polarization and Lowest-Order Correlation

By means of *first-order* one- and two-particle functions, it is possible to evaluate *third*-order energy diagrams. This was demonstrated in connection with the electrostatic calculations described in the previous chapter, and a similar procedure can be applied to the hyperfine interaction in order to evaluate the effect of second-order polarization and first-order correlation. The radial parts of the perturbations can be obtained by means of one- and two-particle Coulomb functions combined with one-particle hyperfine functions.

14.4.1 The Second-Order Polarization

The second-order polarization effect is represented by third-order effective-operator diagrams with single excitations in the intermediate states (see Fig. (14.5).

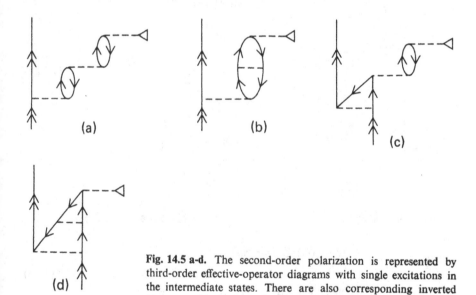

Fig. 14.5 a-d. The second-order polarization is represented by third-order effective-operator diagrams with single excitations in the intermediate states. There are also corresponding inverted diagrams with the hyperfine interaction at the bottom

We assume again for simplicity that there is only one open shell in the atom. Then diagram (a) has the following algebraic expression:

$$= a_m^\dagger a_n \sum \frac{\langle b|t_Q^{(\kappa\lambda)K}|s\rangle \langle as|r_{12}^{-1}|rb\rangle \langle mr|r_{12}^{-1}|na\rangle}{(\varepsilon_b - \varepsilon_s)(\varepsilon_a - \varepsilon_r)}$$

(14.63)

This can be separated into the radial and spin-angular parts

$$a_m^\dagger a_n \sum \frac{\langle b||t^{(\kappa\lambda)K}||s\rangle X^{k_2}(as, rb) X^{k_1}(mr, na)}{(\varepsilon_b - \varepsilon_s)(\varepsilon_a - \varepsilon_r)}$$

$$\times \sum_{\mu\nu} \langle \kappa\mu, \lambda\nu|KQ\rangle G_o G_s ,$$

(14.64)

where G_o and G_s are the orbital and spin-angular-momentum graphs

$$= \delta(k_1, \lambda)\delta(k_2, \lambda)[\lambda]^{-2}$$

(14.65)

$$= 4 \delta(\kappa, 0)$$

By means of first-order functions of the type (14.53) and (14.57)

$$\rho_d^{k_1}(a \to l_r; r_1) = \sum_{n_r}^{\text{part}} \frac{P_r(r_1) X^{k_1}(mr, na)}{\varepsilon_a - \varepsilon_r}$$

$$\rho_h(b \to l_s; r_2) = \sum_{n_s}^{\text{part}} \frac{P_s(r_2)\langle s||t^{(\kappa\lambda)K}||b\rangle}{\varepsilon_b - \varepsilon_s}$$

(14.66)

we can express the radial part of (14.64) as

$$\sum X^{k_1}(a\rho_h(b \to l_s), \rho_d^{k_1}(a \to l_r)b) .$$

(14.67)

The summation runs here over n_a, l_a, n_b, l_b, l_r, l_s, k_1 and k_2. The complete diagram then becomes

$$a_m^\dagger a_n 4\delta(\kappa, 0)\delta(k_1, \lambda)\delta(k_2, \lambda)[\lambda]^{-2}$$
$$\times \sum X^{k_2}(a\rho_h(b \rightarrow l_s), \rho_d^{k_1}(a \rightarrow l_r)b)F \,, \tag{14.68}$$

where F is the spin-angular factor (14.39a,c). In a similar way we can evaluate the remaining diagrams in Fig. 14.5 (see Problem 14.2).

We make here again the important observation that the spin-angular structure of these diagrams is the same as the first-order diagrams (14.50) appearing in the zeroth-order hyperfine constants (14.34). Therefore, we can treat these contributions in the same way as the first-order polarizations, yielding additional terms to the effective hyperfine parameters. (14.51).

Problem 14.2. a) Evaluate diagram (b) in Fig. 14.5 in terms of single-particle functions in analogy with diagram (a) treated in the text. b) Express the contributions to the radial parameters (14.51) due to diagrams (a) and (b).

14.4.2 The Lowest-Order Correlation

The remaining third-order hyperfine diagrams involve at least one double excitation and thus by the definitions we employ here contribute to the correlation effect. A few examples of such diagrams are shown in Fig. 14.6.

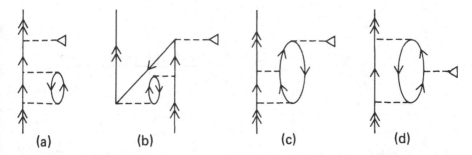

(a)   (b)   (c)   (d)

Fig. 14.6 a–d. Examples of third-order effective-operator diagrams which contribute to the effect of pure correlation upon the hyperfine interaction. There are 66 diagrams of this kind!

In evaluating diagrams of this kind using the same procedure as before we need access also to first-order Coulomb pair functions (13.98)

$$\rho^k(ab \rightarrow l_r l_s; r_1 r_2) = \sum_{n_r n_s}^{\text{part}} \frac{P_r(r_1)P_s(r_2)X^k(rs, ab)}{\varepsilon_a + \varepsilon_b - \varepsilon_r - \varepsilon_s} . \tag{14.69}$$

As a first illustration we consider diagram (a), which has the algebraic expression

$$= a_m^\dagger a_n \sum \frac{\langle m | t_Q^{(\kappa\lambda)K} | t \rangle \langle ta | r_{12}^{-1} | rs \rangle \langle rs | r_{12}^{-1} | na \rangle}{(\varepsilon_m - \varepsilon_t)(\varepsilon_n + \varepsilon_a - \varepsilon_r - \varepsilon_s)}, \qquad (14.70)$$

still assuming that there is a single open shell ($\varepsilon_m = \varepsilon_n$). This has the radial part

$$\sum \frac{\langle m || t^{(\kappa\lambda)K} || t \rangle X^{k_2}(ta, rs) X^{k_1}(rs, na)}{(\varepsilon_m - \varepsilon_t)(\varepsilon_n + \varepsilon_a - \varepsilon_r - \varepsilon_s)}. \qquad (14.71)$$

By means of the first-order hyperfine function (14.57) and the first-order pair function (14.69), this can be expressed as

$$\sum X^{k_2}(\rho_h(nl \to l_t)a, \rho^{k_1}(nl\, a \to l_r l_s)), \qquad (14.72)$$

where nl are the quantum numbers of the open shell.

As a second illustration we consider diagram (d) in Fig. 14.6

$$= a_m^\dagger a_n \sum \frac{\langle ma | r_{12}^{-1} | rt \rangle \langle t | t_Q^{(\kappa\lambda)K} | s \rangle \langle rs | r_{12}^{-1} | na \rangle}{(\varepsilon_m + \varepsilon_a - \varepsilon_r - \varepsilon_t)(\varepsilon_n + \varepsilon_a - \varepsilon_r - \varepsilon_s)}. \qquad (14.73)$$

The radial part of this diagram can be expressed by means of two first-order pair functions as

$$\sum \langle \rho^{k_2}(nl\, a \to l_r l_t) || t^{(\kappa\lambda)K} || \rho^{k_1}(nl\, a \to l_r l_s) \rangle. \qquad (14.74)$$

It is easy to see that all possible third-order diagrams of one-body type can be evaluated by means of first-order Coulomb and hyperfine functions. This is the procedure used by *Garpman* et al. [1975] on the alkali atoms. Equivalent results can be obtained, of course, by generating a complete set of single-particle states and performing the summations and integrations over the virtual states explicitly. This procedure was introduced by *Kelly* [*Kelly* 1968, 1969a,b; *Kelly* and *Ron* 1970] and has also been applied by *Das* and his co-workers [*Lyons* et al. 1969; *Dutta* et al. 1969; *Lee* et al. 1970].

It should now be quite obvious that all one-body effective-operator diagrams—with a single pair of free valence lines—have the same spin-angular structure as the first-order diagram (14.32). This can be proved in a rigorous way by means of theorem JLV3 in Chap.4. The spin and angular structure of an arbitrary one-body diagram can be represented by the diagram to the left in Fig. 14.7 with no free lines other than those shown. Then according to (4.8) this can be transformed into the product of diagrams shown to the right. But the first of these is closed and represents a number, and hence the diagram to the left is proportional to the 3-j symbol appearing in first order. These arguments hold for the spin part and angular part separately and obviously to any order of perturbation theory.

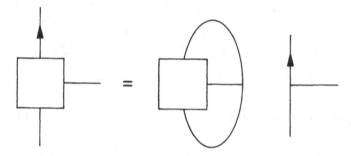

Fig. 14.7. By means of theorem JLV3, the angular-momentum diagram associated with an arbitrary one-body effective hyperfine diagram can be reduced to a closed diagram (which represents a number) times the 3-j symbol appearing in the zeroth-order hyperfine constant. An equivalent result holds for the spin part. This explains why all effective one-body effects can be included in the effective one-body operator (14.44)

14.5 All-Order Polarization

We have seen in the previous section that core polarization, which often has a drastic effect on the hyperfine interaction, may converge slowly in an order-by-order expansion. It can be included to all orders, however, by solving coupled single-particle equations in a self-consistent way, as we shall now demonstrate. Together with the first-order pair functions, which yield the first-order correlation, this procedure can in many cases reproduce the hyperfine interaction in an efficient and accurate way. In special cases the convergence of the perturbation expansion may be so poor that it is necessary to include also the correlation effects to higher orders. This can be done by iterating also the pair equations to self-consistency, as will be further discussed in the following chapter. For the moment we shall see how the polarization effect can be treated to all orders.

14.5.1 The Effective Hyperfine Interaction

Consider an "effective" hyperfine interaction defined by the graphical equation shown in Fig. 14.8. This is a self-consistent equation of the Dyson type (9.54), equivalent to the order-by-order expansion shown in Fig. 14.9.

The function defined in these figures represents a generalization to all orders

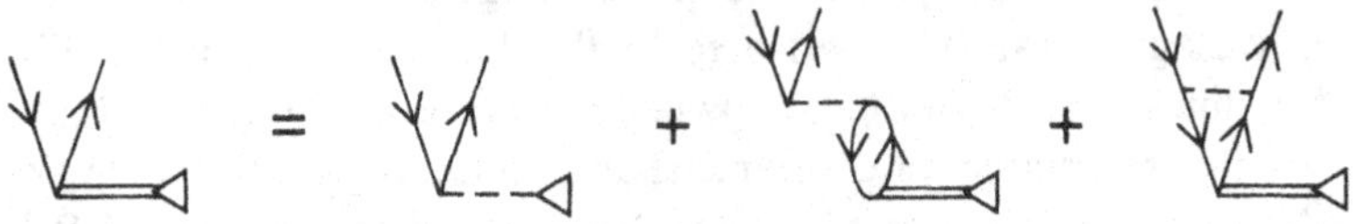

Fig. 14.8. An "effective" hyperfine interaction can be defined by means of a self-consistent equation of the Dyson type. Solving this equation is equivalent to generating the order-by-order expansion indicated in Fig. 14.9

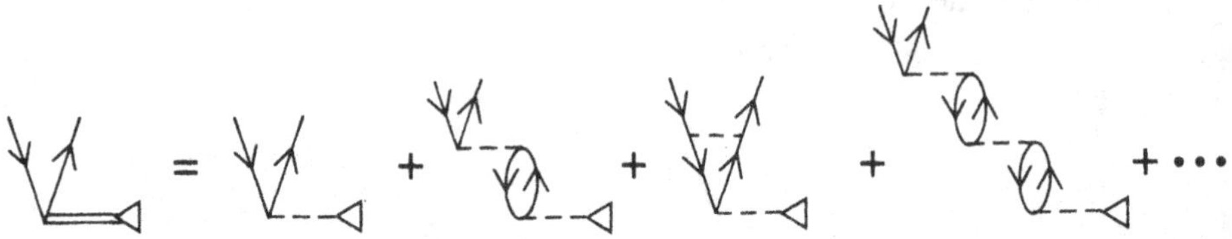

Fig. 14.9. Order-by-order expansion equivalent to the self-consistent equation in Fig. 14.8

of the hyperfine function introduced in the previous section (Fig. 14.2c). By evaluating the diagrams

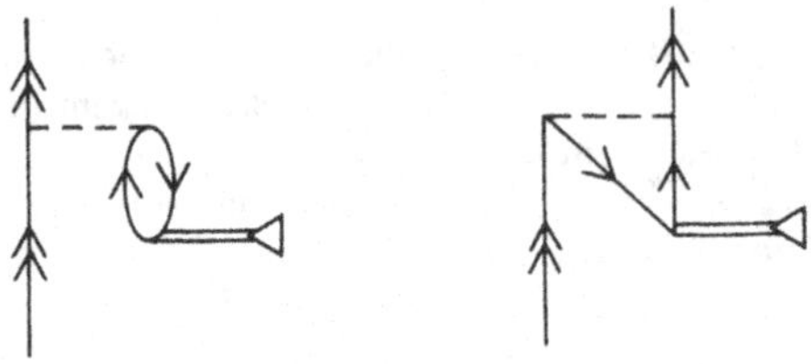

$$(14.75)$$

all polarization diagrams with the hyperfine interaction at the bottom are obtained (cf Fig. 14.5).

In order to convert the equation in Fig. 14.8 to algebraic form, we define single-particle coefficients by

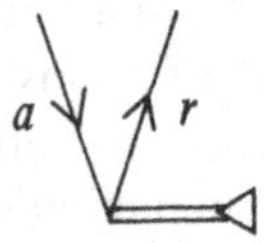

$$= a_r^\dagger a_a h_a^r . \qquad (14.76)$$

The diagrams on the right-hand side then become

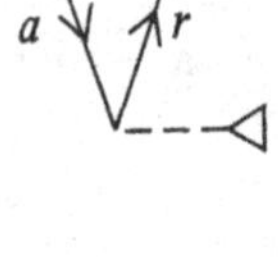

$$= a_r^\dagger a_a \frac{\langle r|h|a \rangle}{\varepsilon_a - \varepsilon_r} \qquad (14.77a)$$

$$= a_r^\dagger a_a \sum_{bs} \frac{\langle rb|r_{12}^{-1}|as\rangle h_b^s}{\varepsilon_a - \varepsilon_r} \qquad (14.77b)$$

$$= -a_r^\dagger a_a \sum_{bs} \frac{\langle br|r_{12}^{-1}|as\rangle h_b^s}{\varepsilon_a - \varepsilon_r}, \qquad (14.77c)$$

where h represents the hyperfine operator. After multiplying both sides in Fig. 14.8 with $(\varepsilon_a - \varepsilon_r)$ we get the following equation:

$$(\varepsilon_a - \varepsilon_r)h_a^r = \langle r|h|a\rangle + \sum_{bs}[\langle rb|r_{12}^{-1}|as\rangle - \langle br|r_{12}^{-1}|as\rangle]h_b^s . \qquad (14.78)$$

This represents a *set of coupled, linear, single-particle equations*, one equation for each excitation $a \to r$.

Using the coefficients (14.76), the diagrams (14.75) become

$$= a_n^\dagger a_m \sum_{ar} \langle na|r_{12}^{-1}|mr\rangle h_a^r \qquad (14.79a)$$

$$= -a_n^\dagger a_m \sum_{ar} \langle an|r_{12}^{-1}|mr\rangle h_a^r, \qquad (14.79b)$$

Solving the coupled equations (14.78) self-consistently and evaluating the two diagrams above is equivalent to evaluating to all orders the polarization diagrams with the hyperfine interaction at the bottom. If there is only one open shell, then these diagrams yield the same results as the corresponding inverted diagrams with the hyperfine interaction at the top, as we have mentioned before. In the case of several open shells it is necessary to solve additional differential equations with modified left-hand sides, corresponding to the modified denominators.

14.5.2 The All-Order Radial Single-Particle Equation

The single-particle equation (14.78) can be separated into radial and spin-angular parts in the same way as the diagrams considered before. It follows immediately from the argument illustrated in Fig. 14.7 that all diagrams (14.77) have the same spin-angular structure, and we can separate the coefficient (14.76) into

$$h_a^r = \bar{h}_a^r F, \tag{14.80}$$

where F is the spin-angular factor (14.39a, c) and $\bar{h}_a^r$ has the character of a reduced matrix element, being independent of all magnetic quantum numbers.

The diagrams (14.77) can now be separated in the following way:

$$= a_r^\dagger a_a \frac{\langle r \| t^{(\kappa\lambda)K} \| a \rangle}{\varepsilon_a - \varepsilon_r} F \tag{14.81a}$$

$$= a_r^\dagger a_a \sum_{bs} \frac{X^k(rb, as)\bar{h}_b^s}{\varepsilon_a - \varepsilon_r} G_1 F \tag{14.81b}$$

$$= -a_r^\dagger a_a \sum_{bs} \frac{X^k(br, as)\bar{h}_b^s}{\varepsilon_a - \varepsilon_r} G_2 F, \tag{14.81c}$$

where the spin-angular factors G_1 and G_2 are obtained by reducing graphs similar to those appearing in the first-order polarization (14.46, 47)

$$G_1 = 2\delta(\kappa, 0)\delta(k, \lambda)[\lambda]^{-1} \tag{14.82a}$$

$$G_2 = (-1)^\lambda \begin{Bmatrix} l_a & \lambda & l_r \\ l_s & k & l_b \end{Bmatrix}. \tag{14.82b}$$

Substituting the expressions above into (14.78), we obtain the following *radial* single-particle equation:

$$(\varepsilon_a - \varepsilon_r)\bar{h}_a^r = \langle r \| t^{(\kappa\lambda)K} \| a \rangle + \sum_{bsk} [G_1 X^k(rb, as) - G_2 X^k(br, as)]\bar{h}_b^s. \tag{14.83}$$

In analogy with the first-order function (14.57) we now introduce the *all-order hyperfine function*

$$\rho_h(a \to l_r; r_1) = \sum_{n_r}^{\text{part}} P_r(r_1) \bar{h}_a^r .\tag{14.84}$$

The differential equation satisfied by this function is obtained by multiplying (14.83) by $P_r(r_1)$ and summing over n_r. Observing that

$$h_0^{l_r} P_r(r_1) = \varepsilon_r P_r(r_1) ,$$

[see (13.81)], we then get

$$(\varepsilon_a - h_0^{l_r})\rho_h(a \to l_r; r_1) = \sum_{n_r}^{\text{part}} P_r(r_1) \langle r \| t^{(\kappa\lambda)K} \| a \rangle$$

$$+ \sum_{n_r}^{\text{part}} P_r(r_1) \sum_{bsk} [G_1 X^k(rb, as) - G_2 X^k(br, as)]\bar{h}_b^s \tag{14.85}$$

or after using the closure property (13.84) and recalling the definition of the reduced matrix element (14.39b) and the Y_k integrals (13.88)

$$(\varepsilon_a - h_0^{l_r})\rho_h(a \to l_r; r_1) = \langle s \| s^\kappa \| s \rangle \langle l_r \| v^\lambda \| l_a \rangle f_{(\kappa\lambda)K}(r_1)P_a(r_1)$$

$$+ \sum_{bsk} [G_1 X(k, l_r l_b l_a l_s)\frac{1}{r_1} Y_k(b, \rho_h(b \to l_s); r_1)P_a(r_1) \tag{14.86}$$

$$-G_2 X(k, l_b l_r l_a l_s)\frac{1}{r_1} Y_k(b, a; r_1)\rho_h(b \to l_s; r_1)] - \text{orthogonality terms.}$$

The radial parts of the diagrams (14.79) are then

$$\sum_{n_r}^{\text{part}} X^k(na, mr) \, \bar{h}_a^r = X^k(na, m\rho_h(a \to l_r))$$

$$\sum_{n_r}^{\text{part}} X^k(an, mr) \, \bar{h}_a^r = X^k(an, m\rho_h(a \to l_r)) \tag{14.87}$$

for the direct and exchange diagram, respectively. This is identical to the first-order polarization results (14.58), the only difference being that the first-order hyperfine function is replaced by the all-order one. Since the spin-angular structure is the same as for first-order polarization, it follows that all polarization corrections can be included in the form (14.51), if we substitute the expressions (14.87) for the X factors.

14.5.3 Ground-State Correlations

The single-particle equations considered above can easily be modified to include another class of diagrams, as we shall now demonstrate. Consider the following diagrams:

$$m \qquad = a_m^\dagger a_n \sum \frac{\langle ma|r_{12}^{-1}|nr\rangle\langle rs|r_{12}^{-1}|ab\rangle\langle b|h|s\rangle}{(\varepsilon_a - \varepsilon_r)(\varepsilon_a + \varepsilon_b - \varepsilon_r - \varepsilon_s)} \qquad (14.88a)$$

$$m \qquad = a_m^\dagger a_n \sum \frac{\langle ma|r_{12}^{-1}|nr\rangle\langle rs|r_{12}^{-1}|ab\rangle\langle b|h|s\rangle}{(\varepsilon_b - \varepsilon_s)(\varepsilon_a + \varepsilon_b - \varepsilon_r - \varepsilon_s)}, \qquad (14.88b)$$

where it is assumed that $\varepsilon_m = \varepsilon_n$. These diagrams differ only in their denominators, and using the identity (12.48)

$$\frac{1}{A(A + B)} + \frac{1}{B(A + B)} = \frac{1}{AB} \qquad (14.89)$$

we see that their sum can be factorized into

$$a_m^\dagger a_n \sum \frac{\langle ma|r_{12}^{-1}|nr\rangle\langle rs|r_{12}^{-1}|ab\rangle\langle b|h|s\rangle}{(\varepsilon_a - \varepsilon_r)(\varepsilon_b - \varepsilon_s)}. \qquad (14.90)$$

This is identical to the polarization diagram (14.63), except for the fact that a and r are interchanged. It follows from the definition of the electrostatic integrals (3.42), however, that

$$\langle ab|r_{12}^{-1}|cd\rangle = \langle ad|r_{12}^{-1}|cb\rangle = \langle cb|r_{12}^{-1}|ad\rangle \qquad (14.91)$$

so (14.90) is identical to (14.63).

Strictly speaking, the diagrams (14.88) represent correlation effects, since double excitations are involved. These kinds of diagrams are sometimes referred to as "ground-state correlations" [Wendin 1971, 1972; Amusia et al. 1972, 1974]. Since they can be evaluated in the same way as the polarization diagrams, however, we shall include them here in the polarization effect.

The exchange variants of (14.88) can be treated similarly. Since the exchange of (14.91) does not hold, however, this leads to a modification of the exchange of (14.63).

We can now include diagrams of the type considered here into the single-particle equation (14.83) by adding the corresponding terms to the right-hand side. This leads to

$$(\varepsilon_a - \varepsilon_r)\bar{h}_a^r = \langle r\|t^{(\kappa\lambda)K}\|a\rangle$$
$$+ \sum_{bsk} [2G_1 X^k(rb, as) - G_2 X^k(br, as) - G_2' X^k(rs, ba)]\bar{h}_b^s , \tag{14.92}$$

where G_2' is obtained from G_2 in (14.82b) by interchanging a and r (or b and s). The corresponding differential equation is obtained in analogy with (14.86).

Solving the coupled equations (14.92)—or the corresponding integral equations—and evaluating the diagrams (14.79) as before is equivalent to generating all diagrams of the kind illustrated in Fig. 14.10.

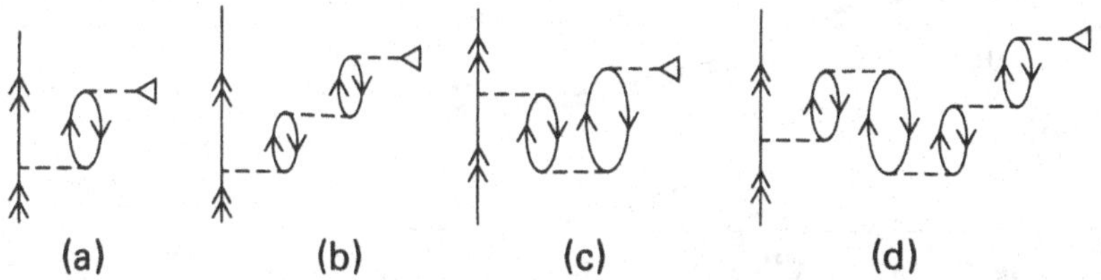

(a)   (b)   (c)   (d)

Fig. 14.10 a-d. Solving (14.92) is equivalent to generating all "bubble" diagrams (with their exchange variants) similar to those appearing in the random-phase approximation (RPA)

The first two diagrams (a,b) represent the polarization effect considered before, while the last two diagrams (c,d) represent effects of the ground-state correlation. These diagrams are quite similar to the diagrams appearing in the "random-phase approximation" (with exchange), which is frequently used in calculations of photoionization cross-sections (see references given above).

The technique described here was developed at Chalmers University and has been applied in many hyperfine and fine-structure calculations [*Garpman* et al. 1976; *Holmgren* et al. 1976; *Lindgren* et al. 1976; *Belin* et al. 1975, 1976a,b; *Lundberg* et al. 1977; *Mårtensson* 1978; *Lindgren* 1984]. Equations analogous to (14.83, 92) have also been derived by *Chang* and *Fano* [1976], particularly for application to photoionization calculations. In recent years similar calculations have also been performed of the *specific* (isotopic) *mass shift* [*Mårtensson* and *Salomonson* 1982; *Lindroth* and *Mårtensson-Pendrill* 1983].

14.5.4 Results of Some All-Order Calculations

As a first illustration of the all-order procedure developed here, we consider the hyperfine interaction of Li-like systems in their ground $2s\ ^2S$ state. Besides the Li atom, experimental values are available for Be$^+$[*Vetter* et al. 1976] and for F^{6+} [*Randolph* et al. 1975].

Calculations including polarization effects to all orders in combination with correlation effects to lowest order have been performed by *Garpman* et al. [1976]. The results are shown in Fig. 14.11, where the experimental values are also indicated. The Hartree-Fock values are used as reference. We then see that the effect of perturbations is of the order of 40% for Li and reduces to about

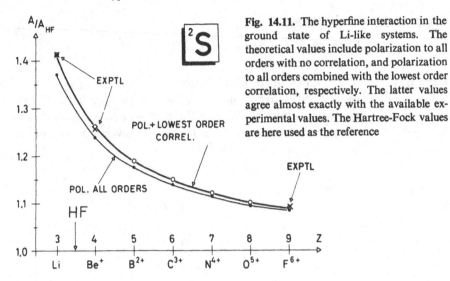

Fig. 14.11. The hyperfine interaction in the ground state of Li-like systems. The theoretical values include polarization to all orders with no correlation, and polarization to all orders combined with the lowest order correlation, respectively. The latter values agree almost exactly with the available experimental values. The Hartree-Fock values are here used as the reference

10% for F^{6+}. The (spin) polarization is here very dominating and is responsible for about 90% of the perturbation for Li and even more for the ionized systems.

Qualitatively we can understand these results. Of course, one expects the independent-particle model to work better for more highly ionized systems, where the central field from the nucleus is more dominant. The effective nuclear charge, felt by a valence electron, is in the inner region three times—and in the outer region seven times—larger in F^{6+} than in Li, which should be compared with the ratio 4:1 of the relative magnitude of the perturbations in the two cases.

In a simple hydrogenic model one finds that the zeroth-order interaction is proportional to the third power of some "effective" nuclear charge, while the polarization and correlation effects varies as the second and first power, respectively, of this charge. This explains why the relative importance of higher-order effects decreases approximately linearly with the nuclear charge and also why the relative importance of the correlation effect decreases even more rapidly.

We see from the examples considered here that such simple systems as the alkali atoms exhibit strong perturbations compared to the central-field model. This is even more conspicuous when we go to high-angular-momentum states of these atoms. Nowadays, experimental data are available for a large number of such states, mainly due to laser work at Columbia University by *Happer* and by *Gupta* et al. and at Chalmers University by *Svanberg* et al. (for a review, see [*Arimondo* et al. 1977]).

In the S and P states of the alkali atoms we have seen that the perturbations are typically of the order of 20–40%. The situation is extremely different for the ^2D states, where it is found that the hyperfine interaction is inverted in all $^2D_{5/2}$ states where information is available. The most accurate experimental determination of this kind has been performed for the $4d\,^2D_{5/2}$ of Rb by the Columbia group [*Liao* et al. 1974; *Lam* et al. 1980]. Their results are given in Table 14.5 together with the Hartree-Fock results.

Table 14.5. Experimental hyperfine data for the $4d\,^2D$ states of Rb compared with Hartree-Fock values. The A factors are given in MHz and the $\langle r^{-3}\rangle$ parameters in atomic units

	$A(^2D_{3/2})$	$A(^2D_{5/2})$	$\langle r^{-3}\rangle_l$	$\langle r^{-3}\rangle_{sd}$	$\langle r^{-3}\rangle_c$
Experimental	25.1(9)	−16.9(6)	0.029(3)	−0.047(12)	−0.625(15)
Hartree-Fock	11.6	+ 5.0	0.041	+0.041	0

We see that the dipole interaction constant is negative for the $^2D_{5/2}$ state in contrast to the Hartree-Fock value. One might expect that this is caused primarily by spin polarization, as in the Li $2p\,^2P_{3/2}$ case considered above. However, the orbital and spin-dipole parameters are drastically different, showing that other perturbations must be of unusual strength. It would, of course, be a challenge to see if the many-body technique available at present would be capable of reproducing these strongly perturbed results. Calculations of this kind have been performed by *Lindgren* et al. [1976, 1977] and by *Mårtensson* [1978], and they will be summarized below.

It is found that the procedure successfully applied to the alkali S and P states, including polarization to all orders and correlation to the lowest order, does not work for the ^2D state in Rb. This is illustrated in the first part of Table 14.6.

The values in the second row of the table include the first-order polarization described in Sect. 14.3. The values of the next row include also the *diagonal* part of the polarization in higher orders, that is, the effect due to excitations from the same nl shell to the same l value in all orders. This represents a simple calculation, since there is no coupling between the single-particle equations. This approximation corresponds to the single-channel RPA approximation, and it can also be reproduced in a first-order polarization calculation using a potential of the "V^{N-1} type" [*Kelly* and *Simons* 1973, *Chang* and *Kelly* 1975]. Therefore, the difference between these two results gives an indication of the sensitivity on the potential of the first-order polarization. The complete all-order polarization, on the other hand, is independent of the choice of potential.

The results obtained when polarization is included to all orders and correlation to the lowest orders are given in the fifth row of the table. We see that these results do not at all reproduce the experimental values. The spin-dipole parameter is positive, in contrast to experiment, and the orbital and contact parameters are off by almost a factor of two. The hyperfine constant in the state $^2D_{3/2}$ happens to be close to experiment but the corresponding value in $^2D_{5/2}$ is off by a factor of three.

Table 14.6. The hyperfine interaction in the $4d\,{}^2D$ state of ^{87}Rb. The A factors are given in MHz and the $\langle r^{-3}\rangle$ parameters in atomic units

	$\langle r^{-3}\rangle_l$	$\langle r^{-3}\rangle_{sd}$	$\langle r^{-3}\rangle_c$	$\langle r^{-3}\rangle_q$	$A({}^2D_{3/2})$	$A({}^2D_{5/2})$
Hartree-Fock	0.041	0.041		0.041	11.6	5.0
With first-order polarization	0.011	−0.024	−0.202	0.115	7.7	−5.0
With all-order polarization (diag.)	−0.006	−0.064	−0.278	0.134	4.0	−9.3
With complete polarization	−0.009	−0.077	−0.447	0.129	8.4	−15.4
Including first-order correl.	0.052	0.011	−0.376	0.161	24.8	−6.1
Approximate Brueckner orbitals	0.083	0.083	0	0.083	23.2	9.9
With all-order pol. + first-order correl.	0.041	−0.024	−0.623	0.174	28.6	−15.6
Incl. higher-order correl.	0.028	−0.043	−0.625	0.170	24.7	−17.1
Experimental	0.029(3)	−0.047(12)	−0.625(15)	0.12(8)	25.1(9)	−16.9(6)

In analyzing third-order calculations of hyperfine interactions, one finds that some of the most dominating diagrams are of the "self-energy" type in which the two Coulomb interactions are connected to a single line. Generally, we can represent such interactions as shown in Fig. 14.12, where the box represents a diagram part with no lines connected to it other than those shown. Those kinds of diagrams can be regarded as *orbital modifications,* diagrams (a) of a core orbital and diagrams (b) of a valence orbital.

It follows from the Brillouin-Brueckner condition (12.33) that all diagrams of this kind vanish if Brueckner orbitals are used. By evaluating diagrams of the type shown in Fig. 14.12 and adding them to the corresponding core or valence orbital, approximate Brueckner orbitals are generated for these shells (provided that the perturbation expansion is properly converging). By iterating this procedure until the diagrams vanish, exact Brueckner orbitals are obtained.

It follows directly from theorem JLV2 in Chap.4 [(4.5,6)], that the angular-momentum graphs associated with the diagrams in Fig. 14.12 vanish unless the l, m_l, and m_s quantum numbers are the same for the free lines. This implies that the spin-angular part of the orbital is unaffected by this type of modification and the corrections can be added to the *radial* functions.

It is not difficult to see that the second-order diagrams in Fig. 14.12 can be generated by solving single-particle equations with pair functions on the right-hand sides. The solutions can then be added directly to the radial part of the

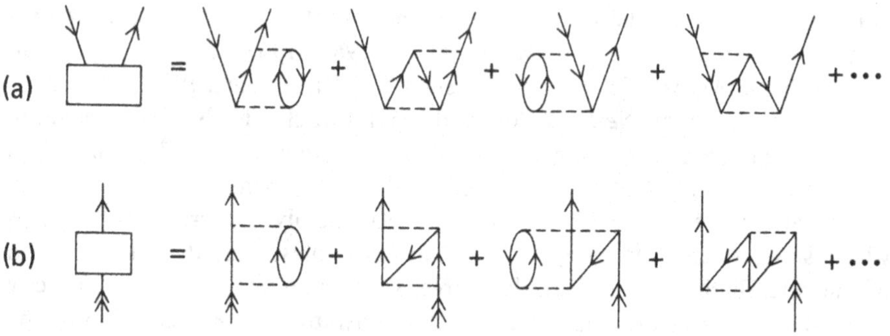

Fig. 14.12 a,b. Correlation diagrams of the "self-energy" type. Diagrams of this type can be used to modify the orbitals of the core (a) and the valence orbitals (b). They vanish when Brueckner orbitals are used

corresponding orbital. Then it is found that diagrams (g-i) in Fig. 13.7 are automatically contained in the zeroth-order diagram (14.32), evaluated by means of the modified orbitals. By reevaluating the polarization and the correlation effects with the new orbitals, important higher-order diagrams are automatically included. If desired, this procedure can be repeated to modify the orbitals several times.

The procedure described here has been applied to the Rb problem in order to improve the accuracy, and the result is given in the sixth and seventh rows in Table 14.6. The sixth row gives the values obtained in zeroth order using the modified orbitals. This shows that the modification changes the hyperfine interaction by a factor of two over the Hartree-Fock values! When the polarization is evaluated to all orders and the correlation to the lowest order the results are in reasonable agreement with experiments (row 7), even if the spin-dipole constant is still off by a factor of two. The last row gives the results of the most complete calculation for his state, performed by *Mårtensson* [1978]. In this calculation the pair-correlation effect is also included to all orders by solving coupled pair equations iteratively. These results are in excellent agreement with the experimental results. Since no other calculation of this kind has been performed, it is not possible to tell to what extent these results are fortuitous. They do indicate, however, that the hyperfine interaction in the highly perturbed alkali ^2D states can be well represented by a wave function containing polarization effects (single excitations) and pair-correlation effects (double excitations) to all orders, implying that genuine three-, four-body effects—which cannot be represented by means of single and double excitations in any orbital basis set—are relatively insignificant. Furthermore, the calculations have clearly demonstrated that these effects have to be carried to high orders in the perturbation expansion and that a low-order expansion may fail completely.

The numerical examples we have studied in the last few sections are based on the perturbative approach in order to illustrate how different kinds of physical effects appear in this scheme. It should be observed, however, that essentially equivalent results can be obtained by nonperturbative methods, such as configuration interaction (CI) [*Schaefer* et al. 1968, 1969], multiconfiguration Hartree-Fock (MCHF) [*Bagus* and *Bauche* 1973; *Bauche* et al. 1974], or "Bethe-Goldstone" equations [*Nesbet* 1970]. At this point we should also like to mention the Green's function approach employed by *Cederbaum* et al. [1975], which to a large extent is equivalent to the procedure developed here.

So far, we have also in this chapter been mainly concerned with atoms with a single valence electron, such as the alkali atoms. Here the interpretation of the perturbations is particularly simple, since the interaction can be exactly expressed by means of a one-body effective operator of the same kind as the ordinary hyperfine operator. The large amount of experimental information which is now available for the alkali atoms can be regarded as well understood in terms of such an effective operator. In the next section we shall briefly consider the hyperfine interaction for atoms with several valence electrons, such as alkaline earths. Here the experimental as well as the theoretical information is much more scarce, but new data are presently being produced at an increased rate.

14.6 Effective Two-Body Hyperfine Interactions

As we have mentioned, all effective one-body contributions to the hyperfine interaction, represented by diagrams with a single pair of free lines, can be included in an effective hyperfine operator of the one-body type (14.44). If there are several open shells, one set of parameters is needed for each shell. Effective two-body diagrams, on the other hand—like diagrams (e) and (f) in Fig. 13.8—cannot be treated in the same way, since their spin-angular dependence will be quite different from that of the first-order diagrams. In order to represent such diagrams one would need effective two-body operators of the type

$$\{t^{\kappa\lambda_1}(i)t^{0\lambda_2}(j)\}^{\kappa K} \tag{14.93}$$

in analogy with the effective two-body Coulomb interaction (13.59, 67). This has been discussed in detail by *Bauche-Arnoult* [1971, 1973]. To cover the two-body effects completely in this way would require a large number of parameters, but it is usually possible to select a few dominating ones. An alternative approach is to evaluate the two-body contributions explicitly for each particular *LS* term, and this is the procedure we shall follow here. We shall see that we can then use the same operator as before, the only difference being that the parameters are now *term dependent*.

Before we evaluate the higher-order two-body contributions, we shall reconsider the zeroth-order hyperfine interaction for a specific *LS* term. We assume that there are two valence electrons, and the zeroth-order hyperfine interaction (which we get from first-order perturbation theory) can then be expressed in analogy with the corresponding Coulomb interaction (13.106) as

$$= \langle n \| t^{(\kappa\lambda)K} \| q \rangle G \, . \tag{14.94}$$

Here the reduced matrix element is given by (14.39b) and G is the spin-angular factor obtained by coupling the spin- and orbital-angular-momentum diagrams of the hyperfine operator to S and L, respectively. The angular part is

$$G_o = \quad = (-1)^{l_p + l_q + L}(2L + 1) \begin{Bmatrix} L & \lambda & L \\ l_q & l_p & l_n \end{Bmatrix} + \quad \tag{14.95}$$

and the spin part is given by an analogous diagram. The spin and orbital parts are coupled by means of vector-coupling coefficients, and we may represent the complete interaction with a combined diagram, as in the previous treatment (14.39c).

Higher-order contributions can be evaluated in a similar way, and we consider as an example diagram (e) in Fig. 13.8. Coupling the valence electrons to L and S, we can express this diagram as

$$= \sum_{l, n, k} X^k(mr, pq) \langle n \| t^{(\kappa\lambda)K} \| r \rangle \, G'/\Delta E \, , \tag{14.96}$$

where G' is the spin-angular part, $\langle n \| t^{(x\lambda)K} \| r \rangle$ is the reduced matrix element (14.39b) and ΔE the energy dominator. The spin part of G' is the same as that of the zeroth-order diagram (14.94), since the Coulomb interaction is spin independent. The angular part is given by the angular-momentum diagram

$$G'_0 = \quad = (2L+1)\begin{Bmatrix} l_m & L & l_r \\ l_q & k & l_p \end{Bmatrix} \begin{Bmatrix} L & \lambda & L \\ l_n & l_m & l_r \end{Bmatrix} + \qquad \text{(14.97)}$$

Thus, we see that the spin-angular structure of diagram (14.96) is the same as that of the zeroth-order diagram (14.94), and we can include the effect of this perturbation by modifying the radial parameters appearing in zeroth order. It is not difficult to see that all effective two-body diagrams can be treated in the same way. In contrast to the effective one-body diagrams, however, these diagrams lead to corrections that depend on the term considered.

In evaluating diagrams of the kind considered here it should be remembered that the coupling represented by joining the free valence lines does not contain any antisymmetrization (see Sect. 4.4). In the case of two equivalent electrons we know from the treatment in Chap. 4 that we can formally ignore the exclusion principle and use LS-coupled functions of the direct product type provided that $L + S$ is even (4.28). For nonequivalent electrons, on the other hand, we have to consider the direct and exchange contributions separately with the appropriate phase factor according to (4.29).

We consider explicitly a configuration $nl\,n'l'$ with an excitation $nl \rightarrow n''l''$. The angular part (14.95) of the zeroth-order diagram then becomes

$$(-1)^{l+l'+L}(2L+1)\begin{Bmatrix} L & \lambda & L \\ l & l' & l \end{Bmatrix}, \tag{14.98}$$

leaving out the diagram of the 3-j symbol. The corresponding part of the first-order diagram (14.97) is

$$(2L+1)\begin{Bmatrix} l' & L & l'' \\ l & k & l' \end{Bmatrix}\begin{Bmatrix} L & \lambda & L \\ l & l' & l'' \end{Bmatrix} \tag{14.99a}$$

for the direct term and

$$(2L + 1) \begin{Bmatrix} l' & L & l'' \\ l' & k & l \end{Bmatrix} \begin{Bmatrix} L & \lambda & L \\ l & l' & l'' \end{Bmatrix} \tag{14.99b}$$

for the exchange term. In addition, there is a phase factor $(-1)^{l+l''+L+S}$ associated with the exchange term according to (4.29). Thus, we can include the effect of diagram (14.96) by modifying the zeroth-order radial parameters by

$$\frac{(-1)^{l+l''+L} \sum_{n''l''k} \langle l||v^\lambda||l''\rangle \langle nl|f_{(\kappa\lambda)\kappa}|n''l''\rangle X^k(n'l'\,n''l'', n'l\,nl) \begin{Bmatrix} l' & L & l'' \\ l & k & l' \end{Bmatrix} \begin{Bmatrix} L & \lambda & L \\ l & l' & l'' \end{Bmatrix}}{\langle l||v^\lambda||l\rangle \begin{Bmatrix} L & \lambda & L \\ l & l' & l \end{Bmatrix} \Delta E} \tag{14.100a}$$

for the direct contribution and by

$$\frac{(-1)^{S} \sum_{n''l''k} \langle l||v^\lambda||l''\rangle \langle nl|f_{(\kappa\lambda)\kappa}|n''l''\rangle X^k(n'l'\,n''l'', nl\,n'l') \begin{Bmatrix} l' & L & l'' \\ l' & k & l \end{Bmatrix} \begin{Bmatrix} L & \lambda & L \\ l & l' & l'' \end{Bmatrix}}{\langle l||v^\lambda||l\rangle \begin{Bmatrix} L & \lambda & L \\ l & l' & l \end{Bmatrix} \Delta E} \tag{14.100b}$$

for the exchange contribution. Other higher-order diagrams can be treated in the same way. From this result we see that the modification of the radial parameters depends on the term considered. The exchange contribution is different for the singlet and triplet states, while the direct contribution is independent of the total spin.

In principle, any central-field approximation can be used as the starting point for the procedure indicated here. Provided that the perturbation expansion is brought to sufficient convergence, the final result should be independent of the central-field used in zeroth order. In practice, however, the choice of central-field potential can be quite critical for the convergence and thus for the success of the calculation.

As an illustration we consider the $3s3p$ configuration in Mg. Here, the configurational-average Hartree-Fock procedure yields the $\langle r^{-3} \rangle$ value

$$\langle r^{-3} \rangle_{\text{conf.av.HF}} = 0.501.$$

From experimental hyperfine data one can deduce the following values of the orbital parameter for the ^1P term and of the spin-dipole parameter of the ^3P term:

$$\langle r^{-3} \rangle_l(^1P)_{\text{exp}} = 0.257(21)$$
$$\langle r^{-3} \rangle_{sd}(^3P)_{\text{exp}} = 0.809(9) .$$

These values indicate strong term dependence of the radial wave functions, and therefore one can expect convergence difficulty in a perturbation expansion starting from term-independent zeroth-order functions.

In principle, one can start the perturbation expansion from LS-dependent Hartree-Fock (LSHF), where the orbitals are optimized by minimizing the expectation energy for a particular term, rather than for the configurational average. Then one will find that the second-order diagrams of the kind (14.96) to a large extent cancel the corresponding potential interaction (diagrams a, g in Fig. 13.8) for this particular term.

The LSHF values of $\langle r^{-3} \rangle$ for the two terms in Mg are

$$\langle r^{-3} \rangle (^1P)_{LSHF} = 0.124$$

$$\langle r^{-3} \rangle (^3P)_{LSHF} = 0.607 ,$$

which shows that LSHF actually *exaggerates* the term dependence in this case.

A many-body calculation, starting from LSHF and including core-polarization to all orders, has recently been performed by *Olsson* and *Salomonson* [1982] on corresponding states in the Ca atom. Later, *Salomonson* [1984] has performed a more complete calculation, including also pair-correlation effects. That calculation started from HF of the Ca^{2+} core, and the pair-correlation between the valence electrons were then treated to self-consistency, *before* the core-polarization was evaluated. In that way it was possible to reproduce the hyperfine data for the different LS terms with good accuracy.

An equivalent way of evaluating LS-dependent hyperfine parameters is to use multi-configurational Hartree-Fock (MCHF). Such calculations have been performed on the above-mentioned $3s\,3p$ configuration in Mg by *Bauche* and co-workers [1974]. Starting from LSHF and adding certain singly and doubly excited configurations, they have obtained the following values for Mg:

$$\langle r^{-3} \rangle_l (^1P)_{MCHF} = 0.248$$

$$\langle r^{-3} \rangle_{sd} (^3P)_{MCHF} = 0.792 ,$$

in very good agreement with the experimental values. A similar calculation has recently been performed on the corresponding states in the Ca atom [*Fonseca* and *Bauche* 1984], also with good agreement with the experimental results.

14.7 Relativistic Effects

As we have mentioned in Chap. 5, there exists today no consistent many-body procedure which includes relativistic effects in a complete and rigorous way. Essentially two approaches are presently available for practical applications. One way is to treat the relativistic interactions as perturbations to the nonrelativistic Hamiltonian. This is normally done in the Pauli approximation, where relativistic effects are treated to order α^2. For light and medium-heavy elements this is a reasonably good approximation, and the additional terms obtained in this way, the mass correction, the spin-orbit interaction, etc., can be treated in the same way as the hyperfine interaction. In this section we shall consider some fine-structure calculations using this procedure.

A more accurate way of treating the relativistic many-body problem is to approximate the Hamiltonian by a sum of single-particle Dirac Hamiltonians h_D and the (instantaneous) Coulomb interaction between the electrons, often referred to as the *"Dirac-Coulomb Hamiltonian"*,

$$H = \sum_j h_D(i) + \sum_{i<j} r_{ij}^{-1} . \tag{14.101}$$

This is the operator normally used in relativistic Hartree-Fock (Dirac-Fock) calculations [*Grant* 1970; *Lindgren* and *Rosén* 1974; *Desclaux* 1975], and it can be used as a basis for a relativistic many-body procedure, very much in the same way as the standard Hamiltonian (10.1) is used in nonrelativistic calculations. The Breit interactions, which represent magnetic two-body effects in lowest order and which in the Pauli approximation correspond to the spin-other-orbit, orbit-orbit and spin-spin interactions are usually omitted but can be added to the Hamiltonian (14.101). All relativistic single-particle effects, namely the mass correction, the Darwin term and the spin-(own-)orbit interaction, are then treated to all orders and the relativistic two-body effects to lowest order − or neglected entirely.

14.7.1 Hyperfine Structure Calculations

Many-body calculations of the hyperfine interaction, based on the Hamiltonian (14.101) have been performed by *Andriessen* et al. [1976, 1978], *Vajed-Samii* et al. [1979, 1981] and by *Dzuba* et al. [1982]. These calculations are relativistic analogues of corresponding nonrelativistic calculations performed by *Kelly* [1968, 1969a], *Kelly* and *Ron* [1979] and by *Das* and co-workers [*Lyons* et al. 1969; *Dutta* et al. 1969; *Lee* et al. 1970] and include second-order core-polarization and the dominating lowest-order correlations (see Figs. 14.5 and 14.6). Relativistic hyperfine calculations, which include core-polarization to *all* orders (see Sect. 14.5) but no correlation, have been performed by *Heully* and *Mårtensson-Pendrill* [1983a, b]. More complete calculations have recently been performed by *Dzuba* et al. [1984]. Here the hyperfine interaction is included in the self-consistent procedure [*Sandars* 1977] − which is equivalent to evaluating polarization effects to all orders − and in addition lowest-order correlation effects are evaluated. In Table 14.7 we show the results obtained by *Dzuba* et al. with this procedure for some states in Cs.

Table 14.7. Hyperfine constants for some states in ^{133}Cs. (In 10^{-3} cm^{-1}, from [*Dzuba* et al. 1984])

Level	RHF[a]	RHFH[b]	RHFH + correlation	Experimental[c]
$6s$	47.3	56.8	78.3	76.66
$7s$	13.0	15.6	18.7	18.22(10)
$6p_{1/2}$	5.37	6.67	9.54	9.737(4)
$7p_{1/2}$	1.92	2.37	3.15	3.147(1)
$6p_{3/2}$	0.797	1.427	1.738	1.679(2)
$7p_{3/2}$	0.288	0.512	0.573	0.5539(2)

[a] Relativistic Hartree-Fock (Dirac-Fock)
[b] Relativistic Hartree-Fock with the hyperfine interaction in the self-consistent procedure
[c] From [*Arimondo* et al. 1977]

Instead of using the four-component Dirac spinors as basis functions, it is possible to work in the *Pauli approximation* and use two-component spinors (see, for instance, [*Bethe* and *Salpeter* 1957]). *Cowan* and *Griffin* [1976] have suggested an improvement of the conventional procedure, where the small component is eliminated without any approximation, yielding an *exact* equation for the large component. *Wood* and *Bohring* [1978] have shown that a self-consistent-field procedure based on that scheme, treating the spin-orbit interaction only to first order, is capable of reproducing relativistic effects on the one-electron binding energies within 10% even for heavy elements like uranium. *Heully* [1982] has used a similar scheme for hyperfine-structure calculations and found that this works remarkably well for s electrons, while for non-s electrons it is necessary to include the spin-orbit interaction at least for the valence electron in the SCF procedure. Such a scheme has also been used by *Heully* and *Salomonson* [1982] in an all-order core-polarization calculation, which was found to reproduce very well the corresponding results of a considerably more time-consuming fully relativistic calculation.

The "improved Pauli approximation" might possibly be used also as a basis for calculating the pair correlation in analogy with the non-relativistic procedure described in Sect. 13.5.2 (see also Sects. 15.2.3 and 15.4.2). One principal difficulty, however, is to handle negative-energy solutions of the Dirac equation in a correct way [*Brown* and *Ravenhall* 1951; *Sucher* 1980]. It is instructive in this connection to consider a *diagonalization* of the single-electron Dirac operator [*Foldy* and *Wouthuysen* 1950]. This will lead to two "effective" operators of Pauli type (with two-component eigenfunctions), which reproduce exactly the positive and negative eigenvalues, respectively, of the Dirac Hamiltonian. In lowest order, however, it is sufficient to eliminate the negative energy solutions by means of *projection operators*, and a relativistic pair program based on this idea is presently being developed at Chalmers.

14.7.2 Fine-Structure Calculations on Alkali-like Systems

It is well known that the fine structure is inverted for many high-angular-momentum states of alkali atoms, and it would be of interest to see to what extent this inversion can be reproduced in a many-body calculation, where the spin-orbit interaction is treated as a perturbation. As an illustration we consider the lowest ^2D state of sodium, where calculations involving polarization effects to all orders have been performed by *Holmgren* et al. [1976] and more accurately by *Mårtensson* [1978, *Lindgren* and *Mårtensson* 1982]. The results of these calculations are given in Table 14.8. Included in the table are also the first-order results obtained by *Sternheimer* et al. [*Foley* and *Sternheimer* 1975; *Sternheimer* et al. 1978] and results of relativistic self-cosistent-field calculations by *Luc-Koenig* [1976] and by *Pyper* and *Marketos* [1981]. In the calculation of *Luc-Koenig* the valence orbital is generated in the Hartree-Fock potential of the core, neglecting the core-valence exchange, which is treated as a perturbation. It can be shown that this treatment is equivalent to a nonrelativistic calculation, where the spin-orbit interaction is treated as a perturbation. In the calculations of *Holmgren* et al. and of *Mårtensson* the spin-other-orbit interaction is also considered, which is not the case in the calculation of *Luc-Koenig*, since this is based on the Hamiltonian (14.101). From the agreement of the two sets of results it can be concluded that this effect is fairly small. Both calculations agree quite well with the experimental results, considering the fact that the nonrelativistic Hartree-Fock model predicts a normal (not inverted) fine structure and that no correlation effects are considered. Since the spin-orbit interaction is spin dependent, it can be shown exactly as in the hyperfine case that only the exchange polarization contributes. Furthermore, it can be shown that it is in particular the polarization of the outermost p shell which is responsible for the inversion. Most of this effect appears already in first order, as can be seen from the results of *Sternheimer* et al.

In Fig. 14.13 we show the results for the $3d$ state of the Na isoelectronic sequence [*Lindgren* and *Mårtensson* 1982]. It is interesting to note that the fine structure becomes normal for higher members of the sequence, which is evidently due to the fact that the central field becomes more dominant and consequently the perturbations less significant at higher degrees of ioniza-

Table 14.8. The fine-structure intervals in some ^2D states of sodium. All values are in $10^{-3}\,\mathrm{cm}^{-1}$. References are given in the text

	Hartree-Fock	Mårtensson	Sternheimer et al.	Luc-Koenig	Pyper-Marketos	Exptl.
$3d$	33.46	−44.93	−36.4	−43.79	−35.89	−49.4
$4d$	13.68	−31.58	−26.38	−30.49	−25.77	−34.31
$5d$	6.90	−19.18	−16.12	−18.43	−15.63	−20.6
$6d$	3.97	−12.09		−11.55	− 9.79	−12.41

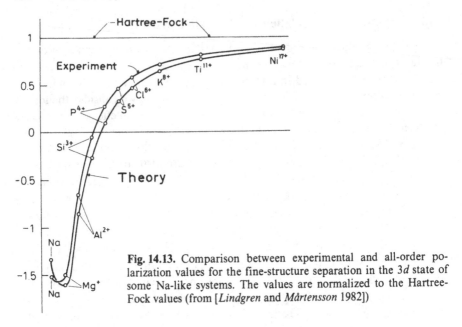

Fig. 14.13. Comparison between experimental and all-order polarization values for the fine-structure separation in the $3d$ state of some Na-like systems. The values are normalized to the Hartree-Fock values (from [*Lindgren* and *Mårtensson* 1982])

tion. From the figure it can be seen that the theoretical values follow the experimental ones quite closely and that the predicted transitions from inverted to normal fine structure appear at approximately the correct place.

For heavier elements the standard Pauli approximation is not sufficient, and it is necessary to use the Dirac-Coulomb procedure, based on (14.101), or some of the "improved Pauli schemes" mentioned above. *Dzuba* et al. [1983 a, b] have calculated the electron binding energies for Cs and Fr, using the relativistic Hartree-Fock procedure and including lowest order correlation effects. As mentioned previously, the RHF procedure is equivalent to a non-relativistic procedure where the core-polarization due to the relativistic perturbations is included to all orders. In contrast to the situation for the sodium-like systems, however, it is found that the correlation plays an important role for the heavier alkali systems. In Table 14.9 we show the results of *Dzuba* et al. [1983 a] for some states in Cs.

Table 14.9. The fine-structure intervals in some states of Cs. (In cm^{-1}, from [*Dzuba* et al. 1983 a])

State	RHF	RHF + correlation	Experimental
$6p$	404	556	554
$7p$	151	185	181
$5d$	−22	63	98
$6d$	0	48	43

15. The Pair-Correlation Problem and the Coupled-Cluster Approach

In the Chapters 12 and 13 we have seen how the electron correlation can be treated order by order for closed-shell and open-shell systems, respectively. It is quite obvious, however, that this approach rapidly gets very cumbersome, and —at least for open-shell systems—it is hardly feasible to go beyond the third-order energy in this way. As demonstrated in Chap. 13, higher-order terms can have quite an appreciable effect on quantities like the term splitting. Therefore, it would be highly desirable to find a more practical way of evaluating higher-order contributions. In the present chapter we shall develop such a procedure, which is based on the solution of a system of coupled differential equations of the type considered in Sect. 13.5. By considering one- and two-particle equations, it is possible to treat the dominating correlation effect, namely the *pair correlation*, in a complete way. This appears to be a very promising approach for an accurate treatment of the electron correlation for atomic and molecular systems.

15.1 Introductory Comments

The basis for our many-body treatment is the generalized Bloch equation (9.73)

$$[\Omega, H_0]P = Q V \Omega P - \chi P V \Omega P, \tag{15.1}$$

where Ω is the wave operator and

$$\Omega = 1 + \chi .$$

Expanding (15.1) order by order leads to the Rayleigh-Schrödinger perturbation expansion, and transformed into graphical form this gives rise to the linked-diagram expansion. The latter expansion can also be generated from an analogous equation (13.47)

$$[\Omega, H_0] = (Q V \Omega - \chi P V \Omega)_{\text{linked}} . \tag{15.2}$$

As an alternative to the order-by-order expansion, the equations above can be separated into a hierarchy of coupled one-, two-, ... particle equations, which can be solved by means of an iterative procedure. Such a separation can be made by transforming the equation into second-quantized or diagrammatic form and identifying terms of one-, two-, ... body type. This leads to a set of coupled equations, and solving them self-consistently is equivalent to evaluating certain terms of the perturbation expansion to all orders. In the previous chapters we have treated the order-by-order approach in detail, and we shall now consider the alternative iterative approach.

The question which immediately arises here is whether an iterative procedure should be based on the original Bloch equation (15.1) or on its linked-diagram version (15.2). If all n-particle equations are considered with $n \leq N$, where N is the number of particles in the system, then, of course, both procedures are inherently exact and should yield the same results. This is not the case, however, if the set of equations is truncated at some point $n < N$. Then the two procedures lead to different results, so the question raised here might be of practical importance.

15.1.1 The Configuration-Interaction and the Linked-Diagram Procedures

In order to illustrate the relative importance of different classes of excitations, we shall consider the Be atom for which very accurate configuration-interaction (CI) calculations have been performed by *Bunge* [1968, 1976a,b]. The results, related to the Hartree-Fock model as the zeroth-order approximation, are given in Table 15.1.

Table 15.1. Relative contributions to the correlation energy for the Be atom derived from [*Bunge* 1976b]. Hartree-Fock orbitals are used

Single excitations	0.35%
Double excitations	95.2%
Triple excitations	1.1%
Quadruple excitations	3.4%

From this table several interesting observations can be made, which seem to be of general validity, at least for closed-shell systems. First of all, it is seen that double excitations dominate very heavily, when Hartree-Fock orbitals are used. Secondly, we see that quadruples have a larger effect than triples, both being more important than single excitations.

Qualitatively, we can understand these results by considering the order-by-order expansion of the energy. For a closed-shell system we know from the treatment in Chap. 12 that the second- and third-order energy contributions contain only double excitations in the intermediate states when Hartree-Fock orbitals are used (see Figs. 12.4 and 12.5). Single, triple and quadruple excitations appear first in fourth order. This explains why double excitations dominate so heavily in the correlation energy and why single, triple and quadruple excitations give contributions of comparable size.

As we have mentioned in Chap. 13, it is not always possible to choose the potential for open-shell systems in such a way that the effective potential (13.7) vanishes. Therefore, single excitations can be important for open-shell systems, since they may appear already in the first-order wave operator.

The calculations by *Bunge* on Be mentioned above are complete in the sense that single, double, triple and quadruple excitations are considered for this four-electron system. This means that the entire configuration space is included in the variational procedure, and the accuracy depends exclusively on the "completeness" of the single-particle functions. Of course, complete calculations of this kind can be performed only on very small systems. For larger systems it is necessary to truncate the procedure at some point. Owing to recent progress in computer technology and program development, it is now possible to treat single and double excitations in a very complete way for quite large systems (see, for instance, [*Roos* and *Siegbahn* 1977, *Siegbahn* 1980], but to go beyond that point in any systematic way is not feasible at the present time.

The Bloch equation (15.1) is equivalent to the Schrödinger equation. Therefore, solving the Bloch equation self-consistently with single, double, ... n-fold excitations in the wave operator is equivalent to solving the Schrödinger equation in the corresponding configuration space. This is also identical to a CI calculation, where the expectation value of the energy is minimized within such a space. If instead of the Bloch equation we treat the linked-diagram equation (15.2) in the same way, however, we shall get a different result for the following reason. In the linked-diagram expansion (LDE) unlinked diagrams of $\chi P V \Omega$ are cancelled by unlinked diagrams of $Q V \Omega$. We can see from the treatment of closed-shell systems in Chap. 12, however, that quadruply excited terms of Ω in $Q V \Omega$ are required in order to cancel unlinked diagrams of $\chi P V \Omega$ with double excitations in the χ and Ω operators. Therefore, if the Bloch equation—or the Schrödinger equation—is truncated after double excitations, all unlinked diagrams with double excitations will *not* be cancelled. Hence, CI with singles and doubles is not identical to LDE restricted to singles and doubles in final and intermediate states.

We have mentioned previously that unlinked diagrams are "unphysical" in the sense that their energy contribution increases nonlinearly with the size of the system. An unlinked energy diagram for a closed-shell system can be separated into two or more diagram parts, each of which is an allowed (linked) energy diagram. In each such part there is an independent summation over *all* electrons of the system. For a system of noninteracting atoms this would lead to "cross-terms", where different parts of the diagram refer to different atoms. Obviously, such nonlinear terms have no physical relevance: and, in particular, they can cause a considerable error in calculations of quantities like dissociation energies. Such nonlinear effects are retained in truncated CI, while they are eliminated in LDE. Of course, CI with singles and doubles is exact for a two-electron system, while the corresponding LDE is not. In order to obtain the exact result with LDE for such a system, it is necessary to include EPV diagrams of three- and four-body type, and we shall see that they can easily be included in an iterative scheme. Then the procedure combines the advantages of truncated CI and LDE of being exact for two-electron systems and free from nonlinear energy contributions. We have then answered the question raised at the beginning of this chapter, and found that an iterative procedure should be based on the linked-diagram formula (15.2) rather than on the Bloch equation (15.1).

15.1.2 The Separability Condition and the Coupled-Cluster Approach

A procedure which is free from nonlinear energy contributions has the important property of yielding the correct separation energy, when a system of interacting subsystem is separated into noninteracting ones. Such a procedure is said to fulfill the *separability condition* for the energy [*Primas* 1965; *Kutzelnigg* 1977] or to be *"size-consistent"* [*Pople* et al. 1976]. From the discussion above it follows that truncated LDE has this property—at least for closed-shell systems—while truncated CI has not. (For a system separating into open-shell fragments, the situation is more complex as discussed by *Bartlett* and *Purvis* [1978, 1980].)

So far we have only considered the separation *energy*. It turns out, however, that a proper energy separation does not necessarily imply that the *wave function* separates in a correct way when the system is separated into its component parts. We have seen above that pair excitations dominate heavily in the wave function (at least for closed-shell systems). If we consider a system of noninteracting subsystems, then the wave function would separate correctly (within the pair approximation) only if it contains *quadruple* excitations, corresponding to simultaneous pair excitations on different subsystems [*Primas* 1965]. Such excitations are represented by the *disconnected* diagram, given in Fig. 12.3m, which gives rise to the linked diagrams shown in Fig. 12.10. Such excitations would not be included in LDE truncated after doubles, since there is a quadruple excitation in the intermediate state. It has been found that such "independent" pair excitations represent the most important quadruple excitations also for interacting systems [*Sinanoğlu* 1962, 1964]. By a slight modification of LDE it is possible to include such effects in the pair approximation, and this leads to the $\exp(S)$ *formalism* or the *coupled-cluster approach*. Such a procedure will then be more accurate than ordinary LDE—truncated at the same point—and it would have the important property that the wave function—as well as the energy—separates correctly when a system dissociates into smaller fragments.

The $\exp(S)$ formalism was introduced long ago in nuclear physics [*Hubbard* 1957, 1958; *Bloch* 1958b; *Coester* 1958; *Coester* and *Kümmel* 1960, *Kümmel* 1962] following a well-known procedure in statistical mechanics [*Mayer* and *Mayer* 1977]. (For a recent review of nuclear applications, see [*Kümmel* et al. 1978].) Later a similar formalism was introduced into quantum chemistry by

Čížek [1966, 1969] under the name "coupled-pair many-electron theory" (CP MET). In recent years this formalism has frequently been used in this field, and it is commonly known as the coupled-cluster approach.

In the present chapter we shall first develop an iterative procedure based on the linked-diagram formula (15.2). Carried to self-consistency, this procedure is equivalent to evaluating linked one-, two-, ... body diagrams to all orders. In the second part of the chapter we shall then consider the coupled-cluster approach in a form valid for closed- as well as open-shell systems. Finally, we shall compare different pair-correlation approaches on a few atomic systems. Our comparison is mainly restricted to closed-shell systems, since so far only very limited calculations of this kind have been performed for open-shell systems.

15.2 Hierarchy of *n*-Particle Equations

15.2.1 General

Instead of expanding the wave operator order by order as in the previous chapters, we shall now separate this operator into one-, two-, ... body parts

$$\Omega = 1 + \chi = 1 + \Omega_1 + \Omega_2 + \Omega_3 + \cdots. \tag{15.3}$$

It should be observed that we have some freedom in making this separation for open-shell systems, which could be of importance in practical applications. For instance, a diagram like

$$(15.4)$$

represents a single excitation out of the model space and can therefore be regarded as a single-particle (one-body) effect, as in the treatment of the polarization effect in Sect. 14.3. Since the diagram has two pairs of free lines, on the other hand, we may also treat it as a two-body effect. Similarly, diagrams representing a double excitation out of the model space with an unexcited valence line—as type 3 in Fig. 13.7—can be included in the two-body effect, although three electrons are involved. In this way one may avoid considering three-body effects in general, which may otherwise be necessary. In the present formal treatment, however, we shall for simplicity assume that the *n*-body part, Ω_n, is defined as the part of Ω which is represented by diagrams with *n* pairs of free lines in the order-by-order expansion we have developed in the previous chapters. Modifications of this separation can then easily be made in the actual calculation.

With the definition we have adopted here we can illustrate the separation (15.3) as in Fig. 15.1, which is the open-shell analogue of the representation in Fig. 11.7. The lines at the bottom represent incoming lines and the lines at the top outgoing lines. We have left out the arrows from the graphs in order to indicate that the incoming lines can be of the core or valence type. As usual, the outgoing lines can represent virtual and valence orbital lines (see Fig. 13.1).

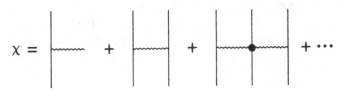

Fig. 15.1. Graphical separation of the correlation operator into one-, two-, ⋯ body terms in the open-shell case. The lines at the bottom represent incoming lines, which can be of core or valence type. The lines at the top represent outgoing lines, which can be of virtual or valence type. All free lines cannot be valence lines. This is a generalization of the closed-shell representation in Fig. 11.7

By considering the n-body parts of the the left- and right-hand sides of the linked-diagram formula (15.2), we get

$$
\begin{cases}
[\Omega_n, H_0] = (Q V \Omega - \chi W)_{n,\text{linked}} \\
W = (V\Omega)_{\text{closed}}
\end{cases}
\tag{15.5}
$$

The operator W appearing here is apart from $P H_0 P$ identical to the effective Hamiltonian (9.82), and we refer to it as the *effective interaction*. It can be separated into zero-, one-, two-, ... body parts in analogy with (15.3)

$$
W = W_0 + W_1 + W_2 + \cdots \\
W_n = (V\Omega)_{n,\text{closed}}
\tag{15.6}
$$

depending on the number of free valence-line pairs. This is represented schematically in Fig. 15.2. Here the diagrams are closed, which means that all free lines are valence lines.

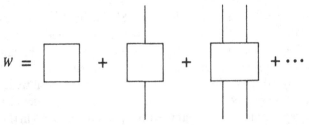

Fig. 15.2. Graphical separation of the effective interaction W into zero-, one-, two-, ⋯ body terms. All free lines are valence lines

The n-particle equations of the wave operator and of the effective interaction are coupled together and have to be solved iteratively. Solving these equations self-consistently is equivalent to evaluating the corresponding diagrams to all orders. The sequence of equations has to be terminated at some point, of course. As mentioned in Sect. 15.1, double excitations dominate in the correlation energy, but we have argued that single excitations can also be of importance for open-shell systems. Single excitations of the type (15.4) can be included in the two-body part, while effects of the type considered in Fig. 14.12 can not. Therefore, it might be a reasonable compromise between accuracy and complexity for open-shell systems to keep the one- and two-body parts of the wave operator but omit the remaining parts. As we have pointed out before, such a calculation is *not* equivalent to a CI calculation, for which the corresponding excitations are considered. The procedure proposed here, being based on the linked-diagram formalism, includes important effects which in the CI procedure require triple and quadruple excitations. In Sect. 15.3 we shall extend this procedure further by means of the exponential ansatz, leading to the coupled-cluster approach or the exp(S) formalism, where additional classes of quadruple excitations are included.

15.2.2 The All-Order Single-Particle Equation

We shall now restrict ourselves to one- and two-particle effects, which means that the wave operator and the effective interaction are approximated by

$$\Omega = 1 + \Omega_1 + \Omega_2$$
$$W = W_1 + W_2 . \tag{15.7}$$

With the definition we use here this is represented by the diagrams in Figs. 15.1 and 15.2 with one or two pairs of free lines.

The single-particle equation is obtained in this approximation by substituting (15.7) into the general equations (15.5, 6), setting $n = 1$,

$$[\Omega_1, H_0] = Q(V + V\Omega_1 + V\Omega_2 - \Omega_1 W_1)_{1,\text{linked}} \tag{15.8a}$$
$$W_1 = (V + V\Omega_1 + V\Omega_2)_{1,\text{closed}} . \tag{15.8b}$$

The graphical form of these equations is obtained directly by means of the standard procedure, based on Wick's theorem. Diagrams of the first three terms on the right-hand side of (15.8a) are shown in Fig. 15.3. The last term represents folded diagrams. We recall that folded diagrams are obtained by joining the lines at the bottom of the wave operator with lines at the top of the effective interaction (see Fig. 13.12). Therefore, the folded diagram has at least as many free lines at the bottom as the effective-interaction diagram. Hence, only W_1 can contribute in the single-particle equation. This diagram is shown in Fig. 15.4. By comparing (15.8a) and (15.8b) we find that the diagrams of W_1 are quite analogous to those of Ω_1, the only difference being that all free lines are valence lines and that there are no folded diagrams.

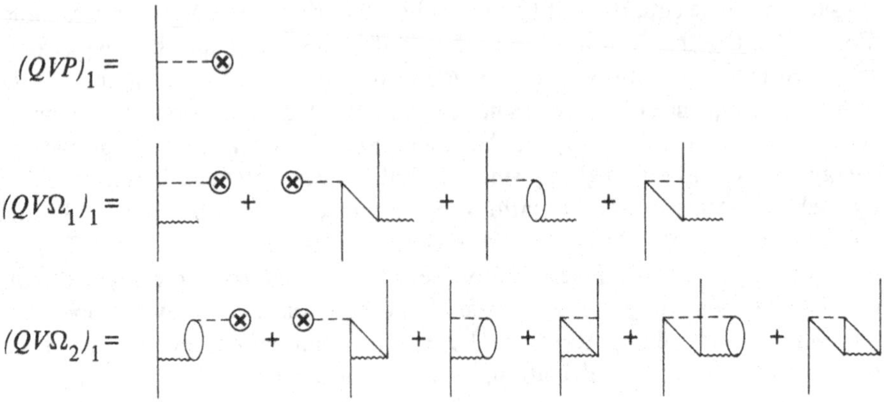

$(QVP)_1 =$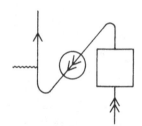

$(QV\Omega_1)_1 =$

$(QV\Omega_2)_1 =$

Fig. 15.3. Diagrams corresponding to the first three terms in the single-particle equation (15.8) are obtained in the standard way by means of the representation of the correlation operator in Fig. 15.1 and the graphical form of Wick's theorem

Fig. 15.4. Only one type of folded diagram can appear in the single-particle equation, namely by combining the diagrams of Ω_1 and W_1

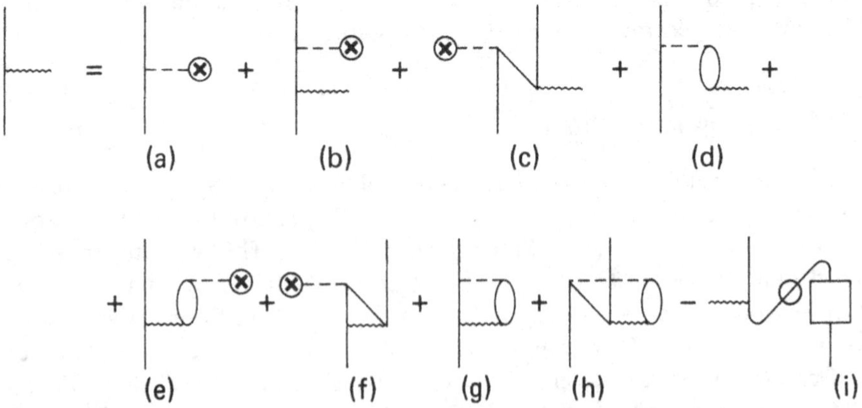

(a)   (b)   (c)   (d)

(e)   (f)   (g)   (h)   (i)

Fig. 15.5 a-i. Graphical form of the all-order single particle equation (15.8a). Exchange variants of the Coulomb interaction are omitted. The line at the bottom represents an incoming core or valence line. The folded diagram (i) appears only if the incoming line is a valence line. It should be observed that this diagram appears with a minus sign in the equation

We can now give the complete single-particle equation in the approximation (15.7) as shown in Fig. 15.5. For simplicity, exchange variants of the Coulomb interaction are omitted. The same equation holds also if the incoming line at the bottom represents a core orbital, with the exception that there can then be no folded diagram.

The algebraic form of the single-particle equation (15.8a) is obtained by evaluating the diagrams in Fig. 15.5 essentially as for the order-by-order procedure. The Goldstone rules, given in Sect. 12.4, are still valid with rather obvious generalizations.

We begin by expressing the wave operator in Fig. 15.1 in second quantized form in analogy with the closed-shell case (11.64)

$$= a_i^\dagger a_j x_j^i \tag{15.9a}$$

$$= \frac{1}{2} a_i^\dagger a_j^\dagger a_l a_k \, x_{kl}^{ij} \tag{15.9b}$$

$$= \frac{1}{3!} a_i^\dagger a_j^\dagger a_k^\dagger a_n a_m a_l \, x_{lmn}^{ijk} \tag{15.9c}$$

and so on. As mentioned before, the lines at the bottom represent incoming core or valence lines and the lines at the top outgoing virtual or valence lines. We assume that the coefficients have the symmetry property

$$x_{kl}^{ij} = x_{lk}^{ji}$$
$$x_{lmn}^{ijk} = x_{mnl}^{jki} = x_{nlm}^{kij} = x_{nml}^{kji} = \cdots . \tag{15.10}$$

which means that all vertices are equivalent. This is in analogy with the definition of the first-order coefficients (12.12) and with the usual convention of expressing the operators in second-quantized form. In corresponding molecular works (see, for instance, [*Paldus* and *Čížek* 1975]) the coefficients are often expressed in antisymmetric form, so that

$$x_{kl}^{ij} = - x_{kl}^{ji} \tag{15.11}$$

and so on. We shall find, however, that the nonantisymmetrized form is more convenient when the graphical technique is used to evaluate the angular part of the diagrams.

The diagrams of W can be expressed in a similar way

$$= a_m^\dagger a_n w_n^m \tag{15.12a}$$

$$= \frac{1}{2} a_m^\dagger a_n^\dagger a_q a_p \; w_{pq}^{mn} \; . \tag{15.12b}$$

If all incoming lines of a wave-operator diagram are not of the same kind, then several equivalent diagrams can be formed, such as

$$\tag{15.13a}$$

and

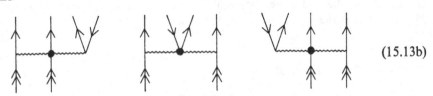

$$\tag{15.13b}$$

When only distinct diagrams are considered, we shall then replace the diagrams shown above by

$$= a_r^\dagger a_s^\dagger a_a a_m \; x_{ma}^{rs} \tag{15.14a}$$

$$= \frac{1}{2} a_r^\dagger a_s^\dagger a_t^\dagger a_a a_n a_m \, x_{mna}^{rst} \, .$$

(15.14b)

The first diagram represents two and the second diagram three equivalent diagrams, which together with the original factors in (15.9) give the weight factors above. These factors are still consistent with rule f') given in Chap. 12, since in (15.14b) two vertices are interchangeable, while in (15.14a) no such exchange can be made.

In principle, we can have more than two interchangeable vertices in this procedure, as in (15.9c), and this leads to the following generalization of rule f') for the weight factors given before:

> f'') *There is a factor of* $(n!)^{-1}$ *for each set of n vertices or n groups of vertices on the same horizontal level which are interchangeable in the sense that any permutation of them leads to the same diagram or to any other diagram appearing in the expansion.*

The interpretation of this rule will become clearer as we proceed with the applications.

As an illustration of the evaluation of the diagrams in the single-particle equation in Fig. 15.5 we consider diagrams (c) and (g) and the folded diagram (i)

$$= -a_r^\dagger a_m \sum_c \langle c|v|m\rangle x_c^r$$

(15.15c)

$$= a_r^\dagger a_m \sum_{cst} \langle rc|r_{12}^{-1}|st\rangle x_{mc}^{st}$$

(15.15g)

$$= a_r^\dagger a_m \sum_n w_m^n x_n^r \, .$$

(15.15i)

No diagrams in the equation have any interchangeable vertices, which means that all weight factors are equal to one. It should be observed that the folded diagram (i) appears with a minus sign in the equation.

The exchange variants of the diagrams containing the two-body Coulomb interaction can be included by replacing the Coulomb interaction in the matrix elements with the Coulomb interaction with exchange

$$\langle ij|R|kl\rangle = \langle ij|r_{12}^{-1}|kl\rangle - \langle ji|r_{12}^{-1}|kl\rangle. \tag{15.16}$$

As mentioned, the diagrams of W_1 are quite analogous to those of Ω_1, the only difference being that all free lines of W_1 are valence lines and that there are no folded diagrams. The evaluation is then performed exactly as for the wave operator, for instance

$$= -a_n^\dagger a_m \sum_c \langle c|v|m\rangle x_c^n \tag{15.17c}$$

$$= a_n^\dagger a_m \sum_{csr} \langle nc|r_{12}^{-1}|rs\rangle x_{mc}^{rs}. \tag{15.17g}$$

By evaluating all diagrams in Fig. 15.5 in this way, we obtain the following single-particle equation in the approximation (15.7):

$$(\varepsilon_m - \varepsilon_r)x_m^r = \underset{\text{(a)}}{\langle r|v|m\rangle} + \underset{\text{(b)}}{\langle r|v|s\rangle x_m^s} - \underset{\text{(c)}}{\langle c|v|m\rangle x_c^r} + \underset{\text{(d)}}{\langle rc|R|ms\rangle x_c^s}$$

$$+ \underset{\text{(e)}}{\langle c|v|s\rangle x_{mc}^{rs}} - \underset{\text{(f)}}{\langle c|v|s\rangle x_{cm}^{rs}} + \underset{\text{(g)}}{\langle rc|R|st\rangle x_{mc}^{st}} - \underset{\text{(h)}}{\langle cd|R|ms\rangle x_{cd}^{rs}} - \underset{\text{(i)}}{w_m^n x_n^r} \tag{15.18}$$

$$w_m^n = \langle n|v|m\rangle + \langle n|v|s\rangle x_m^s - \langle c|v|m\rangle x_c^n + \langle nc|R|ms\rangle x_c^s$$

$$+ \langle c|v|s\rangle x_{mc}^{ns} - \langle c|v|s\rangle x_{cm}^{ns} + \langle nc|R|st\rangle x_{mc}^{st} - \langle cd|R|ms\rangle x_{cd}^{ns}. \tag{15.19}$$

On the right-hand sides of these equations summations are performed over all indices occurring twice.

As we have mentioned before (Sect. 12.2.3), single excitations vanish if Brueckner orbitals are used. Therefore, the solutions of the single-particle equation can be used to construct approximate orbitals of this kind. This was demonstrated in connection with the hyperfine calculations described in the previous chapter.

15.2.3 The All-Order Pair Equation

The two-particle or pair equation in the approximation (15.7) is obtained directly from the general equation (15.5) in the same way as the single-particle equation (15.8)

$$[\Omega_2, H_0] = Q(V + V\Omega_1 + V\Omega_2 - \Omega_1 W_2 - \Omega_2 W_1 - \Omega_2 W_2)_{2,\text{linked}} \quad (15.20\text{a})$$

$$W_2 = (V + V\Omega_1 + V\Omega_2)_{2,\text{closed}} . \quad (15.20\text{b})$$

The construction of the diagrams of this equation is to a large extent analogous to the construction of the second-order diagrams discussed in Chap.13. The result is shown in Fig. 15.6. Folded diagrams can be formed in three different ways, corresponding to the three last terms in (15.20a).

Fig. 15.6 a-m. Graphical form of the all-order pair equation (15.20a). Exchange variants of the Coulomb interaction are omitted. The lines at the bottom represent incoming core or valence lines. The folded diagram (l) appears only if at least one of the lines at the bottom is a valence line and diagrams (k) and (m) appear only if both lines at the bottom are valence lines

The one-body diagrams of W have been discussed before, and the two-body diagrams are obtained in exactly the same way from those of Ω_2 by removing the folded diagrams and replacing all free lines by open-shell lines. Again, we see the analogy with the second-order diagrams (Fig. 13.7).

As an illustration of the evaluation of the diagrams of the pair equation, we consider diagrams (c, g, m)

$$= a_r^\dagger a_s^\dagger a_n a_m \sum_t \langle rs | r_{12}^{-1} | tn \rangle \, x_m^t \qquad (15.21c)$$

$$= \frac{1}{2} \, a_r^\dagger a_s^\dagger a_n a_m \sum_{tu} \langle rs | r_{12}^{-1} | tu \rangle x_{mn}^{tu} \qquad (15.21g)$$

$$= \frac{1}{2} \, a_r^\dagger a_s^\dagger a_n a_m \sum_{pq} w_{mn}^{pq} x_{pq}^{rs} \, . \qquad (15.21m)$$

In the last two diagrams the vertices associated with two orbital paths can be interchanged, which gives the weight of $1/2$ according to rule f″) above. In the first diagrams, on the other hand, no such interchange can be made. Therefore, if only distinct diagrams are considered, this diagram will have the weight of one. As mentioned before, the negative sign associated with the folded diagrams appears explicitly in the equations.

We have assumed that the coefficients of the wave operator are symmetrical with respect to permutations of the vertices (15.10). In order for the solution of the pair equation to fulfill this requirement, we have to symmetrize the right-hand side. This means that diagram (c) above should be replaced by

$$(15.22)$$

Then, of course, each diagram gets the weight of 1/2. This weight is in agreement with rule f″) given above, since an exchange of the vertices associated with the two paths (that is, a reflection in a vertical plane) transforms the diagrams into each other.

By evaluating all diagrams in Fig. 15.6 after this kind of symmetrization we get the following pair equation in the approximation (15.7):

$$
\begin{aligned}
(\varepsilon_m + \varepsilon_n - \varepsilon_r - \varepsilon_s)x_{mn}^{rs} &= \underset{\text{(a)}}{\langle rs|r_{12}^{-1}|mn\rangle} + \underset{\text{(b)}}{\langle s|v|n\rangle x_m^r} + \underset{\text{(b')}}{\langle r|v|m\rangle x_n^s} \\
&+ \underset{\text{(c)}}{\langle rs|r_{12}^{-1}|tn\rangle x_m^t} + \underset{\text{(c')}}{\langle rs|r_{12}^{-1}|mt\rangle x_n^t} - \underset{\text{(d)}}{\langle cs|r_{12}^{-1}|mn\rangle x_c^r} - \underset{\text{(d')}}{\langle rc|r_{12}^{-1}|mn\rangle x_c^s} \\
&+ \underset{\text{(e)}}{\langle r|v|t\rangle x_{mn}^{ts}} + \underset{\text{(e')}}{\langle s|v|t\rangle x_{mn}^{rt}} - \underset{\text{(f)}}{\langle c|v|m\rangle x_{cn}^{rs}} - \underset{\text{(f')}}{\langle c|v|n\rangle x_{mc}^{rs}} \\
&+ \underset{\text{(g)}}{\langle rs|r_{12}^{-1}|tu\rangle x_{mn}^{tu}} + \underset{\text{(h)}}{\langle rc|R|mt\rangle x_{cn}^{ts}} + \underset{\text{(h')}}{\langle cs|R|tn\rangle x_{mc}^{rt}} - \underset{\text{(i)}}{\langle rc|R|tn\rangle x_{mc}^{ts}} \\
&- \underset{\text{(i')}}{\langle cs|R|mt\rangle x_{cn}^{rt}} + \underset{\text{(j)}}{\langle cd|r_{12}^{-1}|mn\rangle x_{cd}^{rs}} - \underset{\text{(k)}}{w_{mn}^{ps}x_p^r} - \underset{\text{(k')}}{w_{nm}^{pr}x_p^s} \\
&- \underset{\text{(l)}}{w_n^p x_{mp}^{rs}} - \underset{\text{(l')}}{w_m^p x_{np}^{sr}} - \underset{\text{(m)}}{w_{mn}^{pq} x_{pq}^{rs}}.
\end{aligned}
$$

$$(15.23)$$

As before, a summation is performed over all indices on the right-hand side occurring twice.

An analogous equation for the w_{mn}^{pq} coefficients is obtained by removing the energy factor to the left as well as the folded terms to the right and replacing the virtual orbitals r, s by the valence orbitals p, q.

15.2.4 The All-Order Radial Equations

For atomic systems the diagrams appearing in the all-order equations can be separated into radial and spin-angular parts in the same way as the ordinary Goldstone diagrams discussed in the previous chapters. In order to obtain the general structure of these diagrams, we shall see how they are obtained in the order-by-order expansion.

The diagrams of the single-particle equation have a single pair of free orbital lines, like the second-order diagrams (12.63, 64) evaluated before. The angular-momentum graph corresponding to such a diagram can always be reduced to a single line. This is a consequence of theorem JLV2 (4.6), saying that if an angular-momentum graph has only two free lines, these must have the same quantum numbers [see also (13.52)]. Therefore, all excitations in the single-particle equation must occur between states with the same l, m_l and m_s. This means that the single-particle interaction on the left-hand side of the equation has the same spin-angular structure as the interaction with a central potential. We can then replace the one-body coefficients of the correlation operator (15.9a) by

$$x_j^i = \delta(l_i,\, l_j)\delta(m_l^i,\, m_l^j)\delta(m_s^i,\, m_s^j)\bar{x}_j^i = \bar{x}_j^i \qquad \begin{array}{c} l_i \\ \\ l_j \end{array} \Bigg| \qquad \begin{array}{c} s_i \\ \\ s_j \end{array} \Bigg| \qquad (15.24)$$

where $\bar{x}_j^i$ is a pure *radial* factor. As before, we let l_i and s_i in the graph represent $l_i m_l^i$ and $s m_s^i$, respectively.

As an illustration we consider the diagram (15.15c). The corresponding angular-momentum diagram is obtained by removing the potential interaction and the single-particle interaction at the bottom, which reduces the graph directly to a single line

$$= -a_r^\dagger a_m \langle a|v|m\rangle_{\mathrm{rad}}\, \bar{x}_a^r \qquad \begin{array}{c} l_r \\ \\ l_m \end{array} \Bigg| \qquad \begin{array}{c} s_r \\ \\ s_m \end{array} \Bigg| \qquad (15.25)$$

Here $\langle a|v|m\rangle_{\mathrm{rad}}$ is the radial part of the matrix element of the effective potential defined before (12.60).

Obviously the one-body diagrams of W have the same spin-angular structure as those of Ω, so we can make a similar separation there. This means that the folded diagrams (15.15i) can be reduced to

$$= a_r^\dagger a_m \sum_n w_m^n \bar{x}_n^r \qquad \begin{array}{c} l_r \\ \\ l_m \end{array} \Bigg| \qquad \begin{array}{c} s_r \\ \\ s_m \end{array} \Bigg| \qquad (15.26)$$

In order to analyze the two-body diagrams of Ω and W, we consider again the order-by-order expansion. One diagram of this kind is evaluated in Chap. 12 (12.65) and two diagrams in Chap. 13 (13.62, 65). It can be shown that the angular-momentum graph of such a diagram can always be reduced to the graph of the ordinary Coulomb interaction (3.42). Therefore, we can make the following separation:

$$x_{mn}^{ij} = \sum_k \bar{x}_{mn}^{ij}(k) \qquad (15.27)$$

leaving out the spin part. Again, $\bar{x}_{mn}^{ij}$ is a pure *radial* factor.

As shown in Chap. 13 (13.65), the k value in (15.27) may take *any* value consistent with the triangular conditions, not only those appearing in the Coulomb interaction, where the vertex sums must be even. For that reason the vertex signs are important. (We recall that similar arguments led to the Trees parameters in treating the effective Coulomb interaction in Chap. 13.)

Applying this separation to the exchange variant of diagram (15.15 g) as an illustration, we get

$$= -a_r^\dagger a_m \sum_{ast} \langle ra | r_{12}^{-1} | st \rangle x_{am}^{st}$$

$$(15.28)$$

$$= -a_r^\dagger a_m \sum_{astk_1k_2} X^{k_1}(ra, st)\bar{x}_{am}^{st}(k_1)$$

All diagrams of the single-particle equation can be evaluated in the same way, and their angular-momentum graphs can always be reduced to a single line. The same line appears on the left-hand side after the separation (15.24) has been made. This line can then be removed, and we get an equation for the radial coefficients alone. The complete equation is rather lengthy, however, and we shall not reproduce it here.

The diagrams of the pair equation can be separated in a similar way, and we consider diagram (15.21 g) as an illustration,

$$= \frac{1}{2} a_r^\dagger a_s^\dagger a_n a_m \sum_{tu} \langle rs | r_{12}^{-1} | tu \rangle x_{mn}^{tu}$$

(15.29)

$$= \frac{1}{2} a_r^\dagger a_s^\dagger a_n a_m \sum_{tukk_1k_2} X^{k_2}(rs, tu) \tilde{x}_{mn}^{tu}(k_1) \, G$$

where G is the factor appearing in the evaluation of diagram (12.65). By evaluating all diagrams in this way, a radial pair equation is obtained. Again, this equation is too lengthy to be given here.

The all-order radial equations can be transformed into *differential equations* in a way similar to that of the first-order equations discussed in Chap. 13 and the all-order hyperfine equation in Chap. 14. This is done by introducing the one- and two-particle functions

$$\rho(m \to l_r; r_1) = \sum_{n_r}^{\text{part}} \tilde{x}_m^r P_r(r_1)$$

(15.30)

$$\rho(mn \to l_r l_s; r_1 r_2) = \sum_{n_r n_s}^{\text{part}} \tilde{x}_{mn}^{rs} P_r(r_1) P_s(r_2) .$$

The corresponding differential equations are obtained in the same way as before, namely by multiplying the radial equations by $P_r(r_1)$ and $P_r(r_1) P_s(r_2)$, respectively, and summing over the principal quantum numbers of the particle states. The infinite summations are then eliminated by means of the closure property (13.84). A complete pair equation of this kind, without single excitations, was given by Mårtensson [1978].

15.3 The Exponential Ansatz

In the preceeding section we have considered an iterative procedure based on the linked-diagram formula (15.2), where all linked one- and two-body diagrams are included in all orders. Such a procedure is usually capable of reproducing the correlation energy for light closed-shell systems to about 95%, as well as predicting properties like the hyperfine interaction with high accuracy. As discussed in Sect. 15.1, this procedure has the important advantage over to the corresponding CI procedure that it does not contain any non-linear (unlinked) energy contributions. Therefore it should be suitable also for heavier systems, where—in contrast to the corresponding CI procedure—it can be expected to give an accuracy comparable to that achieved for light systems, apart from relativistic effects.

As also discussed in Sect. 15.1, it is possible, however, to increase the accuracy further with only a minor modification of the procedure discussed so far, namely by also considering *independent* pair excitations represented by *disconnected* wave-operator diagrams. This leads to the coupled-cluster approach or the exp(S) formalism, where the wave operator is expressed in exponential form. In the present section we shall develop such a procedure, valid for closed- as well as open-shell systems, and we shall begin by considering the factorization of disconnected wave-operator diagrams.

15.3.1 Factorization of Closed-Shell Diagrams

In order to derive the exponential form of the wave operator, we shall consider the factorization of disconnected diagrams, and we choose the closed-shell diagrams in Fig. 12.2e as a first illustration. These diagrams can be factorized in the same way as the corresponding unlinked diagrams (Fig. 12.7). This is illustrated in Fig. 15.7 (top line). The diagrams to the left have the algebraic expressions

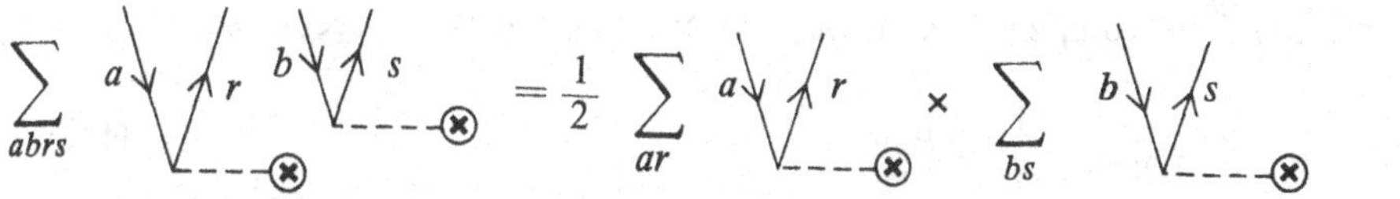

Fig. 15.7. Illustration of the factorization of disconnected wave-operator diagrams (top line). When the summaton is performed over all free lines independently, all diagrams appear twice, which explains the factor of 1/2 in the bottom line

$$a_r^\dagger a_s^\dagger a_b a_a \frac{\langle s|v|b\rangle\langle r|v|a\rangle}{(\varepsilon_a + \varepsilon_b - \varepsilon_r - \varepsilon_s)\,(\varepsilon_a - \varepsilon_r)}$$

$$a_r^\dagger a_s^\dagger a_b a_a \frac{\langle s|v|b\rangle\langle r|v|a\rangle}{(\varepsilon_a + \varepsilon_b - \varepsilon_r - \varepsilon_s)\,(\varepsilon_b - \varepsilon_s)}$$

$$\tag{15.31}$$

the sum of which factorizes to

$$a_r^\dagger a_s^\dagger a_b a_a \frac{\langle s|v|b\rangle\langle r|v|a\rangle}{(\varepsilon_b - \varepsilon_s)\,(\varepsilon_a - \varepsilon_r)}.$$

$$\tag{15.32}$$

The two disconnected parts, *considered as separate diagrams*, have the expressions

$$a_r^\dagger a_a \frac{\langle r|v|a\rangle}{\varepsilon_a - \varepsilon_r} \quad \text{and} \quad a_s^\dagger a_b \frac{\langle s|v|b\rangle}{\varepsilon_b - \varepsilon_s}$$

with the product

$$a_r^\dagger a_a a_s^\dagger a_b \frac{\langle r|v|a\rangle\langle s|v|b\rangle}{(\varepsilon_a - \varepsilon_r)(\varepsilon_b - \varepsilon_s)}. \tag{15.33}$$

The sum (15.32) is identical to this product, apart from the order of the operators. The operators in (15.32) are normal ordered, while those in (15.33) are not. However, the operators appearing here anticommute, so the normal ordering does not lead to any contraction in the present closed-shell case, and we see that the phases of (15.32, 33) are also the same. This verifies the factorization as it is expressed in the top line of Fig. 15.7. If we let "*abrs*" run over all possible combinations, then we see that all diagrams will appear twice on the left-hand side. Therefore, the sum of all diagrams is equal to one-half of the product of the disconnected pieces summed independently, as illustrated in the bottom line of Fig. 15.7. A similar factorization can be made for disconnected second-order diagrams containing two-body parts. All disconnected second-order wave-operator diagrams can then be expressed

$$\Omega^{(2)}_{\text{disconn}} = \frac{1}{2}(\Omega^{(1)})^2, \tag{15.34}$$

where $\Omega^{(1)}$ represents the sum of all first-order diagrams (which are all connected). The complete second-order wave operator then becomes

$$\Omega^{(2)} = \Omega^{(2)}_{\text{conn}} + \frac{1}{2}(\Omega^{(1)}_{\text{conn}})^2, \tag{15.35}$$

where $\Omega^{(2)}_{\text{conn}}$ represents the sum of all connected diagrams in second order.

15.3.2 Factorization of Open-Shell Diagrams

In the open-shell case we have to observe that disconnected parts of a wave-operator diagram do not necessarily commute, and hence the factorization demonstrated above has to be somewhat modified. As an illustration, we consider the following diagram, regarded as a *single* diagram:

$$= a_r^\dagger a_n^\dagger a_s^\dagger a_q a_p a_m \frac{M_1 M_2}{D_1(D_1 + D_2)}$$

and the same diagram where the order between the two interactions is reversed

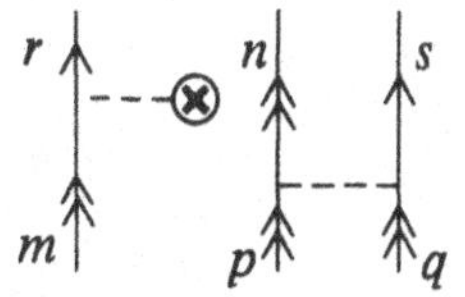

$$\begin{array}{c} \text{(diagram)} \end{array} = a_r^\dagger a_n^\dagger a_s^\dagger a_q a_p a_m \frac{M_1 M_2}{D_2(D_1 + D_2)}.$$

Here, M_1 and M_2 represent the two matrix elements and D_1 and D_2 are the energy expressions

$$D_1 = \varepsilon_m - \varepsilon_r, \qquad D_2 = \varepsilon_p + \varepsilon_q - \varepsilon_n - \varepsilon_s.$$

The sum of the two diagrams factorize as before, and it is equal to

$$a_r^\dagger a_n^\dagger a_s^\dagger a_q a_p a_m \frac{M_1 M_2}{D_1 D_2}. \tag{15.36}$$

On the other hand, the product of the two parts, regarded as separate diagrams, is

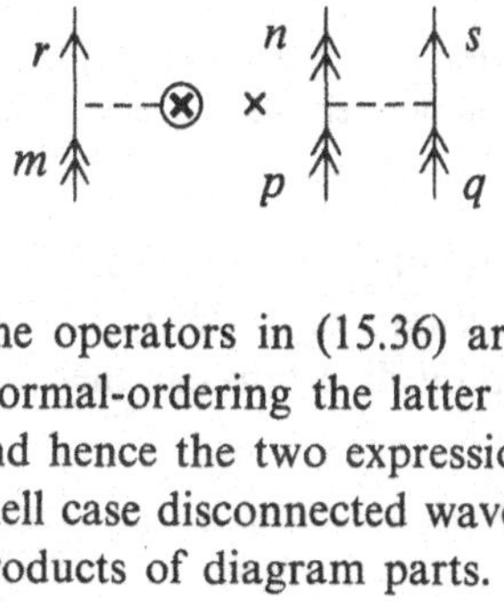

$$\begin{array}{c} \text{(diagram)} \end{array} = (a_r^\dagger a_m)(a_n^\dagger a_s^\dagger a_q a_p) \frac{M_1 M_2}{D_1 D_2}. \tag{15.37}$$

The operators in (15.36) are in normal form, while those in (15.37) are not. Normal-ordering the latter leads to additional terms due to the contractions, and hence the two expressions are *not* identical. This means that in the open-shell case disconnected wave-operator diagrams do not factorize into ordinary products of diagram parts. Instead we find that

   disconnected diagrams factorize into the normal product of the separate diagram parts.

This rule is illustrated in Fig. 15.8 for the example studied here.

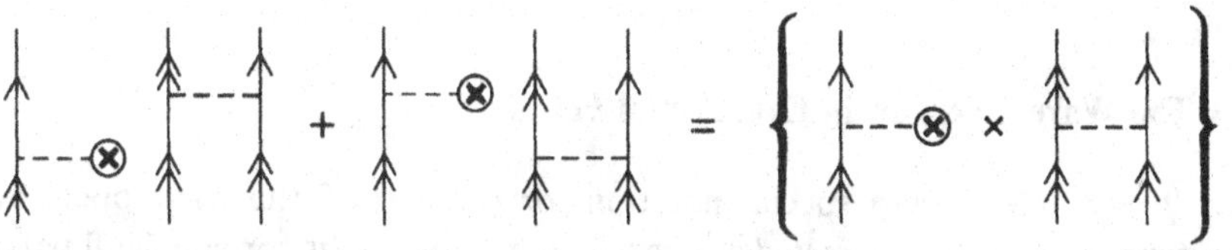

Fig. 15.8. Illustration of the factorization of disconnected diagrams in the open-shell case. The curly brackets denote normal form (without contractions) in the same sense as in Wick's theorem (11.57)

Considering all possible disconnected diagrams of $\Omega^{(2)}$ in the open-shell case we find that they can be expressed in analogy with (15.34) by means of the corresponding *normal* product

$$\Omega^{(2)}_{\text{disconn}} = \frac{1}{2} \{(\Omega^{(1)})^2\} . \tag{15.38}$$

It is not difficult to show that a similar factorization can be performed in higher orders. As an illustration we consider some third-order diagrams in Fig. 15.9, and it is left as an exercise to verify the factorization in this case (Problem 15.1).

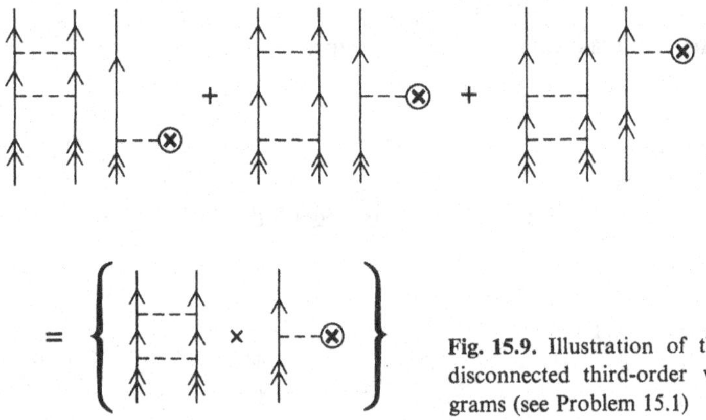

Fig. 15.9. Illustration of the factorization of disconnected third-order wave-operator diagrams (see Problem 15.1)

Disconnected third-order wave-operator diagrams consist either of one first-order and one (connected) second-order piece or of three first-order pieces. All possible combinations and orderings will appear, and the diagrams can therefore be factorized in the following way:

$$\Omega^{(3)}_{\text{disconn}} = \{\Omega^{(2)}_{\text{conn}}\Omega^{(1)}\} + \frac{1}{3!} \{(\Omega^{(1)})^3\} . \tag{15.39}$$

The factor of $1/3!$ in the last term is due to the fact that there are three pieces of the same kind. Here any permutation leads to equivalent diagrams, and hence each distinct diagram will appear $3!$ times in the factorized term. This is in agreement with rule f'') given above.

15.3.3 The Wave Operator in Exponential Form

We shall now introduce a special notation for the part of the wave operator which is represented by *connected* diagrams, that is, diagrams for which all parts are connected to each other either by orbital lines or by interaction lines. This part of the wave operator will be denoted by S with the order-by-order expansion

$$S = S^{(1)} + S^{(2)} + S^{(3)} + \cdots \tag{15.40}$$

The terms in the expansion are defined by

$$S^{(n)} = \Omega^{(n)}_{\text{conn}} \tag{15.41}$$

and symbolically we can write the S operator

$$\boxed{S = \Omega_{\text{conn}}} \tag{15.42}$$

after summing to all orders. The diagrams of S will in the following be referred to as diagram *clusters*, and the S operator as the *cluster operator*.

The factorization demonstrated in the previous subsection can now be expressed

$$\begin{cases} \Omega^{(1)} = S^{(1)} \\[2mm] \Omega^{(2)} = S^{(2)} + \dfrac{1}{2}\,\{(S^{(1)})^2\} \\[2mm] \Omega^{(3)} = S^{(3)} + \{S^{(2)}S^{(1)}\} + \dfrac{1}{3!}\,\{(S^{(1)})^3\} \\[2mm] \cdots\cdots\cdots\cdots\cdots \end{cases} \tag{15.43}$$

We can then see that this expansion is in fact the order-by-order expansion of the operator

$$\{\exp(S)\} = 1 + S + \frac{1}{2}\{S^2\} + \frac{1}{3!}\{S^3\} + \cdots = \sum_{n=0}^{\infty} \frac{1}{n!}\{S^n\}, \tag{15.44}$$

which can be verified by substituting the expansion (15.40) into this expression and identifying terms order by order. The interpretation of this result is that if S represents all *connected* wave-operator diagrams, then (15.44) represents the complete wave operator, that is, connected as well as disconnected wave-operator diagrams,

$$\boxed{\Omega = \{\exp(S)\}}. \tag{15.45}$$

This is a generalization of the ordinary exponential form of *Coester, Kümmel* and others, mentioned before,

$$\Omega = \exp(S) = 1 + S + \frac{1}{2}S^2 + \frac{1}{3!}S^3 + \cdots, \tag{15.46}$$

the only difference being that *normal* instead of ordinary products are used.

Problem 15.1. Verify the factorization illustrated in Fig. 15.9.

15.4 The Coupled-Cluster Equations

15.4.1 General

Instead of an order-by-order expansion, we now separate the cluster operator S (15.40) into one-, two-, ... body parts, in the same way as we have separated the wave operator (15.3)

$$S = S_1 + S_2 + S_3 + \cdots, \tag{15.47}$$

Graphically, S_n is represented by all connected wave-operator diagrams of n-body type, that is

$$S_n = (\Omega_n)_{\text{conn}} . \tag{15.48}$$

It should be observed that the freedom in defining Ω_n, discussed above, obviously is transferred to S_n.

Substituting the expansion (15.47) into the exponential form of the wave operator (15.44, 45) and identifying the one-, two-, ... body terms, we get in analogy with the order-by-order expansion (15.43) the following identities:

$$\begin{cases} \Omega_1 = S_1 \\[2mm] \Omega_2 = S_2 + \frac{1}{2}\{S_1^2\} \\[2mm] \Omega_3 = S_3 + \{S_2 S_1\} + \frac{1}{3!}\{S_1^3\} \\[2mm] \Omega_4 = S_4 + \{S_3 S_1\} + \frac{1}{2}\{S_2^2\} + \frac{1}{2}\{S_2 S_1^2\} + \frac{1}{4!}\{S_1^4\} \\[2mm] \cdots\cdots\cdots\cdots\cdots . \end{cases} \tag{15.49}$$

Note the difference between $S^{(n)}$ and S_n. The symbol $S^{(n)}$ represents all connected wave-operator diagrams in order n, while S_n represents the connected n-body diagrams (with n pairs of free lines) in all orders of perturbation theory.

We have derived the expansion above by first considering the factorization of wave-operator diagrams order by order, which led to the exponential form (15.45), and then separating this form into one-, two-, ... body parts. Alternatively, we can derive the expansion (15.49) more directly by considering disconnected n-body diagrams to all orders, using the general factorization theorem mentioned in Sect. 12.5. From the way the wave-operator diagrams are constructed, one finds that a disconnected diagram will always occur with all possible relative orderings between the interactions of the disconnected parts.

According to the factorization theorem the denominators of the disconnected parts can then be treated separately. Then one can show, for instance, that all disconnected two-body diagrams can be expressed as one-half times the normal product of all pairs of one-body diagrams, which is in keeping with the second relation in (15.49). In a similar way the remaining identities in (15.49) can be obtained.

Graphically we represent the cluster operator as in Fig. 15.10. The double bar is used to distinguish these diagrams from those of the wave operator. The S diagrams are strictly connected in an order-by-order expansion, while the corresponding wave-operator diagrams also contain disconnected parts. The relations between the two kinds of diagrams are illustrated in Fig. 15.11, which is a graphical form of the identities (15.49). When we use rule f″) given in Sect. 15.2.2 for the weight factors, we see that we automatically include the numerical factors in (15.49). For instance, the last diagram in the bottom line contains three interchangeable pieces, which gives the weight of $1/3! = 1/6$.

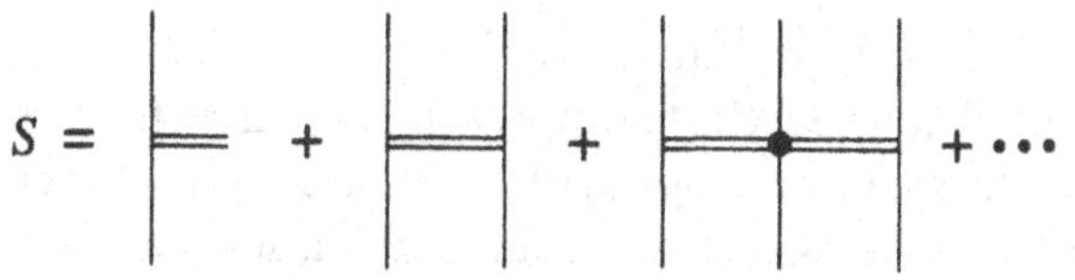

Fig. 15.10. Graphical separation of the cluster operator into one-, two-, ... body terms, in analogy with the corresponding separation of the correlation operator in Fig. 15.1. The vertices connected by double bars are always connected when an order-by-order expansion is performed

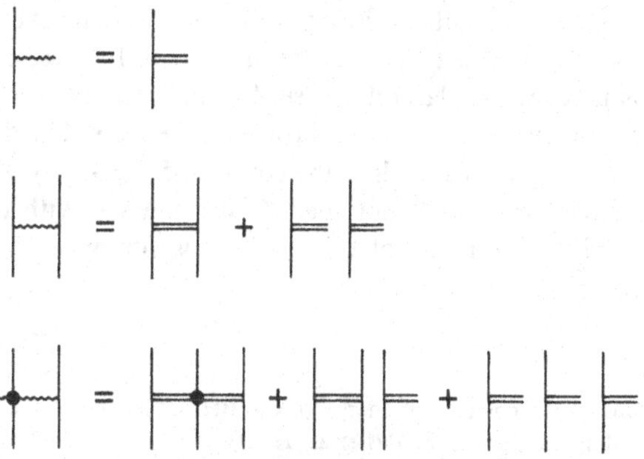

Fig. 15.11. Relations between the correlation-operator diagrams in Fig. 15.1 and the cluster-operator diagrams in Fig. 15.10. The factors appearing in (15.49) are automatically included in the evaluation, when rule f″) above for the weight factors is used

The idea is now to evaluate S_1, S_2, ... rather than Ω_1, Ω_2, ... and once these components are found with the desired accuracy, the wave operator can be constructed by means of (15.49). This procedure has the advantage that by evaluating S_1 and S_2, for instance, important parts of Ω_3 and Ω_4 are automatically included, in particular the term $1/2\{S_2^2\}$, which represents two independent pair excitations.

A general equation for the cluster operator S is obtained from the linked-diagram formula (15.2) by considering the *connected* parts of both sides

$$\boxed{[S, H_0] = (QV\Omega - \chi W)_{conn}}.$$
(15.50)

When this equation is expanded order by order, we evidently generate all connected diagrams of Ω, which by definition represent the cluster operator S.

This equation can be separated into one-, two-, ... body equations in analogy with (15.5)

$$\boxed{[S_n, H_0] = (QV\Omega - \chi W)_{n,conn}}.$$
(15.51)

This will be referred to as the general cluster equation.

The treatment of the open-shell coupled-cluster approach made here follows mainly the procedure recently suggested by *Lindgren* [1978]. In recent years other generalizations of the well-known closed-shell procedure have also been suggested [*Offermann* et al. 1976; *Offermann* 1976; *Ey* 1978; *Mukherje* et al. 1975, 1977, 1979; *Paldus* et al. 1978], which to some extent give equivalent results, but expressed in a different form.

As an illustration of the relative importance of the different types of cluster diagrams for a closed-shell system, we shall consider the BH_3 molecule for which accurate calculations have been performed by *Paldus* et al. [1972]. In these calculations a finite basis set is used, with a large basis set for one- and two-body effects and a smaller set for estimating the three- and four-body effects. The disconnected two- and three-body diagrams as well as the connected four-body diagrams were not calculated directly but were obtained by comparison with the CI results for the same basis set. The result of this analysis is summarized in Table 15.2.

Table 15.2. Relative contributions to the correlation energy of the BH_3 molecule from connected and disconnected n-body diagrams (from [*Paldus* et al. 1972])

	Total	Connected	Disconnected
One-body	0.1%	0.1%	—
Two-body	97.2	97.2	≤ 0.1
Three-body	0.8	0.8	≤ 0.01
Four-body	1.9	≤ 0.01	1.9

As in the Be case discussed before, we see that two-body perturbations dominative very heavily. We see also that these perturbations are almost exclusively described by means of *connected* diagrams. The disconnected part, $1/2\{S_1^2\}$, is quite insignificant. The situation is similar for the three-body part; the connected diagrams dominate heavily over the disconnected ones. For the four-body effects, on the other hand, this situation is reversed. Here the connected part is insignificant compared to the disconnected ones. The latter part is almost exclusively represented by double-pair excitations, that is, by the term $1/2\{S_2^2\}$.

These results suggest that a good approximation for closed-shell systems would be

$$S = S_2 ,$$

which corresponds to the wave-operator approximation

$$\Omega = 1 + S_2 + \frac{1}{2}\{S_2^2\} . \tag{15.52}$$

We know that single excitations can be more important for open-shell systems than for closed-shell systems, and for that reason we shall make the following approximation here:

$$S = S_1 + S_2 \tag{15.53}$$

in analogy with the approximation (15.7) made previously. If the potential is properly chosen, however, we still expect S_1 to be relatively small, so that we may neglect higher powers of S_1. In that case the wave operator is approximated by

$$\Omega = 1 + S_1 + S_2 + \frac{1}{2}\{S_2^2\} . \tag{15.54}$$

This relation is shown in Fig. 15.12. If the single excitations are found to be more important, then it may be necessary to include also other terms involving S_1, like $\{S_1 S_2\}$, but for simplicity we shall assume that this is not the case here.

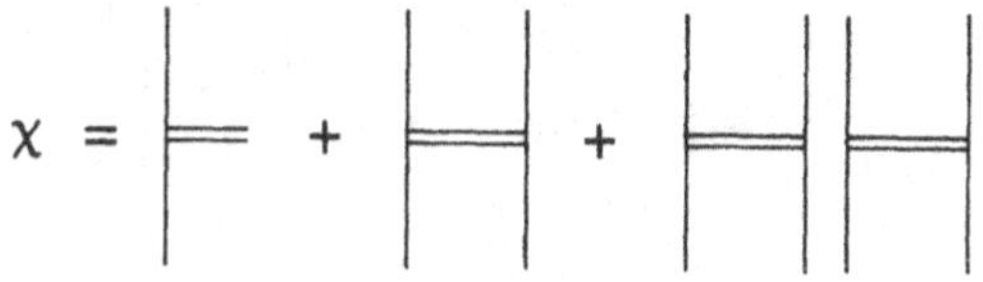

Fig. 15.12. Graphical representation of the approximation (15.54)

15.4.2 The One- and Two-Particle Coupled-Cluster Equations

By substituting the approximation (15.54) into the general cluster equaiton (15.51), we get the following equation for the one-body clusters:

$$[S_1, H_0] = Q(V + VS_1 + VS_2 - S_1 W_1)_{1,\text{conn}} , \tag{15.55a}$$

where

$$W_1 = (V + VS_1 + VS_2)_{1,\text{closed}} . \tag{15.55b}$$

We observe that single-particle diagrams cannot be formed by operating with V on $\{S_2^2\}$. This has the effect that the coupled-cluster single-particle equation is almost identical to the previous single-particle equation (15.8), the only difference being that Ω_n is replaced by S_n. Therefore, we get the graphical form of (15.55) simply by replacing the wavy lines (representing Ω_n) in Fig. 15.5 by the double bars (representing S_n).

Similarly, we get the equation for the two-body clusters

$$[S_2, H_0] = Q(V + VS_1 + VS_2 + \frac{1}{2} V\{S_2^2\}$$
$$- S_1 W_2 - S_2 W_1 - S_2 W_2)_{2,\text{conn}} , \tag{15.56a}$$

where

$$W_2 = (V + VS_1 + VS_2)_{2,\text{closed}} . \tag{15.56b}$$

This differs from the previous pair equation (15.20) primarily by the presence of the coupled-cluster term $1/2 (V\{S_2^2\})_{2,\text{conn}}$. The diagrams of this term are constructed in the same way as the third-order diagrams in Fig. 12.10. The result is shown in Fig. 15.13. [It should be remembered that the disconnected wave-operator part $1/2 \{S_2^2\}$ contains also exclusion-principle-violating (EPV) diagrams, as in the ordinary linked-diagram expansion.]

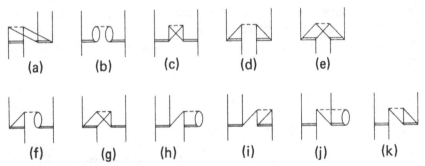

Fig. 15.13 a-k. The coupled-cluster diagrams in the pair equation (15.56a) are obtained by operating with V on the disconnected diagrams of $\{S_2^2\}$

The remaining diagrams of the coupled-pair equation are obtained from the diagrams of the previous pair equation in Fig. 15.6 by removing the disconnected diagram (b) and replacing the wave-operator diagrams by the cluster diagrams. This gives the full coupled-pair equation in the approximation (15.54) shown in Fig. 15.14.

(a) (b) (c) (d) (e)

(f) (g) (h) (i)

(j) (k) (l) (m) (n) (o)

(p) (q) (r)

Fig. 15.14 a-r. Graphical form of the coupled-pair equation (15.56a). Diagrams (j-o) are the coupled-pair diagrams shown in Fig. 15.13. As before, all exchange variants of the Coulomb interaction are omitted

For the evaluation of the coupled-cluster diagrams we introduce the following coefficients in analogy with the corresponding correlation-operator coefficients (15.9):

$$= a_i^\dagger a_j s_j^i \tag{15.57a}$$

$$= \frac{1}{2} a_i^\dagger a_j^\dagger a_l a_k s_{kl}^{ij} \tag{15.57b}$$

$$= \frac{1}{3!} a_i^\dagger a_j^\dagger a_k^\dagger a_n a_m a_l \, s_{lmn}^{ijk} \,. \tag{15.57c}$$

Here we need consider only the coupled-cluster diagrams. The remaining diagrams are evaluated in the same way as before. As illustrations we consider the following coupled-cluster diagrams:

$$= \frac{1}{2} a_r^\dagger a_s^\dagger a_n a_m \sum \langle ab | r_{12}^{-1} | tu \rangle s_{mn}^{tu} s_{ab}^{rs} \tag{15.58j}$$

$$= \frac{1}{2} a_r^\dagger a_s^\dagger a_n a_m \sum \langle ab | r_{12}^{-1} | tu \rangle s_{ma}^{rt} s_{bn}^{us} \tag{15.58k}$$

$$= - a_r^\dagger a_s^\dagger a_n a_m \sum \langle ab | r_{12}^{-1} | tu \rangle s_{ma}^{rs} s_{nb}^{tu} \,. \tag{15.58n}$$

These diagrams differ from the analogous third-order diagrams considered in Chap.12 primarily in the respect that the two clusters are initially equivalent, which was not the case with the electrostatic interactions at the bottom of the third-order diagrams. This gives rise to an additional symmetry consideration, namely whether the two clusters are equivalent or not in the sense of that word used in the rule for the weight factor. If an interchange of the two clusters gives back the same diagram or gives another diagram in the expansion, then the factor of 1/2, originally associated with the coupled-pair term $(1/2)V\{S_2^2\}$ is kept, and if this is not the case, this factor is removed.

Diagram (k) is symmetric with respect to an interchange of the two clusters, which gives a factor of 1/2. On the other hand, the vertices within the clusters cannot be interchanged, so the total weight is 1/2.

In diagram (j) we get back the same diagram when the two open paths are interchanged, in analogy with diagram (12.51c). Since the two clusters cannot be interchanged, the weight is 1/2.

Diagram (n) has no symmetry, and if only distinct diagrams are considered it will then have the weight of unity. In order to satisfy the symmetry condition (15.10), on the other hand, a symmetric diagram has to be added, which gives each of the diagrams the weight of 1/2. For the diagrams (j) and (k) no such symmetrization is required.

In this way one finds that all coupled-cluster diagrams of the pair equation have the weight of 1/2 after symmetrization, and we can remove this factor from the equation as in the previous case (15.23). The contribution to the right-hand side of the equation due to the coupled-cluster diagrams is then

$$
\begin{aligned}
&\langle ab|r_{12}^{-1}|tu\rangle s_{mn}^{tu}s_{ab}^{rs} + \langle ab|R|tu\rangle \, [s_{ma}^{rt}s_{bn}^{us} + s_{am}^{rt}s_{nb}^{us} - s_{am}^{rt}s_{bn}^{us} - s_{ma}^{rt}s_{nb}^{us} \\
&\quad\;\;\text{(j)} \qquad\qquad\qquad\qquad\quad\;\; \text{(k)} \qquad\quad \text{(l)} \qquad\;\; \text{(m)} \qquad\;\; \text{(m')} \\[4pt]
&\quad - s_{ma}^{rs}s_{nb}^{tu} - s_{am}^{tu}s_{bn}^{rs} - s_{mn}^{rt}s_{ab}^{su} - s_{ab}^{tr}s_{mn}^{us}] \\
&\qquad \text{(n)} \qquad\;\; \text{(n')} \qquad\;\; \text{(o)} \qquad\;\; \text{(o')}
\end{aligned}
\tag{15.59}
$$

using the same sum convention as before. The remaining terms of the coupled-pair equation are obtained by replacing the x coefficients of the corresponding terms of the linear equation (15.23) by the s coefficients and leaving out the disconnected diagram (b).

We have then obtained the complete coupled-cluster equations in the approximation (15.54). This represents a generalization to the open-shell case of the "coupled-pair many-electron theory" (CPMET) developed by Čížek [1966, 1969] for closed-shell systems. The procedure used by Čížek is to a large extent similar to that employed here, the main difference being that it is based on the ordinary exponential ansatz (15.46) rather than the modification (15.44).

15.5 Comparison Between Different Pair-Correlation Approaches

The coupled-cluster approach in the pair approximation—or the coupled-pair many-electron theory (CPMET)—developed in the previous section seems to be the best procedure that is presently available for systems with more than a few electrons. The most serious approximation in this scheme is the neglect of three-body clusters, which are very difficult to handle. Even in the pair approximation, however, this scheme is quite time consuming, and only a few complete calculations have been performed.

The most complicated part of the coupled-pair equation is, of course, the "quadratic" terms (15.59), which involve fourfold summations/integrations.

Several approximate schemes have therefore been suggested in order to simplify this part, and we shall now consider some of them. The discussion will be mainly restricted to closed-shell systems, for which most calculations have been performed so far.

For closed-shell systems there are no folded diagrams, and the pair equation (15.56) reduces to

$$[S_2, H_0] = Q\left(V + VS_2 + \frac{1}{2}VS_2^2\right)_{2,\text{conn}},\tag{15.60}$$

omitting single excitations. The quadratic term $(1/2)VS_2^2$ also contains an EPV part, which can be quite large. We shall now consider this part in some detail, since it plays an important role in certain approximation schemes.

By operating with V on S_2^2, we get linked as well as unlinked diagrams, as in the third-order wave operator treated in Sect. 12.5. The EPV diagrams violate the exclusion principle in the intermediate state, and hence they do not contribute to S_2^2. Consequently, all diagrams obtained by operating with V on these diagrams—linked and unlinked—must cancel. The unlinked diagrams are eliminated in LDE, while the linked ones are retained. We are interested in the linked EPV diagrams, and we can obtain their contribution most easily by evaluating the unlinked EPV diagrams and reversing the sign.

The unlinked diagrams of $(1/2)VS_2^2$ are of the type

$$\tag{15.61}$$

The EPV contribution is obtained by summing all combinations for which the same orbital appears in the two diagram parts. For simplicity, we shall disregard the exclusion principle as far as virtual orbitals are concerned, since the effect there is quite small. Hence, we consider only the cases where c and/or d is equal to a and/or b.

The open diagram in (15.61) represents the two-body cluster parameter s_{ab}^{rs} defined in (15.57). The closed diagram—together with its exchange variant—represents the correlation energy for the system, apart from single excitations. Keeping the core orbital pair (c, d) fixed and summing over the virtual orbitals (t, u), this part of the diagram yields the pair-correlation energy ε_{cd}. After this partial summation (15.61) becomes $\varepsilon_{cd}s_{ab}^{rs}$, and the EPV part of (15.61), that is, the unlinked EPV part of $(1/2)VS_2^2$, can then be expressed as $W_{ab}s_{ab}^{rs}$, where

$$W_{ab} = \varepsilon_{ab} + \sum_d^{d\neq b} \varepsilon_{ad} + \sum_c^{c\neq a} \varepsilon_{cb} .\tag{15.62}$$

Here, the term ε_{ab} is excluded from the sums, since it should appear only once.

According to the argument above, the linked EPV part of $(1/2)VS_2^2$ is equal to the negative of the unlinked part, or $-W_{ab}s_{ab}^{rs}$.

If we now retain only the EPV part of $(1/2)VS_2^2$ in (15.60), then we can replace the coupled-cluster expression in (15.59) simply by $-W_{ab}s_{ab}^{rs}$. Due to the factorization in this case, we can move this term to the left-hand side of the equation, and we obtain an equation quite similar to the previous linear equation (15.23)

$$(\varepsilon_a + \varepsilon_b - \varepsilon_r - \varepsilon_s + W_{ab})s_{ab}^{rs} = \langle rs|r_{12}^{-1}|ab\rangle$$
$$\text{(a)}$$

$$+ \langle rs|r_{12}^{-1}|tu\rangle s_{ab}^{tu} + \langle rc|R|at\rangle s_{cb}^{ts} + \langle cs|R|tb\rangle s_{ac}^{rt}$$
$$\text{(g)} \qquad\qquad \text{(h)} \qquad\qquad \text{(h')}$$

$$- \langle rc|R|tb\rangle s_{ac}^{ts} - \langle cs|R|at\rangle s_{cb}^{rt} + \langle cd|r_{12}^{-1}|ab\rangle s_{cd}^{rs}.$$
$$\text{(i)} \qquad\qquad \text{(i')} \qquad\qquad \text{(j)}$$

$$(15.63)$$

The corresponding graphical representation is shown in Fig. 15.15.

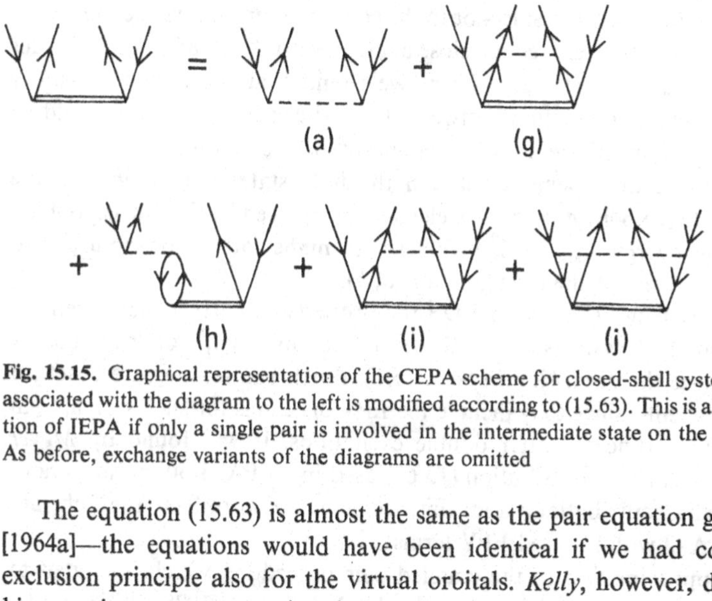

Fig. 15.15. Graphical representation of the CEPA scheme for closed-shell systems. The energy associated with the diagram to the left is modified according to (15.63). This is also a representation of IEPA if only a single pair is involved in the intermediate state on the right-hand side. As before, exchange variants of the diagrams are omitted

The equation (15.63) is almost the same as the pair equation given by *Kelly* [1964a]—the equations would have been identical if we had considered the exclusion principle also for the virtual orbitals. *Kelly*, however, did not derive his equation as an approximation to the coupled-cluster equation, as we have done here, but included the EPV diagrams for more intuitive reasons. *Kelly* has not made any calculation with this scheme.

Other schemes, similar to that of *Kelly*, have been introduced and applied to several molecular systems by *Meyer* [1973, 1976], and these schemes are nowadays referred to as the *coupled-electron-pair approximation* (CEPA). The schemes of *Meyer* differ from that of *Kelly* only in the form of W_{ab}. Two schemes, known as CEPA–1 and CEPA–2 [*Kutzelnigg* 1977; *Ahlrichs* 1979], have been used, namely

$$W_{ab} = \frac{1}{2} \sum_d \varepsilon_{ad} + \frac{1}{2} \sum_c \varepsilon_{cb} \qquad (15.64a)$$

and

$$W_{ab} = \varepsilon_{ab} . \qquad (15.64b)$$

The form (15.62) is sometimes referred to as CEPA-3.

We mentioned in Sect. 15.1 that LDE, truncated after double excitations, does not yield exact results for a two-electron system, since the EPV diagrams needed to cancel the unlinked diagrams are partly of three- and four-body type. In the CEPA-3 scheme (15.62) the EPV diagrams are explicitly included, and therefore this scheme is exact for two-electron systems—provided that single excitations are also considered. Furthermore, we see that for two-electron systems the energy factor W_{ab} is equal to the correlation energy ε_{ab} in all schemes discussed here, so there is no difference between the different CEPA schemes for such systems.

The CEPA schemes differ from the truncated LDE scheme only in the energy factor on the left-hand side—or in the corresponding energy denominator of the diagrams. We can here see the close analogy with the summation of ladder diagrams, discussed in Sect. 12.7. There we found that the hole-hole ladder representing repeated interactions between the particles in core states, could be included by modifying the energy denominator of the second-order diagram by the Coulomb matrix element between the hole states $\langle ab|R|ab \rangle$. In the CEPA-2 scheme (15.64b), this matrix element is replaced by the corresponding pair-correlation energy ε_{ab}, which includes also higher-order contributions to the interaction energy of the electron pair (a, b).

The CEPA schemes share with LDE the property of correct energy separation for closed-shell systems, since there are no unlinked, nonlinear energy contributions (see the discussion in Sect. 15.1). Furthermore, they are exact for two-electron systems, since they include the EPV diagrams for such systems—in contrast to LDE truncated after double excitations. It was found by *Meyer* [1973] that the full EPV contribution (15.62) used in CEPA-3 often exaggerates the coupled-pair contribution, and this led him to introduce the schemes CEPA-1 CEPA-2 with reduced EPV terms.

Another approximation to the coupled-pair procedure, which is related to the CEPA schemes, has recently been used by *Mårtensson* [1978, 1979] and by *Adams* et al. [1979] (see also [*Jankowski* and *Paldus* 1980]). This can be derived in the following way. The EPV diagrams for two-electron systems are of the same kind as the coupled-cluster diagrams. It turns out, however, that only three of these diagrams contribute for such systems, namely diagrams (a, h, i) in Fig. 15.13—or diagrams (j, n) in Fig. 15.14. These diagrams have a particularly simple form also for more general systems. Writing out the summations over the virtual orbitals in (15.59 j, n, n') explicitly, we get the following expressions for these three diagrams:

$$[\sum_{tu} \langle ab | r_{12}^{-1} | tu \rangle s_{mn}^{tu}] \times s_{ab}^{rs} \tag{15.65j}$$

$$- [\sum_{tu} \langle ab | R | tu \rangle s_{nb}^{tu}] \times s_{ma}^{rs} \tag{15.65n}$$

$$- [\sum_{tu} \langle ab | R | tu \rangle s_{am}^{tu}] \times s_{bn}^{rs} . \tag{15.65n'}$$

The remaining terms in (15.59) cannot be factorized in this way. The contributions (15.65) are somewhat more complicated than the EPV part (15.62) derived before, but they are still quite easy to evaluate. This scheme has the advantage over the CEPA schemes that it is more well-defined in terms of diagrams, which makes it easier to estimate the effect of the omitted part. This scheme, which will be referred to as "approximate CPMET", contains the full EPV contribution as does CEPA–3, and in addition the EP allowed part of the corresponding diagrams. This is identical to the scheme ACP of *Adams* et al. [1979] and the scheme ACP–45 of *Jankowski* and *Paldus* [1980], who have found this scheme to be closer to the full CPMET than the CEPA schemes in case of quasidegeneracy (near degeneracy), except for very strong quasidegeneracy where CEPA–3 is somewhat more accurate.

In the schemes described above the different pair excitations are coupled in the sense that all possible excitations may appear in the intermediate state. A simpler scheme has for a long time been employed by *Nesbet* [1958, 1965, 1969] and *Sinanoğlu* [1962, 1964, 1969], where the coupling between excitations from different pairs of core orbitals is neglected. *Nesbet* called his equation the atomic "Bethe-Goldstone" equation, in analogy with the corresponding equation in nuclear physics [*Bethe* and *Goldstone* 1957; *Fetter* and *Walečka* 1971, p. 322], and *Sinanoğlu* named his procedure the "exact pair theory". At this point we would also like to mention the pair-correlation works of *Brenig* [1957] and *Szasz* [1959, 1968], which are closely related to those of *Nesbet* and *Sinanoğlu* but not so widely known. Nowadays, all schemes of this kind are referred to as *independent-electron-pair approximations* (IEPA) [*Kutzelnigg* 1977].

The algebraic form of the IEPA equation is obtained from the CEPA equation (15.63) by replacing (c, d) by (a, b) and excluding the summation over the core orbitals. The graphical form is the same as in Fig. 15.15 with the restriction that only a single pair of core orbitals (a, b) is involved in the intermediate state. In IEPA the parameter W_{ab} is usually put equal to the pair-correlation energy ε_{ab}, that is, the same as in CEPA–2 (15.64b).

Early calculations by *Nesbet* and *Sinanoğlu* using the IEPA procedure yielded very accurate correlation energies, often within 1 % of the experimental value. It was shown later, however, that this was partly due to cancellations between the effects of a limited basis set and of the neglect of pair-pair couplings [*Barr* and *Davidson* 1970; *Viers* et al. 1970]. In the case of the Ne each of these effects was estimated to be of the order of 15 %.

It should be observed that in the IEPA scheme only a single pair of core orbitals is involved in all intermediate states. However, when angular-momentum graphs are used to evaluate the diagrams, a summation over the magnetic quantum numbers (m_s, m_l) of the core lines will automatically be included. In this way there will be a coupling between all pair excitations from a given pair of nl shells. This will be referred to as *intrashell pair coupling* (ISPC). For a pair of s shells, this is identical to the ordinary IEPA, since the Coulomb interaction is diagonal with respect to m_s. If non-s shells are involved, on the other hand, the situation is different. ISPC then includes coupling between all excitations from pairs of core orbitals which differ in the m_l values. It will be demonstrated below that the intrashell pair coupling dominates over the remaining *intershell* coupling, at least for atoms. (For molecules one can expect that all couplings within the same *main* shells, rather than the subshells, are of importance.) Therefore, the ISPC scheme, which is intermediate between IEPA and CEPA, provides us with an additional approximation, which combines simplicity with good accuracy.

15.5.1 Application to Helium

As a first example of a pair-correlation calculation we consider the He atom in its ground state $1s^2\ {}^1S$. Pair-correlation calculations on this system have been performed by *Winter* et al. [1970], using a partial-wave expansion of the Coulomb interaction. The resulting two-dimensional radial equations are solved numerically in different point grids and extrapolations to infinite grid made by means of the Richardson extrapolation procedure [*Richardson* and *Gaunt* 1927]. Similar calculations have been performed by *Mårtensson* [1978, 1979] as a test of a general pair-correlation program. The results of these calculations are shown in Table 15.3, where also the multiconfiguration Hartree-Fock results of *Froese-Fischer* [1973] and the accurate CI results of *Bunge* [1970] are given. The results for a certain "L-limit" are obtained by considering all excitations $1s^2 \rightarrow l^2$ for which $l < L$, including couplings between them. (Spectroscopic notations are used in the table, which means that $S, P, D, \ldots$ represent $L = 0,$ $1, 2, \ldots$.)

Table 15.3. L-limits for the He atom

Limit	*Winter* et al.	*Mårtensson*	*Froese-Fischer*	*Bunge*
HF				—2.86168
S	−2.87903	−2.87902	−2.87899	−2.87903
P	−2.90051	−2.90050	−2.90015	−2.90051
D	−2.90276	−2.90274	−2.90252	−2.90275
F	−2.90331	−2.90328	−2.90291	−2.90331
G	−2.90351	−2.90347	−2.90303	−2.90347
∞				−2.90372

The correlation energy, that is, the difference between the exact and the Hartree-Fock energies, is in this case $-2.90372 + 2.86168 = -0.04204$ H. From the expansion above, we see that the S limit ($1s^2 \rightarrow s^2$) represents about 41% and the P limit ($1s^2 \rightarrow s^2, p^2$) about 93% of the correlation energy. For larger L values one finds that the contribution generally decreases as the fourth power of L [*Bunge* 1976a].

15.5.2 Application to Beryllium

As a second example we consider the Be atom in its ground state $1s^2 2s^2$ ^1S, for which several accurate calculations have been performed. The results of some of these calculations are summarized in Table 15.4. The CI results of *Bunge* [1976a] represent the most accurate calculation performed so far on this system. The total correlation energy obtained in this calculation agrees very well with the estimated experimental value, corrected for relativistic and other effects which are not included in the ab initio calculations discussed here.

Table 15.4. Calculations of the ground-state correlation energy of the Be atom (unit 10^{-4} H, reversed sign)

	$1s^2$	$1s2s$	$2s^2$	Total	% of experimental value
Kelly	421	50	449	920	97.5
Nesbet	418	59	444	921	97.6
Szasz and *Byrne*	423		445		
Byron and *Joachain*	425	52	448	925	98.1
Sims and *Hagstrom*				935	99.1
Kaldor	421	54	447	922	97.8
Froese-Fischer and *Saxena*	421	41	466	928	98.4
Lindgren and *Salomonson*	426	55	448	930	98.6
Bunge	425	54	464	943	100.0
Experimental (nonrelativistic)				943.0	

The calculation of *Kelly* [1963, 1964b] has been described in Chap.12. It is obtained by means of the linked-diagram expansion in second order, with certain higher-order effects included by means of energy-denominator modifications. The calculation of *Nesbet* [1967] is carried out by means of his "Bethe-Goldstone" equation, which is a form of IEPA as described above. Kelly's and Nesbet's calculations are carried out to the D limit, which means that excitations to states with $l = 0$, 1, and 2 are considered. The "three-body" effects reported by *Kelly* and *Nesbet* represent mainly what we call pair-pair interactions and they have been included in the $2s^2$ correlation energy in the Table 15.4.

The calculations of *Szasz* and *Byrne* [1967], *Byron* and *Joachain* [1967] and *Sims* and *Hagstrom* [1971] are performed using so-called Hylleraas wave functions, where the interelectronic distance r_{ij} is introduced explicitly. The calculation of *Sims* and *Hagstrom*, which uses a 107-configuration wave function, is the most accurate calculation of this kind. Unfortunately, the contributions are not separated into pair-correlation energies, so no comparison with other calculations can be made on that level.

The calculation of *Kaldor* [1973] is similar to that of *Kelly* but more accurate. The third-order contributions are calculated more completely and excitations to f states are included. The results of *Froese-Fischer* and *Saxena* [1974] are obtained by means of a numerical multiconfiguration Hartree-Fock procedure with 52 configurations and with the $2s^2 2s^2$ and the $1s^2 2p^2$ configurations in the zeroth-order approximation. The results of *Lindgren* and *Salomonson* [1980] are obtained with a numerical coupled-cluster procedure in the pair approximation (CPMET), neglecting single and triple excitations,

From Table 15.4 we see that all calculations mentioned here reproduce the experimental correlation energy within 2–3%. Even if this is in some cases partly fortuitous, it is quite satisfactory in view of the variety of methods that has been employed. A more detailed comparison between the results obtained in these calculations, however, is difficult to perform due to the differences in the techniques employed. Therefore, in order to be able to see the effect of various contributions to the correlation energy more directly, we have performed a series of calculations on the Be ground state with different pair approximations using the same numerical procedure [*Lindgren* and *Salomonson* 1980]. The results are summarized in Table 15.5. (At this point we should also like to call attention to a recent comparison between different CPMET and CEPA schemes performed by *Adams* et al. [1979].)

The first row represents the linked-diagram expansion to all orders using double excitations only, that is, no EPV diagrams or any other coupled-cluster

Table 15.5. Comparison between various pair approximations for the correlation energy of the Be ground state (unit 10^{-4} H, reversed sign)

	$1s^2$	$1s2s$	$2s^2$	Total	% of experimental value
LDE double } L-CPMET }	425	55	503	984	104.3
CI double	424	56	415	894	94.8
IEPA, ISPC	427	60	456	944	100.1
CEPA, CPMET	426	55	448	930	98.6
Complete CI	425	54	464	943	100.0

contributions. This scheme, which sometimes is referred to as "linear CPMET" (L-CPMET), is obtained by setting $W_{ab} = 0$ in (15.63). The second row represents the corresponding configuration-interaction result, which is obtained from the CEPA equation (15.63) by setting W_{ab} equal to the entire correlation energy. The CI result is consistent with the result shown in Table 15.1, namely that in Be double excitations are responsible for approximately 95% of the correlation energy, and single, triple and quadruple excitations for the remaining 5%. When only double excitations are considered in LDE, on the other hand, we see that we "overshoot" the correlation energy by almost the same amount. Here, certain quadruple excitations are implicitly included in the elimination of the unlinked diagrams. The error is mainly due to the fact that EPV diagrams with quadrupole excitations are omitted.

The results in the fourth row in Table 15.5 are obtained with the approximate CPMET scheme (15.65). As we have mentioned, this includes three coupled-cluster diagrams, which are identical to the EPV diagrams appearing in LDE. These diagrams contain also an EP allowed part, which, however, is comparatively small in the present case. Therefore, this is essentially identical to CEPA-3. The difference between the L-CPMET and the CEPA results gives the effect of the EPV diagrams in the present case. The remaining eight coupled-cluster diagrams, which are not included in the approximate scheme (15.65), have a very small effect for Be. Therefore, with the number of digits given in the table, this result represents also CPMET.

By again comparing with Table 15.1, we see that the coupled-cluster results in Table 15.5 agree almost exactly with the CI result with double *and* quadruple excitations. It then follows that the quadruple excitations are well represented by the coupled-cluster diagrams—which in the present case are mainly of the EPV type. This is also in agreement with the results shown in Table 15.2, namely that the *connected* four-body wave-operator diagrams are almost negligible compared to the disconnected ones.

The third row in Table 15.5 gives the result in the approximation where the intershell pair coupling (ISPC) is neglected. In the present case there is very little pair coupling within the shells, since only two s shells are involved with one pair each. Such coupling appears only in the $1s2s$ pair. Therefore, in the present case the ISPC approximation is almost identical to IEPA, where pair-pair coupling is neglected completely. The correlation energy obtained in this approximation agrees almost exactly with the experimental result. Obviously this is fortuitous, and due to cancellation between the intershell pair coupling and the effects of single and triple excitations. By comparing with the complete CI results of *Bunge* in the last row, we see that the agreement is not as good for the individual pair-correlation energies. Nevertheless, it is found that ISPC represents a good approximation in view of its relative simplicity. The Be atom ground state is rather special, however, since only s shells are involved in the unperturbed state. Therefore, we shall also consider the Ne atom where a full p shell is involved.

15.5.3 Application to Neon

The results of various calculations for the Ne ground state $1s^2 2s^2 2p^6 \, ^1S$ are shown in Table 15.6. The first row gives the result of *Nesbet* [1967], which represents the first accurate calculation on this system. It was performed with the IEPA scheme, where couplings between excitations from different pairs of core orbitals are neglected. The early result of *Nesbet* agrees almost exactly with the experimental correlation energy, which is found to be in the range 0.385–0.390 H [*Bunge* and *Peixoto* 1970; *Sasaki* and *Yoshimine* 1974]. As we have mentioned, however, it was argued by several groups that this agreement was largely accidental and due to cancellation between the pair-pair coupling and the effect due to an incomplete basis set. A later calculation by *Nesbet* et al. [1969] with a larger basis set gave a correlation energy which was about 6% larger than the experimental value. A further analysis by *Barr* and *Davidson* [1970] and by *Viers* et al. [1970] indicated that the effect of pair-pair coupling would be of the order of 15%. In other words, an IEPA calculation with a very large basis set is expected to yield about 115% of the experimental correlation energy.

Table 15.6. Pair-correlation energies for neon (unit 10^{-4} H, reversed sign)

		$1s^2$	$1s2s$	$2s^2$	$1s2p$	$2s2p$	$2p^2$	Total
Nesbet		399	51	108	199	815	2249	3822
Nesbet, Barr, and *Davidson*					117		906	2431	4103
Lee, Dutta and *Das*		273	40	120	235	908	2333	3891
Jankowski-	Second-order	402	56	120	222	872	2208	3879
Malinowski	Third-order							3855
Lindgren-	Second-order	402	56	120	222	873	2210	3883
Salomonson	Third-order	410	57	124	224	882	2151	3848
	Appr. CPMET, ISPC	411	57	125	231	894	2114	3832
	Appr. CPMET complete coupling	410	56	124	223	890	2091	3794
	CPMET	410	56	124	223	894	2104	3812

The results of *Lee* et al. [1970] in the third row in Table 15.6 were obtained by means of low-order LDE, using energy-denominator modifications to include higher-order effects in an approximate way. The total correlation energy agrees well with the experimental value, but this is partly due to cancellation of errors, which can be seen by comparing the pair-correlation energies with those obtained in other calculations.

Recently, very accurate second-order correlation-energy calculations for Ne have been performed by *Jankowski* and *Malinowski* [1980], using analytical basis functions. The results are given in the fourth row in Table 15.6. These results agree very well with the accurate second-order results of *Lindgren* and *Salomonson* [1980], using numerical integrations, shown in the sixth row. *Jankowski* and *Malinowski* have also calculated the third-order energy-contribution, using a smaller basis set. The corresponding result of *Lindgren* and *Salomonson* is about 50% larger and probably more accurate, since numerical integration is used.

The schemes represented by the last three rows in Table 15.6 are the same as used in the Be calculation discussed above. Here the pair-correlation is carried to all orders of perturbation theory, using the iterative scheme described previously. In the ISPC scheme, based on formulas (15.65), the couplings between pair excitations involving the same *nl* shells are considered. Since a *p* shell is involved here, this scheme is in the present case considerably different from the IEPA scheme, where no pair-pair coupling is included. As mentioned, an accurate IEPA calculation is expected to yield about 115% of the experimental correlation energy, while the ISPC scheme used here yields 98–99% of the experimental value. When the remaining intershell pair coupling is included (ninth row), the result is reduced by approximately 1%. The last row in Table 15.6 gives the complete CPMET results, that is, by adding the effect of the remaining eight coupled-cluster diagrams (see Fig. 15.13) to the previous result. This effect is about 0.5% of the correlation energy.

In the CPMET calculations only two-body clusters are considered. *Bunge* [1980] has estimated the effect of triple excitations to be about 1% and, assuming the singles to contribute 0.5%, we would from the CPMET result estimate the total correlation energy of neon to be about 0.387 H, in good agreement with the experimental estimate.

In comparing the results for Be and Ne one finds that the difference between the different approximations is much smaller in the latter case, which reflects the fact that the coupled-cluster contribution is there less important. In third order there are no diagrams of this kind, and since the third-order contribution is quite small for Ne, this result is almost equivalent to LDE (double) or linear CPMET, where the diagrams appearing in third order are carried to all orders. Therefore, the difference between the third-order and CPMET results gives the coupled-cluster contribution in this case. This is only about 1% of the correlation energy, as compared to 5–6% for Be. This is in agreement with the conclusion of *Jankowski* and *Paldus* [1980] that the coupled-cluster contribution is more important in the case of quasidegeneracy, which is quite pronounced for Be.

15.5.4 Examples of Open-Shell Coupled-Cluster Calculations

Two coupled-cluster calculations for atomic open-shell systems have recently been reported by *Morrison* and *Salomonson* [1980] on carbon and by *Salomonson* et al. [1980] on Be-like systems. Here we shall consider the former calculation as a final illustration.

As described in Sect. 13.4, the energy levels of the $1s^2 2s^2 2p^2$ configuration of carbon can be represented by the effective operator

$$H_{eff} = E_0 + a u^1(1) \cdot u^1(2) + F^2 C^2(1) \cdot C^2(2) , \tag{15.66}$$

where E_0 is a constant energy term, F^2 is a Slater parameter and a is the coefficient of an additional pair of tensor operators of odd rank, corresponding to one of the Trees parameters introduced in Sect. 13.4. This coefficient is zero in first order, but certain higher-order effects contribute to this parameter, as we have shown in Chap. 13.

The $1s^2 2s^2 2p^2$ configuration has three LS terms, ^1S, ^1D, and ^3P. Since the relative positions of the ^1D and ^1S levels of carbon are well known and the total (nonrelativistic) energy of the ^3P ground state has been accurately estimated [A. Bunge and C. Bunge 1970], the "experimental" values of the parameters in (15.66) can be determined

$$E_0 = -37.807, \quad a = 0.0173, \quad F^2 = 0.169.$$

The corresponding Hartree-Fock values of these parameters are

$$E_0 = -37.528, \quad a = 0 \quad F^2 = 0.239.$$

In the Hartree-Fock model the Slater integral F^2 is more than 40% larger than the experimental value, and the Hartree-Fock value of the parameter a is zero, since this is a first-order theory. The two parameters a and F^2 are responsible for the splitting of the energy levels. E_0 is a constant term, which is the same for all of the states of the configuration, and therefore affects only the average energy.

In Table 15.7 we give the results of Morrison and Salomonson [1980] obtained with the second-order perturbation theory and with pair-correlation effects included to all orders. These results were obtained by solving the coupled pair equations in a self-consistent way, including folded diagrams and the factorizable coupled-cluster diagrams (a, h, i) in Fig. 15.13.

Table 15.7. The Slater and Trees parameter for carbon in different approximations

	F^2	a
Hartree-Fock	0.2387	0
Second order	0.1640	0.0040
Coupled-cluster	0.1694	0.0107
Experimental	0.1690	0.0173

As we can see, the coupled-cluster value of the Slater parameter agrees well with the experimental value, while the value of the a parameter is 40% too small. This discrepancy is due mainly to additional coupled-cluster diagrams, not included in the calculation, and to another family of diagrams, mentioned in Sect. 13.3, for which an additional valence line polarizes a core or valence orbital.

The coupled-pair procedure discussed in this chapter can be readily extended to include these two additional families of diagrams. Furthermore, the calculation presented here was based on orbitals generated in the Hartree-Fock potential of the C^{2+} core. It would be interesting to repeat the calculation with a potential which includes the monopole part of the interaction between the valence electrons. A potential of this kind has the right asymptotic behavior and the perturbation theory can be expected to converge more rapidly with this potential.

Appendix

A. Hartree Atomic Units

In this book we employ a system of atomic units, usually referred to as Hartree's system. This is based on the following units:

		Dimension
e	the absolute value of the electric charge of the electron	[A s]
m	the rest mass of the electron	[kg]
$\hbar$	Planck's constant divided by 2π	[kg m²/s]
$4\pi\varepsilon_0$	4π times the permittivity of vacuum	[A²s⁴/kg m³]

The dimensions in SI units of the basic quantities are shown in square brackets. Other atomic units can be obtained by combining these quantities to give the correct dimension.

The first Bohr radius

$$a_0 = \frac{4\pi\varepsilon_0 \hbar^2}{me^2} \tag{A.1}$$

has the dimension [m], and, hence, it is the atomic unit of length.

The fine-structure constant

$$\alpha = \frac{e^2}{4\pi\varepsilon_0 c\hbar}, \tag{A.2}$$

where c is the velocity of light in vacuum, is a pure number ($\approx 1/137$). This means that αc is the atomic unit of velocity, or that the velocity of light is approximately 137 atomic units.

In spectroscopy, the energy is often expressed in terms of

$$hcR_\infty = \frac{me^4}{2(4\pi\varepsilon_0)^2\hbar^2}, \tag{A.3}$$

where R_∞ is the Rydberg constant for infinite nuclear mass. In atomic units this has the value of 1/2, so the atomic energy unit is

$$\boxed{1\text{H} = 2hcR_\infty \approx 27.2 \text{ eV}}. \tag{A.4}$$

This unit is nowadays often referred to as the *Hartree unit*.

The Rydberg constant itself has the dimension $[m^{-1}]$, and its value in atomic units is $\alpha/4\pi$ (h has the value of 2π). This leads to the relation

$$R_\infty = \frac{\alpha}{4\pi a_0} , \tag{A.5}$$

which is easily checked by means of the complete expressions.

The Bohr magneton

$$\mu_B = \frac{e\hbar}{2m}$$

has the value $1/2$ in atomic units, so the unit for magnetic moment is $2\mu_B$. (It should be observed that the expression $e\hbar/2mc$ used in older literature for the Bohr magneton is based on a *mixed* unit system and therefore does not have the correct dimension.)

From the relation

$$\mu_0\varepsilon_0 = 1/c^2 ,$$

where μ_0 is the permeability of vacuum, we find that

$$\frac{\mu_0}{4\pi} = \frac{1}{4\pi\varepsilon_0 c^2}$$

has the value α^2 in atomic units.

B. States and Operators

B.1 Representation of Physical States

The state of a physical system is in the time-independent Schrödinger formalism represented by a wave function $\Psi^a(r)$, where r stands for the space and spin coordinates of all the particles. In the *Dirac notation* such a state is represented by a *ket*, or *ket vector*, $|\Psi^a\rangle$. The corresponding complex conjugate function $\Psi^a(r)^*$ is represented in Dirac's notation by a *bra*, or *bra vector*, $\langle\Psi^a|$.

The *scalar product* of two functions, $\Psi^a(r)$ and $\Psi^b(r)$, is in the Dirac and Schrödinger notations

$$\langle\Psi^a|\Psi^b\rangle = \int \Psi^a(r)^*\Psi^b(r)dr . \tag{B.1}$$

This represents an *integration* over all space coordinates and a *summation* over all spin coordinates. If the scalar product is zero, the functions are *orthogonal*. The function Ψ^a is *normalized*, if $\langle\Psi^a|\Psi^a\rangle = 1$. It follows directly from (B.1) that

$$\langle \Psi^b | \Psi^a \rangle = \langle \Psi^a | \Psi^b \rangle^* \, . \tag{B.2}$$

Let $\{\phi_i\}$ represent a set of *orthonormal* wave functions. This implies that

$$\langle \phi_i | \phi_j \rangle = \delta(i,j) \, , \tag{B.3}$$

where $\delta(i,j)$ is the Kronecker delta symbol

$$\delta(i,j) = \begin{cases} 1 & \text{for} \quad i = j \\ 0 & \text{for} \quad i \neq j \, . \end{cases}$$

[The symbol $\delta(i,j)$ is used for typographical reasons instead of the more common symbol δ_{ij}·]. This set is *complete* if any wave function representing a possible state of the system considered can be expanded as

$$\Psi^a = \sum_i c_i \phi_i \, , \tag{B.4}$$

where c_i are ordinary (complex) numbers. The number of basis functions $\{\phi_i\}$ may not be enumerable. The subscript i will then be a *continuous* variable, and the sum replaced by an *integral*. This formal difficulty can be avoided by enclosing the system in a finite box, in which case the basis functions can always be numbered (but still infinite in number). The real situation is approached by letting the box increase without limits. For simplicity, we shall in the following assume that such a procedure is used.

It follows directly from (B.2–4) that

$$c_i = \int \phi_i^* \Psi^a d\tau = \langle \phi_i | \Psi^a \rangle \, . \tag{B.5}$$

The expansion (B.4) then becomes

$$\Psi^a(r) = \sum_i \phi_i(r) \int \phi_i^*(r') \Psi^a(r') dr' \, . \tag{B.6}$$

Since $\Psi^a(r)$ is an arbitrary wave function, this leads to the so-called *closure property* [*Schiff* 1968, p. 51; *Jackson* 1975, p. 66].

$$\sum_i \phi_i(r)\phi_i(r') = \delta(r - r') \, . \tag{B.7}$$

Here, $\delta(r - r')$ is the Dirac delta function, which has the property

$$\int f(r')\delta(r - r')dr' = f(r) \tag{B.8}$$

for any function $f(r)$. The closure property is a consequence of the fact that the set $\{\phi_i\}$ is complete. In Dirac's notation (B.6) becomes

$$| \Psi^a \rangle = \sum_i | \phi_i \rangle \langle \phi_i | \Psi^a \rangle \, ,$$

which leads to the important formal identity

$$\boxed{\sum_i |\phi_i\rangle \langle \phi_i| \equiv 1}.$$

(B.9)

This is also referred to as the *completeness relation* or the *resolution of the identity*.

The function

$$c_i |\phi_i\rangle = |\phi_i\rangle\langle\phi_i| \Psi^a\rangle$$

(B.10)

is called the *projection* of the function Ψ^a onto the function ϕ_i. The operator

$$P_i = |\phi_i\rangle\langle\phi_i|.$$

(B.11)

is the corresponding *projection operator*. Equation (B.9) can then be written

$$\sum_i P_i \equiv 1,$$

(B.12)

which simply means that all the projections add up to the function itself. The coefficients c_i define the state Ψ^a completely and form the *vector representation* of it.

Another useful representation of a state is the *coordinate representation* (see, for instance, [*Rodberg* and *Thaler* 1967] p. 168). Here the basis functions are eigenfunctions of position and denoted by $|r\rangle$. The representation of a state Ψ^a is given by the scalar product with the basis functions $\langle r | \Psi^a\rangle$. This represents the amplitude for finding the system at position r, and hence it is identical to the Schrödinger wave function

$$\langle r | \Psi^a\rangle \equiv \Psi^a(r).$$

B.2 Representation of Physical Quantities

Each physical quantity is in quantum mechanics represented by a *linear operator*. An operator α is linear, if

$$\alpha(c_1\phi_1 + c_2\phi_2) = c_1\alpha\phi_1 + c_2\alpha\phi_2.$$

(B.13)

ϕ_1 and ϕ_2 are here arbitrary functions and c_1 and c_2 arbitrary numbers. Quantum-mechanical operators are, in general, noncommutative, which means that the *commutator*

$$[\alpha, \beta] = \alpha\beta - \beta\alpha$$

(B.14)

is generally nonvanishing.

A quantum-mechanical operator acting on a wave function always produces a wave function, which represents a possible state for the system considered,

$$\alpha |\Psi\rangle = |\Psi'\rangle \, .$$

If

$$\alpha |\Psi\rangle = a |\Psi\rangle \, , \tag{B.15}$$

where a is a number, $|\Psi\rangle$ is an *eigenfunction* of the operator α and a is the corresponding *eigenvalue*. The eigenvalues represent possible values of the corresponding physical quantity.

The operator is completely defined if we know the result of the operation on all possible wave functions. Since the operator is linear, it is sufficient to know the result of the operation on the basis functions,

$$\phi_k' = \alpha \phi_k \, . \tag{B.16}$$

Using the identity (B.9), this becomes

$$|\phi_k'\rangle = \sum_i |\phi_i\rangle\langle\phi_i|\alpha\phi_k\rangle \, . \tag{B.17}$$

The coefficient is here usually written $\langle\phi_i|\alpha|\phi_i\rangle$, and it corresponds in the Schrödinger representation to the integral

$$\langle\phi_i|\alpha|\phi_k\rangle = \int \phi_i(r)^* \alpha\phi_k(r) dr \, . \tag{B.18}$$

These coefficients form the *matrix representation* of the operator α.

The operator α in (B.18) is assumed to operate to the right. An operator $\alpha^\dagger$, which yields the same result operating to the left as α operating to the right, is called the *hermitian adjoint* of α,

$$\langle\overset{\frown}{\phi_i|\alpha^\dagger}|\phi_k\rangle = \langle\phi_i|\overset{\frown}{\alpha|\phi_k}\rangle \tag{B.19}$$

or

$$\int (\alpha^\dagger\phi_i)^* \phi_k dr = \int \phi_i^* \alpha\phi_k dr \, . \tag{B.20}$$

By taking the complex conjugate of the last equation, we get

$$\int \phi_k^* \alpha^\dagger\phi_i dr = [\int \phi_i^* \alpha\phi_k dr]^* \, . \tag{B.21}$$

or, in Dirac notation,

$$\langle\phi_k|\alpha^\dagger|\phi_i\rangle = \langle\phi_i|\alpha|\phi_k\rangle^* \, . \tag{B.22}$$

Thus, the hermitian adjoint operator $\alpha^\dagger$ is represented by a matrix, which is the hermitian adjoint of the matrix representation of α.

An operator, which is equal to its hermitian adjoint,

$$\alpha^\dagger = \alpha\,, \tag{B.23}$$

is said to be *self-adjoint* or *hermitian*. Such an operator is represented by a hermitian matrix,

$$\langle \phi_k | \alpha | \phi_l \rangle = \langle \phi_l | \alpha | \phi_k \rangle^*\,. \tag{B.24}$$

B.3 Theorems Concerning Hermitian Operators and Observables

Theorem B.1. *A hermitian operator has real eigenvalues.*

Proof:

$$\alpha \Psi = a \Psi$$

gives

$$\langle \Psi | \alpha | \Psi \rangle = a \langle \Psi | \Psi \rangle = a\,,$$

if Ψ is normalized. But if α is hermitian, (B.24) also gives

$$\langle \Psi | \alpha | \Psi \rangle = a^*\,,$$

which implies that a must be real.

Theorem B.2. *Eigenfunctions of a hermitian operator corresponding to* different *eigenvalues are orthogonal.*

Proof:

$$\alpha \phi_1 = a_1 \phi_1$$
$$\alpha \phi_2 = a_2 \phi_2$$

gives

$$\langle \phi_2 | \alpha | \phi_1 \rangle = a_1 \langle \phi_2 | \phi_1 \rangle$$
$$\langle \phi_1 | \alpha | \phi_2 \rangle = a_2 \langle \phi_1 | \phi_2 \rangle\,.$$

Since α is hermitian, the second equation can be rewritten as

$$\langle \phi_2 | \alpha | \phi_1 \rangle = a_2^* \langle \phi_1 | \phi_2 \rangle^* = a_2 \langle \phi_2 | \phi_1 \rangle\,,$$

using Theorem B.1 and equation (B.2). Subtracting this from the first equation, we get

$$(a_1 - a_2)\langle\phi_2|\phi_1\rangle = 0 .$$

This implies that the scalar product is zero, if the eigenvalues are different.

Theorem B.3. *An operator representing a measurable quantity (observable) has a complete set of eigenfunctions.*

This theorem is stated here without proof. Interested readers are referred to the book by *Dirac* [1958, p. 37].

Theorem B.4. *If the eigenfunctions of an observable α form the basis for a matrix representation, then the matrix of β is diagonal with respect to the eigenvalues of α, if α and β commute.*

Proof: The set $\{\phi_{ik}\}$ is an orthonormal set of eigenfunctions of α,

$$\alpha|\phi_{ik}\rangle = a_i|\phi_{ik}\rangle , \tag{B.25}$$

or, more briefly,

$$\alpha|ik\rangle = a_i|ik\rangle .$$

(The second index is used to distinguish eigenfunctions with the same eigenvalue.) The set $\{\phi_{ik}\}$ is complete according to Theorem B.3. Thus, the identity (B.9) can be written

$$\sum_{ik} |ik\rangle\langle ik| \equiv 1 . \tag{B.26}$$

Then,

$$\beta|ik\rangle = \sum_{i'k'} |i'k'\rangle\langle i'k'|\beta|ik\rangle$$

and

$$\alpha\beta|ik\rangle = \sum_{i'k'} a_{i'}|i'k'\rangle\langle i'k'|\beta|ik\rangle ,$$

using (B.25). But since α and β commute, this must be equal to

$$\beta\alpha|ik\rangle = a_i \sum_{i'k'} |i'k'\rangle\langle i'k'|\beta|ik\rangle .$$

Since the functions $|i'k'\rangle$ are mutually orthogonal, all the coefficients of the two expressions must be equal,

$$a_{i'}\langle i'k'|\beta|ik\rangle = a_i\langle i'k'|\beta|ik\rangle .$$

But this implies that

$$\langle i'k'|\beta|ik\rangle = 0 \quad \text{if} \quad a_i \neq a_{i'} \tag{B.27}$$

or, in other words, the matrix of β is diagonal with respect to i.

Theorem B.5. *If α and β are commuting observables, then the simultaneous eigenfunctions of α and β form a complete set.*

Proof: We use the same set $\{\phi_{ik}\}$ as in the proof of previous theorem,

$$\alpha|ik\rangle = a_i|ik\rangle \,.$$

Any linear combination

$$\Psi = \sum_k c_k|ik\rangle$$

is an eigenfunction of α with the eigenvalue a_i. Assume that Ψ is an eigenfunction also of β,

$$\beta\Psi = b\Psi$$

or

$$\beta\sum_k c_k|ik\rangle = b\sum_k c_k|ik\rangle \,.$$

Using the identity (B.26), we get

$$\sum_{kk'} c_k|ik'\rangle\langle ik'|\beta|ik\rangle = k\sum_{k'} c_k|ik'\rangle \,,$$

utilizing the fact that the matrix of β is diagonal with respect to i (Theorem B.4) and that the functions $|ik\rangle$ are orthonormal. This is a system of homogeneous, linear equations, which has nontrivial solutions only for certain values of b. These are the eigenvalues of β, which can be combined with a particular eigenvalue a_i of α. The corresponding eigenvectors represent the simultaneous eigenfunctions. We see that there are as many solutions as there are basis functions $|ik\rangle$ for that particular i. Since β is hermitian and, thus, has a complete set of eigenfunctions, the solutions must be linearly independent. It then follows that the simultaneous eigenfunctions span the entire function space.

Theorem B.6. *If α is a hermitian operator that commutes with* all *components of the angular-momentum operator, $\mathbf{j}$, then the matrix elements of α in a scheme where $\mathbf{j}^2$ and j_z are diagonal (jm representation) is independent of the eigenvalue (m) of j_z.*

Proof: The basis functions $|\gamma jm\rangle$ are eigenfunctions of $\mathbf{j}^2$ and j_z,

$$\mathbf{j}^2|\gamma jm\rangle = j(j+1)|\gamma jm\rangle$$
$$j_z|\gamma jm\rangle = m|\gamma jm\rangle \,.$$

It follows from Theorem B.4 that the matrix of α is diagonal in this representation. Thus the completeness relation (B.9) gives

$$\alpha|\gamma jm\rangle = \sum_{\gamma'} |\gamma'jm\rangle \langle \gamma'jm|\alpha|\gamma jm\rangle.$$

Let us operate on this equation with $j_+ = j_x + ij_y$, using (2.19a),

$$j_+\alpha|\gamma jm\rangle = [j(j+1) - (m+1)]^{1/2} \sum_{\gamma'} |\gamma'\, jm + 1\rangle \langle \gamma'jm|\alpha|\gamma jm\rangle$$

Similarly,

$$\alpha j_+|\gamma jm\rangle = \alpha[j(j+1) - m(m+1)]^{1/2}|\gamma jm + 1\rangle$$
$$= [j(j+1) - m(m+1)]^{1/2} \sum_{\gamma'} |\gamma'jm + 1\rangle \langle \gamma'jm + 1|\alpha|\gamma jm + 1\rangle.$$

Since j_+ and α commute, these expressions are identical. It then follows that

$$\langle \gamma'jm + 1|\alpha|\gamma jm + 1\rangle = \langle \gamma'jm|\alpha|\gamma jm\rangle,$$

or, in other words, that the matrix elements of α are independent of m.

Appendix C. Formulas of 3-j and 6-j Symbols

Reprint from: *Edmonds* [1957]

$$\begin{pmatrix} j_1 & j_2 & j_3 \\ 0 & 0 & 0 \end{pmatrix} = (-1)^{\frac{1}{2}J}\left[\frac{(j_1+j_2-j_3)!(j_1+j_3-j_2)!(j_2+j_3-j_1)!}{(j_1+j_2+j_3+1)!}\right]^{\frac{1}{2}}$$

$$\frac{(\frac{1}{2}J)!}{(\frac{1}{2}J-j_1)!(\frac{1}{2}J-j_2)!(\frac{1}{2}J-j_3)!}$$

if J is even.

$$\begin{pmatrix} j_1 & j_2 & j_3 \\ 0 & 0 & 0 \end{pmatrix} = 0 \quad \text{if} \quad J \text{ is odd} \quad \text{where} \quad J = j_1 + j_2 + j_3$$

$$\begin{pmatrix} J+\frac{1}{2} & J & \frac{1}{2} \\ M & -M-\frac{1}{2} & \frac{1}{2} \end{pmatrix} \qquad (-1)^{J-M-\frac{1}{2}}\left[\frac{J-M+\frac{1}{2}}{(2J+2)(2J+1)}\right]^{\frac{1}{2}}$$

$$\begin{pmatrix} J+1 & J & 1 \\ M & -M-1 & 1 \end{pmatrix} \qquad (-1)^{J-M-1}\left[\frac{(J-M)(J-M+1)}{(2J+3)(2J+2)(2J+1)}\right]^{\frac{1}{2}}$$

$$\begin{pmatrix} J+1 & J & 1 \\ M & -M & 0 \end{pmatrix} \qquad (-1)^{J-M-1}\left[\frac{(J+M+1)(J-M+1)\cdot 2}{(2J+3)(2J+2)(2J+1)}\right]^{\frac{1}{2}}$$

$$\begin{pmatrix} J & J & 1 \\ M & -M-1 & 1 \end{pmatrix} \qquad (-1)^{J-M}\left[\frac{(J-M)(J+M+1)\cdot 2}{(2J+2)(2J+1)(2J)}\right]^{\frac{1}{2}}$$

$$\begin{pmatrix} J & J & 1 \\ M & -M & 0 \end{pmatrix} \qquad (-1)^{J-M}\frac{M}{[(2J+1)(J+1)J]^{\frac{1}{2}}}$$

$$\begin{Bmatrix} a & b & c \\ 1 & c-1 & b-1 \end{Bmatrix} = (-1)^s\left[\frac{s(s+1)(s-2a-1)(s-2a)}{(2b-1)2b(2b+1)(2c-1)2c(2c+1)}\right]^{\frac{1}{2}}$$

$$\begin{Bmatrix} a & b & c \\ 1 & c-1 & b \end{Bmatrix} = (-1)^s\left[\frac{2(s+1)(s-2a)(s-2b)(s-2c+1)}{2b(2b+1)(2b+2)(2c-1)2c(2c+1)}\right]^{\frac{1}{2}}$$

$$\begin{Bmatrix} a & b & c \\ 1 & c-1 & b+1 \end{Bmatrix} = (-1)^s\left[\frac{(s-2b-1)(s-2b)(s-2c+1)(s-2c+2)}{(2b+1)(2b+2)(2b+3)(2c-1)2c(2c+1)}\right]^{\frac{1}{2}}$$

$$\begin{Bmatrix} a & b & c \\ 1 & c & b \end{Bmatrix} = (-1)^{s+1}\frac{2[b(b+1)+c(c+1)-a(a+1)]}{[2b(2b+1)(2b+2)2c(2c+1)(2c+2)]^{\frac{1}{2}}}$$

where $s = a + b + c$.

Appendix D. Table of 3-*j* and 6-*j* Symbols

Reprint from: *Rotenberg* et al. [1959]

Use of the Tables

Use of the Tables of 3-*j* Symbols

The squares of the 3-*j* symbols are rational fractions. For this reason, it is the *squares* of the 3-*j* symbols which are tabulated. If the negative radical is to be used, the number is preceded by an asterisk.

The notation adopted is, in essence, that of Sharp and co-workers.[†]

Any integer may be expressed in terms of products of prime factors. If it is specified that the primes be listed in increasing order with the smallest prime at the left, the representation is unique. For example,

$$20 = 2 \times 2 \times 5 = 2^2 \times 3^0 \times 5^1$$

$$375 = 2^0 \times 3^1 \times 5^3.$$

A unique but much more compact notation is to list only the powers of the prime factors:

$$20 \longrightarrow 201$$

$$375 \longrightarrow 013.$$

Rational fractions are represented in the same notation, with the convention that negative exponents are underscored. Thus,

$$12/49 = 2^2 \times 3^1/7^2 \longrightarrow 210\underline{2}.$$

The following conventions have been adopted:

1. If a power should exceed 10, then only the excess over 10 is written, with a bar *above* it. For example,

$$6144 = 2^{11} \times 3^1 \longrightarrow \bar{1}1.$$

2. A number is preceded by an asterisk if the negative radical is to be used.

[†] W. T. Sharp, J. M. Kennedy, B. J. Sears, and M. G. Hoyle, *Tables of Coefficients for Angular Distribution Analysis*, CRT 556, AECL No. 97, Atomic Energy of Canada Limited, Chalk River, Ontario (1954).

the 3-j and 6-j symbols

3. The first 11 primes are 2, 3, 5, 7, 11, 13, 17, 19, 23, 29, 31. If a number cannot be completely factored by the first 11 primes, the *remainder* appears in parentheses immediately after the prime factors, in conventional arithmetic notation. Example:

$$1(37)^2 \text{ means } 2^1 \times 37^2 = 2738.$$

Sometimes a number will appear with two sets of parentheses, such as *09$(41)^2 \times$ $(10657)^2$. The first parentheses contain the product of all the prime factors between, and including, 37 and 97. The second parentheses contain the product of all factors greater than 97. Note that if only one set of parentheses appears it means either *all* factors contained in it are greater than 97 or *all* are 97 or less. (It turns out that the numbers in parentheses are usually prime.)

For convenience, a comma is used to set the primes off in groups of four. Thus, all the primes in the first group are less than 10, and in the second group all primes are between 10 and 20. (In the tables of special 3-j symbols, where 15 prime factors are used, the third group includes all the primes between 20 and 40.)

4. Coefficients that are zero for physical reasons are omitted. "Accidental" zeros are represented by a single 0.

Advantage has been taken of the symmetry properties expressed in Eqs. 1.6, 1.7, and 1.8: Only those coefficients for which $j_1 \geq j_2 \geq j_3$ are tabulated. By interchanging columns according to Eqs. 1.6 and 1.7, any 3-j symbol may be brought to this standard form. By using Eq. 1.8, the sign of m_2 may be changed. Consequently, only the symbols for $m_2 \leq 0$ have been tabulated.

To summarize, the following look-up procedure for the 3-j symbols is recommended:

Rule 1.† Interchange the columns so that $j_1 \geq j_2 \geq j_3$.

Rule 2.† If necessary, change the signs of all the magnetic quantum numbers so that $m_2 \leq 0$.

The parameters are listed in the order $j_1, j_2, j_3, m_1, m_2, m_3$. The parameters j_1, j_2, j_3, m_1, m_2 are listed in a true odometer manner; that is, j_1 varies most slowly, m_2 most quickly; m_3 simply follows the rule that $m_1 + m_2 = -m_3$.

Use of the Table of 6-j Symbols

Like the squares of the 3-j symbols, the squares of the 6-j symbols are rational fractions, which is the reason the *squares* of the 6-j symbols are tabulated. If the negative radical is to be used, the number is preceded by an asterisk.

---

† The possible sign change indicated by Eq. 1.7 or Eq. 1.8 must always be kept in mind.

use of the table of 6-j symbols

The "power-of-prime-factor" notation is used again. See page 434 under "Use of the Tables of 3-j Symbols" for a guide to this notation.

The following symmetry properties have been used to shorten the table:

(a) The columns of a 6-j symbol may be rearranged in any order.

(b) Any *two* numbers in the bottom row of a 6-j symbol may be interchanged with the *corresponding* two numbers in the top row.

Denote a 6-j symbol as

$$\begin{Bmatrix} j_1 & j_2 & j_3 \\ l_1 & l_2 & l_3 \end{Bmatrix}.$$

To find any 6-j symbol in the table:

Rule 1. Use symmetry properties (a) and (b) to place the *largest* of the six parameters in the *upper left corner* (the j_1 position).

Rule 2. Use the symmetry properties again to place the *largest of the remaining four* parameters in the *middle of the top row* (the j_2 position).

Rule 3. If $j_1 = j_2$, then, if necessary, use symmetry property (a) to make $l_1 > l_2$.

These rules may be summarized in a simple diagram. Let an arrow ($\longrightarrow$) stand for the symbol $\geq$ (greater than or equal). The symbol that is listed is the one for which

$$\begin{Bmatrix} j_1 \longrightarrow j_2 \longrightarrow j_3 \\ l_1 \dashrightarrow l_2 \quad\ l_3 \end{Bmatrix}.$$

The dotted arrow holds if $j_1 = j_2$.

Examples. In the foregoing, assume that $A > B > C > D > E > F$. Suppose that we are confronted with the symbol $\begin{Bmatrix} F & B & C \\ E & A & D \end{Bmatrix}$. We wish to get A in the j_1 position since it is the largest parameter. Use symmetry (a), $\begin{Bmatrix} B & F & C \\ A & E & D \end{Bmatrix}$. Then use symmetry (b), $\begin{Bmatrix} A & E & C \\ B & F & D \end{Bmatrix}$. Since, of the remaining four parameters $E, F, C,$ and D, C is the largest, use symmetry (a) to place it in the j_2 position: $\begin{Bmatrix} A & C & E \\ B & D & F \end{Bmatrix}$. This coefficient will be found in the tables.

the 3-j and 6-j symbols

As another example, let the 6-j symbol first appear as $\begin{Bmatrix} D & C & F \\ A & A & E \end{Bmatrix}$. Symmetry (b) allows this to become $\begin{Bmatrix} A & A & F \\ D & C & E \end{Bmatrix}$, but Rule 3 requires that $l_1 > l_2$. Use of symmetry (a) gives $\begin{Bmatrix} A & A & F \\ C & D & E \end{Bmatrix}$, which will be found in the tables.

We list a few more examples. In each case, the last symbol is the one that can be found in the tables:

$$\begin{Bmatrix} 1 & 2 & 3 \\ 4 & 5 & 6 \end{Bmatrix} = \begin{Bmatrix} 3 & 2 & 1 \\ 6 & 5 & 4 \end{Bmatrix} = \begin{Bmatrix} 6 & 5 & 1 \\ 3 & 2 & 4 \end{Bmatrix}$$

$$\begin{Bmatrix} 1 & 2 & 6 \\ 6 & 5 & 3 \end{Bmatrix} = \begin{Bmatrix} 1 & 6 & 2 \\ 6 & 3 & 5 \end{Bmatrix} = \begin{Bmatrix} 6 & 6 & 5 \\ 1 & 3 & 2 \end{Bmatrix} = \begin{Bmatrix} 6 & 6 & 5 \\ 3 & 1 & 2 \end{Bmatrix}$$

$$\begin{Bmatrix} 1 & 6 & 3 \\ 2 & 6 & 4 \end{Bmatrix} = \begin{Bmatrix} 6 & 3 & 1 \\ 6 & 4 & 2 \end{Bmatrix} = \begin{Bmatrix} 6 & 4 & 1 \\ 6 & 3 & 2 \end{Bmatrix}.$$

The parameters are listed in the table in the order $j_1, j_2, j_3, l_1, l_2, l_3$. Odometer ordering is used throughout; that is, l_3 varies most rapidly, j_1 most slowly.

3-j Symbols

j_1	j_2	j_3	m_1	m_2	m_3	
1	1	0	0	0	0	*01
2	1	1	0	0	0	111
2	2	0	0	0	0	001
2	2	2	0	0	0	*1011,
3	2	1	0	0	0	*0111,
3	3	0	0	0	0	*0001,
3	3	2	0	0	0	2111,
4	2	2	0	0	0	1011,
4	3	1	0	0	0	2201,
4	3	3	0	0	0	*1001,1
4	4	0	0	0	0	02
4	4	2	0	0	0	*2211,1
4	4	4	0	0	0	1201,11
5	3	2	0	0	0	*1111,1
5	4	1	0	0	0	*0210,1
5	4	3	0	0	0	2011,11
5	5	0	0	0	0	*0000,1
5	5	2	0	0	0	1110,11
5	5	4	0	0	0	*1000,11
6	3	3	0	0	0	2121,11
6	4	2	0	0	0	0010,11
6	4	4	0	0	0	*2210,11
6	5	1	0	0	0	1100,11
6	5	3	0	0	0	*0101,11
6	5	5	0	0	0	4110,111
6	6	0	0	0	0	0000,01
6	6	2	0	0	0	*1011,11
6	6	4	0	0	0	2001,111
6	6	6	0	0	0	*4020,1111,
7	4	3	0	0	0	*0211,11
7	5	2	0	0	0	*0111,11
7	5	4	0	0	0	3211,111
7	6	1	0	0	0	*0111,01
7	6	3	0	0	0	3111,111
7	6	5	0	0	0	*2111,1111,
7	7	0	0	0	0	*011
7	7	2	0	0	0	3111,011
7	7	4	0	0	0	*2411,1111,
7	7	6	0	0	0	3130,1111,
8	4	4	0	0	0	1212,111
8	5	3	0	0	0	3001,111
8	5	5	0	0	0	*1012,1111,
8	6	2	0	0	0	2011,011
8	6	4	0	0	0	*3201,1111,
8	6	6	0	0	0	1021,1111,

j_1	j_2	j_3	m_1	m_2	m_3	
8	7	1	0	0	0	3110,001
8	7	3	0	0	0	*2211,0111,
8	7	5	0	0	0	3210,1111,
8	7	7	0	0	0	*1131,0111,1
8	8	0	0	0	0	0000,001
8	8	2	0	0	0	*3110,0011,
8	8	4	0	0	0	2200,0111,
8	8	6	0	0	0	*3120,0111,1
8	8	8	0	0	0	1012,0111,1
9	5	4	0	0	0	*1202,1111,
9	6	3	0	0	0	*2101,0111,
9	6	5	0	0	0	2111,1111,
9	7	2	0	0	0	*2210,0011,
9	7	4	0	0	0	3010,0111,
9	7	6	0	0	0	*1211,0111,1
9	8	1	0	0	0	*0200,0011,
9	8	3	0	0	0	3101,0011,
9	8	5	0	0	0	*2110,1111,1
9	8	7	0	0	0	3201,0111,1
9	9	0	0	0	0	*0000,0001,
9	9	2	0	0	0	1111,0011,
9	9	4	0	0	0	*2201,1011,1
9	9	6	0	0	0	4100,1111,1
9	9	8	0	0	0	*1102,0011,1
10	5	5	0	0	0	2301,1111,
10	6	4	0	0	0	1011,0111,
10	6	6	0	0	0	*2301,0111,1
10	7	3	0	0	0	3011,0011,
10	7	5	0	0	0	*1011,1111,1
10	7	7	0	0	0	3521,0111,1
10	8	2	0	0	0	0211,0011,
10	8	4	0	0	0	*3011,1011,1
10	8	6	0	0	0	2211,1111,1
10	8	8	0	0	0	*3111,0011,1
10	9	1	0	0	0	1111,0001,
10	9	3	0	0	0	*0211,1011,1
10	9	5	0	0	0	4211,1011,1
10	9	7	0	0	0	*2201,1011,1
10	9	9	0	0	0	2113,0011,11
10	10	0	0	0	0	0101,
10	10	2	0	0	0	*1111,1001,1
10	10	4	0	0	0	1411,1011,1
10	10	6	0	0	0	*4201,1111,1
10	10	8	0	0	0	1203,1011,11
10	10	10	0	0	0	*2413,0011,111

3-j Symbols

j_1	j_2	j_3	m_1	m_2	m_3	
1/2	1/2	0	1/2	-1/2	0	1
1	1/2	1/2	0	-1/2	1/2	11
1	1/2	1/2	1	-1/2	-1/2	*01
1	1	0	0	0	0	*01
1	1	0	1	-1	0	01
1	1	1	-1	0	1	11
1	1	1	0	-1	1	*11
1	1	1	0	0	0	0
1	1	1	1	-1	0	11
1	1	1	1	0	-1	*11
3/2	1	1/2	-1/2	0	1/2	*11
3/2	1	1/2	1/2	-1	1/2	21
3/2	1	1/2	1/2	0	-1/2	11
3/2	1	1/2	3/2	-1	-1/2	*2
3/2	3/2	0	1/2	-1/2	0	*2
3/2	3/2	0	3/2	-3/2	0	2
3/2	3/2	1	-1/2	-1/2	1	111
3/2	3/2	1	1/2	-3/2	1	*101
3/2	3/2	1	1/2	-1/2	0	*211
3/2	3/2	1	3/2	-3/2	0	211
3/2	3/2	1	3/2	-1/2	-1	*101
2	1	1	-1	0	1	*101
2	1	1	0	-1	1	111
2	1	1	0	0	0	111
2	1	1	1	-1	0	*101
2	1	1	1	0	-1	*101
2	1	1	2	-1	-1	001
2	3/2	1/2	0	-1/2	1/2	*101
2	3/2	1/2	1	-3/2	1/2	201
2	3/2	1/2	1	-1/2	-1/2	211
2	3/2	1/2	2	-3/2	-1/2	*001
2	3/2	3/2	-1	-1/2	3/2	101
2	3/2	3/2	0	-3/2	3/2	*201
2	3/2	3/2	0	-1/2	1/2	*201
2	3/2	3/2	1	-3/2	1/2	101
2	3/2	3/2	1	-1/2	-1/2	0
2	3/2	3/2	2	-3/2	-1/2	*101
2	3/2	3/2	2	-1/2	-3/2	101
2	2	0	0	0	0	001
2	2	0	1	-1	0	*001
2	2	0	2	-2	0	001
2	2	1	-1	0	1	*101
2	2	1	0	-1	1	101
2	2	1	0	0	0	0
2	2	1	1	-2	1	*011
2	2	1	1	-1	0	*111
2	2	1	1	0	-1	101
2	2	1	2	-2	0	111
2	2	1	2	-1	-1	*011
2	2	2	-2	0	2	1011,
2	2	2	-1	-1	2	*0111,
2	2	2	-1	0	1	1011,
2	2	2	0	-2	2	1011,
2	2	2	0	-1	1	1011,
2	2	2	0	0	0	*1011,
2	2	2	1	-2	1	*0111,
2	2	2	1	-1	0	1011,
2	2	2	1	0	-1	1011,
2	2	2	2	-2	0	1011,
2	2	2	2	-1	-1	*0111,
2	2	2	2	0	-2	1011,
5/2	3/2	1	-1/2	-1/2	1	*201
5/2	3/2	1	1/2	-3/2	1	211
5/2	3/2	1	1/2	-1/2	0	101
5/2	3/2	1	3/2	-3/2	0	*011
5/2	3/2	1	3/2	-1/2	-1	*101
5/2	3/2	1	5/2	-3/2	-1	11
5/2	2	1/2	-1/2	0	1/2	101
5/2	2	1/2	1/2	-1	1/2	*011
5/2	2	1/2	1/2	0	-1/2	*101
5/2	2	1/2	3/2	-2	1/2	111
5/2	2	1/2	3/2	-1	-1/2	111
5/2	2	1/2	5/2	-2	-1/2	*11
5/2	2	3/2	-3/2	0	3/2	*0111,
5/2	2	3/2	-1/2	-1	3/2	2211,
5/2	2	3/2	-1/2	0	1/2	1011,
5/2	2	3/2	1/2	-2	3/2	*0011,
5/2	2	3/2	1/2	-1	1/2	*2111,
5/2	2	3/2	1/2	0	-1/2	1011,
5/2	2	3/2	3/2	-2	1/2	3111,
5/2	2	3/2	3/2	-1	-1/2	·1111,
5/2	2	3/2	3/2	0	-3/2	*0111,
5/2	2	3/2	5/2	-2	-1/2	*1101,
5/2	2	3/2	5/2	-1	-3/2	1001,
5/2	5/2	0	1/2	-1/2	0	11
5/2	5/2	0	3/2	-3/2	0	*11
5/2	5/2	0	5/2	-5/2	0	11
5/2	5/2	1	-1/2	-1/2	1	*0111,
5/2	5/2	1	1/2	-3/2	1	3111,
5/2	5/2	1	1/2	-1/2	0	1111,

3-j Symbols

j_1	j_2	j_3	m_1	m_2	m_3	
5/2	5/2	1	3/2	-5/2	1	*0101,
5/2	5/2	1	3/2	-3/2	0	*1111,
5/2	5/2	1	3/2	-1/2	-1	3111,
5/2	5/2	1	5/2	-5/2	0	1111,
5/2	5/2	1	5/2	-3/2	-1	*0101,
5/2	5/2	2	-3/2	-1/2	2	2211,
5/2	5/2	2	-1/2	-3/2	2	*2211,
5/2	5/2	2	-1/2	-1/2	1	0
5/2	5/2	2	1/2	-5/2	2	2001,
5/2	5/2	2	1/2	-3/2	1	0011,
5/2	5/2	2	1/2	-1/2	0	*2111,
5/2	5/2	2	3/2	-5/2	1	*1001,
5/2	5/2	2	3/2	-3/2	0	2111,
5/2	5/2	2	3/2	-1/2	-1	0011,
5/2	5/2	2	5/2	-5/2	0	2111,
5/2	5/2	2	5/2	-3/2	-1	*1001,
5/2	5/2	2	5/2	-1/2	-2	2001,
3	3/2	3/2	-1	-1/2	3/2	*0011,
3	3/2	3/2	0	-3/2	3/2	2011,
3	3/2	3/2	0	-1/2	1/2	2211,
3	3/2	3/2	1	-3/2	1/2	*0011,
3	3/2	3/2	1	-1/2	-1/2	*0111,
3	3/2	3/2	2	-3/2	-1/2	1001,
3	3/2	3/2	2	-1/2	-3/2	1001,
3	3/2	3/2	3	-3/2	-3/2	*0001,
3	2	1	-1	0	1	1011,
3	2	1	0	-1	1	*0011,
3	2	1	0	0	0	*0111,
3	2	1	1	-2	1	0111,
3	2	1	1	-1	0	3111,
3	2	1	1	0	-1	1011,
3	2	1	2	-2	0	*0101,
3	2	1	2	-1	-1	*1101,
3	2	1	3	-2	-1	0001,
3	2	2	-2	0	2	*1001,
3	2	2	-1	-1	2	1111,
3	2	2	-1	0	1	0011,
3	2	2	0	-2	2	*1011,
3	2	2	0	-1	1	*1011,
3	2	2	0	0	0	0
3	2	2	1	-2	1	1111,
3	2	2	1	-1	0	0011,
3	2	2	1	0	-1	*0011,
3	2	2	2	-2	0	*1001,
3	2	2	2	-1	-1	0

j_1	j_2	j_3	m_1	m_2	m_3	
3	2	2	2	0	-2	1001,
3	2	2	3	-2	-1	1001,
3	2	2	3	-1	-2	*1001,
3	5/2	1/2	0	-1/2	1/2	1001,
3	5/2	1/2	1	-3/2	1/2	*0101,
3	5/2	1/2	1	-1/2	-1/2	*1101,
3	5/2	1/2	2	-5/2	1/2	1101,
3	5/2	1/2	2	-3/2	-1/2	1111,
3	5/2	1/2	3	-5/2	-1/2	*0001,
3	5/2	3/2	-1	-1/2	3/2	*2211,
3	5/2	3/2	0	-3/2	3/2	1111,
3	5/2	3/2	0	-1/2	1/2	0011,
3	5/2	3/2	1	-5/2	3/2	*3001,
3	5/2	3/2	1	-3/2	1/2	*3111,
3	5/2	3/2	1	-1/2	-1/2	2111,
3	5/2	3/2	2	-5/2	1/2	2111,
3	5/2	3/2	2	-3/2	-1/2	2101,
3	5/2	3/2	2	-1/2	-3/2	*1001,
3	5/2	3/2	3	-5/2	-1/2	*3011,
3	5/2	3/2	3	-3/2	-3/2	3101,
3	5/2	5/2	-2	-1/2	5/2	2111,
3	5/2	5/2	-1	-3/2	5/2	*0101,
3	5/2	5/2	-1	-1/2	3/2	*1111,
3	5/2	5/2	0	-5/2	5/2	2211,
3	5/2	5/2	0	-3/2	3/2	2211,
3	5/2	5/2	0	-1/2	1/2	*2211,
3	5/2	5/2	1	-5/2	3/2	*0101,
3	5/2	5/2	1	-3/2	1/2	*1111,
3	5/2	5/2	1	-1/2	-1/2	2111,
3	5/2	5/2	2	-5/2	1/2	2111,
3	5/2	5/2	2	-3/2	-1/2	*2101,
3	5/2	5/2	2	-1/2	-3/2	*2101,
3	5/2	5/2	3	-5/2	-1/2	*1211,
3	5/2	5/2	3	-3/2	-3/2	2201,
3	5/2	5/2	3	-1/2	-5/2	*1211,
3	3	0	0	0	0	*0001,
3	3	0	1	-1	0	0001,
3	3	0	2	-2	0	*0001,
3	3	0	3	-3	0	0001,
3	3	1	-1	0	1	1001,
3	3	1	0	-1	1	*1001,
3	3	1	0	0	0	0
3	3	1	1	-2	1	2111,
3	3	1	1	-1	0	2101,
3	3	1	1	0	-1	*1001,

j_1	j_2	j_3	m_1	m_2	m_3	
3	3	1	2	-3	1	*2001,
3	3	1	2	-2	0	*0101,
3	3	1	2	-1	-1	2111,
3	3	1	3	-3	0	2101,
3	3	1	3	-2	-1	*2001,
3	3	2	-2	0	2	*0101,
3	3	2	-1	-1	2	1011,
3	3	2	-1	0	1	*1111,
3	3	2	0	-2	2	*0101,
3	3	2	0	-1	1	*1111,
3	3	2	0	0	0	2111,
3	3	2	1	-3	2	1101,
3	3	2	1	-2	1	2001,
3	3	2	1	-1	0	*2111,
3	3	2	1	0	-1	*1111,
3	3	2	2	-3	1	*2111,
3	3	2	2	-2	0	0
3	3	2	2	-1	-1	2001,
3	3	2	2	0	-2	*0101,
3	3	2	3	-3	0	2111,
3	3	2	3	-2	-1	*2111,
3	3	2	3	-1	-2	1101,
3	3	3	-3	0	3	1101,
3	3	3	-2	-1	3	*0101,
3	3	3	-2	0	2	1101,
3	3	3	-1	-2	3	0101,
3	3	3	-1	-1	2	0
3	3	3	-1	0	1	*1101,
3	3	3	0	-3	3	*1101,
3	3	3	0	-2	2	*1101,
3	3	3	0	-1	1	1101,
3	3	3	0	0	0	0
3	3	3	1	-3	2	0101,
3	3	3	1	-2	1	0
3	3	3	1	-1	0	*1101,
3	3	3	1	0	-1	1101,
3	3	3	2	-3	1	*0101,
3	3	3	2	-2	0	1101,
3	3	3	2	-1	-1	0
3	3	3	2	0	-2	*1101,
3	3	3	3	-3	0	1101,
3	3	3	3	-2	-1	*0101,
3	3	3	3	-1	-2	0101,
3	3	3	3	0	-3	*1101,
7/2	2	3/2	-3/2	0	3/2	2001,

j_1	j_2	j_3	m_1	m_2	m_3	
7/2	2	3/2	-1/2	-1	3/2	*1011,
7/2	2	3/2	-1/2	0	1/2	*2211,
7/2	2	3/2	1/2	-2	3/2	3011,
7/2	2	3/2	1/2	-1	1/2	1111,
7/2	2	3/2	1/2	0	-1/2	2211,
7/2	2	3/2	3/2	-2	1/2	*3001,
7/2	2	3/2	3/2	-1	-1/2	*1001,
7/2	2	3/2	3/2	0	-3/2	*2001,
7/2	2	3/2	5/2	-2	-1/2	3101,
7/2	2	3/2	5/2	-1	-3/2	1001,
7/2	2	3/2	7/2	-2	-3/2	*3
7/2	5/2	1	-1/2	-1/2	1	2001,
7/2	5/2	1	1/2	-3/2	1	*3001,
7/2	5/2	1	1/2	-1/2	0	*1001,
7/2	5/2	1	3/2	-5/2	1	3101,
7/2	5/2	1	3/2	-3/2	0	2111,
7/2	5/2	1	3/2	-1/2	-1	2111,
7/2	5/2	1	5/2	-5/2	0	*2001,
7/2	5/2	1	5/2	-3/2	-1	*3011,
7/2	5/2	1	7/2	-5/2	-1	3
7/2	5/2	2	-3/2	-1/2	2	*0101,
7/2	5/2	2	-1/2	-3/2	2	3211,
7/2	5/2	2	-1/2	-1/2	1	2211,
7/2	5/2	2	1/2	-5/2	2	*1201,
7/2	5/2	2	1/2	-3/2	1	*3211,2
7/2	5/2	2	1/2	-1/2	0	*1111,
7/2	5/2	2	3/2	-5/2	1	3111,
7/2	5/2	2	3/2	-3/2	0	2001,
7/2	5/2	2	3/2	-1/2	-1	*2101,
7/2	5/2	2	5/2	-5/2	0	*2111,
7/2	5/2	2	5/2	-3/2	-1	*3201,
7/2	5/2	2	5/2	-1/2	-2	2201,
7/2	5/2	2	7/2	-5/2	-1	321
7/2	5/2	2	7/2	-3/2	-2	*12
7/2	3	1/2	-1/2	0	1/2	*1001,
7/2	3	1/2	1/2	-1	1/2	3101,
7/2	3	1/2	1/2	0	-1/2	1001,
7/2	3	1/2	3/2	-2	1/2	*2001,
7/2	3	1/2	3/2	-1	-1/2	*3011,
7/2	3	1/2	5/2	-3	1/2	3001,
7/2	3	1/2	5/2	-2	-1/2	2101,
7/2	3	1/2	7/2	-3	-1/2	*3
7/2	3	3/2	-3/2	0	3/2	2111,
7/2	3	3/2	-1/2	-1	3/2	*0101,
7/2	3	3/2	-1/2	0	1/2	*2101,

3-j Symbols

j_1	j_2	j_3	m_1	m_2	m_3	
7/2	3	3/2	1/2	-2	3/2	3111,
7/2	3	3/2	1/2	-1	1/2	2001,
7/2	3	3/2	1/2	0	-1/2	*2101,
7/2	3	3/2	3/2	-3	3/2	*2101,
7/2	3	3/2	3/2	-2	1/2	*3101,
7/2	3	3/2	3/2	-1	-1/2	0
7/2	3	3/2	3/2	0	-3/2	2111,
7/2	3	3/2	5/2	-3	1/2	0101,
7/2	3	3/2	5/2	-2	-1/2	3001,
7/2	3	3/2	5/2	-1	-3/2	*2111,
7/2	3	3/2	7/2	-3	-1/2	*21
7/2	3	3/2	7/2	-2	-3/2	31
7/2	3	5/2	-5/2	0	5/2	*0101,
7/2	3	5/2	-3/2	-1	5/2	0101,
7/2	3	5/2	-3/2	0	3/2	0
7/2	3	5/2	-1/2	-2	5/2	*1201,
7/2	3	5/2	-1/2	-1	3/2	*0201,
7/2	3	5/2	-1/2	0	1/2	1101,
7/2	3	5/2	1/2	-3	5/2	2101,
7/2	3	5/2	1/2	-2	3/2	1211,
7/2	3	5/2	1/2	-1	1/2	*3201,
7/2	3	5/2	1/2	0	-1/2	*1101,
7/2	3	5/2	3/2	-3	3/2	*2001,
7/2	3	5/2	3/2	-2	1/2	*2101,
7/2	3	5/2	3/2	-1	-1/2	3111,
7/2	3	5/2	3/2	0	-3/2	0
7/2	3	5/2	5/2	-3	1/2	3101,
7/2	3	5/2	5/2	-2	-1/2	*2201,
7/2	3	5/2	5/2	-1	-3/2	*2211,
7/2	3	5/2	5/2	0	-5/2	0101,
7/2	3	5/2	7/2	-3	-1/2	*31
7/2	3	5/2	7/2	-2	-3/2	12
7/2	3	5/2	7/2	-1	-5/2	*22
7/2	7/2	0	1/2	-1/2	0	*3
7/2	7/2	0	3/2	-3/2	0	3
7/2	7/2	0	5/2	-5/2	0	*3
7/2	7/2	0	7/2	-7/2	0	3
7/2	7/2	1	-1/2	-1/2	1	2201,
7/2	7/2	1	1/2	-3/2	1	*2111,
7/2	7/2	1	1/2	-1/2	0	*3201,
7/2	7/2	1	3/2	-5/2	1	0101,
7/2	7/2	1	3/2	-3/2	0	3001,
7/2	7/2	1	3/2	-1/2	-1	*2111,
7/2	7/2	1	5/2	-7/2	1	*22
7/2	7/2	1	5/2	-5/2	0	*3221,

j_1	j_2	j_3	m_1	m_2	m_3	
7/2	7/2	1	5/2	-3/2	-1	0101,
7/2	7/2	1	7/2	-7/2	0	3201,
7/2	7/2	1	7/2	-5/2	-1	*22
7/2	7/2	2	-3/2	-1/2	2	*0101,
7/2	7/2	2	-1/2	-3/2	2	0101,
7/2	7/2	2	-1/2	-1/2	1	0
7/2	7/2	2	1/2	-5/2	2	*2001,
7/2	7/2	2	1/2	-3/2	1	*2101,
7/2	7/2	2	1/2	-1/2	0	3111,
7/2	7/2	2	3/2	-7/2	2	211
7/2	7/2	2	3/2	-5/2	1	2111,
7/2	7/2	2	3/2	-3/2	0	*3111,
7/2	7/2	2	3/2	-1/2	-1	*2101,
7/2	7/2	2	5/2	-7/2	1	*201
7/2	7/2	2	5/2	-5/2	0	*3111,
7/2	7/2	2	5/2	-3/2	-1	2111,
7/2	7/2	2	5/2	-1/2	-2	*2001,
7/2	7/2	2	7/2	-7/2	0	3111,
7/2	7/2	2	7/2	-5/2	-1	*201
7/2	7/2	2	7/2	-3/2	-2	211
7/2	7/2	3	-5/2	-1/2	3	3101,1
7/2	7/2	3	-3/2	-3/2	3	*1111,1
7/2	7/2	3	-3/2	-1/2	2	0101,1
7/2	7/2	3	-1/2	-5/2	3	3101,1
7/2	7/2	3	-1/2	-3/2	2	0101,1
7/2	7/2	3	-1/2	-1/2	1	*1001,1
7/2	7/2	3	1/2	-7/2	3	*1100,1
7/2	7/2	3	1/2	-5/2	2	*2201,1
7/2	7/2	3	1/2	-3/2	1	1111,1
7/2	7/2	3	1/2	-1/2	0	3101,1
7/2	7/2	3	3/2	-7/2	2	2110,1
7/2	7/2	3	3/2	-5/2	1	1101,1
7/2	7/2	3	3/2	-3/2	0	*3101,1
7/2	7/2	3	3/2	-1/2	-1	1111,1
7/2	7/2	3	5/2	-7/2	1	*1000,1
7/2	7/2	3	5/2	-5/2	0	3121,1
7/2	7/2	3	5/2	-3/2	-1	1101,1
7/2	7/2	3	5/2	-1/2	-2	*2201,1
7/2	7/2	3	7/2	-7/2	0	3101,1
7/2	7/2	3	7/2	-5/2	-1	*1000,1
7/2	7/2	3	7/2	-3/2	-2	2110,1
7/2	7/2	3	7/2	-1/2	-3	*1100,1
4	2	2	-2	0	2	1101,
4	2	2	-1	-1	2	*1201,
4	2	2	-1	0	1	*0101,

3-*j* Symbols

j_1	j_2	j_3	m_1	m_2	m_3		j_1	j_2	j_3	m_1	m_2	m_3	
4	2	2	0	-2	2	1211,	4	3	1	1	-1	0	*2111,
4	2	2	0	-1	1	3211,	4	3	1	1	0	-1	*1211,
4	2	2	0	0	0	1011,	4	3	1	2	-3	1	2201,
4	2	2	1	-2	1	*1201,	4	3	1	2	-2	0	0101,
4	2	2	1	-1	0	*0101,	4	3	1	2	-1	-1	2111,
4	2	2	1	0	-1	*0101,	4	3	1	3	-3	0	*22
4	2	2	2	-2	0	1101,	4	3	1	3	-2	-1	*21
4	2	2	2	-1	-1	2201,	4	3	1	4	-3	-1	02
4	2	2	2	0	-2	1101,	4	3	2	-2	0	2	0101,
4	2	2	3	-2	-1	*12	4	3	2	-1	-1	2	*1201,
4	2	2	3	-1	-2	*12	4	3	2	-1	0	1	*1101,
4	2	2	4	-2	-2	02	4	3	2	0	-2	2	0201,
4	5/2	3/2	-1	-1/2	3/2	2211,	4	3	2	0	-1	1	1211,
4	5/2	3/2	0	-3/2	3/2	*1201,	4	3	2	0	0	0	0
4	5/2	3/2	0	-1/2	1/2	*0101,	4	3	2	1	-3	2	*1111,
4	5/2	3/2	1	-5/2	3/2	3201,	4	3	2	1	-2	1	*2211,
4	5/2	3/2	1	-3/2	1/2	3111,	4	3	2	1	-1	0	*2101,
4	5/2	3/2	1	-1/2	-1/2	2111,	4	3	2	1	0	-1	1101,
4	5/2	3/2	2	-5/2	1/2	*2101,	4	3	2	2	-3	1	2111,
4	5/2	3/2	2	-3/2	-1/2	*2111,	4	3	2	2	-2	0	2111,
4	5/2	3/2	2	-1/2	-3/2	*1211,	4	3	2	2	-1	-1	*2201,
4	5/2	3/2	3	-5/2	-1/2	31	4	3	2	2	0	-2	*0101,
4	5/2	3/2	3	-3/2	-3/2	321	4	3	2	3	-3	0	*201
4	5/2	3/2	4	-5/2	-3/2	*02	4	3	2	3	-2	-1	*221
4	5/2	5/2	-2	-1/2	5/2	*2001,	4	3	2	3	-1	-2	12
4	5/2	5/2	-1	-3/2	5/2	0201,	4	3	2	4	-3	-1	011
4	5/2	5/2	-1	-1/2	3/2	1211,	4	3	2	4	-2	-2	*121
4	5/2	5/2	0	-5/2	5/2	*2201,	4	3	3	-3	0	3	*1000,1
4	5/2	5/2	0	-3/2	3/2	*2001,	4	3	3	-2	-1	3	0101,1
4	5/2	5/2	0	-1/2	1/2	*0201,	4	3	3	-2	0	2	1101,1
4	5/2	5/2	1	-5/2	3/2	0201,	4	3	3	-1	-2	3	*0111,1
4	5/2	5/2	1	-3/2	1/2	1211,	4	3	3	-1	-1	2	*4201,1
4	5/2	5/2	1	-1/2	-1/2	0	4	3	3	-1	0	1	1111,1
4	5/2	5/2	2	-5/2	1/2	*2001,	4	3	3	0	-3	3	1001,1
4	5/2	5/2	2	-3/2	-1/2	*2211,	4	3	3	0	-2	2	1201,1
4	5/2	5/2	2	-1/2	-3/2	2211,	4	3	3	0	-1	1	1201,1
4	5/2	5/2	3	-5/2	-1/2	12	4	3	3	0	0	0	*1001,1
4	5/2	5/2	3	-3/2	-3/2	0	4	3	3	1	-3	2	*0111,1
4	5/2	5/2	3	-1/2	-5/2	*12	4	3	3	1	-2	1	*4201,1
4	5/2	5/2	4	-5/2	-3/2	*12	4	3	3	1	-1	0	1111,1
4	5/2	5/2	4	-3/2	-5/2	12	4	3	3	1	0	-1	1111,1
4	3	1	-1	0	1	*1211,	4	3	3	2	-3	1	0101,1
4	3	1	0	-1	1	1101,	4	3	3	2	-2	0	1101,1
4	3	1	0	0	0	2201,	4	3	3	2	-1	-1	*2211,1
4	3	1	1	-2	1	*2101,	4	3	3	2	0	-2	1101,1

6-j Symbols

j_1	j_2	j_3	l_1	l_2	l_3	
1/2	1/2	0	0	0	1/2	*1
1/2	1/2	0	1/2	1/2	0	*2
1	1/2	1/2	0	1/2	1/2	2
1	1/2	1/2	1	1/2	1/2	22
1	1	0	0	0	1	01
1	1	0	1/2	1/2	1/2	11
1	1	0	1	1	0	02
1	1	0	1	1	1	*02
1	1	1	1/2	1/2	1/2	*02
1	1	1	1	0	1	*02
1	1	1	1	1	0	*02
1	1	1	1	1	1	22
3/2	1	1/2	0	1/2	1	*11
3/2	1	1/2	1/2	1	1/2	*02
3/2	1	1/2	1	1/2	1	*22
3/2	1	1/2	3/2	1	1/2	*42
3/2	3/2	0	0	0	3/2	*2
3/2	3/2	0	1/2	1/2	1	*3
3/2	3/2	0	1	1	1/2	*21
3/2	3/2	0	1	1	3/2	21
3/2	3/2	0	3/2	3/2	0	*4
3/2	3/2	0	3/2	3/2	1	4
3/2	3/2	1	1/2	1/2	1	321
3/2	3/2	1	1	0	3/2	21
3/2	3/2	1	1	1	1/2	321
3/2	3/2	1	1	1	3/2	*121
3/2	3/2	1	3/2	1/2	1	22
3/2	3/2	1	3/2	3/2	0	4
3/2	3/2	1	3/2	3/2	1	*4220,2
2	1	1	0	1	1	02
2	1	1	1	1	1	22
2	1	1	2	1	1	222
2	3/2	1/2	0	1/2	3/2	3
2	3/2	1/2	1/2	1	1	21
2	3/2	1/2	1	1/2	3/2	301
2	3/2	1/2	1	3/2	1/2	4
2	3/2	1/2	1	3/2	3/2	*201
2	3/2	1/2	3/2	1	1	311
2	3/2	1/2	2	3/2	1/2	402
2	3/2	1/2	2	3/2	3/2	*202
2	3/2	3/2	0	3/2	3/2	*4
2	3/2	3/2	1/2	1	1	*31
2	3/2	3/2	1	1/2	3/2	*201
2	3/2	3/2	1	3/2	1/2	*201
2	3/2	3/2	1	3/2	3/2	402
2	3/2	3/2	3/2	1	1	*112
2	3/2	3/2	2	1/2	3/2	*202
2	3/2	3/2	2	3/2	1/2	*202
2	3/2	3/2	2	3/2	3/2	422
2	2	0	0	0	2	001
2	2	0	1/2	1/2	3/2	101
2	2	0	1	1	1	011
2	2	0	1	1	2	*011
2	2	0	3/2	3/2	1/2	201
2	2	0	3/2	3/2	3/2	*201
2	2	0	2	2	0	002
2	2	0	2	2	1	*002
2	2	0	2	2	2	002
2	2	1	1/2	1/2	3/2	*201
2	2	1	1	0	2	*011
2	2	1	1	1	1	*201
2	2	1	1	1	2	221
2	2	1	3/2	1/2	3/2	*301
2	2	1	3/2	3/2	1/2	*322
2	2	1	3/2	3/2	3/2	102
2	2	1	2	1	1	*202
2	2	1	2	1	2	2121,
2	2	1	2	2	0	*002
2	2	1	2	2	1	22
2	2	1	2	2	2	*202
2	2	2	1	1	1	2121,
2	2	2	1	1	2	2121,
2	2	2	3/2	1/2	3/2	3021,
2	2	2	3/2	3/2	1/2	3021,
2	2	2	3/2	3/2	3/2	0
2	2	2	2	0	2	002
2	2	2	2	1	1	2121,
2	2	2	2	1	2	*202
2	2	2	2	2	0	002
2	2	2	2	2	1	*202
2	2	2	2	2	2	*2222,
5/2	3/2	1	0	1	3/2	*21
5/2	3/2	1	1/2	3/2	1	*4
5/2	3/2	1	1	1	3/2	*301
5/2	3/2	1	3/2	3/2	1	*202
5/2	3/2	1	2	1	3/2	*312
5/2	3/2	1	5/2	3/2	1	*422
5/2	2	1/2	0	1/2	2	*101
5/2	2	1/2	1/2	1	3/2	*011
5/2	2	1/2	1	1/2	2	*021

6-*j* Symbols

j_1	j_2	j_3	l_1	l_2	l_3	
5/2	2	1/2	1	3/2	1	*201
5/2	2	1/2	1	3/2	2	2211,
5/2	2	1/2	3/2	1	3/2	*311
5/2	2	1/2	3/2	2	1/2	*002
5/2	2	1/2	3/2	2	3/2	3021,
5/2	2	1/2	2	3/2	1	*212
5/2	2	1/2	2	3/2	2	202
5/2	2	1/2	5/2	2	1/2	*222
5/2	2	1/2	5/2	2	3/2	022
5/2	2	3/2	0	3/2	2	201
5/2	2	3/2	1/2	1	3/2	4111,
5/2	2	3/2	1/2	2	3/2	422
5/2	2	3/2	1	1/2	2	2211,
5/2	2	3/2	1	3/2	1	3021,
5/2	2	3/2	1	3/2	2	*322
5/2	2	3/2	3/2	1	3/2	2121,
5/2	2	3/2	3/2	2	1/2	3021,
5/2	2	3/2	3/2	2	3/2	*202
5/2	2	3/2	2	1/2	2	202
5/2	2	3/2	2	3/2	1	3121,
5/2	2	3/2	2	3/2	2	*3001,
5/2	2	3/2	5/2	1	3/2	402
5/2	2	3/2	5/2	2	1/2	022
5/2	2	3/2	5/2	2	3/2	*4222,(47)^2
5/2	5/2	0	0	0	5/2	*11
5/2	5/2	0	1/2	1/2	2	*21
5/2	5/2	0	1	1	3/2	*12
5/2	5/2	0	1	1	5/2	12
5/2	5/2	0	3/2	3/2	1	*31
5/2	5/2	0	3/2	3/2	2	31
5/2	5/2	0	2	2	1/2	*111
5/2	5/2	0	2	2	3/2	111
5/2	5/2	0	2	2	5/2	*111
5/2	5/2	0	5/2	5/2	0	*22
5/2	5/2	0	5/2	5/2	1	22
5/2	5/2	0	5/2	5/2	2	*22
5/2	5/2	1	1/2	1/2	2	2211,
5/2	5/2	1	1	0	5/2	12
5/2	5/2	1	1	1	3/2	2211,
5/2	5/2	1	1	1	5/2	*0211,
5/2	5/2	1	3/2	1/2	2	021
5/2	5/2	1	3/2	3/2	1	3021,
5/2	5/2	1	3/2	3/2	2	*3221,02
5/2	5/2	1	2	1	3/2	202
5/2	5/2	1	2	1	5/2	*5221,

j_1	j_2	j_3	l_1	l_2	l_3	
5/2	5/2	1	2	2	1/2	0221,
5/2	5/2	1	2	2	3/2	*2221,2
5/2	5/2	1	2	2	5/2	0021,
5/2	5/2	1	5/2	3/2	1	022
5/2	5/2	1	5/2	3/2	2	*1021,
5/2	5/2	1	5/2	5/2	0	22
5/2	5/2	1	5/2	5/2	1	*2222,0000,002
5/2	5/2	1	5/2	5/2	2	2222,0000,2
5/2	5/2	2	1	1	3/2	*1221,
5/2	5/2	2	1	1	5/2	*5221,
5/2	5/2	2	3/2	1/2	2	*112
5/2	5/2	2	3/2	3/2	1	*2121,
5/2	5/2	2	3/2	3/2	2	*2121,
5/2	5/2	2	2	0	5/2	*111
5/2	5/2	2	2	1	3/2	*102
5/2	5/2	2	2	1	5/2	0021,
5/2	5/2	2	2	2	1/2	*112
5/2	5/2	2	2	2	3/2	1102,
5/2	5/2	2	2	2	5/2	1102,
5/2	5/2	2	5/2	1/2	2	*202
5/2	5/2	2	5/2	3/2	1	*1021,
5/2	5/2	2	5/2	3/2	2	4022,
5/2	5/2	2	5/2	5/2	0	*22
5/2	5/2	2	5/2	5/2	1	2222,0000,2
5/2	5/2	2	5/2	5/2	2	*422
3	3/2	3/2	0	3/2	3/2	4
3	3/2	3/2	1	3/2	3/2	422
3	3/2	3/2	2	3/2	3/2	402
3	3/2	3/2	3	3/2	3/2	4022,
3	2	1	0	1	2	011
3	2	1	1/2	3/2	3/2	201
3	2	1	1	1	2	021
3	2	1	1	2	1	002
3	2	1	1	2	2	*112
3	2	1	3/2	3/2	3/2	202
3	2	1	2	1	2	0121,
3	2	1	2	2	1	022
3	2	1	2	2	2	*1021,
3	2	1	5/2	3/2	3/2	2121,
3	2	1	3	2	1	0222,
3	2	1	3	2	2	*0022,
3	2	2	0	2	2	*002
3	2	2	1/2	3/2	3/2	*102
3	2	2	1	1	2	*112
3	2	2	1	2	1	*112

6-j Symbols

j_1	j_2	j_3	l_1	l_2	l_3			j_1	j_2	j_3	l_1	l_2	l_3	
3	2	2	1	2	2	0		3	5/2	3/2	5/2	2	1	*6222,
3	2	2	3/2	3/2	3/2	*102		3	5/2	3/2	5/2	2	2	2122,02
3	2	2	2	1	2	*1021,		3	5/2	3/2	3	3/2	3/2	*0022,
3	2	2	2	2	1	*1021,		3	5/2	3/2	3	3/2	5/2	4422,
3	2	2	2	2	2	4022,		3	5/2	3/2	3	5/2	1/2	*0202,
3	2	2	5/2	3/2	3/2	*1222,		3	5/2	3/2	3	5/2	3/2	6222,(71)^2
3	2	2	3	1	2	*0022,		3	5/2	3/2	3	5/2	5/2	*202
3	2	2	3	2	1	*0022,		3	5/2	5/2	0	5/2	5/2	22
3	2	2	3	2	2	2002,		3	5/2	5/2	1/2	2	2	102
3	5/2	1/2	0	1/2	5/2	21		3	5/2	5/2	1	3/2	3/2	202
3	5/2	1/2	1/2	1	2	12		3	5/2	5/2	1	3/2	5/2	0121,
3	5/2	1/2	1	1/2	5/2	2211,		3	5/2	5/2	1	5/2	3/2	0121,
3	5/2	1/2	1	3/2	3/2	31		3	5/2	5/2	1	5/2	5/2	*2222,2
3	5/2	1/2	1	3/2	5/2	*1201,		3	5/2	5/2	3/2	1	2	0121,
3	5/2	1/2	3/2	1	2	1201,		3	5/2	5/2	3/2	2	1	0121,
3	5/2	1/2	3/2	2	1	111		3	5/2	5/2	3/2	2	2	1022,
3	5/2	1/2	3/2	2	2	*0011,		3	5/2	5/2	2	1/2	5/2	2111,
3	5/2	1/2	2	3/2	3/2	3011,		3	5/2	5/2	2	3/2	3/2	2422,
3	5/2	1/2	2	3/2	5/2	*0111,		3	5/2	5/2	2	3/2	5/2	*1122,
3	5/2	1/2	2	5/2	1/2	22		3	5/2	5/2	2	5/2	1/2	2111,
3	5/2	1/2	2	5/2	3/2	*3211,		3	5/2	5/2	2	5/2	3/2	*1122,
3	5/2	1/2	2	5/2	5/2	2111,		3	5/2	5/2	2	5/2	5/2	*4222,0000,02
3	5/2	1/2	5/2	2	1	1211,		3	5/2	5/2	5/2	1	2	1322,
3	5/2	1/2	5/2	2	2	*1111,		3	5/2	5/2	5/2	2	1	1322,
3	5/2	1/2	3	5/2	1/2	2202,		3	5/2	5/2	5/2	2	2	*3422,
3	5/2	1/2	3	5/2	3/2	*0202,		3	5/2	5/2	3	1/2	5/2	2002,
3	5/2	1/2	3	5/2	5/2	2002,		3	5/2	5/2	3	3/2	3/2	4422,
3	5/2	3/2	0	3/2	5/2	*31		3	5/2	5/2	3	3/2	5/2	*202
3	5/2	3/2	1/2	1	2	*021		3	5/2	5/2	3	5/2	1/2	2002,
3	5/2	3/2	1/2	2	2	*102		3	5/2	5/2	3	5/2	3/2	*202
3	5/2	3/2	1	1/2	5/2	*1201,		3	5/2	5/2	3	5/2	5/2	4422,(79)^2
3	5/2	3/2	1	3/2	3/2	*112		3	3	0	0	0	3	0001,
3	5/2	3/2	1	3/2	5/2	3221,		3	3	0	1/2	1/2	5/2	1001,
3	5/2	3/2	1	5/2	3/2	*202		3	3	0	1	1	2	0101,
3	5/2	3/2	1	5/2	5/2	0121,		3	3	0	1	1	3	*0101,
3	5/2	3/2	3/2	1	2	*5221,		3	3	0	3/2	3/2	3/2	2001,
3	5/2	3/2	3/2	2	1	*112		3	3	0	3/2	3/2	5/2	*2001,
3	5/2	3/2	3/2	2	2	0021,		3	3	0	2	2	1	0011,
3	5/2	3/2	2	1/2	5/2	*0111,		3	3	0	2	2	2	*0011,
3	5/2	3/2	2	3/2	3/2	*1021,		3	3	0	2	2	3	0011,
3	5/2	3/2	2	3/2	5/2	4121,2		3	3	0	5/2	5/2	1/2	1101,
3	5/2	3/2	2	5/2	1/2	*3211,		3	3	0	5/2	5/2	3/2	*1101,
3	5/2	3/2	2	5/2	3/2	2222,0000,2		3	3	0	5/2	5/2	5/2	1101,
3	5/2	3/2	2	5/2	5/2	*1122,		3	3	0	3	3	0	0002,
3	5/2	3/2	5/2	1	2	*1021,		3	3	0	3	3	1	*0002,

6-*j* Symbols

j_1	j_2	j_3	l_1	l_2	l_3	
3	3	0	3	3	2	0002,
3	3	0	3	3	3	*0002,
3	3	1	1/2	1/2	5/2	*1201,
3	3	1	1	0	3	*0101,
3	3	1	1	1	2	*1201,
3	3	1	1	1	3	3201,
3	3	1	3/2	1/2	5/2	*2211,
3	3	1	3/2	3/2	3/2	*0011,
3	3	1	3/2	3/2	5/2	4211,
3	3	1	2	1	2	*0111,
3	3	1	2	1	3	3001,
3	3	1	2	2	1	*3211,
3	3	1	2	2	2	1011,
3	3	1	2	2	3	*3011,
3	3	1	5/2	3/2	3/2	*1111,
3	3	1	5/2	3/2	5/2	3111,
3	3	1	5/2	5/2	1/2	*1212,
3	3	1	5/2	5/2	3/2	3212,002
3	3	1	5/2	5/2	5/2	*1012,
3	3	1	3	2	1	*0202,
3	3	1	3	2	2	1112,
3	3	1	3	2	3	*1002,
3	3	1	3	3	0	*0002,
3	3	1	3	3	1	4202,2
3	3	1	3	3	2	*4202,
3	3	1	3	3	3	2002,
3	3	2	1	1	2	1021,
3	3	2	1	1	3	3001,
3	3	2	3/2	1/2	5/2	2111,
3	3	2	3/2	3/2	3/2	0121,
3	3	2	3/2	3/2	5/2	4121,
3	3	2	2	0	3	0011,
3	3	2	2	1	2	0121,
3	3	2	2	1	3	*3011,
3	3	2	2	2	1	3122,
3	3	2	2	2	2	*1122,
3	3	2	2	2	3	*3122,2
3	3	2	5/2	1/2	5/2	0111,
3	3	2	5/2	3/2	3/2	1322,
3	3	2	5/2	3/2	5/2	*3122,002
3	3	2	5/2	5/2	1/2	0002,
3	3	2	5/2	5/2	3/2	*4022,2
3	3	2	5/2	5/2	5/2	*2222,
3	3	2	3	1	2	2022,
3	3	2	3	1	3	*1002,

j_1	j_2	j_3	l_1	l_2	l_3	
3	3	2	3	2	1	1112,
3	3	2	3	2	2	*202
3	3	2	3	2	3	2112,
3	3	2	3	3	0	0002,
3	3	2	3	3	1	*4202,
3	3	2	3	3	2	4222,0002,
3	3	2	3	3	3	2202,
3	3	3	3/2	3/2	3/2	*1112,
3	3	3	3/2	3/2	5/2	*3312,
3	3	3	2	1	2	*0112,
3	3	3	2	1	3	*1002,
3	3	3	2	2	1	*0112,
3	3	3	2	2	2	*2112,
3	3	3	2	2	3	2112,
3	3	3	5/2	1/2	5/2	*1112,
3	3	3	5/2	3/2	3/2	*3312,
3	3	3	5/2	3/2	5/2	3112,
3	3	3	5/2	5/2	1/2	*1112,
3	3	3	5/2	5/2	3/2	3112,
3	3	3	5/2	5/2	5/2	3312,002
3	3	3	3	0	3	*0002,
3	3	3	3	1	2	*1002,
3	3	3	3	1	3	2002,
3	3	3	3	2	1	*1002,
3	3	3	3	2	2	2112,
3	3	3	3	2	3	2202,
3	3	3	3	3	0	*0002,
3	3	3	3	3	1	2002,
3	3	3	3	3	2	2202,
3	3	3	3	3	3	*2002,
7/2	2	3/2	0	3/2	2	*201
7/2	2	3/2	1/2	2	3/2	*002
7/2	2	3/2	1	3/2	2	*102
7/2	2	3/2	3/2	2	3/2	*202
7/2	2	3/2	2	3/2	2	*1021,
7/2	2	3/2	5/2	2	3/2	*0022,
7/2	2	3/2	3	3/2	2	*3022,
7/2	2	3/2	7/2	2	3/2	*6022,
7/2	5/2	1	0	1	5/2	*12
7/2	5/2	1	1/2	3/2	2	*31
7/2	5/2	1	1	1	5/2	*2211,
7/2	5/2	1	1	2	3/2	*111
7/2	5/2	1	1	2	5/2	2111,
7/2	5/2	1	3/2	3/2	2	*0111,
7/2	5/2	1	3/2	5/2	1	*22

6-j Symbols

j_1	j_2	j_3	l_1	l_2	l_3		j_1	j_2	j_3	l_1	l_2	l_3	
7/2	5/2	1	3/2	5/2	3	2111,	7/2	3	1/2	5/2	2	3/2	*4011,
7/2	5/2	1	2	1	5/2	*3201,	7/2	3	1/2	5/2	2	5/2	1111,
7/2	5/2	1	2	2	3/2	*1111,	7/2	3	1/2	5/2	3	1/2	*0002,
7/2	5/2	1	2	2	5/2	3111,	7/2	3	1/2	5/2	3	3/2	4112,
7/2	5/2	1	5/2	3/2	2	*4111,	7/2	3	1/2	5/2	3	5/2	*1112,
7/2	5/2	1	5/2	5/2	1	*0202,	7/2	3	1/2	3	5/2	1	*3102,
7/2	5/2	1	5/2	5/2	2	1112,	7/2	3	1/2	3	5/2	2	3002,
7/2	5/2	1	3	2	3/2	*3112,	7/2	3	1/2	3	5/2	3	*2002,
7/2	5/2	1	3	2	5/2	2012,	7/2	3	1/2	7/2	3	1/2	*6002,
7/2	5/2	1	7/2	5/2	1	*6202,	7/2	3	1/2	7/2	3	3/2	4002,
7/2	5/2	1	7/2	5/2	2	6002,	7/2	3	1/2	7/2	3	5/2	*6202,
7/2	5/2	2	0	2	5/2	111	7/2	3	3/2	0	3/2	3	2001,
7/2	5/2	2	1/2	3/2	2	312	7/2	3	3/2	1/2	1	5/2	3001,
7/2	5/2	2	1/2	5/2	2	222	7/2	3	3/2	1/2	2	5/2	3001,
7/2	5/2	2	1	1	5/2	2111,	7/2	3	3/2	1	1/2	3	4101,
7/2	5/2	2	1	2	3/2	102	7/2	3	3/2	1	3/2	2	2111,
7/2	5/2	2	1	2	5/2	2221,	7/2	3	3/2	1	3/2	3	0
7/2	5/2	2	3/2	3/2	2	0121,	7/2	3	3/2	1	5/2	2	0111,
7/2	5/2	2	3/2	5/2	1	2111,	7/2	3	3/2	1	5/2	3	*4111,
7/2	5/2	2	3/2	5/2	2	*2222,2	7/2	3	3/2	3/2	1	5/2	3001,
7/2	5/2	2	2	1	5/2	3111,	7/2	3	3/2	3/2	2	3/2	2111,
7/2	5/2	2	2	2	3/2	1322,	7/2	3	3/2	3/2	2	5/2	*3011,
7/2	5/2	2	2	2	5/2	*3122,002	7/2	3	3/2	3/2	3	3/2	2002,
7/2	5/2	2	5/2	3/2	2	4422,	7/2	3	3/2	3/2	3	5/2	*1002,
7/2	5/2	2	5/2	5/2	1	1112,	7/2	3	3/2	2	1/2	3	4001,
7/2	5/2	2	5/2	5/2	2	*202	7/2	3	3/2	2	3/2	2	3111,
7/2	5/2	2	3	1	5/2	2012,	7/2	3	3/2	2	3/2	3	*2101,
7/2	5/2	2	3	2	3/2	4322,	7/2	3	3/2	2	5/2	1	0002,
7/2	5/2	2	3	2	5/2	*1222,2	7/2	3	3/2	2	5/2	2	*1012,
7/2	5/2	2	7/2	3/2	2	2022,	7/2	3	3/2	2	5/2	3	4102,
7/2	5/2	2	7/2	5/2	1	6002,	7/2	3	3/2	5/2	1	5/2	4101,
7/2	5/2	2	7/2	5/2	2	*6420,002	7/2	3	3/2	5/2	2	3/2	1112,
7/2	3	1/2	0	1/2	3	*1001,	7/2	3	3/2	5/2	2	5/2	*4212,0002,
7/2	3	1/2	1/2	1	5/2	*0101,	7/2	3	3/2	5/2	3	1/2	4112,
7/2	3	1/2	1	1/2	3	*3001,	7/2	3	3/2	5/2	3	3/2	*4202,
7/2	3	1/2	1	3/2	2	*2001,	7/2	3	3/2	5/2	3	5/2	52
7/2	3	1/2	1	3/2	3	4101,	7/2	3	3/2	3	3/2	2	2012,
7/2	3	1/2	3/2	1	5/2	*5111,	7/2	3	3/2	3	3/2	3	*3112,
7/2	3	1/2	3/2	2	3/2	*0011,	7/2	3	3/2	3	5/2	1	5012,
7/2	3	1/2	3/2	2	5/2	5311,	7/2	3	3/2	3	5/2	2	*5112,02
7/2	3	1/2	2	3/2	2	*3011,	7/2	3	3/2	3	5/2	3	2112,
7/2	3	1/2	2	3/2	3	4001,	7/2	3	3/2	7/2	2	3/2	6002,
7/2	3	1/2	2	5/2	1	*1101,	7/2	3	3/2	7/2	2	5/2	*2102,
7/2	3	1/2	2	5/2	2	2111,	7/2	3	3/2	7/2	3	1/2	4002,
7/2	3	1/2	2	5/2	3	*3001,	7/2	3	3/2	7/2	3	3/2	*6202,2

6-j Symbols

j_1	j_2	j_3	l_1	l_2	l_3	
7/2	3	3/2	7/2	3	5/2	42
7/2	3	5/2	0	5/2	3	*1101,
7/2	3	5/2	1/2	2	5/2	*0201,
7/2	3	5/2	1/2	3	5/2	*2202,
7/2	3	5/2	1	3/2	2	*2011,
7/2	3	5/2	1	3/2	3	*4111,
7/2	3	5/2	1	5/2	2	*5212,
7/2	3	5/2	1	5/2	3	3212,
7/2	3	5/2	3/2	1	5/2	*5101,
7/2	3	5/2	3/2	2	3/2	*0112,
7/2	3	5/2	3/2	2	5/2	*5212,02
7/2	3	5/2	3/2	3	3/2	*1002,
7/2	3	5/2	3/2	3	5/2	42
7/2	3	5/2	2	1/2	3	*3001,
7/2	3	5/2	2	3/2	2	*3312,
7/2	3	5/2	2	3/2	3	4102,
7/2	3	5/2	2	5/2	1	*2102,
7/2	3	5/2	2	5/2	2	3112,
7/2	3	5/2	2	5/2	3	2002,
7/2	3	5/2	5/2	1	5/2	*1002,
7/2	3	5/2	5/2	2	3/2	*5412,
7/2	3	5/2	5/2	2	5/2	2112,
7/2	3	5/2	5/2	3	1/2	*1112,
7/2	3	5/2	5/2	3	3/2	52
7/2	3	5/2	5/2	3	5/2	2202,
7/2	3	5/2	3	1/2	3	*2002,
7/2	3	5/2	3	3/2	2	*3212,
7/2	3	5/2	3	3/2	3	2112,
7/2	3	5/2	3	5/2	1	*4012,
7/2	3	5/2	3	5/2	2	4312,0000,02
7/2	3	5/2	3	5/2	3	*1312,
7/2	3	5/2	7/2	1	5/2	*4002,
7/2	3	5/2	7/2	2	3/2	*2102,
7/2	3	5/2	7/2	2	5/2	4402,0002,
7/2	3	5/2	7/2	3	1/2	*6202,
7/2	3	5/2	7/2	3	3/2	42
7/2	3	5/2	7/2	3	5/2	*6402,$(37)^2$
7/2	7/2	0	0	0	7/2	*3
7/2	7/2	0	1/2	1/2	3	*4
7/2	7/2	0	1	1	5/2	*31
7/2	7/2	0	1	1	7/2	31
7/2	7/2	0	3/2	3/2	2	*5
7/2	7/2	0	3/2	3/2	3	5
7/2	7/2	0	2	2	3/2	*301
7/2	7/2	0	2	2	5/2	301

j_1	j_2	j_3	l_1	l_2	l_3	
7/2	7/2	0	2	2	7/2	*301
7/2	7/2	0	5/2	5/2	1	*41
7/2	7/2	0	5/2	5/2	2	41
7/2	7/2	0	5/2	5/2	3	*41
7/2	7/2	0	3	3	1/2	*3001,
7/2	7/2	0	3	3	3/2	3001,
7/2	7/2	0	3	3	5/2	*3001,
7/2	7/2	0	3	3	7/2	3001,
7/2	7/2	0	7/2	7/2	0	*6
7/2	7/2	0	7/2	7/2	1	6
7/2	7/2	0	7/2	7/2	2	*6
7/2	7/2	0	7/2	7/2	3	6
7/2	7/2	1	1/2	1/2	3	4101,
7/2	7/2	1	1	0	7/2	31
7/2	7/2	1	1	1	5/2	4101,
7/2	7/2	1	1	1	7/2	*2301,
7/2	7/2	1	3/2	1/2	3	3001,
7/2	7/2	1	3/2	3/2	2	5311,
7/2	7/2	1	3/2	3/2	3	*5111,
7/2	7/2	1	2	1	5/2	4001,
7/2	7/2	1	2	1	7/2	*0201,
7/2	7/2	1	2	2	3/2	2111,
7/2	7/2	1	2	2	5/2	*4311,02
7/2	7/2	1	2	2	7/2	2111,
7/2	7/2	1	5/2	3/2	2	1111,
7/2	7/2	1	5/2	3/2	3	*3211,
7/2	7/2	1	5/2	5/2	1	4112,
7/2	7/2	1	5/2	5/2	2	*4312,$(37)^2$
7/2	7/2	1	5/2	5/2	3	4332,
7/2	7/2	1	3	2	3/2	3002,
7/2	7/2	1	3	2	5/2	*3302,
7/2	7/2	1	3	2	7/2	3102,1
7/2	7/2	1	3	3	1/2	5302,
7/2	7/2	1	3	3	3/2	*1102,
7/2	7/2	1	3	3	5/2	5302,0002,
7/2	7/2	1	3	3	7/2	*1102,
7/2	7/2	1	7/2	5/2	1	4002,
7/2	7/2	1	7/2	5/2	2	*4222,
7/2	7/2	1	7/2	5/2	3	2202,1
7/2	7/2	1	7/2	7/2	0	6
7/2	7/2	1	7/2	7/2	1	*6402,$(59)^2$
7/2	7/2	1	7/2	7/2	2	6202,002
7/2	7/2	1	7/2	7/2	3	*6202,02
7/2	7/2	2	1	1	5/2	*4001,
7/2	7/2	2	1	1	7/2	*0201,

6-j Symbols

j_1	j_2	j_3	l_1	l_2	l_3	
7/2	7/2	2	3/2	1/2	3	*3001,
7/2	7/2	2	3/2	3/2	2	*5101,
7/2	7/2	2	3/2	3/2	3	*5101,
7/2	7/2	2	2	0	7/2	*301
7/2	7/2	2	2	1	5/2	*4111,
7/2	7/2	2	2	1	7/2	2111,
7/2	7/2	2	2	2	3/2	*2102,
7/2	7/2	2	2	2	5/2	4102,
7/2	7/2	2	2	2	7/2	4122,
7/2	7/2	2	5/2	1/2	3	*4001,
7/2	7/2	2	5/2	3/2	2	*1002,
7/2	7/2	2	5/2	3/2	3	3102,
7/2	7/2	2	5/2	5/2	1	*5022,
7/2	7/2	2	5/2	5/2	2	52
7/2	7/2	2	5/2	5/2	3	3202,
7/2	7/2	2	3	1	5/2	*1102,
7/2	7/2	2	3	1	7/2	3102,1
7/2	7/2	2	3	2	3/2	*4012,
7/2	7/2	2	3	2	5/2	2112,
7/2	7/2	2	3	2	7/2	*2112,1
7/2	7/2	2	3	3	1/2	*5022,
7/2	7/2	2	3	3	3/2	3222,
7/2	7/2	2	3	3	5/2	*5002,
7/2	7/2	2	3	3	7/2	*1202,
7/2	7/2	2	7/2	3/2	2	*4002,
7/2	7/2	2	7/2	3/2	3	4102,1
7/2	7/2	2	7/2	5/2	1	*4222,
7/2	7/2	2	7/2	5/2	2	42
7/2	7/2	2	7/2	5/2	3	*2202,1
7/2	7/2	2	7/2	7/2	0	*6
7/2	7/2	2	7/2	7/2	1	6202,002
7/2	7/2	2	7/2	7/2	2	*6222,
7/2	7/2	2	7/2	7/2	3	6202,
7/2	7/2	3	3/2	3/2	2	5112,1
7/2	7/2	3	3/2	3/2	3	5112,1
7/2	7/2	3	2	1	5/2	3102,1
7/2	7/2	3	2	1	7/2	3102,1
7/2	7/2	3	2	2	3/2	4112,1
7/2	7/2	3	2	2	5/2	2112,1
7/2	7/2	3	2	2	7/2	*2112,1
7/2	7/2	3	5/2	1/2	3	4002,1
7/2	7/2	3	5/2	3/2	2	2012,1
7/2	7/2	3	5/2	3/2	3	0
7/2	7/2	3	5/2	5/2	1	5112,1
7/2	7/2	3	5/2	5/2	2	5312,1
7/2	7/2	3	5/2	5/2	3	*3312,1
7/2	7/2	3	3	0	7/2	3001,
7/2	7/2	3	3	1	5/2	3102,1
7/2	7/2	3	3	1	7/2	*1102,
7/2	7/2	3	3	2	3/2	3102,1
7/2	7/2	3	3	2	5/2	*3202,1
7/2	7/2	3	3	2	7/2	*1202,
7/2	7/2	3	3	3	1/2	4002,1
7/2	7/2	3	3	3	3/2	*4202,1
7/2	7/2	3	3	3	5/2	*4202,1
7/2	7/2	3	3	3	7/2	2202,1
7/2	7/2	3	7/2	1/2	3	2002,
7/2	7/2	3	7/2	3/2	2	4102,1
7/2	7/2	3	7/2	3/2	3	*42
7/2	7/2	3	7/2	5/2	1	2202,1
7/2	7/2	3	7/2	5/2	2	*2202,1
7/2	7/2	3	7/2	5/2	3	2202,
7/2	7/2	3	7/2	7/2	0	6
7/2	7/2	3	7/2	7/2	1	*6202,02
7/2	7/2	3	7/2	7/2	2	6202,
7/2	7/2	3	7/2	7/2	3	6002,2000,002
4	2	2	0	2	2	002
4	2	2	1	2	2	222
4	2	2	2	2	2	2022,
4	2	2	3	2	2	2022,
4	2	2	4	2	2	2422,
4	5/2	3/2	0	3/2	5/2	31
4	5/2	3/2	1/2	2	2	111
4	5/2	3/2	1	3/2	5/2	3001,
4	5/2	3/2	1	5/2	3/2	22
4	5/2	3/2	1	5/2	5/2	*0201,
4	5/2	3/2	3/2	2	2	0111,
4	5/2	3/2	2	3/2	5/2	4101,
4	5/2	3/2	2	5/2	3/2	2002,
4	5/2	3/2	2	5/2	5/2	*1002,
4	5/2	3/2	5/2	2	2	2012,
4	5/2	3/2	3	3/2	5/2	4202,
4	5/2	3/2	3	5/2	3/2	6002,
4	5/2	3/2	3	5/2	5/2	*2102,
4	5/2	3/2	7/2	2	2	4212,
4	5/2	3/2	4	5/2	3/2	6402,
4	5/2	3/2	4	5/2	5/2	*2402,
4	5/2	5/2	0	5/2	5/2	*22
4	5/2	5/2	1/2	2	2	*121
4	5/2	5/2	1	3/2	5/2	*0201,

References

Abragam, A., Horowitz, J., Pryce, M.H.L. (1955): Proc. R. Soc. London A**230**, 169
Abragam, A., Pryce, M.H.L. (1951): Proc. R. Soc. London A**205**, 135
Adams, B.G., Jankowski, K., Paldus, J. (1979): Chem. Phys. Lett. **67**, 144
Ahlenius, T., Larsson, S. (1973): Phys. Rev. A**8**, 1
Ahlrichs, R. (1979): Comput. Phys. Commun. **17**, 31
Amusia, M.Ya., Ivanov, V.K., Cherepkov, N.A., Chernysheva, L.V. (1972): Phys. Lett. A**40**, 361
Amusia, M.Ya., Cherepkov, N.A., Janev, R.K., Živanović, Dj. (1974): J. Phys. B**7**, 1435
Andriessen, J., Ray, S.N., Lee, T., Das, T.P., Ikenberry, D. (1976): Phys. Rev. A**13**, 1669
Andriessen, J., van Ormondt, D., Ray, S.N., Das, T.P. (1978): J. Phys. B**11**, 2601
Arfken, G. (1970): *Mathematical Methods for Physicists*, 2nd ed. (Academic, New York)
Arimondo, E., Inguscio, M., Violino, P. (1977): Rev. Mod. Phys. **49**, 31
Armstrong, L. Jr. (1971): *Theory of Hyperfine Structure of Free Atoms* (Wiley-Interscience, New York)
— (1978): "Relativistic Effects in Highly Ionized Atoms" in *Structure and Collisions of Ions and Atoms*, ed. by I.A. Sellin, Topics in Current Physics, Vol. 5 (Springer, Berlin, Heidelberg, New York) p. 69
Armstrong, L. Jr., Feneuille, S. (1974): in *Advances in Atomic and Molecular Physics*, Vol. 10, ed. by D.R. Bates, B. Bederson (Academic, New York) p. 1
Avery, J. (1976): *Creation and Annihilation Operators* (McGraw-Hill, New York)

Bagus, P.S., Bauche, J. (1973): Phys. Rev. A**8**, 734
Baker, G.A. (1971): Rev. Mod. Phys. **43**, 479
Barnes, R.G., Smith, W.V. (1954): Phys. Rev. **93**, 95
Barr, T.L., Davidson, E.R. (1970): Phys. Rev. A**1**, 644
Barrett, B.R., Kirson, M.W. (1973): Adv. Nucl. Phys. **6**, 219
Bartlett, R.J., Purvis, G.D. (1978): Int. J. Quantum Chem. **14**, 561
Bartlett, R.J., Purvis, G.D. (1980): Phys. Scr. **21**, 255
Bauche-Arnoult, C. (1971): Proc. R. Soc. London A**322**, 361
— (1973): J. Phys. Paris **34**, 301
Bauche, J., Couarraze, G., Labarthe, J.-J. (1974): Z. Phys. **270**, 211
Belin, G., Holmgren, L., Lindgren, I., Svanberg, S. (1975): Phys. Scr. **12**, 287
Belin, G., Holmgren, L., Svanberg, S. (1976a): Phys. Scr. **13**, 351
— (1976b): Phys. Scr. **14**, 39
Berezin, F.A. (1966): *The Method of Second Quantization* (Academic, New York)
Bessis, N., Lefebvre-Brion, H., Moser, C.M. (1962): Phys. Rev. **128**, 213
Bethe, H.A. (1956): Phys. Rev. **103**, 1353
Bethe, H.A., Goldstone, J. (1957): Proc. R. Soc. London A**238**, 551
Bethe, H.A., Salpeter, E.E. (1957): *Quantum Mechanics of One- and Two-Electron Atoms* (Springer, Berlin, Göttingen, Heidelberg)
Bloch, C. (1958a): Nucl. Phys. **6**, 329
— (1958b): Nucl. Phys. **7**, 451
Blume, M., Freeman, A.J., Watson, R.E. (1964): Phys. Rev. A**134**, 320
Blume, M., Watson, R.E. (1962): Proc. R. Soc. London A**270**, 127
— (1963): Proc. R. Soc. London A**271**, 565

Bogoliubov, N.N., Shirkov, D.V. (1959): *Introduction of the Theory of Quantized Fields* (Wiley, New York) [English transl.]

Bohr, N. (1922): Z. Phys. **9**, 1

Brandow, B.H. (1966): In *Proceedings of the International School of Physics "Enrico Fermi", Course 36,* ed. C. Bloch (Academic, New York)

—   (1967): Rev. Mod. Phys. **39**, 771

—   (1975): "Perturbation Theory of Effective Hamiltonians", in *Effective Interactions and Operators in Nuclei,* ed. by B.R. Barrett, Lecture Notes in Physics. Vol. 40 (Springer, Berlin, Heidelberg, New York)

—   (1977): Adv. Quantum Chem. **10**, 187

Breit, G. (1929): Phys. Rev. **34**, 553

—   (1930): Phys. Rev. **36**, 383

—   (1932): Phys. Rev. **39**, 616

Brenig, W. (1957): Nucl. Phys. **4**, 363

Briggs, J.S. (1971): Rev. Mod. Phys. **43**, 189

Brillouin, L. (1933): Actual Sci. Ind. no. **71**

—   (1934): Actual. Sci. Ind. no. **159**

Brink, D.M., Satchler, G.R. (1968): *Angular Momentum,* 2nd ed. (Clarendon, Oxford)

Brown, G.E. (1967): *Unified Theory of Nuclear Models and Forces* (North-Holland, Amsterdam)

Brown, G.E., Ravenhall, D.G. (1951): Proc. Roy. Soc. London A **208**, 552

Brueckner, K.A. (1955): Phys. Rev. **100**, 36

Brueckner, K.A., Levinson, C.A. (1955): Phys. Rev. **97**, 1344

Bunge, A., Bunge, C.F. (1970): Phys. Rev. A **1**, 1599

Bunge, C.F. (1968): Phys. Rev. **168**, 92

—   (1970): Theor. Chim. Acta Berlin **16**, 126

—   (1976a): Phys. Rev. A **14**, 1965

—   (1976b): At. Data Nucl. Data Tables **18**, 293

—   (1980): Phys. Scr. **21**, 328

Bunge, C.F., Peixoto, E.M.A. (1970): Phys. Rev. A **1**, 1277

Byron, F.W., Joachain, C.J. (1967): Phys. Rev. **157**, 7

Carnall, W.T., Fields, P.R., Sarup, R. (1969): J. Chem. Phys. **51**, 2587

Carter, S.L., Kelly, H.P. (1977): Phys. Rev. A **16**, 1525

Cederbaum, L.S., Matschke, F.E.P., von Niessen, W. (1975): Phys. Rev. A **12**, 6

Chang, J.-J., Kelly, H.P. (1975): Phys. Rev. A **12**, 92

Chang, T.N., Fano, U. (1976): Phys. Rev. A **13**, 263

Chang, T.N., Poe, R.T. (1975): Phys. Rev. A **11**, 191

Čížek, J. (1966): J. Chem. Phys. **45**, 4256

—   (1969): Adv. Chem. Phys. **14**, 35

Coester, F. (1958): Nucl. Phys. **7**, 421

Coester, F., Kümmel, H. (1960): Nucl. Phys. **17**, 477

Condon, E.U., Shortley, G.H. (1935): *The Theory of Atomic Spectra* (Cambridge University Press, Cambridge)

Cowan, R.D., Andrew, K.L. (1965): J. Opt. Soc. Am. **55**, 502

Cowan, R.D., Griffin, D.C. (1976): J. Opt. Soc. Am. **66**, 1010

Dalgarno, A. (1959): Proc. R. Soc. London **251**, 282

—   (1962): Adv. Phys. **11**, 281

Dalgarno, A., Lewis, J.T. (1955): Proc. R. Soc. London A **233**, 70

Desclaux, J.-P. (1975): Comput. Phys. Commun. **9**, 31

Dirac, P.A.M. (1927): Proc. R. Soc. London A **114**, 243

Dirac, P.A.M. (1958): *The Principles of Quantum Mechanics,* 4th ed. (Oxford University Press, Oxford)

Dolan, V.L. (1970): Techn. Rep. AFML-TR-70-249, U.S. Air Force Materials Lab.

Dutta, N.C., Matsubara, C., Pu, R.T., Das, T.P. (1969): Phys. Rev. **177**, 33
Dyson, F.J. (1949): Phys. Rev. **75**, 486, 1736
Dzuba, V.A., Flambaum, V.V., Sushkov, O.P. (1982): Reported at the 8th International Conf. on Atomic Physics, Göteborg, August, 1982
– (1983a): Phys. Lett. **95A**, 230
– (1983b): J. Phys. B **16**, 715
– (1984): J. Phys. B **17**, 1953

Eckart, C. (1930): Rev. Mod. Phys. **2**, 305
Eden, R.J., Francis, N.C. (1955): Phys. Rev. **97**, 1366
Edmonds, A.R. (1957): *Angular Momentum in Quantum Mechanics* (Princeton University Press, Princeton)
El Baz, E., Castel, B. (1972): *Graphical Methods of Spin Algebras in Atomic, Nuclear, and Particle Physics* (Dekker, New York)
Ellis, P.J., Osnes, E. (1977): Rev. Mod. Phys. **49**, 777
Ey, W. (1978): Nucl. Phys. A **296**, 189

Fano, U., Racah, G. (1959): *Irreducible Tensor Sets* (Academic, New York)
Fassett, J.D. et al. (1985): *Europhysics Conference Abstracts*, Vol. 9 B
Feshbach, H. (1958a): Ann. Phys. N.Y. **5**, 357
– (1958b): Ann. Rev. Nucl. Sci. **8**, 49
– (1962): Ann. Phys. N.Y. **19**, 287
Fetter, A.L., Walečka, J.D. (1971): *Quantum Theory of Many-Particle Systems* (McGraw-Hill, New York)
Feynman, R.P. (1949): Phys. Rev. **76**, 749, 769
Foldy, L.L., Wouthuysen, S.A. (1950): Phys. Rev. **78**, 29
Foley, H.M., Sternheimer, R.M. (1975): Phys. Lett. A **55**, 276
Fonseca, A.L.A., Bauche, J. (1984): Z. Phys. A **318**, 13
Francis, N.C., Watson, K.M. (1953): Phys. Rev. **92**, 291
Frantz, L.M., Mills, R.L. (1960): Nucl. Phys. **15**, 16
Freeman, A.J., Watson, R.E. (1965): In *Magnetism*, Vol. IIA ed. by G.T. Rado, H. Suhl (Academic, New York)
Froese-Fischer, C. (1973): J. Comput. Phys. **13**, 502
– (1977): *The Hartree-Fock Method for Atoms* (Wiley, New York)
Froese-Fischer, C., Saxena, K.M.S. (1974): Phys. Rev. A **9**, 1498

Garpman, S., Lindgren, I., Lindgren, J., Morrison, J. (1975): Phys. Rev. A **11**, 758
– (1976): Z. Phys. A **276**, 167
Glass, R., Hibbert, A. (1976): J. Phys. B **9**, 875
Goldstein, H. (1950): *Classical Mechanics* (Addison-Wesley, London)
Goldstone, J. (1957): Proc. R. Soc. London A **239**, 267
Grant, I.P. (1970): Adv. Phys. **19**, 747

Hartree, D.R. (1957): *The Calculation of Atomic Structures* (Wiley, New York)
Haque, A., Mukherjee, D. (1984): Pramāna **23**, 651
Harvey, J.S.M. (1965): Proc. R. Soc. London A **285**, 581
Heully, J.-L. (1982): J. Phys. B **15**, 4079
Heully, J.-L., Salomonson, S. (1982): J. Phys. B **15**, 4093
Heully, J.-L., Mårtensson-Pendrill, A.-M. (1983a): Phys. Rev. A **27**, 3332
– (1983b): Phys. Scripta **27**, 291
Holmgren, L., Lindgren, I., Morrison, J., Mårtensson, A.-M. (1976): Z. Phys. A **276**, 179
Hose, G., Kaldor, U. (1979): J. Phys. B **12**, 3827
– (1980): Phys. Scr. **21**, 357
– (1982): J. Phys. Chem. **86**, 2133

454   References

van Hove, L. (1955): Physica **21**, 901
− (1956): Physica **22**, 343
Huang, K.-N., Starace, A.F. (1978): Phys. Rev. A**18**, 354
Hubbard, J. (1957): Proc. R. Soc. London A**240**, 539
− (1958): Proc. R. Soc. London A**243**, 336
Hugenholtz, N.M. (1957): Physica **23**, 481

Jackson, J.D. (1975): *Classical Electrodynamics* 2nd ed. (Wiley, New York)
Jankowski, K., Malinowski, P. (1980): Phys. Rev. A**21**, 45
Jankowski, K., Paldus, J. (1980): Int. J. Quantum Chem. **18**, 1243
Jeziorski, B., Monkhorst, H.J. (1981): Phys. Rev. A**24**, 1668
Johnson, M.B., Baranger, M. (1971): Ann. Phys. N.Y. **62**, 172
Jordan, P., Klein, O. (1927): Z. Phys. **45**, 751
Jordan, P., Wigner, E.P. (1928): Z. Phys. **47**, 631
Jucys (Yutsis), A.P., Levinson, I.B., Vanagas, V.V. (1962): *Mathematical Apparatus of the Theory of Angular Momentum* (Israel Program for Scientific Translations, Jerusalem) [English transl.], see also
Jucys, A.P., Bandzaitis, A.A. (1977): *The Theory of Angular Momentum in Quantum Mechanics*, 2nd ed. (Academy of Sciences of the Lithuanian SSR, Institute of Physics, Vilnius)
Judd, B.R. (1962): Phys. Rev. **125**, 613
− (1963a): *Operator Techniques in Atomic Spectroscopy* (McGraw-Hill, New York)
− (1963b): Proc. Phys. Soc. London **82**, 874
− (1967): *Second Quantization and Atomic Spectroscopy* (Johns Hopkins University Press, Baltimore)
− (1969): Adv. Chem. Phys. **14**, 91
Judd, B.R., Elliott, J.P. (1970): *Topics in Atomic and Nuclear Theory* (University of Canterbury, New Zealand, The Caxton Press, Christchurch)

Kaldor, U. (1973): Phys. Rev. A**7**, 427
Kato, T. (1949): Progr. Theor. Phys. Jpn. **4**, 514
− (1966): *Perturbation Theory for Linear Operators*, Grundlehren der mathematischen Wissenschaften, Vol. 132 (Springer, Berlin, Heidelberg, New York)
Kelly, H.F. (1963): Phys. Rev. **131**, 684
− (1964a): Phys. Rev. A**134**, 1450
− (1964b): Phys. Rev. B**136**, 896
− (1968): Phys. Rev. **173**, 142
− (1969a): Phys. Rev. **180**, 55
− (1969b): Adv. Chem. Phys. **14**, 129
− (1976): In *Photoionization and Other Probes of Many-Electron Interactions*, ed. by F.S. Wuilleumier (Plenum, New York)
Kelly, H.P., Ron, A. (1970): Phys. Rev. A**2**, 1261
Kelly, H.P., Simons, R.L. (1973): Phys. Rev. Lett. **30**, 529
Kobe, D.H. (1971): Phys. Rev. C**3**, 417
Koopmans, T. (1933): Physica **1**, 104
Kopfermann, H. (1958): *Nuclear Moments* (Academic, New York) [English transl.]
Kümmel, H. (1962): In *Lecture Notes on the Many-Body Problem*, ed. by E.R. Caianiello (Academic, New York)
Kümmel, H., Lührmann, K.H., Zabolitzky, J.G. (1978): Phys. Rep. **36C**, 1
Kuo, T.T.S., Lee, S.Y., Ratcliff, K.F. (1971): Nucl. Phys. A**176**, 65
Kutzelnigg, W. (1977): In *Methods in Electronic Structure Theory*, ed. by H.F. Schaefer (Plenum, New York)
Kvasnička, V. (1974): Czech. J. Phys. B**24**, 605
− (1977a): Czech. J. Phys. B**27**, 599
− (1977b): Adv. Chem. Phys. **36**, 345

Lam, L.K., Gupta, R., Happer, W. (1980): Phys. Rev. A **21,** 1225
Larsson, S. (1970): Phys. Rev. A **2,** 1248
- (1971): Chem. Phys. Lett. **10,** 162
Larsson, S., Brown, R.E., Smith, V.H. (1972): Phys. Rev. A **6,** 1375
Lee, T., Dutta, N.C., Das, T.P. (1970): Phys. Rev. A **1,** 995
Liao, K.H., Lam, L.K., Gupta, R., Happer, W. (1974): Phys. Rev. Lett. **32,** 1340
Linderberg, J., Öhrn, Y. (1973): *Propagators in Quantum Chemistry* (Academic, New York)
Lindgren, I. (1974): J. Phys. B **7,** 2441
- (1978): Int. J. Quant. Chem. S **12,** 33
- (1984): Rep. Prog. Phys. **47,** 345
- (1985a): Phys. Rev. A **31,** 1273
- (1985b): Phys. Scripta **32,** 291,611
Lindgren, I., Lindgren, J., Mårtensson, A.-M. (1976): Z. Phys. A **279,** 113
- (1977): Phys. Rev. A **15,** 2123
Lindgren, I., Mårtensson, A.-M. (1982): Phys. Rev. A **26,** 3249
Lindgren, I., Rosén, A. (1974): Case Stud. At. Phys. **4,** 93
Lindgren, I., Salomonson, S. (1980): Phys. Scr. **21,** 335
Lindroth, E., Mårtensson-Pendrill, A.-M. (1983): Z. Phys. A **309,** 277
Lippman, B.A., Schwinger, J. (1950): Phys. Rev. **79,** 469
Löwdin, P.O. (1962): J. Math. Phys. **3,** 969, 1171
- (1965): Phys. Rev. A **139,** 357
- (1966): In *Perturbation Theory and its Applications in Quantum Mechanics,* ed. by C.H. Wilcox (Wiley, New York)
- (1968): Int. J. Quantum Chem. **2,** 867
- (1969): Adv. Phys. Chem. **14,** 283
Luc-Koenig, E. (1976): Phys. Rev. A **13,** 2114
- (1980): J. Physique (Paris) **41,** 1273
Lundberg, H., Mårtensson, A.-M., Svanberg, S. (1977): J. Phys. B **10,** 1971
Lyons, J.D., Pu, R.T., Das, T.P. (1969): Phys. Rev. **178,** 103

McKoy, V., Winter, N.W. (1968): J. Chem. Phys. **48,** 5514
March, N.H., Young, W.H., Sampanthar, S. (1967): *The Many-Body Problem in Quantum Mechanics* (Cambridge University Press, Cambridge)
Mårtensson, A.-M. (1978): Thesis, Department of Physics, Chalmers University of Technology/University of Gothenburg
- (1979): J. Phys. B **12,** 3995
Mårtensson, A.-M., Salomonson, S. (1982): J. Phys. B **15,** 2115
Martin, W.C., Zalubas, R., Hagan, L. (1978): *Atomic Energy Levels – The Rare-Earth Elements* (National Bureau of Standards, Washington DC)
Mayer, J.E., Goeppert-Mayer, M. (1977): *Statistical Mechanics,* 2nd ed. (Wiley, New York)
Merzbacher, E. (1961): *Quantum Mechanics* (Wiley, New York)
Messiah, A. (1961): *Quantum Mechanics* (North-Holland, Amsterdam)
Meyer, W. (1973): J. Chem. Phys. **58,** 1017
- (1976): J. Chem. Phys. **64,** 2901
Møller, C. (1945): K. Dan. Vidensk. Selsk. **22,** no. 1
- (1946): K. Dan. Vidensk. Selsk. **23,** no. 19
Morita, T. (1963): Prog. Theor. Phys. **29,** 351
Morrison, J. (1972): Phys. Rev. A **6,** 643
- (1973): J. Phys. B **6,** 2205
Morrison, J., Rajnak, K. (1971): Phys. Rev. A **4,** 536
Morrison, J., Rajnak, K., Wilson, M. (1970): J. Phys. Paris Colloq. C **4,** 167
Morrison, J., Salomonson, S. (1980): Phys. Scr. **21,** 343
Mukherjee, D. (1979): Pramāna **12,** 203
Mukherjee, D., Moitra, R.K., Mukhopadhyay, A. (1975): Pramāna **4,** 247

Mukherjee, D., Moitra, R.K., Mukhopadhyay, A. (1977): Mol. Phys. **33**, 955
Musher, J.I., Schulman, J.M. (1968): Phys. Rev. **173**, 93

Nesbet, R.K. (1958): Phys. Rev. **109**, 1632
− (1965): Adv. Chem. Phys. **9**, 321
− (1967): Phys. Rev. **155**, 51
− (1969): Adv. Chem. Phys. **14**, 1
− (1970): Phys. Rev. A**2**, 661
Nesbet, R.K., Barr, T.L., Davidson, E.R. (1969): Chem. Phys. Lett. **4**, 203
Nielson, C.W., Koster, G.F. (1963): *Spectroscopic Coefficients for p^n, d^n and f^n Configurations* (MIT Press, Cambridge, MA)

Oberlechner, G., Owono-N-Guema, F., Richert, J. (1970): Nouvo Cimento B**68**, 23
Offermann, R. (1976): Nucl. Phys. A**273**, 368
Offermann, R., Ey, W., Kümmel, H. (1976): Nucl. Phys. A**273**, 349
Olsson, G., Salomonson, S. (1982): Z. Phys. A**307**, 99

Paldus, J. (1977): J. Chem. Phys. **67**, 303
Paldus, J., Čížek, J. (1975): Adv. Quantum Chem. **9**, 105
Paldus, J., Čížek, J., Saute, M., Laforgue, A. (1978): Phys. Rev. A**17**, 805
Paldus, J., Čížek, J., Shavitt, I. (1972): Phys. Rev. A**5**, 50
Pauli, W. (1926): Z. Phys. **31**, 765
Pyper, N.C., Marketos, P. (1981): J. Phys. B**14**, 4469
Pople, J.A., Binkley, J.S., Seeger, R. (1976): Int. J. Quantum Chem. S**10**, 1
Primas, H. (1965): In *Modern Quantum Chemistry*, ed. by O. Sinanoğlu (Academic, New York)

Racah, G. (1942): Phys. Rev. **62**, 438
− (1943): Phys. Rev. **63**, 367
− (1949): Phys. Rev. **76**, 1352
− (1952): Phys. Rev. **85**, 381
Racah, G., Stein, J. (1967): Phys. Rev. **156**, 58
Rajnak, K., Wybourne, B.G. (1963): Phys. Rev. **132**, 280
Randolph, W.L., Asher, J., Kroen, J.W., Rowe, P., Matthias, E. (1975): Hyperfine Interact. **1**, 145
Richardson, L., Gaunt, J. (1927): Trans. R. Soc. London A**226**, 299
Rodberg, L.S., Thaler, R.M. (1967): *Introduction to the Quantum Theory of Scattering* (Academic, New York)
Roman, P. (1965): *Advanced Quantum Theory* (Addison-Wesley, London)
Roos, B., Siegbahn, P. (1977): In *Electronic Structure Theory*, ed. by H.F. Schaefer (Plenum, New York)
Rose, M.E. (1957): *Elementary Theory of Angular Momentum* (Wiley, New York)
Rotenberg, M., Bevins, R., Metropolis, M., Wooten, J.K. (1959): *The 3-j and 6-j Symbols* (MIT Press, Cambridge, MA)

Salomonson, S. (1984): Z. Phys. A**316**, 135
Salomonson, S., Lindgren, I., Mårtensson, A.-M. (1980): Phys. Scr. **21**, 351
Sandars, P.G.H. (1969): Adv. Chem. Phys. **14**, 365
− (1971): In *Atomic Physics and Astrophysics*, ed. by M. Chrestien, E. Lipworth (Gordon and Breach, London)
− (1977): J. Phys. B**10**, 2983
Sasaki, F., Yoshimine, M. (1974): Phys. Rev. A**9**, 17
Schaefer, H.F. (1972): *The Electronic Structure of Atoms and Molecules* (Addison-Wesley, London)
− (ed.) (1977): *Methods of Electronic Structure Theory* (Plenum, New York)

Schaefer, H.F., Klemm, R.A. (1969): Phys. Rev. **188**, 152
Schaefer, H.F., Klemm, R.A., Harris, F.E. (1968): Phys. Rev. **176**, 49
– (1969): Phys. Rev. **181**, 137
Schäfer, L., Weidenmüller, H.A. (1971): Nucl. Phys. A **174**, 1
Schiff, L.I. (1968): *Quantum Mechanics*, 3rd ed. (McGraw-Hill, New York)
Schucan, T.H., Weidenmüller, H.A. (1972): Ann. Phys. NY **73**, 108
– (1973): Ann. Phys. NY **76**, 483
Schulman, J.M., Lee, W.S. (1972): Phys. Rev. A **5**, 13
Schwartz, C. (1955): Phys. Rev. **97**, 380
Shortley, G.H. (1936): Phys. Rev. **50**, 1072
Siegbahn, P.E.M. (1980): J. Chem. Phys. **72**, 1647
Sims, J.S., Hagstrom, S. (1971): Phys. Rev. A **4**, 908
Sinanoğlu, O. (1962): J. Chem. Phys. **36**, 706, 3198
– (1964): Adv. Chem. Phys. **6**, 315
– (1969): Adv. Chem. Phys. **14**, 237
Slater, J.C. (1960): *Quantum Theory of Atomic Structure* (McGraw-Hill, New York)
– (1968): Phys. Rev. **165**, 655
Sternheimer, R.M. (1950): Phys. Rev. **80**, 102
– (1951): Phys. Rev. **84**, 244
– (1972): Phys. Rev. A **6**, 1702
Sternheimer, R.M., Peierls, R.F. (1971 a): Phys. Rev. A **3**, 837
– (1971 b): Phys. Rev. A **4**, 1722
Sternheimer, R.M., Rodgers, J.E., Das, T.P. (1978): Phys. Rev. A **17**, 505
Sucher, J. (1980): Phys. Rev. A **22**, 348
Szász, L. (1959): Z. Naturforschung A **14**, 1014
– (1968): J. Chem. Phys. **49**, 679
Szász, L., Byrne, J. (1967): Phys. Rev. **158**, 34

Thomas, L. H. (1926): Nature **117**, 514
Thouless, D.J. (1972): *The Quantum Mechanics of Many-Body Systems* (Academic, New York)
Tolmachev, V.V. (1969): Adv. Chem. Phys. **14**, 421, 471
Trees, R.E. (1951 a): Phys. Rev. **83**, 756
– (1951 b): Phys. Rev. **84**, 1089
– (1952): Phys. Rev. **85**, 382

Vajed-Samii, M., Ray, S.N., Das, T.P., Andriessen, J. (1979): Phys. Rev. A **20**, 1787
– (1981): Phys. Rev. A **24**, 1204
Vetter, J., Ackermann, H., zu Putlitz, G., Weber, E.W. (1976): Z. Phys. A **276**, 161
Viers, J.M., Harris, F.E., Schaefer, H.F. (1970): Phys. Rev. A **1**, 24

Watson, K.M. (1953): Phys. Rev. **89**, 575
Wendin, G. (1971): J. Phys. B **4**, 1080
– (1972): J. Phys. B **5**, 110
Wick, G.C. (1950): Phys. Rev. **80**, 268
Wigner, E.P. (1931): *Gruppentheorie* (Vieweg, Braunschweig); [English transl.: *Group Theory and Its Application to Quantum Mechanics of Atomic Spectra* (Academic, New York)]
Winter, N.W., Laferrière, A., McKoy, V. (1970): Phys. Rev. A **2**, 49
Wood, J.H., Bohring, A.M. (1978): Phys. Rev. B **18**, 2701
Wybourne, B.G. (1965 a): *Spectroscopic Properties of Rare Earths* (Wiley, New York)
– (1965 b): Phys. Rev. A **137**, 364

Author Index

Numbers in *italics* refer to the list of references

Subject Index

Numbers in *italics* refer to page where the subject is defined or explained in more detail